Spring Boot 实战

从0开始动手搭建企业级项目

十三 / 著

电子工业出版社
Publishing House of Electronics Industry
北京·BEIJING

内 容 简 介

本书首先从 Spring Boot 基础知识部分讲起，涉及学习方法、技术趋势、开发环境和基础实践；然后是进阶应用部分，涉及核心源码、自动配置、模板引擎、数据库等分析，以及分页功能、文件上传、验证码等功能的编码实现；最后是实战开发部分，即从 0 开始动手搭建一个企业级的商城项目，涉及后台管理系统和前端页面所有功能模块的技术开发。

本书重在引导读者进入真实的项目开发体验，围绕 Spring Boot 技术栈全面展开，兼顾相关技术的知识拓展，由浅入深，步步为营，使读者既能学习基础知识，又能掌握一定的开发技巧。本书的目标是让读者拥有一个完整且高质量的学习体验，远离"Hello World 项目"，为技术深度的挖掘和薪水、职位的提升提供保障。

本书适合 Spring Boot 和 Java Web 的开发者、对大型商城项目开发感兴趣的技术人员，以及对 Spring Boot 技术栈感兴趣的读者。

未经许可，不得以任何方式复制或抄袭本书之部分或全部内容。
版权所有，侵权必究。

图书在版编目（CIP）数据

Spring Boot 实战：从 0 开始动手搭建企业级项目 / 十三著. —北京：电子工业出版社，2021.7
ISBN 978-7-121-41424-4

Ⅰ. ①S… Ⅱ. ①十… Ⅲ. ①JAVA 语言－程序设计 Ⅳ. ①TP312.8

中国版本图书馆 CIP 数据核字(2021)第 122523 号

责任编辑：陈　林
印　　刷：三河市良远印务有限公司
装　　订：三河市良远印务有限公司
出版发行：电子工业出版社
　　　　　北京市海淀区万寿路 173 信箱　邮编：100036
开　　本：787×980　1/16　印张：45.5　字数：1019.2 千字
版　　次：2021 年 7 月第 1 版
印　　次：2022 年 10 月第 2 次印刷
定　　价：138.00 元

凡所购买电子工业出版社图书有缺损问题，请向购买书店调换。若书店售缺，请与本社发行部联系，联系及邮购电话：(010) 88254888, 88258888。

质量投诉请发邮件至 zlts@phei.com.cn，盗版侵权举报请发邮件至 dbqq@phei.com.cn。
本书咨询联系方式：010-51260888-819，faq@phei.com.cn。

自　　序

大家好，我是韩帅，也是程序员十三。

非常感谢你阅读本书，在技术道路上，从此我们不再独行。

1. 写作背景

2017 年 2 月 24 日，笔者正式开启技术写作之路，同时也开始在 GitHub 网站上做开源项目，由于一直坚持更新文章和开源项目，慢慢被越来越多的人所熟悉。

从 2018 年开始，有不少出版社的编辑向笔者邀约写书。2018 年 6 月 7 日，电子工业出版社的陈林编辑通过邮件联系笔者并邀请笔者写书。从此，笔者与电子工业出版社结缘。

对于图书写作的邀约，一开始笔者都会婉拒，因为笔者对自己有比较清楚的认知。博客文章的写作要求相较于图书出版的要求还是有很大差别的，笔者觉得当时的可行性太低，于是逐一拒绝了出书的邀约。可是邀约多了，心态也有了改变。就像当初写第一篇博客抱着"试一试"的心态，笔者慢慢也有了"试一试"写书的念头。

生活中总有意外的惊喜。

2018 年，笔者也被不同的平台邀请制作付费专栏和课程。自 2018 年 9 月开始，笔者陆陆续续在 CSDN 图文课、实验楼、蓝桥云课、掘金小册、极客时间等平台上线了多个付费专栏和课程。其中，与 Spring Boot 技术栈相关的付费专栏就有 7 个。同时，笔者也会将付费专栏中的实战项目开源到 GitHub 和 Gitee 两个开源代码平台上。本书中的实战项目 newbee-mall 新蜂商城也是在这段时期内开发的，并于 2019 年开源。

基于这些开源项目的维护经验和多个付费专栏的制作经验，笔者逐渐觉得自己已经有能力完成一本技术书的写作，并且做到言之有物了。所以，笔者打定主意要写一本实体技术书。这也是一个开发者长久以来的梦想。于是笔者与陈林编辑联系并沟通了写作事宜，签订了图书出版合同。笔者写作的初衷就是希望自己把对 Spring Boot 技术栈的理解及实战项目开发的经验分享给读者。

笔者过去几年的经历可以整理成一张图，"免费文章→付费专栏→付费视频→实体书"，从 0 到 1，从无到有，都是一步一步走过来的。这些也是笔者的写作背景。

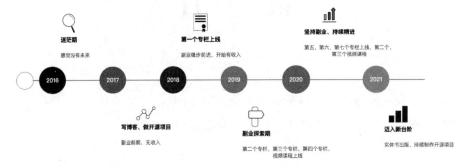

如果本书能够帮助读者学会 Spring Boot 并且用 Spring Boot 开发一些实际项目，那么笔者就非常满足了，这一次的写作也变得意义非凡。另外，笔者也在开始整理第二本书的底稿，内容是关于 Spring Boot + Vue3 前后端分离的实战项目开发。项目预览图如下所示。

2. 你会学到什么

本书的代码基于 Spring Boot 2.3.7-RELEASE 版本。笔者用 25 章全面深入地讲解 Spring Boot 技术栈的技术原理、功能点开发和项目实战。

工欲善其事，必先利其器。本书注重基础环境的搭建和开发工具的使用，以帮助读者少走弯路，快速掌握 Spring Boot 项目的开发技能。

读者学习本书，会有以下收获。

- Spring Boot 技术栈的基础使用和开发技巧
- Spring Boot 的进阶知识，自动配置特性的源码解读
- Spring Boot 项目的实战开发
- 为在校学生的毕业设计提供思路
- Thymeleaf 模板引擎的整合及运用
- AdminLTE3、Bootstrap4、SweetAlert、JqGrid、JQuery 等前端框架组件及控件的使用
- Spring Boot 企业级商城项目的全流程开发实践
- 大型技术项目的开发、设计和统筹

3. 适宜人群

本书定位 Spring Boot 项目的实战和进阶，资深开发人员可按需要选择对应章节阅读。为了照顾初学者，本书也设置了入门章节。以下读者非常适合学习本书。

- 从事 Java Web 开发的技术人员
- 对 Spring Boot 感兴趣、想要改变原有开发模式的开发人员
- 希望将 Spring Boot 技术实际运用到项目中的开发人员
- 需要使用 Spring Boot 进行完整项目学习的开发人员
- 想要独立完成一个 Java Web 项目的开发人员
- 需要大型商城项目实践的开发人员

- 想要将自己的项目上线到互联网的开发人员

4．源码

本书每个实战章节都会有对应的源码并提供下载，读者可以在本书封底扫码获取。

最终的实战项目是笔者的开源项目 newbee-mall，源码在开源网站 GitHub 和 Gitee 上都能搜索并下载到最新的源码。

- github.com/newbee-ltd/newbee-mall

- gitee.com/newbee-ltd/newbee-mall

5．致谢

感谢本书编辑陈林老师。从第一封邮件开始，他就展现了出版社编辑的专业性和耐心。在写作中，陈老师对本书的内容脉络做了非常多的指导工作，也给予笔者非常多的帮助和鼓励。在书稿整理完成后，陈老师不断调整和优化稿件中的内容，以确保图书质量。感谢电子工业出版社的美术编辑李玲和其他老师，本书能够顺利出版离不开他们的奉献，感谢他们辛苦、严谨的工作。

感谢 newbee-mall 开源仓库的各位用户及笔者专栏文章的所有读者。他们提供了非常多的修改和优化意见，使 Spring Boot 商城项目变得更加完善，也为笔者提供了持续写作的动力。

感谢掘金社区及其运营负责人优弧和运营人员 Captain。本书大部分内容是对掘金小册《Spring Boot 大型线上商城项目实战教程》的系统化升级和优化，本书能顺利出版也得到了掘金社区的大力支持。

特别感谢一下家人，没有他们的默默付出和巨大的支持，笔者不可能有如此多的时间和精力专注于本书的写作。

感谢每一位没有提及名字，但是曾经帮助过笔者的贵人。

韩帅

2021 年 7 月 1 日 于杭州

目 录

第 1 章 如何系统地学习 Spring Boot ························· 1

1.1 学习 Spring Boot 遇到的问题 ························· 1
1.2 学习 Spring Boot 的系统性建议 ························· 3
    1.2.1 基础实践、源码分析、项目开发 ························· 3
    1.2.2 如何开发和统筹一个完整的项目 ························· 4
    1.2.3 其他建议 ························· 6

第 2 章 Spring Boot：Java 开发人员的选择 ························· 7

2.1 认识 Spring Boot ························· 7
    2.1.1 越来越流行的 Spring Boot ························· 7
    2.1.2 Java 开发人员必备的技术栈 ························· 8
2.2 选择 Spring Boot ························· 9
    2.2.1 Spring Boot 的理念 ························· 9
    2.2.2 Spring Boot 可以简化开发 ························· 10
    2.2.3 Spring Boot 的特性 ························· 11

第 3 章 Spring Boot 基础开发环境的安装和配置 ························· 14

3.1 JDK 的安装和配置 ························· 14
    3.1.1 下载安装包 ························· 14
    3.1.2 安装 JDK ························· 16
    3.1.3 配置环境变量 ························· 17
    3.1.4 JDK 环境变量验证 ························· 18

- 3.2 Maven 的安装和配置 ·· 18
  - 3.2.1 下载安装包 ·· 18
  - 3.2.2 安装并配置 Maven ·· 19
  - 3.2.3 Maven 环境变量验证 ·· 20
  - 3.2.4 配置国内 Maven 镜像 ·· 20
- 3.3 开发工具 IDEA 的安装和配置 ·· 22
  - 3.3.1 下载 IDEA 安装包 ·· 22
  - 3.3.2 安装 IDEA 及其功能介绍 ·· 24
  - 3.3.3 配置 IDEA 的 Maven 环境 ·· 25

第 4 章 Spring Boot 项目搭建及快速上手 ·· 28

- 4.1 Spring Boot 项目创建 ·· 28
  - 4.1.1 认识 Spring Initializr ·· 28
  - 4.1.2 Spring Boot 项目初始化配置 ·· 29
  - 4.1.3 使用 Spring Initializr 初始化一个 Spring Boot 项目 ·· 30
  - 4.1.4 其他方式创建 Spring Boot 项目 ·· 32
- 4.2 Spring Boot 项目目录结构介绍 ·· 33
- 4.3 启动 Spring Boot 项目 ·· 34
  - 4.3.1 在 IDEA 编辑器中启动 Spring Boot 项目 ·· 34
  - 4.3.2 Maven 插件启动 ·· 36
  - 4.3.3 java-jar 命令启动 ·· 37
  - 4.3.4 Spring Boot 项目启动日志 ·· 38
- 4.4 开发第一个 Spring Boot 项目 ·· 39

第 5 章 Spring Boot 核心详解及源码分析 ·· 42

- 5.1 约定优于配置 ·· 42
- 5.2 Spring Boot 之依赖管理 ·· 43
- 5.3 @SpringBootApplication 注解与分析 ·· 49
  - 5.3.1 @SpringBootApplication 注解 ·· 49
  - 5.3.2 @SpringBootConfiguration 注解 ·· 51
  - 5.3.3 @EnableAutoConfiguration 注解 ·· 51
  - 5.3.4 @AutoConfigurationPackage 源码解析 ·· 52
  - 5.3.5 EnableAutoConfigurationImportSelector 类的源码解析 ·· 54
  - 5.3.6 @ComponentScan 注解 ·· 61
- 5.4 SpringApplication 启动流程解析 ·· 62

目 录

第 6 章　Spring Boot 之 DispatcherServlet 自动配置及源码解读 …………… 78

- 6.1　Spring MVC 的核心分发器 DispatcherServlet ………………………… 78
  - 6.1.1　核心分发器 DispatcherServlet 介绍 …………………………… 78
  - 6.1.2　DispatcherServlet 自动配置的日志输出 ……………………… 80
- 6.2　DispatcherServlet 自动配置的源码调试记录 …………………………… 81
- 6.3　自动配置类 DispatcherServletAutoConfiguration ……………………… 86
  - 6.3.1　DispatcherServletAutoConfiguration 类的讲解 ……………… 86
  - 6.3.2　DispatcherServletAutoConfiguration 在 IOC 容器中的注册 … 91
- 6.4　DispatcherServlet 自动配置流程 ………………………………………… 91
  - 6.4.1　注册至 IOC 容器 ………………………………………………… 92
  - 6.4.2　创建并启动嵌入式的 Tomcat 对象 …………………………… 94
  - 6.4.3　装载至 Servlet 容器 …………………………………………… 99

第 7 章　Spring Boot 之 Web 开发和 Spring MVC 自动配置分析 ………… 106

- 7.1　Spring MVC 自动配置内容 ……………………………………………… 106
- 7.2　WebMvcAutoConfiguration 源码分析 …………………………………… 107
- 7.3　ViewResolver 视图解析器的自动配置 ………………………………… 109
- 7.4　自动注册 Converter、Formatter ………………………………………… 111
- 7.5　消息转换器 HttpMessageConverter ……………………………………… 113
- 7.6　Spring Boot 对静态资源的映射规则 …………………………………… 116
- 7.7　welcomePage 和 favicon 配置 …………………………………………… 120
  - 7.7.1　welcomePage 配置 ……………………………………………… 120
  - 7.7.2　favicon 配置 …………………………………………………… 122

第 8 章　Thymeleaf 模板引擎使用详解 ……………………………………… 126

- 8.1　模板引擎技术介绍 ……………………………………………………… 126
- 8.2　Thymeleaf 模板引擎 …………………………………………………… 127
  - 8.2.1　Thymeleaf 模板引擎介绍 ……………………………………… 127
  - 8.2.2　Thymeleaf 并非 Spring Boot 默认的模板引擎 ……………… 128
- 8.3　Spring Boot 之 Thymeleaf 整合 ………………………………………… 129
  - 8.3.1　引入 Thymeleaf 依赖 …………………………………………… 130
  - 8.3.2　创建模板文件 …………………………………………………… 132
  - 9.3.3　编辑 Controller 代码 …………………………………………… 132
  - 8.3.4　Thymeleaf 模板引擎使用注意事项 …………………………… 134
- 8.4　Thymeleaf 属性值讲解 ………………………………………………… 135

ix

		8.4.1 Thymeleaf 模板解读·····135

- 8.4.1 Thymeleaf 模板解读······135
- 8.4.2 Thymeleaf 属性值的设置······136
- 8.4.3 修改属性值实践······137

8.5 Thymeleaf 语法讲解······139
- 8.5.1 Thymeleaf 语法······139
- 8.5.2 Thymeleaf 简单语法实践······141
- 8.5.3 Thymeleaf 表达式······143

第 9 章 Spring Boot 操作 MySQL 数据库······151

9.1 Spring Boot 连接 MySQL 实践······151
- 9.1.1 Spring Boot 对数据库连接的支持······151
- 9.1.2 Spring Boot 整合 spring-boot-starter-jdbc······152
- 9.1.3 Spring Boot 连接 MySQL 数据库验证······155

9.2 Spring Boot 数据源自动配置源码详解······157
- 9.2.1 Spring Boot 默认数据源 HikariDataSource······157
- 9.2.2 数据源自动配置类 DataSourceAutoConfiguration······158
- 9.2.3 属性绑定······160
- 9.2.4 DataSourceConfiguration 源码解析······160

9.3 使用 JdbcTemplate 进行数据库的增、删、改、查······164
- 9.3.1 JdbcTemplate 介绍······165
- 9.3.2 详解 JdbcTemplate 对数据库的增、删、改、查······165

9.4 Spring Boot 项目中 MyBatis 相关组件的自动配置讲解······170
- 9.4.1 MyBatis 简介······170
- 9.4.2 mybatis-springboot-starter 介绍······171
- 9.4.3 MyBatis 自动配置详解······171

9.5 Spring Boot 整合 MyBatis 的过程······179
- 9.5.1 添加依赖······179
- 9.5.2 application.properties 的配置······182
- 9.5.3 启动类增加 Mapper 扫描······182

9.6 Spring Boot 整合 MyBatis 进行数据库的增、删、改、查······183
- 9.6.1 新建实体类和 Mapper 接口······183
- 9.6.2 创建 Mapper 接口的映射文件······185
- 9.6.3 新建 MyBatisController······187

第 10 章 分页功能的讲解和编码实现······192

10.1 分页功能介绍······192

         10.1.1 百度分页功能演示 ····································· 192
         10.1.2 GitHub 分页功能演示 ·································· 194
         10.1.3 商城后台管理系统分页功能演示 ·························· 195
         10.1.4 商品搜索页分页功能演示 ································ 196
   10.2 分页功能的作用 ··· 197
   10.3 分页功能的设计 ··· 198
         10.3.1 前端分页功能设计 ······································ 198
         10.3.2 后端分页功能设计 ······································ 199
   10.4 分页功能的编码实现 ··· 199
         10.4.1 新增分页测试数据 ······································ 200
         10.4.2 分页功能返回结果的封装 ································ 202
         10.4.3 分页功能代码的具体实现 ································ 207
         10.4.4 分页功能测试 ·· 210
   10.5 jqGrid 分页插件 ··· 212
         10.5.1 jqGrid 分页插件介绍 ···································· 212
         10.5.3 导入 jqGrid 分页插件资源到项目中 ························ 214
         10.5.3 使用 jqGrid 实现分页的步骤 ····························· 214
         10.5.4 分页数据格式详解 ······································ 217
   10.6 整合 jqGrid 实现分页功能 ···································· 218
         10.6.1 前端页面制作 ·· 218
         10.6.2 jqGrid 初始化 ·· 219
         10.6.3 整合 jqGrid 实现分页功能测试 ··························· 221

第 11 章 Spring Boot 文件上传功能的实现 ······························· 223
   11.1 Spring MVC 处理文件上传的源码分析 ··························· 223
         11.1.1 文件上传功能源码调用链 ································ 223
         11.1.2 文件上传功能源码分析 ·································· 224
         11.1.3 Spring Boot 中 MultipartResolver 的自动配置 ··············· 227
   11.2 Spring Boot 文件上传功能的实现案例 ··························· 229
         11.2.1 Spring Boot 文件上传配置项 ····························· 229
         11.2.2 新建文件上传页面 ······································ 230
         11.2.3 新建文件上传处理 Controller 类 ························· 231
         11.2.5 文件上传功能测试 ······································ 232
   11.3 Spring Boot 文件上传路径回显 ································ 234
   11.4 Spring Boot 多文件上传功能的实现 ···························· 236

11.4.1　文件名相同时的多文件上传处理·················236
　　11.4.2　文件名不同时的多文件上传处理·················239

第 12 章　Spring Boot 实现验证码生成及验证功能·················245

12.1　验证码介绍·················245
12.1.1　什么是验证码·················245
12.1.2　验证码的形式·················246

12.2　Spring Boot 整合 easy-captcha 生成验证码·················248
12.2.1　添加 easy-captcha 依赖·················249
12.2.2　验证码格式·················251
12.2.3　验证码字符类型·················251
12.2.4　字体设置·················252
12.2.5　验证码图片输出·················253

12.3　生成并显示验证码·················254
12.3.1　后端逻辑实现：生成并输出验证码·················254
12.3.2　前端逻辑实现：在页面中展示验证码·················255

12.4　验证码的输入验证·················256
12.4.1　后端逻辑实现·················256
12.4.2　前端逻辑实现·················257

第 13 章　商城项目需求分析与功能设计·················260

13.1　选择开发商城系统的原因·················260
13.1.1　什么是商城系统·················260
13.1.2　为什么要做商城系统·················261

13.2　认识新蜂商城系统·················262
13.2.1　新蜂商城系统介绍·················262
13.2.2　新蜂商城开发背景·················263
13.2.3　新蜂商城开源过程·················264
13.2.4　新蜂商城运行预览图·················266

13.3　新蜂商城功能详解·················272
13.3.1　商城端功能整理·················272
13.3.2　后台管理系统功能整理·················273
13.3.3　新蜂商城架构图·················274

第 14 章　项目初体验：启动和使用新蜂商城·················275

14.1　下载商城项目的源码·················275

		14.1.1　使用 clone 命令下载源码 ································· 275
		14.1.2　通过开源网站下载源码 ································· 276
	14.2　新蜂商城目录结构讲解 ·· 279
	14.3　启动商城项目 ·· 281
		14.3.1　导入数据库 ··· 281
		14.3.2　修改数据库连接配置 ··································· 281
		14.3.3　静态资源目录设置 ····································· 282
		14.3.4　启动并访问商城项目 ··································· 283
	14.4　注意事项 ·· 285
		14.4.1　关于项目地址 ··· 285
		14.4.2　关于账号及密码 ······································· 286
		14.4.3　商城登录和后台管理系统登录演示 ······················· 287

第 15 章　页面设计及商城后台管理系统页面布局的实现 ················· 289
	15.1　前端页面实现的技术选型 ······································ 289
		15.1.1　Bootstrap 产品介绍 ···································· 289
		15.1.2　为什么选择 Bootstrap ·································· 290
		15.1.3　AdminLTE3 产品介绍 ·································· 291
		15.1.4　为什么选择 AdminLTE3 ································ 293
		15.1.5　前端技术选型的 5 个原则 ······························· 294
	15.2　商城页面布局讲解 ·· 294
		15.2.1　后台管理系统页面布局介绍 ····························· 294
		15.2.2　商城端页面布局介绍 ··································· 296
	15.3　后台管理系统页面制作 ·· 297
		15.3.1　AdminLTE3 整合到 Spring Boot 项目中 ··················· 297
		15.3.2　后台管理系统页面制作 ································· 298
		15.3.3　Controller 类处理页面跳转 ····························· 302
		15.3.4　公共页面抽取 ··· 303
		15.3.5　分段表达式传参 ······································· 310

第 16 章　后台管理系统登录功能的实现 ······························· 313
	16.1　登录流程设计 ·· 313
		16.1.1　什么是登录 ··· 313
		16.1.2　用户登录状态 ··· 314
		16.1.3　登录流程设计 ··· 315

16.2 管理员登录功能实践 ··· 316
　　16.2.1 管理员登录页面的实现 ······································ 316
　　16.2.2 管理员表结构设计 ·· 319
　　16.2.3 新建管理员实体类和 Mapper 接口 ······················· 321
　　16.2.4 创建 AdminUserMapper 接口的映射文件 ··············· 323
　　16.2.5 业务层代码的实现 ·· 324
　　16.2.6 管理员登录控制层代码的实现 ····························· 325
　　16.2.7 管理员登录功能演示及注意事项 ·························· 326
16.3 后台管理系统登录拦截器的实现 ································· 328
　　16.3.1 登录拦截器 ·· 328
　　16.3.2 定义拦截器 ·· 329
　　16.3.3 配置拦截器 ·· 331
16.4 管理员模块功能的完善 ·· 333

第 17 章　轮播图管理模块的开发 ·· 343

17.1 轮播图模块介绍 ·· 343
17.2 轮播图管理页面跳转逻辑的实现 ································· 345
　　17.2.1 导航栏中增加"轮播图配置"栏目 ························· 345
　　17.2.2 控制类处理跳转逻辑 ··· 345
　　17.2.3 轮播图管理页面基础样式的实现 ·························· 346
17.3 轮播图管理模块后端功能的实现 ································· 348
　　17.3.1 轮播图表结构设计 ·· 348
　　17.3.2 轮播图管理模块接口介绍 ··································· 349
　　17.3.3 新建轮播图实体类和 Mapper 接口 ······················· 350
　　17.3.4 创建 CarouselMapper 接口的映射文件 ················· 354
　　17.3.5 业务层的代码实现 ·· 360
　　17.3.6 轮播图管理模块控制层的代码实现 ······················ 363
17.4 轮播图管理模块前端功能的实现 ································· 366
　　17.4.1 功能按钮和分页信息展示区域 ····························· 366
　　17.4.2 轮播图管理页面分页功能的实现 ·························· 367
　　17.4.3 添加和修改按钮触发事件及 Modal 框的实现 ········· 368
　　17.4.4 轮播图管理页面添加和编辑功能的实现 ················ 371
　　17.4.5 轮播图管理页面删除功能的实现 ·························· 373
　　17.4.6 功能测试 ·· 375

第 18 章　分类管理模块的开发 379

18.1　分类管理模块介绍 379
18.1.1　商品分类 379
18.1.2　分类层级 380
18.1.3　分类模块的主要功能 381
18.2　商品类目管理模块前端页面的制作 381
18.2.1　在导航栏中增加"分类管理"栏目 381
18.2.2　控制类处理跳转逻辑 382
18.2.3　分类管理页面基础样式的实现 383
18.2.4　功能按钮和分页信息展示区域 386
18.2.5　URL 参数处理 387
18.3　商品分类表的结构设计 388
18.4　分类模块后端功能的实现 389
18.4.1　新建分类实体类和 Mapper 接口 390
18.4.2　创建 GoodsCategoryMapper 接口的映射文件 395
18.4.3　业务层代码的实现 401
18.4.4　分类管理模块控制层的代码实现 405
18.5　商品类目管理模块前端功能的实现 410
18.5.1　分类管理页面分页功能的实现 410
18.5.2　上下级分类页面的跳转逻辑处理 412
18.5.3　分类管理页面添加和修改按钮的触发事件 413
18.5.4　分类管理页面添加和编辑功能的实现 414
18.5.5　分类管理页面删除功能的实现 416
18.5.6　功能测试 417
18.6　分类数据的三级联动功能开发 421
18.6.1　多层级数据联动效果的常见场景 421
18.6.2　多层级的数据联动实现原理和方式 423
18.6.3　分类三级联动页面基础样式的实现 423
18.6.4　数据初始化 426
18.6.5　数据联动后端接口的实现 429
18.6.6　监听选择框的 change 事件并实现联动功能 431

第 19 章　富文本编辑器介绍及整合 435

19.1　富文本编辑器详解 435

19.1.1 如何处理复杂的文本内容 ································· 435
19.1.2 富文本编辑器介绍及其优势 ····························· 437
19.2 富文本编辑器 wangEditor 的介绍 ································· 438
19.3 wangEditor 整合编码案例 ·· 441
19.4 新蜂商城项目 wangEditor 的应用情况 ····························· 444
19.4.1 为什么选择 wangEditor ································ 444
19.4.2 wangEditor 整合过程中的问题 ·························· 445

第 20 章 商品编辑页面及商品管理模块的开发 ·························· 448

20.1 新蜂商城商品管理模块简介 ······································ 448
20.2 新蜂商城商品信息表结构的设计 ·································· 449
20.3 商品编辑页面的制作 ·· 451
 20.3.1 导航栏中增加"商品信息"栏目 ························ 451
 20.3.2 控制类处理跳转逻辑 ································· 452
 20.3.3 商品信息编辑页面的制作 ····························· 453
 20.3.4 初始化插件 ··· 458
 20.3.5 新增控制类处理图片上传 ····························· 462
20.4 商品信息添加接口的开发与联调 ·································· 467
 20.4.1 新建商品实体类和 Mapper 接口 ······················· 467
 20.4.2 创建 NewBeeMallGoodsMapper 接口的映射文件 ········· 472
 20.4.3 业务层的代码实现 ··································· 476
 20.4.4 商品添加接口控制层的代码实现 ······················· 477
 20.4.5 前端调用商品添加接口 ······························· 478
 20.4.6 功能测试 ··· 482
20.5 商品信息编辑页面的完善 ·· 483
 20.5.1 控制类处理跳转的逻辑 ······························· 484
 20.5.2 商品信息编辑页面数据的回显 ························· 486
20.6 商品信息修改的开发与联调 ······································ 489
 20.6.1 数据层代码的实现 ··································· 490
 20.6.2 业务层代码的实现 ··································· 492
 20.6.3 商品添加接口控制层代码的实现 ······················· 493
 20.6.4 前端调用商品修改接口 ······························· 494
20.7 商品信息管理页面的制作 ·· 496
 20.7.1 导航栏中增加"商品管理"按钮 ······················· 496
 20.7.2 控制类处理跳转逻辑 ································· 497

20.7.3　商品管理页面基础样式的实现 ………………………………………… 497
　20.8　商品信息管理模块接口的实现 ……………………………………………………… 499
　　　20.8.1　数据层代码的实现 …………………………………………………… 500
　　　20.8.2　业务层代码的实现 …………………………………………………… 501
　　　20.8.3　控制层代码的实现 …………………………………………………… 502
　20.9　商品管理模块前端功能的实现 ……………………………………………………… 503
　　　20.9.1　商品管理页面功能按钮的设置 ………………………………………… 503
　　　20.9.2　商品管理页面分页功能的实现 ………………………………………… 504
　　　20.9.3　商品添加和修改按钮的触发事件 ……………………………………… 506
　　　20.9.4　商品上架和下架功能的实现 …………………………………………… 506
　　　20.9.5　功能测试 ……………………………………………………………… 509

第 21 章　新蜂商城首页功能的开发 …………………………………………………… 514

　21.1　新蜂商城首页静态页面的制作 ……………………………………………………… 514
　　　21.1.1　商城首页的设计注意事项 …………………………………………… 514
　　　21.1.2　新蜂商城首页的排版设计 …………………………………………… 515
　　　21.1.3　新蜂商城首页基础样式的实现 ………………………………………… 517
　21.2　新蜂商城首页功能的实现 …………………………………………………………… 523
　　　21.2.1　首页跳转逻辑的实现 ………………………………………………… 523
　　　21.2.2　Controller 处理跳转 …………………………………………………… 523
　　　21.2.3　公共页面的抽取 ……………………………………………………… 525
　21.3　商城端首页轮播图功能的实现 ……………………………………………………… 528
　　　21.3.1　Swiper 轮播图插件的介绍 …………………………………………… 528
　　　21.3.2　轮播图插件 Swiper 的整合 …………………………………………… 529
　　　21.3.3　轮播图数据的读取 …………………………………………………… 530
　　　21.3.4　轮播图数据的渲染 …………………………………………………… 531
　　　21.3.5　轮播效果的实现 ……………………………………………………… 532
　21.4　首页分类效果的制作 ………………………………………………………………… 532
　　　21.4.1　首页商品分类数据的读取 …………………………………………… 533
　　　21.4.2　首页商品分类数据的渲染 …………………………………………… 535
　　　21.4.3　首页商品分类联动效果的实现 ………………………………………… 536
　21.5　商城首页推荐商品模块的介绍 ……………………………………………………… 538
　21.6　首页配置管理页面的制作 …………………………………………………………… 539
　　　21.6.1　导航栏中增加首页配置相关栏目 ……………………………………… 539
　　　21.6.2　控制类处理跳转逻辑 ………………………………………………… 540

- 21.6.3 首页配置商品管理页面基础样式的实现 ·············· 541
- 21.7 首页配置管理模块接口的设计及实现 ················· 543
  - 21.7.1 首页配置表结构的设计 ··············· 543
  - 21.7.2 新建首页配置实体类和 Mapper 接口 ··············· 544
  - 21.7.3 创建 IndexConfigMapper 接口的映射文件 ··············· 549
  - 21.7.4 业务层代码的实现 ··············· 555
  - 21.7.5 首页管理模块控制层代码的实现 ··············· 559
- 21.8 首页配置管理模块前端功能的实现 ················· 563
  - 21.8.1 功能按钮和分页信息展示区域 ··············· 563
  - 21.8.2 首页配置管理页面分页功能的实现 ··············· 563
  - 21.8.3 添加和修改按钮触发事件及 Modal 框实现 ··············· 565
  - 21.8.4 首页配置管理页面添加和编辑功能的实现 ··············· 568
  - 21.8.5 首页配置管理页面删除功能的实现 ··············· 570
- 21.9 商城首页功能完善 ················· 571
  - 21.9.1 首页推荐商品数据的读取 ··············· 571
  - 21.9.2 首页推荐商品数据的渲染 ··············· 573

第 22 章 商城端用户登录和注册功能的开发 ·············· 575

- 22.1 商城端用户表结构的设计 ················· 575
- 22.2 商城端用户登录和注册页面的制作 ················· 577
  - 22.2.1 商城端登录页面基础样式的实现 ··············· 577
  - 22.2.2 商城端注册页面基础样式的实现 ··············· 579
  - 22.2.3 控制类处理跳转逻辑 ··············· 580
- 22.3 商城端用户登录和注册模块接口的实现 ················· 580
  - 22.3.1 新建商城端用户实体类和 Mapper 接口 ··············· 580
  - 22.3.2 创建 MallUserMapper 接口的映射文件 ··············· 584
  - 22.3.3 业务层代码的实现 ··············· 587
  - 22.3.4 商城端用户登录和注册控制层代码的实现 ··············· 590
- 22.4 商城端用户登录注册模块前端功能的实现 ················· 593
  - 22.4.1 注册功能的实现 ··············· 593
  - 22.4.2 登录功能的实现 ··············· 595
- 22.5 商城端用户登录拦截器的实现 ················· 596
  - 22.5.1 定义拦截器 ··············· 597
  - 22.5.2 配置拦截器 ··············· 598
- 22.6 功能测试 ················· 600

第 23 章 商城端搜索商品功能的开发·····602

23.1 搜索页面的设计和数据格式的定义·····602
- 23.1.1 搜索页面的设计·····602
- 23.1.2 数据格式的定义·····603

23.2 发起搜索请求·····605
- 23.2.1 商品的关键字搜索·····605
- 23.2.2 商品的分类搜索功能·····607

23.3 商品数据查询的实现代码·····608
- 23.3.1 数据层代码的实现·····609
- 23.3.2 业务层代码的实现·····611

23.4 商品搜索结果页面数据的渲染·····612
- 23.4.1 参数封装及分页数据的获取·····612
- 23.4.2 搜索结果页面渲染的逻辑实现·····614

第 24 章 商品详情页及购物车功能的开发·····619

24.1 商城端商品详情页面的制作·····619
- 24.1.1 商品详情页跳转逻辑的实现·····619
- 24.1.2 商品详情页面数据的渲染·····620

24.2 购物车模块简介及表结构设计·····624
- 24.2.1 购物车模块简介·····624
- 24.2.2 购物车表结构设计·····624

24.3 将商品加入购物车功能的实现·····625
- 24.3.1 新建购物项实体类和 Mapper 接口·····625
- 24.3.2 创建 NewBeeMallShoppingCartItemMapper 接口的映射文件·····628
- 24.3.3 业务层代码的实现·····631
- 24.3.4 将商品加入购物车接口的实现·····633
- 24.3.5 前端功能的实现·····635

24.4 购物车列表功能的实现·····636
- 24.4.1 数据格式的定义·····636
- 24.4.2 购物车列表数据的获取·····638
- 24.4.3 购物车列表数据的渲染·····641

24.5 编辑购物项功能的实现·····645
- 24.5.1 数据层代码的实现·····645
- 24.5.2 业务层代码的实现·····646

- 24.5.3 控制层代码的实现·················648
- 24.5.4 前端调用修改和删除购物项的接口·················650
- 24.6 功能测试·················652

第25章 订单模块功能开发及讲解·················656

- 25.1 订单确认页面的功能开发·················657
  - 25.1.1 商城中的订单确认步骤·················657
  - 25.1.2 订单确认的前置步骤·················658
  - 25.1.3 订单确认页面的数据整合·················659
  - 25.1.4 订单确认页面制作及数据渲染·················660
- 25.2 订单模块中的表结构设计·················667
  - 25.2.1 订单主表和订单项关联表设计·················667
  - 25.2.2 订单项表的设计思路·················670
- 25.3 订单生成功能的实现·················670
  - 25.3.1 新蜂商城订单生成的流程·················670
  - 25.3.2 发起订单生成请求·················671
  - 25.3.3 订单生成请求处理·················672
  - 25.3.4 订单生成逻辑的实现·················674
- 25.4 订单详情页面功能的实现·················677
  - 25.4.1 订单详情跳转处理·················677
  - 25.4.2 订单详情数据的渲染·················678
- 25.5 商城端订单列表功能·················685
  - 25.5.1 订单列表数据格式的定义·················685
  - 25.5.2 订单列表页面数据的获取·················690
  - 25.5.3 订单列表页面渲染的逻辑·················692
- 25.6 订单处理流程及订单状态的介绍·················697
  - 25.6.1 订单处理流程·················697
  - 25.6.2 订单状态的介绍·················698
- 25.7 订单状态转换的讲解·················700
  - 25.7.1 订单支付·················700
  - 25.7.2 订单确认·················703
  - 25.7.3 订单出库·················704
  - 25.7.4 确认收货·················705
  - 25.7.5 取消订单·················705
- 25.8 商城系统的展望·················707

第 1 章

如何系统地学习 Spring Boot

开发人员学习任何一门技术都需要经过如下步骤：了解→入门→实践，直到最终掌握这门技术。笔者会结合个人经验谈一谈在学习 Spring Boot 过程中会遇到的问题和处理方法。开发人员掌握一门新技术的最终目标是能够把它运用到实际的开发项目中，因此笔者也会讲解如何开发和统筹一个完整的大型项目。

1.1 学习 Spring Boot 遇到的问题

近几年笔者一直在做关于 Spring Boot 的技术实践和分享，也不断有朋友与笔者进行技术交流。在交流过程中，很多朋友描述了他们遇到的 Spring Boot 学习困境，如图 1-1 所示。

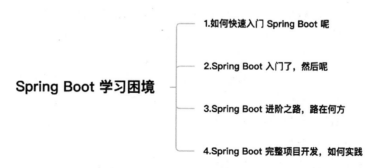

图 1-1　Spring Boot 学习困境

笔者将结合自身的学习和实践经验，对这些问题进行总结和回答。

1. 如何快速入门 Spring Boot 呢

在学习一门新技术时，很多开发人员会在网上找 demo 或者在开源网站上找对应的开源项目，通过阅读源码，学习项目作者的开发思路和解决问题的方法。这一过程对大多数人来说并不容易，要么不知道从何下手，要么由于技术文档的不完善导致被某个环节卡住。这些不利的因素最终可能导致开发人员无法坚持学习。

学习新技术的第一阶段，首先应该了解这门技术的产生背景、理念、发展历程，同时要分析它主要解决什么痛点。然后可以参照官网的案例进行尝试性的编码。

2. Spring Boot 入门了，然后呢

对于开发人员来说，学习任何技术的最终目的都是为了进行实际企业项目的开发。了解一门技术，甚至通过编码实现了一些小功能，但是没有尝试与正在开发或者已经上线的项目进行结合与类比，这是不行的。如果开发人员处于这个状态下，就应该思考一下用 Spring Boot 重构正在开发或者已经上线的项目。笔者认为，原项目中的模块和功能，都可以尝试使用 Spring Boot 实现，并逐一比较该开发模式与平时直接使用 Spring 进行开发有何不同。通过一两个项目的实践，就能更加了解 Spring Boot 技术栈，而且可以整合前期零散的知识点，获得较大提升。

3. Spring Boot 进阶之路，路在何方

随着开发人员对 Spring Boot 技术学习和使用的深入，其需求会由使用转向深入理解和掌握。此时的开发人员往往处于一个比较迷茫的状态，即知其然不知其所以然，主要表现在以下几个方面。

（1）能够使用 Spring Boot 进行功能开发，但是并不知道所写的代码具体是如何生效的。比如最常见的数据源配置，即在配置文件中设置 JDBC 的相关参数就可以直接连接数据库并进行相关操作。此时的痛点就是虽然能够写出一个功能的实现代码，但不知道为什么要这么写。

（2）在面试或者技术交流时，谈到 Spring Boot 的底层实现和设计思想，该阶段的开发人员往往一脸茫然，无法给出自己的回答。他们对 Spring Boot 的了解并不全面，而且 Spring Boot 技术栈涉及的模块很多，知识点的串联难度也就比较高。

比如"约定优于配置"。什么是约定优于配置？它具有什么特点，又能给实际开发工作的效率带来哪些提升？比如自动配置，什么是自动配置？它又是如何实现的？自动配置的机制能够给开发工作带来哪些改变？再比如 Spring Boot 中的各种 starter 是什么？该如何深入了解它？此时的痛点就是对于耳熟能详的特性或者概念不理解其内在含义。

（3）对于阅读和学习源码，该阶段的开发人员往往不知道从哪里看起，也没有坚实的理论基础，看不懂源码的含义，最后就放弃学习。此时的痛点就是对于源码阅读有心无力，没有良好的方法论和指导。

如果正在阅读本书的读者也遇到了相似问题，一方面说明你已经掌握了 Spring Boot 的基本使用方法，另一方面也说明你需要升级相关知识的认知了。

4．Spring Boot 完整项目开发，如何实践

该阶段的开发人员有了自己的项目构想，但是不知道如何开发一个完整的项目。他们经常开发到某一个阶段就会被一些小问题卡住，导致无法继续开发。由于缺少完整的项目源码和系统的知识讲解，他们在开发过程中时常会碰到各种小问题，最终导致开发进度停滞不前甚至被迫终止开发。其实笔者在项目开发的过程中也会遇到各种各样的问题，甚至也遇到过不能继续开发的死局。任何人都会遇到问题，我们应该尝试着解决它，而不是退缩和逃避，通过摸索、学习最终都会解决问题。

1.2 学习Spring Boot的系统性建议

针对前文涉及的问题，笔者会给出一些建议，希望读者在学习本书之后能够掌握 Spring Boot 基础开发知识，同时也能够知晓其源码实现原理，并根据书中的案例掌握实战项目的开发技巧，最终把这些技巧和知识灵活地运用在实际的企业项目开发中。

1.2.1 基础实践、源码分析、项目开发

想要掌握 Spring Boot 技术栈，基础实践、源码分析、项目开发三个步骤都不可或缺，这也是笔者整理资料写作本书的目的。

本书所选择的知识点都比较实用，且源码完备，在理论知识介绍完之后都配有相应的源码案例可以下载到本地。建议读者在阅读时能够参考书中提供的源码，自己动手实现相关功能点，或者直接运行书中给出的源码。不管是选择哪种方式，读者都一定要动手操作，而不是翻一翻书就完事。本书主要目的就是加强读者的动手能力。

本书的章节按照"基础实践、源码分析、项目开发"三个步骤展开，知识结构合理。首先，由 Spring Boot 技术栈介绍和基础环境搭建讲起，有多个章节涉及 Spring Boot 技术栈的相关知识。

其次，理论结合源码讲解，由 Spring Boot 的基础使用讲到 Spring Boot 的源码解读，包括 Spring Boot 整合 Thymeleaf 制作页面、Spring Boot 整合 MyBatis 操作数据库、Spring Boot 启动流程的源码分析、Spring Boot 自动配置流程讲解等。

最后，商城项目的开发实战，主要包括商城后台管理系统的开发和商城端的功能开发。在实践过程中将对项目功能进行拆分，使用 Spring Boot 分别实现各个独立的功能点，比如图片上传功能、分页功能、登录功能、验证码功能等，并整合多个独立的基础功能到一个完整的功能模块中，最终完成各个功能模块的功能和交互，开发出一个完整的商城系统。

1.2.2 如何开发和统筹一个完整的项目

帮助读者获得开发和统筹一个完整项目的能力是笔者写作本书的一个重要目标。笔者将通过技术栈的详细讲解、多个功能模块的开发实践，并结合在实际项目开发中的产品流程来达成此目标。

一个可以实操练手的完整项目，再配备上详细的技术讲解手册，是提高开发人员技术水平最高效的方式。围绕 Spring Boot 技术栈，笔者将给读者呈现一个大型项目十分完整的开发流程。实践项目包含一个内容展示系统和一个后台管理系统，其中功能模块包括登录认证模块、管理员模块、商品发布和管理模块、分类管理模块、搜索模块、订单管理模块、会员管理模块等。本书既能让读者得到一个完整的实操项目，也能让读者加满 Spring Boot 技能点，从而帮助读者提升专业技术能力，为升职加薪提供知识保障。

一个小的 demo 项目做起来并不复杂，它功能单一，也没有过多复杂的知识点需要掌握，跟着教程能很快完成。但是开发和统筹一个完整的项目则不同，它所涉及的知识点庞大而复杂。

1. 功能模块齐全

一个完整的项目一般分为后台管理系统和内容展示系统（前台网站），而一个完整的系统可以纵向拆分出很多个功能模块。后台管理系统包含管理员登录、身份认证、菜单设置、商品管理、订单管理、文件上传及管理、富文本编辑器整合、系统设置、数据统计及常用的交互功能等模块。而前台网站包括首页门户、商品分类、新品上线、首页轮播、商品推荐、商品搜索、商品展示、购物车、订单结算、订单流程、个人订单管理、会员中心、帮助中心等模块。实战项目所涉及的模块和技术如图 1-2 所示。

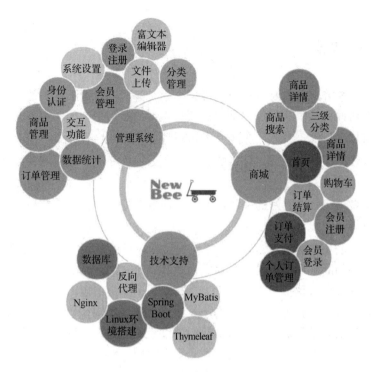

图1-2 实战项目所涉及的模块和技术

2. 涉及的技能点庞杂

开发和统筹一个完整的大型项目需要开发人员了解不同的技术或框架，比如常用的前端页面模板和基本的 Web 开发知识、后端开发技术框架（Spring Boot、模板引擎、ORM 框架）、服务器基础设施（Shell 命令、Nginx、MySQL）等，都需要开发人员全局考虑和布置。

大部分开发人员独自开发一个完整的项目是很费力的。这不仅考验着开发人员的技术储备深度，更考验着开发人员技术使用的熟练度，同时对于开发人员的系统设计能力也有较高要求（系统切分、功能点设计、页面结构和交互优化等）。

虽然开发大型项目比较复杂，但也不是完全无法实现。只要开发人员计划合理、选用有效的解决方案就可以完成这项任务。业内流行的一个解决方案就是"拆"，化繁为简，将大项目拆解成若干个小项目，大系统拆分出若干个功能模块，大功能拆解成若干个小功能，之后再对各个环节或者各个功能做具体的实现和完善。当开发人员将这些各个击破并全部完善的时候，一个完整的项目也就逐渐展现在眼前了。

1.2.3 其他建议

在本书中，笔者尽可能对重点知识进行全面讲解，不过囿于本书的篇幅和定位，在内容上肯定有所取舍。本书的定位是实战项目类型的书，会更偏向实战介绍。这里笔者再给出一些建议，以便读者有一个更好的学习体验。

（1）遇到任何问题，先尝试自己解决，实在不行再寻求帮助，这样有助于提升自己独立解决问题的能力。

（2）善于做笔记，看到好的文章或者解决问题的好办法，一定要做好笔记，避免自己犯同样的错误。

（3）IT 技术的更新迭代非常快，一定要关注行业资讯，及时更新自己的知识。同样，流行的技术框架的版本迭代也很快，要学会查看官方文档，获取最新的知识和材料，这样才能更有效提升自身技术水平。

（4）使用正确的方式进行提问。对于自己无法解决的问题，可以尝试向别人提问，在提问时尽量提供充足的信息，把遇到问题的过程说清楚，可以附上错误日志、页面截图、录屏等内容，千万不要上来就问"在吗？""项目 404 了怎么解决？"

（5）开发人员一定要多动手实践、多写代码、多做练习，看了不等于会了，只有把代码编写出来才算真正掌握了。

第 2 章

Spring Boot：Java 开发人员的选择

为什么 Java 开发人员需要掌握 Spring Boot？因为 Spring Boot 已经成为其在职业道路上"打怪升级"的必备技能包了。本章将通过介绍 Spring Boot 的基本情况、特点和优势展开具体讨论。

2.1 认识Spring Boot

2.1.1 越来越流行的 Spring Boot

Spring Boot 是目前 Java 社区最有影响力的项目之一，也是下一代企业级应用开发的首选技术。Spring Boot 是伴随 Spring 4 而产生的技术框架，具备良好的技术基因。在继承 Spring 框架所有优点的同时，它也为开发人员带来了巨大的便利。与普通的 Spring 项目相比，Spring Boot 可以简化项目的配置和编码，使项目部署更方便，而且它还为开发人员提供了"开箱即用"的良好体验，可以进一步提升开发效率。

Spring Boot 正在成为越来越流行的开发框架。从 Spring Boot 词条的百度指数可以确切地看出，开发人员对 Spring Boot 技术栈的关注度越来越高，如图 2-1 所示。

Spring Boot 以其优雅简单的启动配置和便利的开发模式深受好评，其开源社区也空前的活跃。截至 2020 年年底，Spring Boot 项目在 GitHub 网站上已经有 52.5k 的 stars，32.3k 的 forks（如图 2-2 所示），并且数量仍在高速增长。另外，各种基于 Spring Boot

的项目也如雨后春笋一般出现在开发人员的面前，其受欢迎程度可见一斑。现在，很多技术团队在使用 Spring Boot 进行企业项目的开发。

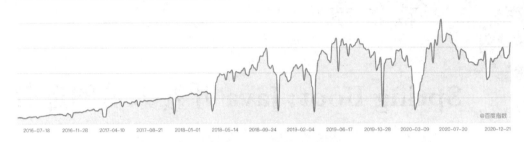

图 2-1　Spring Boot 词条的百度指数

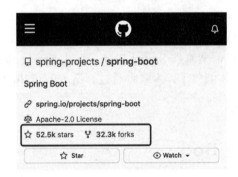

图 2-2　Spring Boot 项目的 GitHub 数据

2.1.2　Java 开发人员必备的技术栈

在五六年前，Java 开发工程师只要掌握 JSP 和 Servlet，并且有一些简单的项目经验，就可以获得很多面试机会。如果面试过程表现良好，拿到一份 offer 并不困难。然而，在现如今的大环境下，这几乎是不可能的。

现在，企业对 Java 开发工程师的要求更高，需要有一些实际开发的项目经验，并且多半是 SSM（Spring+Spring MVC+MyBatis）或者 Spring Boot 相关的项目经验。如果求职者简历中没有足够的项目经验，那么简历投递可能就会杳无音信。

Spring Boot 已经成为企业招聘需求的重要部分了。这也使得 Spring Boot 成为 Java 开发人员必备的技术栈。无论应届毕业生还是有经验的 Java 开发人员，Spring Boot 技术栈及相关项目经验都已经成为他们简历中的必要元素。

除此之外，Java 技术社区和 Spring 官方团队也对 Spring Boot 有非常多的资源倾斜。Spring 官方极力推崇 Spring Boot，后续笔者会向读者介绍 Spring 官方对 Spring Boot 的重视。

Spring Boot 有着非常好的前景，具体如图 2-3 所示。

图 2-3　Spring Boot 的前景

2.2　选择 Spring Boot

2.2.1　Spring Boot 的理念

关于 Spring Boot 框架的理念，可以通过 Spring 官网探知一二，如图 2-4 所示。

图 2-4　Spring Boot 的理念

在该页面中，官方毫不吝啬对于 Spring Boot 的赞美之词，也极力推荐开发人员使用 Spring Boot 升级 Java 项目的代码。同时，也引用了 Netflix 高级开发工程师的话："I'm very proud to say, as of early 2019, we've moved our platform almost entirely over to Spring Boot."其中的含义不言自明。

官方也在不断鼓励开发人员使用 Spring Boot，并使用 Spring Boot "升级" 项目代码进而达到优化 Java 项目的目的。图 2-4 已经是改版后的 Spring 官网，比之前的话术略微有一些收敛。在 2018 年 Spring 的官网中，官方对于 Spring Boot 的描述是 "Spring Boot BUILD ANYTHING！"

翻译过来就是 "用 Spring Boot 构造一切！"

彼时的官网如图 2-5 所示。Spring Boot 位于 Spring 三个重量级产品的第一位，可以看出 Spring 官方也非常重视 Spring Boot 的发展。

使用 Spring Boot 的目的在于用最少的 Spring 预先配置，让开发人员尽快构建和运行应用，最终创建产品级的 Spring 应用和服务。

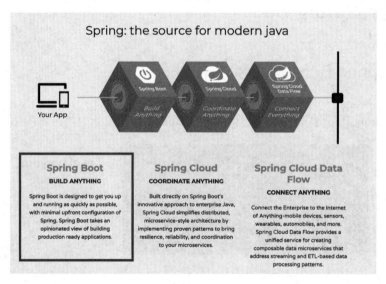

图 2-5　2018 年 Spring 官网对于 Spring Boot 框架的描述文案

2.2.2　Spring Boot 可以简化开发

"当你终于把 Spring 的 XML 配置文件调试完成的时候，我已经用 Spring Boot 开发好 N 个功能了。"

这可不是一句玩笑话，相信熟悉 Spring 开发项目的读者都深有体会。无论 Spring 框架的初学者还是具有经验的开发人员，对 Spring 项目的配置文件多少都会感到头痛，尤其在项目日渐庞大之后，纷繁复杂的 XML 配置文件让开发人员十分头痛。在一个项目开发完成后，这种痛苦也许会消除，但是一旦接手新项目，又要复制、粘贴一些十分雷同的 XML 配置文件，周而复始地进行这种枯燥死板的操作让人不胜其烦。

Spring Boot 解决了这个问题。Spring Boot 通过大量的自动化配置等方式简化了原 Spring 项目开发过程中编码人员的配置步骤。其中，大部分模块的设置和类的装载都由 Spring Boot 预先做好，从而使得开发人员不用再重复地进行 XML 配置，极大地提升了开发人员的工作效率。开发人员可以更加注重业务的实现而不是繁杂的配置工作，从而可以快速地构建应用。

框架的封装和抽象程度的完善，也使得代码的复用性更高、项目的可维护性提高、开发和学习成本降低，能加快开发进度并最终形成行业开发标准。从这个角度来说，越简洁的开发模式就越能减轻开发人员的负担并提升开发效率，行业内普遍认可并接受的框架也会越来越流行，并最终会形成一套读者都认可的开发标准。Spring Boot 就正在逐渐改变原有的开发模式，形成行业认可的开发标准。

2.2.3 Spring Boot 的特性

1. 继承 Spring 的优点

Spring Boot 来自 Spring 家族，因此 Spring 所有具备的功能和优点，Spring Boot 同样拥有。官方还对 Spring Boot 做了大量的封装和优化，从而使开发人员更容易上手和学习。相对于 Spring 来说，使用 Spring Boot 完成同样的功能和效果，开发人员需要操作和编码的工作更少了。

2. 可以快速创建独立运行的 Spring 项目

Spring Boot 简化了基于 Spring 的应用开发，通过少量的代码就能快速构建一个个独立的、产品级别的 Spring 应用。

Spring Initializr 方案是官方提供的创建新 Spring Boot 项目的不错选择。开发人员使用官方的初始化方案创建 Spring Boot 项目能够确保获得经过测试和验证的依赖项，这些依赖项适用于自动配置，能够极大简化项目创建流程。同时，IDEA 和 STS 编辑器也支持这种直接初始化 Spring Boot 项目的方式，使开发人员在一分钟之内就可以完成一个项目的初始化工作。

3. 习惯优于配置

Spring Boot 遵循习惯优于配置的原则，在使用 Spring Boot 后开发人员只需要很少的配置甚至零配置即可完成项目开发，一般使用 Spring Boot 默认配置即可。

4. 拥有大量的自动配置

自动进行 Spring 框架的配置，可以节省开发人员大量的时间和精力，能够让开发人员专注在业务逻辑代码的编写上。

5. starter 自动依赖与版本控制

Spring Boot 通过一些 starter 的定义可以减少开发人员在依赖管理上所花费的时间。开发人员在整合各项功能的时候，不需要自己搜索和查找所需依赖，但可以在 Maven 的 pom 文件中进行定义。starter 可以简单理解为"场景启动器"，开发人员可以在不同的场景和功能中引入不同的 starter。如果需要开发 Web 项目，就在 pom 文件中导入 spring-boot-starter-web。在 Web 项目开发中所需的依赖都已经维护在 spring-boot-starter-web 中，无须再导入 Servlet、Spring MVC 等所需要的 jar 包。项目中如果需要使用 JDBC，在 pom 文件中导入 spring-boot-starter-jdbc 即可。针对其他企业级开发中遇到的各种场景，Spring Boot 都有相关的 starter。如果没有对应的 starter 开发人员也可以自行定义。

使用 Spring Boot 开发项目可以非常方便地进行包的管理，所需依赖以及依赖 jar 包的关系和版本都由 starter 自行维护，在很大程度上减少了维护依赖版本所造成的 jar 包冲突或者依赖的版本冲突。

Spring Boot 官方 stater 的详细内容可以参考"Spring Boot-starter-*"。

6. 使用嵌入式的 Servlet 容器

Spring Boot 直接嵌入 Tomcat、Jetty 或者 Undertow 作为 Servlet 容器，降低了对环境的要求，在开发和部署时都无须安装相关 Web 容器，调试方便。在开发完成后可以将项目打包为 jar 包，并使用命令行直接启动项目，从而简化部署环节打包并发布到 Servlet 容器中的流程。

使用嵌入式的 Servlet 容器使得开发调试环节和部署环节的工作量有所减少，同时开发人员也可以通过 Spring Boot 配置文件修改内置 Servlet 容器的配置，简单又灵活。

7. 对主流框架无配置集成，使用场景全覆盖

Spring Boot 集成的技术栈丰富，不同公司使用的技术框架大部分可以无配置集成，

即使不行，也可以通过自定义 spring-boot-starter 进行快速集成。这就意味着 Spring Boot 的应用场景非常广泛，包括常见的 Web、SOA 和微服务等应用。

在 Web 应用中，Spring Boot 提供了 spring-boot-starter-web 来为 Web 开发予以支持。spring-boot-starter-web 为开发人员提供了嵌入的 Tomcat 和 Spring MVC 的依赖，可以快速构建 MVC 模式的 Web 工程。在 SOA 和微服务中，用 Spring Boot 可以包装每个服务。Spring Cloud 即是一套基于 Spring Boot 实现分布式系统的工具，适用于构建微服务。Spring Boot 提供了 spring-boot-starter-websocket 来快速实现消息推送，同时也可以整合流行的 RPC 框架，提供 RPC 服务接口（只要简单加入对应的 starter 组件即可）。

从以上各个特性可以看出，Spring Boot 可以简化 Spring 项目开发过程中冗余复杂的流程。另外，引入 spring-boot-start-actuator 依赖并进行相应的设置可获取 Spring Boot 进程的运行期性能参数，让运维人员也能体验到 Spring Boot 的魅力。

第 3 章

Spring Boot 基础开发环境的安装和配置

工欲善其事必先利其器。本章介绍如何搭建 Spring Boot 项目的基础开发环境，包括 JDK 的安装和配置、Maven 的安装和配置，以及开发工具 IDEA 的安装和配置。

3.1 JDK的安装和配置

由于 Spring Boot 2.x 版本要求 Java 8 作为最低语言版本，因此需要安装 JDK 8 或者以上版本运行。而目前大部分公司或者 Java 开发人员都在使用 Java 8，因此笔者选择 JDK 8 进行安装和配置。

3.1.1 下载安装包

JDK 的安装包可以在 Oracle 官网免费下载。在下载之前，需要确定所使用电脑的系统信息，这里以 Windows 系统为例。首先在电脑桌面上用鼠标右键点击"计算机"或"此电脑"，然后点开属性面板，之后可以在"属性"栏中查看"系统属性"。如果是 64 位操作系统，则需要下载对应的 64 位 JDK 安装包；如果是 32 位操作系统，则需要下载对应的 32 位 JDK 安装包。

打开浏览器，在 Oracle 官网找到对应的 JDK 下载页面。

如果还没有 Oracle 官网的账号，则需要注册一个账号，否则无法在 Oracle 官网下载 JDK 安装包，注册页面如图 3-1 所示。

图 3-1　Oracle 注册页面

在 JDK 下载页面中查看不同系统的安装包，选择对应 JDK 安装包进行下载，如图 3-2 所示。

图 3-2　选择 JDK 安装包

这里选择 Windows x64 的 JDK 安装包，下载前需要同意 Oracle 的许可协议，否则无法下载。

3.1.2 安装 JDK

在 JDK 安装包下载完成后，首先鼠标左键双击“jdk-8u271-windows-x64.exe”文件进行安装，然后会出现 JDK 安装界面，如图 3-3 所示。

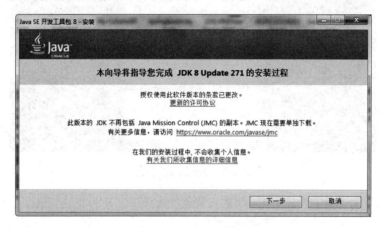

图 3-3 JDK 安装界面

按照 JDK 安装界面的提示，依次点击“下一步”按钮即可完成安装。

需要注意的是，此步骤中 JDK 的安装路径，可以选择安装到 C 盘的默认路径，也可以自行更改安装路径，比如笔者将安装路径修改为 D:\Java\jdk1.8.0_171。另外，因为 JDK 中已经包含 JRE，在安装过程中需要取消公共 JRE 的安装。在安装步骤完成后，可以看到 D:\Java\jdk1.8.0_171 目录下的文件（如图 3-4 所示），这就表示 JDK 安装成功了。

bin	2020/12/23 星期三 下午 21:35	文件夹	
include	2020/12/23 星期三 下午 21:35	文件夹	
jre	2020/12/23 星期三 下午 21:35	文件夹	
legal	2020/12/23 星期三 下午 21:35	文件夹	
lib	2020/12/23 星期三 下午 21:35	文件夹	
COPYRIGHT	2020/9/16 星期三 下午 19:55	文件	4 KB
javafx-src.zip	2020/12/23 星期三 下午 21:35	360压缩 ZIP 文件	5,098 KB
jmc.txt	2020/12/23 星期三 下午 21:35	TXT 文件	1 KB
LICENSE	2020/12/23 星期三 下午 21:35	文件	1 KB
README.html	2020/12/23 星期三 下午 21:35	Chrome HTML D...	1 KB
release	2020/12/23 星期三 下午 21:35	文件	1 KB
src.zip	2020/9/16 星期三 下午 19:55	360压缩 ZIP 文件	20,658 KB
THIRDPARTYLICENSEREADME.txt	2020/12/23 星期三 下午 21:35	TXT 文件	1 KB
THIRDPARTYLICENSEREADME-JAVAF...	2020/12/23 星期三 下午 21:35	TXT 文件	1 KB

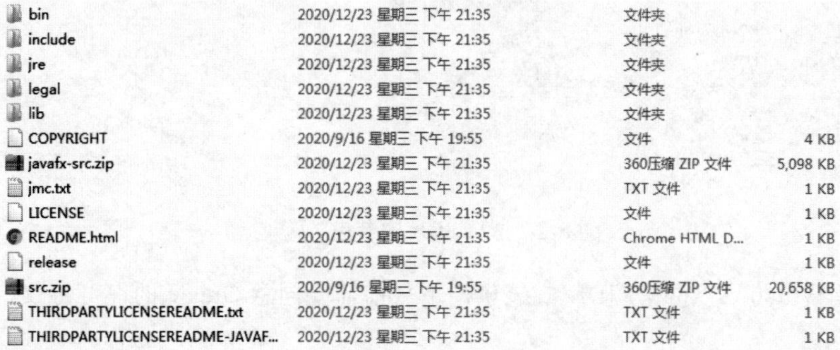

图 3-4 JDK 安装文件

3.1.3 配置环境变量

在安装成功后,还需要配置 Java 的环境变量,具体如下所示。

首先在电脑桌面上鼠标右键点击"计算机"或者"此电脑",然后点开属性面板,点击"高级系统设置",在弹出的"系统属性"面板中点击"高级"选项卡,最后点击"环境变量"按钮。

在环境变量面板中,点击"系统变量"下方的"新建"按钮,在"变量名"输入框中输入"JAVA_HOME";在"变量值"输入框中输入安装步骤中选择的 JDK 安装目录,比如"D:\Java\jdk1.8.0_271",点击"确定"按钮,如图 3-5 所示。

图 3-5 新建 JAVA_HOME 环境变量

编辑 Path 变量,在变量的末尾添加:

```
.;%JAVA_HOME%\bin;%JAVA_HOME%\jre\bin;
```

具体如图 3-6 所示。

图 3-6 编辑 Path 环境变量

增加 CLASS_PATH 变量，与添加 JAVA_HOME 变量的过程一样，变量名为"CLASS_PATH"，变量值为";%JAVA_HOME%\lib;%JAVA_HOME%\lib\tools.jar"。

至此，环境变量设置完成。

3.1.4　JDK 环境变量验证

在完成环境变量配置后，还需要验证是否配置正确。

打开 cmd 命令窗口，输入 java -version 命令。这里演示安装的 JDK 版本为 1.8.0_271，如果环境变量配置正确，命令窗口会输出正确的 JDK 版本号：

```
java version "1.8.0_271"
```

如果验证结果如图 3-7 所示，则表示 JDK 安装成功。如果输入命令后报错，则需要检查在环境变量配置步骤中是否存在路径错误或者拼写错误并进行改正。

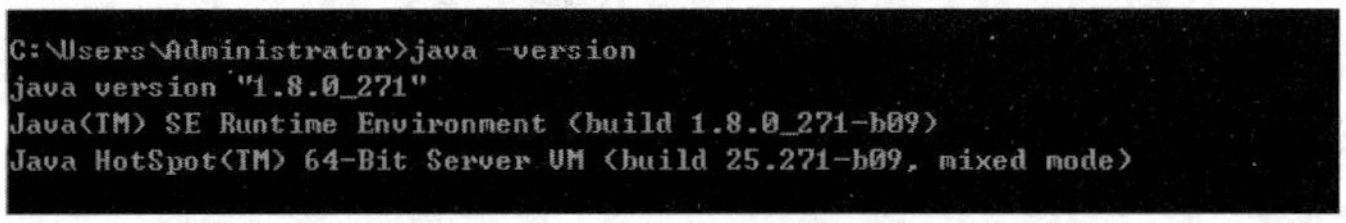

```
C:\Users\Administrator>java -version
java version "1.8.0_271"
Java(TM) SE Runtime Environment (build 1.8.0_271-b09)
Java HotSpot(TM) 64-Bit Server VM (build 25.271-b09, mixed mode)
```

图 3-7　JDK 安装验证结果

3.2　Maven 的安装和配置

Maven 是 Apache 的一个软件项目管理和构建工具，它可以对 Java 项目进行构建和依赖管理。本书中所有源码都选择了 Maven 作为项目依赖管理工具，本节内容将讲解 Maven 的安装和配置。

当然，Gradle 也是目前比较流行的项目管理工具，感兴趣的读者可以尝试使用。

3.2.1　下载安装包

打开浏览器，在 Apache 官网找到 Maven 下载页面，其下载文件列表如图 3-8 所示。点击"apache-maven-3.6.3-bin.zip"即可完成下载。

第 3 章 Spring Boot 基础开发环境的安装和配置

图 3-8 Maven 下载文件列表

3.2.2 安装并配置 Maven

首先安装 Maven 并不像安装 JDK 一样需要执行安装程序，直接将下载的安装包解压到相应的目录下即可。这里笔者解压到 D:\maven\apache-maven-3.6.3 目录下，如图 3-9 所示。

图 3-9 Maven 解压目录

然后需要配置 Maven 命令的环境变量，步骤与配置 JDK 环境变量类似。在环境变量面板中，点击"系统变量"下方的"新建"按钮，在"变量名"输入框中输入"MAVEN_HOME"，在"变量值"输入框中输入目录，比如"D:\maven\apache-maven-3.6.3"，点击"确定"按钮，具体如图 3-10 所示。

图 3-10 新建 Maven_HOME 环境变量

最后修改 Path 环境变量，在末尾增加：

;%MAVEN_HOME%\bin;

3.2.3　Maven 环境变量验证

在 Maven 环境变量配置完成后，同样需要验证是否配置正确。

打开 cmd 命令窗口，输 mvn -v 命令。这里安装的 Maven 版本为 3.6.3，安装目录为 D:\maven\apache-maven-3.6.3。如果环境变量配置正确，在命令窗口会输出如图 3-11 所示的验证结果，表示 Maven 安装成功。

图 3-11　Maven 安装验证结果

如果在输入命令后报错，则需要检查在环境变量配置步骤中是否存在路径错误或者拼写错误并进行改正。

3.2.4　配置国内 Maven 镜像

在完成以上步骤后就可以正常使用 Maven 工具了。为了获得更好的使用体验，建议国内开发人员修改一下 Maven 的配置文件。

国内开发人员在使用 Maven 下载项目的依赖文件时，通常会面临下载速度缓慢的情况，甚至出现"编码 5 分钟，启动项目半小时"的窘境。这是因为 Maven 的中央仓库在国外的服务器中，如图 3-12 所示。

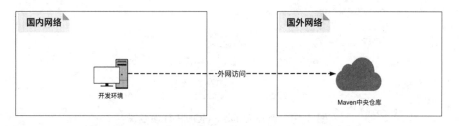

图 3-12　Maven 中央仓库

由于每次下载新的依赖文件都需要通过外网访问 Maven 中央仓库，如果不进行配置的优化处理则会极大地影响开发流程。笔者建议使用国内公司提供的中央仓库镜像，比如阿里云的镜像、华为云的镜像。另外一种做法是自己搭建一个私有的中央仓库，并修改 Maven 配置文件中的 mirror 标签来设置镜像仓库。

这里以阿里云镜像仓库为例，介绍如何配置国内 Maven 镜像加快依赖的访问速度。

进入 Maven 安装目录 D:\maven\apache-maven-3.6.3，在 conf 文件夹中打开 settings.xml 配置文件。添加阿里云镜像仓库的链接，修改后的 settings.xml 配置文件如下：

```xml
<?xml version="1.0" encoding="UTF-8"?>
<settings xmlns="http://maven.apache.org/SETTINGS/1.0.0"
      xmlns:xsi="http://www.w3.org/2001/XMLSchema-instance"
      xsi:schemaLocation="http://maven.apache.org/SETTINGS/1.0.0
http://maven.apache.org/xsd/settings-1.0.0.xsd">

<!-- 本地仓库的路径设置在 D 盘 maven/repo 目录下（自行配置一个文件夹即可，默认是 ~/.m2/repository) -->
<localRepository>D:\maven\repo</localRepository>

<!-- 配置阿里云镜像服务器-->
<mirrors>
  <mirror>
    <id>alimaven</id>
    <name>aliyun maven</name>
    <url>http://maven.aliyun.com/nexus/content/groups/public/</url>
    <mirrorOf>central</mirrorOf>
  </mirror>
</mirrors>

</settings>
```

在配置完成后，可以直接访问国内的镜像仓库，从而使 Maven 下载 jar 包依赖的速度变得更快，可以节省很多时间，如图 3-13 所示。

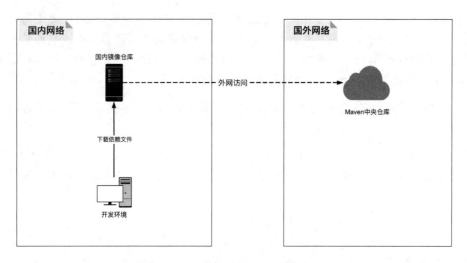

图 3-13 访问 Maven 国内镜像仓库

3.3 开发工具IDEA的安装和配置

Java 开发人员常使用的开发工具包括 Eclipse、MyEclipse 和 IDEA。

关于在 Spring Boot 项目开发时编辑器的选择,笔者推荐 IDEA 作为主要的开发工具。IDEA 对于开发人员非常友好和方便,本书关于项目的开发和演示都会选择使用 IDEA 编辑器。

IDEA 全称 IntelliJ IDEA,是用 Java 语言开发的集成环境(也可用于其他语言)。IntelliJ 在业界被公认为最好的 Java 开发工具之一,尤其在智能代码助手、代码自动提示、重构、J2EE 支持、JUnit 单元测试、CVS 版本控制、代码审查、创新的 GUI 设计等方面的功能可以说是超常的。

3.3.1 下载 IDEA 安装包

打开浏览器,进入 JetBrains 官网。在进入 IDEA 页面后能够查看其基本信息和特性介绍,如图 3-14 所示。感兴趣的读者可以在该页面了解 IDEA 编辑器更多的信息。

第 3 章　Spring Boot 基础开发环境的安装和配置

图 3-14　IDEA 编辑器介绍页面

点击页面中的"Download"按钮，进入 IDEA 编辑器的下载页面，如图 3-15 所示。笔者在整理书稿时，IDEA 编辑器的最新版本为 2020.3。

图 3-15　IDEA 编辑器下载页面

在 IDEA 编辑器的下载页面可以看到两种收费模式的版本。

（1）Ultimate 为商业版本，需要付费购买使用，功能更加强大，插件也更多，使用起来也会更加顺手，可以免费试用 30 天。

（2）Community 为社区版本，可以免费使用，功能和插件相较于付费版本有一定的减少，不过对于项目开发并没有太大的影响。

读者根据所使用的系统版本下载对应的安装包即可，本书将以 Community 社区版本为例进行讲解。

3.3.2 安装 IDEA 及其功能介绍

在下载完成后，双击下载的安装包程序，按照 IDEA 安装界面的提示，依次点击"Next"按钮即可完成安装，如图 3-16 所示。

图 3-16　IDEA 编辑器安装界面

首次打开 IDEA 编辑器可以看到它的欢迎页面，如图 3-17 所示。

图 3-17　IDEA 编辑器欢迎页面

功能区域有三个按钮，功能分别如下所示。

（1）New Project：创建一个新项目。

（2）Open：打开一个计算机中已有的项目。

（3）Get from VCS：通过在版本控制系统中打开项目获取一个项目，比如通过 GitHub、Gitee、GitLab 以及自建的版本控制系统。

在创建或者打开一个项目后，则进入 IDEA 编辑器的主界面。这里以一个基础的 Spring Boot 项目为例进行介绍。在打开项目后，IDEA 编辑器界面如图 3-18 所示。

图 3-18　IDEA 编辑器主界面

由上至下，依次为菜单栏区域、代码操作区域、控制台和终端区域。代码操作区域是开发时主要操作的区域，包括项目结构、代码编辑区、Maven 工具栏。菜单栏区域主要的作用是放功能配置的按钮和增强功能的按钮。控制台和终端区域主要显示项目信息、程序运行日志、代码的版本提交记录、终端命令行等内容。

3.3.3　配置 IDEA 的 Maven 环境

IDEA 编辑器是自带 Maven 环境的，如图 3-19 所示。

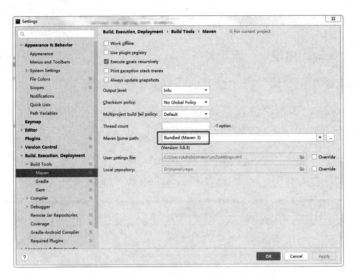

图 3-19　IDEA 编辑器自带 Maven 环境

为了避免一些不必要的麻烦，笔者建议将 IDEA 编辑器中的 Maven 设置为之前已经在全局设置的 Maven 环境。

想要之前安装的 Maven 可以正常在 IDEA 中使用，则需要进行以下配置。依次点击菜单栏中的按钮"File→ Settings→Build,Execution,Deployment→Build Tools→Maven"，在 Maven 设置面板中配置 Maven 目录和 settings.xml 配置文件位置，如图 3-20 所示。

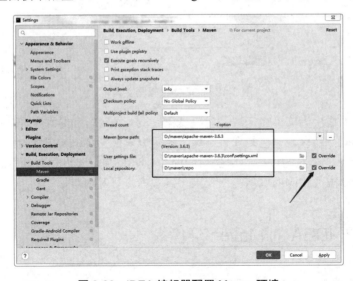

图 3-20　IDEA 编辑器配置 Maven 环境

磨刀不误砍柴工，准备好基础的开发环境和开发工具才有利于后续的编码实践。还要提醒各位读者，如果已经习惯了其他代码编辑工具可以继续使用。这里只是考虑对 Spring Boot 项目的支持，笔者建议使用 IDEA 编辑器。由于本书使用的 MySQL 数据库版本为 5.7，为了避免一些问题，建议读者使用 MySQL 5.7 或以上版本。本书中的所有源码选择的 Spring Boot 版本为 2.3.7，要求 JDK 的最低版本为 JDK 8，建议读者安装 JDK 8 或者以上版本。

第 4 章

Spring Boot 项目搭建及快速上手

本章讲解使用 IDEA 进行 Spring Boot 项目的创建和开发，笔者将和读者一起编写本书的第一个 Spring Boot 项目，希望读者能够尽快上手和体验。

4.1 Spring Boot项目创建

4.1.1 认识 Spring Initializr

Spring 官方提供了 Spring Initializr 来进行 Spring Boot 项目的初始化。这是一个在线生成 Spring Boot 基础项目的工具，可以将其理解为 Spring Boot 项目的"初始化向导"，它可以帮助开发人员快速创建一个 Spring Boot 项目。接下来将讲解如何使用 Spring Initializr 快速初始化一个 Spring Boot 骨架工程。

访问 Spring 官方提供的 Spring Initializr 网站，打开浏览器并输入 Spring Initializr 的网站地址，页面如图 4-1 所示。

从图 4-1 可以看到 Spring Initializr 页面展示的内容。如果想初始化一个 Spring Boot 项目需要提前对其进行简单的配置，直接对页面中的配置项进行勾选和输入即可。在默认情况下相关配置项已经有缺省值，可以根据实际情况进行简单修改。

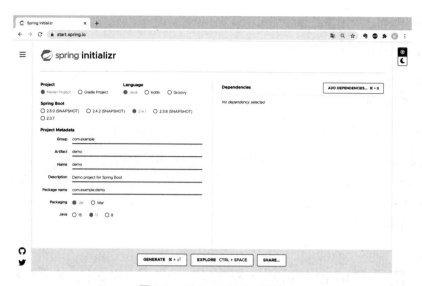

图 4-1　Spring Initializr 网站

4.1.2　Spring Boot 项目初始化配置

需要配置的参数释义如下所示。

Project：表示将要初始化的 Spring Boot 项目类型，可以选择 Maven 构建或者 Gradle 构建，这里选择常用的 Maven 方式。

Language：表示编程语言的选择，支持 Java、Kotlin 和 Groovy。

Spring Boot：表示将要初始化的 Spring Boot 项目所使用的 Spring Boot 版本。由于版本更新迭代较快，Spring Initializr 页面会展示最新的几个 Spring Boot 版本号。其他的版本号虽然不会在这里展示，但是依然可以正常使用。

Project Metada：表示项目的基础设置，包括项目包名的设置、打包方式、JDK 版本选择等。

Group：即 GroupID，表示项目组织的标识符，实际对应 Java 的包结构，是 main 目录里 Java 的目录结构。

Artifact：即 ArtifactId，表示项目的标识符，实际对应项目的名称，也就是项目根目录的名称。

Description：表示项目描述信息。

Package name：表示项目包名。

Packaging：表示项目的打包方式，有 Jar 和 War 两种选择。在 Spring Boot 项目初始化时，如果选用的方式不同，那么导入的打包插件也有区别。

Java：表示 JDK 版本的选择，有 15、11 和 8 三个版本供开发人员选择。

Dependencies：表示将要初始化的 Spring Boot 项目所需的依赖和 starter。如果不选择此项的话，在默认生成的项目中仅有核心模块 spring-boot-starter 和测试模块 spring-boot-starter-test。在这个配置项中可以设置项目所需的 starter，比如 Web 开发所需的依赖、数据库开发所需的依赖等。

4.1.3 使用 Spring Initializr 初始化一个 Spring Boot 项目

在 Spring Initializr 页面中的配置项需要开发人员逐一进行设置，过程非常简单，根据项目情况依次填写即可。

在本次演示中，开发语言选择 Java。因为本地安装的项目管理工具是 Maven，在 Project 项目类型选项中勾选 Maven Project。Spring Boot 版本选择 2.3.7，根据实际开发情况也可以选择其他稳定版本。即使这里已经选择了一个版本号，在初始化成功后也能够在项目中的 pom.xml 文件或者 build.gradle 文件中修改 Spring Boot 版本号。

在项目基础信息中，Group 输入框中填写"ltd.newbee.mall"，Artifact 输入框中填写"newbee-mall"，Name 输入框中填写"newbee-mall"，Description 输入框中填写"NEWBEE 商城"，Package name 输入框中填写"ltd.newbee.mall"，Packaging 打包方式选择 Jar，JDK 版本选择 8。

由于即将开发的是一个 Web 项目，因此需要添加 web-starter 依赖，点击 Dependencies 右侧的"ADD DEPENDENCIES"按钮，在弹出的弹框中输入关键字"web"并选择"Spring Web：Build web, including RESTful, applications using Spring MVC. Uses Apache Tomcat as the default embedded container."如图 4-2 所示。

很明显，该项目将会采用 Spring MVC 开发框架并且使用 Tomcat 作为默认的嵌入式容器。

至此，初始化 Spring Boot 项目的选项配置完成，如图 4-3 所示。

第 4 章　Spring Boot 项目搭建及快速上手

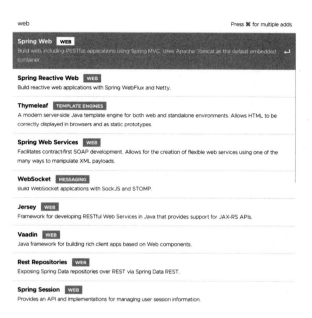

图 4-2　选择 web-starter 依赖

图 4-3　初始化 Spring Boot 项目的选项配置

最后，点击页面底部的"Generate"按钮，即可获取一个 Spring Boot 基础项目的代码压缩包。

4.1.4 其他方式创建 Spring Boot 项目

除了使用官方推荐的 Spring Initializr 方式创建 Spring Boot 项目之外，开发人员也可以选择其他方式创建 Spring Boot 项目。

1. IDEA 编辑器初始化 Spring Boot 项目

在 IDEA 编辑器中内置了初始化 Spring Boot 项目的插件，可以直接新建一个 Spring Boot 项目，创建过程如图 4-4 所示。

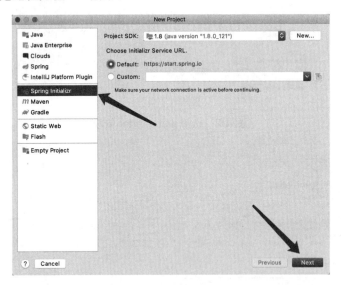

图 4-4　使用 IDEA 编辑器初始化 Spring Boot 项目

需要注意的是，这种方式仅在商业版本的 IDEA 编辑器中可行。IDEA 编辑器社区版本在默认情况下不支持直接生成 Spring Boot 项目。

2. Maven 命令行创建 Spring Boot 项目

使用 Maven 命令也可以创建一个新的项目，操作方式如下所示。

打开命令行工具并将目录切换到对应的文件夹中，运行以下命令：

```
mvn archetype:generate -DinteractiveMode=false -DgroupId=ltd.newbee.mall -DartifactId=newbee-mall -Dversion=0.0.1-SNAPSHOT
```

在构建成功后会生成一个 Maven 骨架项目。但是由于生成的项目仅仅是骨架项目，因此在 pom.xml 文件中需要自己添加依赖，主方法的启动类也需要自行添加。该方法没有前两种方式方便快捷，因此不推荐。

当然，如果计算机中已经存在 Spring Boot 项目则直接打开即可。点击"Open"按钮跳出文件选择框，选择想要导入的项目目录，导入成功就可以进行 Spring Boot 项目开发了。

4.2 Spring Boot项目目录结构介绍

在使用 IDEA 编辑器打开项目之后，可以看到 Spring Boot 项目的目录结构，如图 4-5 所示。

图 4-5　Spring Boot 项目的目录结构图解

Spring Boot 的目录结构主要由以下部分组成：

```
newbee-mall
├── src/main/java
├── src/main/resources
├── src/test/java
    └── pom.xml
```

src/main/java 表示 Java 程序开发目录，开发人员在该目录下进行业务代码的开发。

这个目录对于 Java Web 开发人员来说应该比较熟悉，唯一的不同是 Spring Boot 项目中会多一个主程序类。

src/main/resources 表示配置文件目录，主要用于存放静态文件、模板文件和配置文件。它与普通的 Spring 项目相比有些区别，该目录下有 static 和 templates 两个目录，是 Spring Boot 项目默认的静态资源文件目录和模板文件目录。在 Spring Boot 项目中是没有 webapp 目录的，它默认是使用 static 和 templates 两个文件夹。

static 目录用于存放静态资源文件，如 JavaScript 文件、图片、CSS 文件。

templates 目录用于存放模板文件，如 Thymeleaf 模板文件或者 FreeMarker 文件。

src/test/java 表示测试类文件夹，与普通的 Spring 项目差别不大。

pom.xml 用于配置项目依赖。

以上即为 Spring Boot 项目的目录结构，与普通的 Spring 项目存在一些差异，但是在正常开发过程中这个差异的影响并不大。真正差别较大的地方应该是部署和启动方式的差异，接下来将详细介绍 Spring Boot 项目的启动方式。

4.3 启动Spring Boot项目

4.3.1 在 IDEA 编辑器中启动 Spring Boot 项目

由于 IDEA 编辑器对于 Spring Boot 项目的支持非常友好，在项目导入成功后会被自动识别为 Spring Boot 项目，可以快速进行启动操作。

在 IDEA 编辑器中，有以下三种方式可以启动 Spring Boot 项目。

（1）主类上的启动按钮：打开程序启动类，比如本次演示的 NewBeeMallApplication.java，在 IDEA 代码编辑区域中可以看到左侧有两个绿色的三角形启动按钮，点击任意一个按钮即可启动 Spring Boot 项目。

（2）右键运行 Spring Boot 的主程序类：与普通 Java 类的启动方式类似，在左侧 Project 侧边栏或者类文件编辑器中，执行右键点击操作，可以看到启动 main()方法的按钮，点击 "Run 'NewbeeMallApplication.main()'" 即可启动 Spring Boot 项目，如图 4-6 所示。

第 4 章 Spring Boot 项目搭建及快速上手

图 4-6 右键点击运行 Spring Boot 的主程序类

（3）工具栏中的 Run/Debug 按钮：点击工具栏中的 Run/Debug 按钮也可以启动 Spring Boot 项目，如图 4-7 所示。

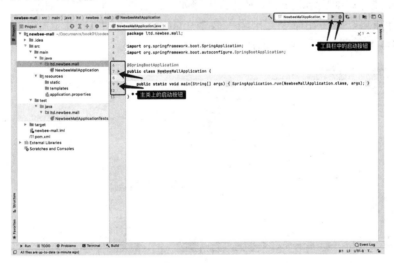

图 4-7 使用工具栏中的按钮启动 Spring Boot 的主程序类

Spring Boot 项目的启动比普通的 Java Web 项目更便捷，减少了几个中间步骤，不用配置 Servlet 容器，也不用打包并且发布到 Servlet 容器再启动，而是直接运行主方法即可启动项目。其开发、调试都十分方便且节省时间。

4.3.2 Maven 插件启动

在项目初始化时，配置项选择的项目类型为 Maven Project，pom.xml 文件中会默认引入 spring-boot-maven-plugin 插件依赖，因此可以直接使用 Maven 命令来启动 Spring Boot 项目，插件配置如下：

```xml
<build>
    <plugins>
        <plugin>
            <groupId>org.springframework.boot</groupId>
            <artifactId>spring-boot-maven-plugin</artifactId>
        </plugin>
    </plugins>
</build>
```

如果在 pom.xml 文件中没有该 Maven 插件配置，是无法通过这种方式启动 Spring Boot 项目的，这一点需要注意。

Maven 插件启动 Spring Boot 项目的步骤如下：首先点击下方工具栏中的 Terminal，打开命令行窗口，然后在命令行中输入命令 mvn spring-boot:run 并执行该命令，即可启动 Spring Boot 项目，如图 4-8 所示。

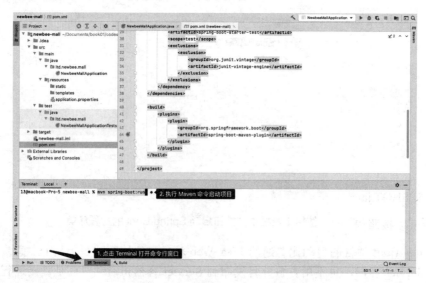

图 4-8　使用 Maven 插件启动 Spring Boot 项目

4.3.3 java-jar 命令启动

在项目初始化时，配置项选择的打包方式为 Jar，那么项目开发完成打包后的结果就是一个 jar 包文件。通过 Java 命令行运行 jar 包的命令为 java -jar xxx.jar，因此可以使用这种方式启动 Spring Boot 项目，如图 4-9 所示。

（1）首先点击下方工具栏中的 Terminal 打开命令行窗口。

（2）然后使用 Maven 命令将项目打包，执行命令为 mvn clean package-Dmaven.test.skip=true，等待打包结果即可。

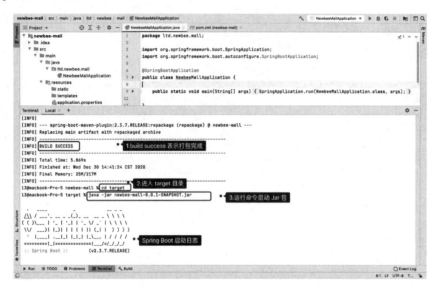

图 4-9 使用 java-jar 命令启动 Spring Boot 项目

（3）在打包成功后进入 target 目录，切换目录的命令为 cd target。

（4）最后是启动已经生成的 jar 包文件，执行命令为 java -jar newbee-mall-0.0.1-SNAPSHOT.jar。

读者可以按照以上步骤练习几次。

需要注意的是，每次在项目启动之前，如果使用了其他方式启动项目工程，则需要将其关掉，否则会因为端口占用导致启动报错，进而无法正常启动 Spring Boot 项目。

4.3.4　Spring Boot 项目启动日志

无论使用以上哪种方式，在 Spring Boot 项目启动时都会在控制台上输出启动日志，如果一切正常则很快就能够启动成功，启动日志如下所示：

```
  .   ____          _            __ _ _
 /\\ / ___'_ __ _ _(_)_ __  __ _ \ \ \ \
( ( )\___ | '_ | '_| | '_ \/ _` | \ \ \ \
 \\/  ___)| |_)| | | | | || (_| |  ) ) ) )
  '  |____| .__|_| |_|_| |_\__, | / / / /
 =========|_|==============|___/=/_/_/_/
 :: Spring Boot ::        (v2.3.7.RELEASE)

2021-01-13 14:50:10.137  INFO 21651 --- [main]
ltd.newbee.mall.NewbeeMallApplication    : Starting NewbeeMallApplication with PID 21651
2021-01-13 14:50:10.143  INFO 21651 --- [main]
ltd.newbee.mall.NewbeeMallApplication    : No active profile set, falling back to default profiles: default
2021-01-13 14:50:12.492  INFO 21651 --- [main]
o.s.b.w.embedded.tomcat.TomcatWebServer  : Tomcat initialized with port(s): 8080 (http)
2021-01-13 14:50:12.517  INFO 21651 --- [main]
o.apache.catalina.core.StandardService   : Starting service [Tomcat]
2021-01-13 14:50:12.518  INFO 21651 --- [main]
org.apache.catalina.core.StandardEngine  : Starting Servlet engine: [Apache Tomcat/9.0.41]
2021-01-13 14:50:12.702  INFO 21651 --- [main]
o.a.c.c.C.[Tomcat].[localhost].[/]       : Initializing Spring embedded WebApplicationContext
2021-01-13 14:50:12.702  INFO 21651 --- [main]
w.s.c.ServletWebServerApplicationContext : Root WebApplicationContext: initialization completed in 2405 ms
2021-01-13 14:50:13.105  INFO 21651 --- [main]
o.s.s.concurrent.ThreadPoolTaskExecutor  : Initializing ExecutorService 'applicationTaskExecutor'
2021-01-13 14:50:13.695  INFO 21651 --- [main]
o.s.b.w.embedded.tomcat.TomcatWebServer  : Tomcat started on port(s): 8080
```

```
(http) with context path ''
2021-01-13 14:50:13.725  INFO 21651 --- [main]
ltd.newbee.mall.NewbeeMallApplication      : Started NewbeeMallApplication in
3.634 seconds (JVM running for 6.557)
```

日志前面部分为 Spring Boot 的启动 Banner 和 Spring Boot 的版本号，中间部分为 Tomcat 启动信息及 ServletWebServerApplicationContext 加载完成信息，后面部分则是 Tomcat 的启动端口和项目启动时间。通过以上日志信息，可以看出 Spring Boot 启动成功共花费 3.634 秒，Tomcat 服务器监听的端口号为 8080。

4.4 开发第一个Spring Boot项目

在项目成功启动后，打开浏览器访问 8080 端口，看到的页面却是一个 Whitelabel Error Page 页面，如图 4-10 所示。

图 4-10 Whitelabel Error Page 页面

这个页面是 Spring Boot 项目的默认错误页面，由页面内容可以看出此次访问的报错为 404 错误。访问其他地址也会出现这个页面。原因是此时在 Web 服务中并没有任何可访问的资源。在生成 Spring Boot 项目之后，由于并没有在项目中增加任何一行代码，就没有接口，也没有页面。

此时，需要自行实现一个 Controller 查看 Spring Boot 如何处理 Web 请求。接下来使用 Spring Boot 实现一个简单的接口，步骤如下所示。

（1）在根目录 ltd.newbee.mall 中点击鼠标右键，在弹出的菜单栏中选择"New→Package"，新建名称为 controller 的 Java 包，如图 4-11 所示。

（2）在 ltd.newbee.mall.controller 中点击鼠标右键，在弹出的菜单栏中选择"New→Java Class"，新建名称为 HelloController 的 Java 类，此时的目录结构如图 4-12 所示。

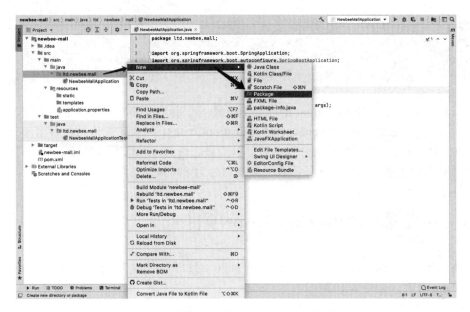

图 4-11　新建名称为 controller 的 Java 包

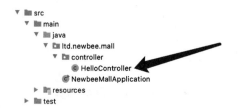

图 4-12　HelloController 目录结构

（3）在 HelloController 类中输入如下代码：

```java
package ltd.newbee.mall.controller;

import org.springframework.stereotype.Controller;
import org.springframework.web.bind.annotation.GetMapping;
import org.springframework.web.bind.annotation.ResponseBody;

@Controller
public class HelloController {

    @GetMapping("/hello")
    @ResponseBody
    public String hello() {
        return "hello,spring boot!";
```

```
    }
}
```

以上这段代码的实现读者应该很熟悉，写法与 Spring 项目开发相同。该段代码的含义是处理请求路径为 /hello 的 GET 请求并返回一个字符串。

在编码完成后，重新启动项目，启动成功后在浏览器中输入以下请求地址：

`http://localhost:8080/hello`

这时页面上展示的内容已经不是错误信息，而是 HelloController 中的正确返回信息，如图 4-13 所示。第一个 Spring Boot 项目实例就完成了！

图 4-13　HelloController 页面效果

本章主要介绍了如何创建一个 Spring Boot 项目，并使用 IDEA 编辑器开发 Spring Boot 项目。

根据笔者的开发经验，在新建 Spring Boot 项目时，建议开发人员使用 Spring Initializr 向导构建。因为该方式生成的代码比较齐全，可避免人为错误，可以直接使用，更加节省时间。而采用 Maven 构建的方式需要进行 pom.xml 文件配置和主程序类的编写。

Spring Boot 项目的启动方式笔者列举了 IDEA 直接启动、Maven 插件启动和命令行启动三种。以上三种方式都很简单，在练习时读者可以自行选择适合自己的启动方式。

在日常开发中通常使用 IDEA 上的按钮或者快捷键直接启动项目，这也比较符合开发人员的开发习惯。Maven 插件启动也是一种 Spring Boot 项目的启动方式，直接运行 Maven 命令即可启动项目。命令行启动项目的方式一般在服务器上部署项目时使用，这是因为项目在上线时通常在生产环境的服务器上直接上传 jar 包文件，再运行 java -jar xxx.jar 命令启动 Spring Boot 项目。

第 5 章

Spring Boot 核心详解及源码分析

Spring Boot 不需要额外安装 Tomcat，也不需要配置 Spring MVC 的 XML 配置文件，就可以快速构建出一个可访问的 Web 工程，启动和运行也十分方便。为何会如此呢？本章将对 Spring Boot 源码进行分析，让读者对 Spring Boot 的特性和启动流程有一个更加清晰、深入的认知。

5.1 约定优于配置

首先讲解一下 Spring Boot 的核心设计思想：约定优于配置。

很多 Java 开发人员在阅读和学习 Spring Boot 教程的时候往往会对自动化配置感到困惑，更会被教程中所谓的 Spring Boot 核心设计思想约定优于配置所迷惑。那么，这个核心设计思想究竟是什么呢？

约定优于配置是由英文 convention over configuration 翻译而来的，也可翻译为惯例优先原则。这并不是一个新潮的概念，相反，这个概念很早就被提出来了。convention 意为惯例、习俗、规矩，因此，可以说 convention over configuration 是一个理念，它并不是 Spring Boot 所独有的特性。

这个理念其实一直都在被遵循、被使用。举一个具体的例子，在 MVC 项目的开发中会把实体类放到 entity 目录下、把数据接口层定义在 dao 目录下、把控制器定义在 controller 目录下。如果在数据库中有一张名称为 tb_mall_user 的表，那么可能在项目中就对应有一个名称为 MallUser 的实体类。这就是开发人员都在遵循的规矩或者说是一种约定。但是所谓的规矩或者约定也需要配置文件来支持。约定优于配置并不是说不需要配置，它其实是一种开发原则，目的是在一定限度内减少需要开发人员手动配置的工作量，用默认的配置达到开发人员的需求。

正因为遵循这个约定，Spring Boot 才让开发人员的开发效率更高。比如在第 4 章中开发的第一个 Spring Boot 项目。在项目中加入 spring-boot-starter-web 后，Spring Boot 会自动导入 Spring MVC 的相关依赖和一个内置的 Tomcat 容器。在这里就有"约定优于配置"的理念。在导入 spring-boot-starter-web 后，就表示开发人员与 Spring Boot 之间约定当前项目是一个 Web 项目。既然是 Web 项目，又进一步约定开发人员会使用 Spring MVC，从而自动装配 DispatchServlet 到 IOC 容器中。同理，开发人员在项目中加入 jdbc-starter 后，就表示开发人员与 Spring Boot 之间约定会对数据库进行操作，Spring Boot 会自动导入 JDBC 相关的依赖并配置一个数据源对象供开发人员调用。

因为相关配置工作由 Spring Boot 主动进行，并且可以达到同样的目的，开发人员的工作量就减少了。这就是 Spring Boot 中的约定优于配置的特性。当然，如果开发人员想要自己完成相关配置，也可以通过编码的方式操作。

这种简洁高效的配置方式源于 Spring Boot 众多的自动化配置类及相关的编码设计。这些自动化配置类在类路径 META-INF/spring.factories 文件中，它通过@EnableAutoConfiguration 注解加载到容器中并发挥作用。在 Spring Boot 中，开发人员甚至可以做到零配置快速构建出所需的应用。

为了读者能更好地理解 Spring Boot 约定优于配置的设计思想，笔者将在后续章节中进行详细的源码剖析和讲解。

5.2 Spring Boot 之依赖管理

打开 IDEA 编辑器，进入在第 4 章中创建的 newbee-mall 项目。打开 pom.xml 文件，可以看到在该 Spring Boot 项目的 pom.xml 文件中配置了一个 parent 属性，其代码如下所示：

```xml
<parent>
    <groupId>org.springframework.boot</groupId>
    <artifactId>spring-boot-starter-parent</artifactId>
    <version>2.3.7.RELEASE</version>
    <relativePath/>
</parent>
```

该项目依赖一个父项目，这个父项目是 spring-boot-starter-parent。为了了解这个父项目的作用，按住 Ctrl 键并点击 spring-boot-starter-parent，如图 5-1 所示。

图 5-1　按住 Ctrl 键并点击 spring-boot-starter-parent

在点击后会进入 spring-boot-starter-parent-2.3.7.RELEASE.pom 文件，查看其源码（节选）：

```
<modelVersion>4.0.0</modelVersion>
<parent>
  <groupId>org.springframework.boot</groupId>
  <artifactId>spring-boot-dependencies</artifactId>
  <version>2.3.7.RELEASE</version>
</parent>
<artifactId>spring-boot-starter-parent</artifactId>
<packaging>pom</packaging>
<name>spring-boot-starter-parent</name>
<description>Parent pom providing dependency and plugin management for applications built with Maven</description>
<properties>
  <java.version>1.8</java.version>
  <resource.delimiter>@</resource.delimiter>
  <maven.compiler.source>${java.version}</maven.compiler.source>
  <maven.compiler.target>${java.version}</maven.compiler.target>
  <project.build.sourceEncoding>UTF-8</project.build.sourceEncoding>
<project.reporting.outputEncoding>UTF-8</project.reporting.outputEncoding>
</properties>
```

通过源码可以看到，在 pom 文件中仅仅定义了 JDK 所需的版本为 1.8、项目的默认编码 UTF-8 和 Maven 项目编译设置。也就是说该 pom 文件定义了 Maven 构建插件等基础配置的属性。继续查看源码会发现，该 pom 文件也依赖了一个父项目 spring-boot-dependencies。

同样的做法，按住 Ctrl 键并点击 spring-boot-dependencies，进入 spring-boot-dependencies-2.3.7.RELEASE.pom 文件中，查看其源码（节选）：

```xml
<modelVersion>4.0.0</modelVersion>
 <groupId>org.springframework.boot</groupId>
 <artifactId>spring-boot-dependencies</artifactId>
 <version>2.3.7.RELEASE</version>
 <packaging>pom</packaging>
 <name>spring-boot-dependencies</name>
 <description>Spring Boot Dependencies</description>
 <url>https://spring.io/projects/spring-boot</url>
 <licenses>
   <license>
     <name>Apache License, Version 2.0</name>
     <url>https://www.apache.org/licenses/LICENSE-2.0</url>
   </license>
 </licenses>
 <developers>
   <developer>
     <name>Pivotal</name>
     <email>info@pivotal.io</email>
     <organization>Pivotal Software, Inc.</organization>
     <organizationUrl>https://www.spring.io</organizationUrl>
   </developer>
 </developers>
 <scm>
   <url>https://github.com/spring-projects/spring-boot</url>
 </scm>
 <properties>
   <activemq.version>5.15.14</activemq.version>
   <antlr2.version>2.7.7</antlr2.version>
   <appengine-sdk.version>1.9.83</appengine-sdk.version>
   <artemis.version>2.12.0</artemis.version>
   <aspectj.version>1.9.6</aspectj.version>
   <assertj.version>3.16.1</assertj.version>
   <atomikos.version>4.0.6</atomikos.version>
   <awaitility.version>4.0.3</awaitility.version>
   <bitronix.version>2.1.4</bitronix.version>
```

```xml
    <build-helper-maven-plugin.version>3.1.0</build-helper-maven-plugin.version>
    <byte-buddy.version>1.10.18</byte-buddy.version>
    <caffeine.version>2.8.8</caffeine.version>
    <cassandra-driver.version>4.6.1</cassandra-driver.version>
    <classmate.version>1.5.1</classmate.version>
    <commons-codec.version>1.14</commons-codec.version>
    <commons-dbcp2.version>2.7.0</commons-dbcp2.version>
    <commons-lang3.version>3.10</commons-lang3.version>
    <commons-pool.version>1.6</commons-pool.version>
    <commons-pool2.version>2.8.1</commons-pool2.version>
    ...省略部分代码

</properties>
    ...省略部分代码
```

其实整个 pom.xml 文件有 3000 多行，在该文件中定义了大量的依赖信息，如 commons 相关依赖、log 相关依赖、数据库相关依赖、Spring 相关依赖、ElasticSearch 搜索引擎相关依赖、消息队列相关依赖等。在 J2EE 项目开发中所涉及的大部分功能场景的依赖都定义在该文件中。

这里就是 Spring Boot 项目依赖的版本管理中心。

版本管理中心已经默认配置好大部分依赖的版本信息。这些版本信息随着 Spring Boot 版本的更新也会随之更改。这种设计让开发人员在今后导入依赖时不需要写版本号，直接使用 Spring Boot 提供的默认版本即可。当然开发人员也可以自行修改版本号。如果不是 dependencies 中默认管理的依赖，就需要声明其版本号。

结合 Spring Boot 约定优于配置的特性，可以这样理解：Spring Boot 为开发人员设置了默认编码、默认 JDK 版本及 Maven 编译的默认设置，同时维护了一套项目依赖的配置，相关依赖可直接导入，并不需要声明其版本号。

这就是 Spring Boot 与开发人员之间做出的一个开发约定。如果认可这个约定就可以减少一些基本配置和很多依赖配置；如果不认可这种方式也可以自行配置，这些配置会将默认配置覆盖。

在 pom.xml 文件中一个需要关注的地方是 spring-boot-starter-web，即在创建 Spring Boot 项目时选择的 Web 模块。spring-boot-starter-web 的版本号已经默认配置好了，查看 spring-boot-dependencies-2.3.7.RELEASE.pom 文件可知其版本号为 2.3.7.RELEASE，如图 5-2 所示。

第 5 章 Spring Boot 核心详解及源码分析

图 5-2　spring-boot-starter-web 的版本号

这个依赖配置有什么作用呢？

接下来，继续按住 Ctrl 键并点击该文件，进入 spring-boot-starter-web- 2.3.7.RELEASE. pom 文件中，查看其源码：

```xml
<?xml version="1.0" encoding="UTF-8"?>
<project xsi:schemaLocation="http://maven.apache.org/POM/4.0.0 http://maven.apache.org/xsd/maven-4.0.0.xsd" xmlns="http://maven.apache.org/POM/4.0.0"
    xmlns:xsi="http://www.w3.org/2001/XMLSchema-instance">
  <modelVersion>4.0.0</modelVersion>
  <groupId>org.springframework.boot</groupId>
  <artifactId>spring-boot-starter-web</artifactId>
  <version>2.3.7.RELEASE</version>
  <name>spring-boot-starter-web</name>
  <description>Starter for building web, including RESTful, applications using Spring MVC. Uses Tomcat as the default embedded container</description>
  <url>https://spring.io/projects/spring-boot</url>
  <organization>
    <name>Pivotal Software, Inc.</name>
    <url>https://spring.io</url>
  </organization>
  <licenses>
    <license>
      <name>Apache License, Version 2.0</name>
      <url>https://www.apache.org/licenses/LICENSE-2.0</url>
    </license>
  </licenses>
  <developers>
    <developer>
      <name>Pivotal</name>
      <email>info@pivotal.io</email>
      <organization>Pivotal Software, Inc.</organization>
      <organizationUrl>https://www.spring.io</organizationUrl>
    </developer>
  </developers>
  <scm>
    <connection>scm:git:git://github.com/spring-projects/spring-boot.git</
```

```xml
connection>
    <developerConnection>scm:git:ssh://git@github.com/spring-projects/spring-boot.git</developerConnection>
    <url>https://github.com/spring-projects/spring-boot</url>
  </scm>
  <issueManagement>
    <system>GitHub</system>
    <url>https://github.com/spring-projects/spring-boot/issues</url>
  </issueManagement>
  <dependencies>
    <dependency>
      <groupId>org.springframework.boot</groupId>
      <artifactId>spring-boot-starter</artifactId>
      <version>2.3.7.RELEASE</version>
      <scope>compile</scope>
    </dependency>
    <dependency>
      <groupId>org.springframework.boot</groupId>
      <artifactId>spring-boot-starter-json</artifactId>
      <version>2.3.7.RELEASE</version>
      <scope>compile</scope>
    </dependency>
    <dependency>
      <groupId>org.springframework.boot</groupId>
      <artifactId>spring-boot-starter-tomcat</artifactId>
      <version>2.3.7.RELEASE</version>
      <scope>compile</scope>
    </dependency>
    <dependency>
      <groupId>org.springframework</groupId>
      <artifactId>spring-web</artifactId>
      <version>5.2.12.RELEASE</version>
      <scope>compile</scope>
    </dependency>
    <dependency>
      <groupId>org.springframework</groupId>
      <artifactId>spring-webmvc</artifactId>
      <version>5.2.12.RELEASE</version>
      <scope>compile</scope>
    </dependency>
  </dependencies>
</project>
```

查看该配置文件可以发现，原来 Spring MVC 所需的依赖和内置的 Tomcat 依赖都已经定义在该文件中了。这也是为什么在开发 HelloController 时，仅仅在 pom.xml 文件中加入 spring-boot-starter-web 场景启动器，就可以进行 Web 项目的开发。

结合 Spring Boot 约定优于配置的特性，可以这样理解：Spring Boot 的场景启动器默认已经维护好了在该场景中所需的依赖 jar 包，导入场景启动器即可进行该场景的功能开发。其他类似的场景启动器，如数据库连接场景启动器、模板引擎场景、缓存整合场景等功能的开发，开发人员都不需要专门维护依赖版本或者解决版本带来的冲突问题。

5.3 @SpringBootApplication注解与分析

Spring Boot 项目的大部分配置和代码写法都和日常开发的项目相同。开发人员甚至可以直接将一些代码直接复制到 Spring Boot 项目中直接使用。唯一比较明显的区别就是在 Spring Boot 项目目录结构中多了一个主程序类 Application.class。

那么这个主程序类是做什么的呢？查看代码如下：

```
/**
 * @SpringBootApplication 标注在一个主程序类上,说明这是一个 Spring Boot 应用
 */
@SpringBootApplication
public class NewbeeMallApplication {
    public static void main(String[] args) {
        //启动 Spring 应用
        SpringApplication.run(NewbeeMallApplication.class, args);
    }
}
```

由这部分代码能够看出，与普通的 Spring 项目对比，二者的区别在 @SpringBootApplication 注解和启动方法 SpringApplication.run() 上。因为未使用 Spring Boot 开发的 Java Web 项目既没有@SpringBootApplication 注解也没有启动方法。

笔者在这一节将结合源码对主程序类进行分析讲解，揭开 Spring Boot 的神秘面纱。

5.3.1 @SpringBootApplication 注解

按住 Ctrl 键并点击@SpringBootApplication 注解，查看其源码：

```
@Target({ElementType.TYPE})
```

```
@Retention(RetentionPolicy.RUNTIME)
@Documented
@Inherited
@SpringBootConfiguration
@EnableAutoConfiguration
@ComponentScan(
    excludeFilters = {@Filter(
    type = FilterType.CUSTOM,
    classes = {TypeExcludeFilter.class}
), @Filter(
    type = FilterType.CUSTOM,
    classes = {AutoConfigurationExcludeFilter.class}
)}
)
public @interface SpringBootApplication {
    …省略部分代码
}
```

由注解源码可以看出，@SpringBootApplication 注解是一个复合注解，其中前面四个注解是 Java 元注解，含义分别如下所示。

@Target(ElementType.TYPE)：类、接口（包括注解类型）和 enum 声明。

@Retention(RetentionPolicy.RUNTIME)：运行时注解。

@Documented：将注解添加到 Java doc 中。

@Inherited：允许继承。

重要的是后面三个注解，含义分别如下所示。

@SpringBootConfiguration：Spring Boot 配置注解。

@EnableAutoConfiguration：启用自动配置注解。

@ComponentScan：组件自动扫描注解。

Java 元注解并无特殊含义，这里可以将@SpringBootApplication 注解理解为@SpringBootConfiguration、@EnableAutoConfiguration 和@ComponentScan 这三个注解的组合注解。如果在主程序类中不使用@SpringBootApplication 注解的话，也可以在主程序类上直接标注 @SpringBootConfiguration、@EnableAutoConfiguration 和@ComponentScan 这三个注解，其效果也是一样的，如下所示：

```
@SpringBootConfiguration
@EnableAutoConfiguration
@ComponentScan
public class NewbeeMallApplication {
```

```
public static void main(String[] args) {
    SpringApplication.run(NewbeeMallApplication.class, args);
}
}
```

@SpringBootApplication 注解的含义:如果 Application 标注在某个类上说明这个类是 Spring Boot 的主配置类,Spring Boot 会运行这个类的 main 方法来启动 Spring Boot 应用。

5.3.2 @SpringBootConfiguration 注解

@SpringBootConfiguration 注解的源码如下所示:

```
@Target({ElementType.TYPE})
@Retention(RetentionPolicy.RUNTIME)
@Documented
@Configuration
public @interface SpringBootConfiguration {
    @AliasFor(
        annotation = Configuration.class
    )
    boolean proxyBeanMethods() default true;
}
```

查看该段代码可以发现,类上声明了@Configuration 注解。这个 Spring 自定义注解读者应该很熟悉,它是从 Spring 3 版本开始就存在的注解,主要用于定义配置类,替代 XML 配置文件。

@SpringBootConfiguration 注解仅仅是对@Configuration 注解进行了包装,本质上依然是@Configuration 注解。@SpringBootConfiguration 注解是 1.4 版本中新增的注解,标注在某个类上表示这是一个 Spring Boot 的配置类。

5.3.3 @EnableAutoConfiguration 注解

@EnableAutoConfiguration 注解表示开启自动配置功能。自动配置是 Spring Boot 最为核心的一个特性,也是"约定大于配置"设计思想的主要体现。而@EnableAutoConfiguration 注解就是这个功能的入口。

@EnableAutoConfiguration 注解的源码如下所示：

```
@Target({ElementType.TYPE})
@Retention(RetentionPolicy.RUNTIME)
@Documented
@Inherited
@AutoConfigurationPackage
@Import({AutoConfigurationImportSelector.class})
public @interface EnableAutoConfiguration {
    String ENABLED_OVERRIDE_PROPERTY = "spring.boot.enableautoconfiguration";

    Class<?>[] exclude() default {};

    String[] excludeName() default {};
}
```

除 Java 元注解以外，这个注解最重要的就是@AutoConfigurationPackage 注解和使用@Import 注解引入的 AutoConfigurationImportSelector 组件。

5.3.4　@AutoConfigurationPackage 源码解析

@AutoConfigurationPackage 注解的源码如下所示：

```
@Target(ElementType.TYPE)
@Retention(RetentionPolicy.RUNTIME)
@Documented
@Inherited
@Import(AutoConfigurationPackages.Registrar.class)
public @interface AutoConfigurationPackage {

    String[] basePackages() default {};

    Class<?>[] basePackageClasses() default {};
}
```

该注解中包含 Spring 框架的@Import 注解，其作用就是将标注了该注解的组件注册到 Spring 的 IOC 容器中，而导入的内容则由 AutoConfigurationPackages. Registrar.class

类指定。也就是说 Spring Boot 会注册自动配置包的名称，默认为当前主程序类所在的包及其子包。这些包中的组件会被加载到容器中。

需要注意的是，加载到其他包中的组件默认是不会被扫描的。这也是为什么有的开发人员在配置了组件后无法被加载的原因。比如本次演示的项目，NewbeeMallApplication 主程序类所在的目录是根目录，其包名为 "ltd.newbee.mall"，所以在该包及子包下的组件都会被扫描进来。如果新建一个 "ltd.newbee.demo" 包，Spring Boot 项目在启动时 demo 包中的组件是无法被加载到 Spring IOC 容器中的。

"默认为当前主程序类所在的包及其子包" 这个结论并不是猜测。接下来笔者结合源码来解释这个结论，读者可以根据以下步骤来验证。

（1）在 IDEA 中打开 org.springframework.boot.autoconfigure.AutoConfigurationPackages 类。在代码的第 124 行打上一个断点。

（2）点击右上角工具栏中的启动按钮，注意，一定要点击右侧的按钮，以 debug 方式启动项目。

（3）启动后会在 register() 方法上阻塞，此时选中 new PackageImports(metadata).getPackageNames().toArray(new String[0]) 并点击右键出现工具栏，点击 "Evaluate Expression"，可以获取该表达式的最终结果，如图 5-3 所示。

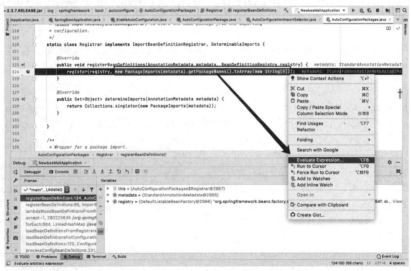

图 5-3　debug 方式获取默认扫描包

整个 debug 过程如图 5-4 所示，最终得出的结果为 "ltd.newbee.mall"，即 "默认会扫描当前主程序类所在的包及其子包"。

图 5-4　debug 方式获取默认扫描包结果

5.3.5　EnableAutoConfigurationImportSelector 类的源码解析

EnableAutoConfigurationImportSelector 类是整个自动配置功能的核心实现，它负责把返回自动配置的相关组件名称注册至 IOC 容器中，源码及源码解释如下所示：

```java
public class AutoConfigurationImportSelector
    implements DeferredImportSelector, BeanClassLoaderAware, Resource
LoaderAware, BeanFactoryAware, EnvironmentAware, Ordered {
    ...省略部分代码

    @Override
    public String[] selectImports(AnnotationMetadata annotationMetadata) {
        if (!this.isEnabled(annotationMetadata)) {
            return NO_IMPORTS;
        } else {
            AutoConfigurationImportSelector.AutoConfigurationEntry auto
ConfigurationEntry = this.getAutoConfigurationEntry(annotationMetadata);
            return
StringUtils.toStringArray(autoConfigurationEntry.getConfigurations());
        }
    }
```

```java
    protected AutoConfigurationImportSelector.AutoConfigurationEntry
getAutoConfigurationEntry(AnnotationMetadata annotationMetadata) {
        if (!this.isEnabled(annotationMetadata)) {
            return EMPTY_ENTRY;
        } else {
            AnnotationAttributes attributes = this.getAttributes(annotationMetadata);
            // 获取自动装配配置项
            List<String> configurations = this.getCandidateConfigurations(annotationMetadata, attributes);
            // 去重
            configurations = this.removeDuplicates(configurations);
            // 获取停用配置项（开发人员自行设置，排除不需要自动配置的组件）
            Set<String> exclusions = this.getExclusions(annotationMetadata, attributes);
            this.checkExcludedClasses(configurations, exclusions);
            // 移除停用配置项
            configurations.removeAll(exclusions);
            configurations = this.getConfigurationClassFilter().filter(configurations);
            this.fireAutoConfigurationImportEvents(configurations, exclusions);
            // 返回所有的自动装配配置
            return new AutoConfigurationImportSelector.AutoConfigurationEntry(configurations, exclusions);
        }
    }

    @Override
    public Class<? extends Group> getImportGroup() {
        return AutoConfigurationGroup.class;
    }

    //该方法判断是否开启自动配置
    protected boolean isEnabled(AnnotationMetadata metadata) {
        //获取 spring.boot.enableautoconfiguration 的值是 true 还是 false，默认为 true
        return this.getClass() == AutoConfigurationImportSelector.class ? (Boolean)this.getEnvironment().getProperty("spring.boot.enableautoconfiguration", Boolean.class, true) : true;
    }

    protected List<String> getCandidateConfigurations(AnnotationMetadata metadata, AnnotationAttributes attributes) {
```

```
        // 从META-INF/spring.factories中获取 EnableAutoConfiguration 的所有配置项
        List<String> configurations = SpringFactoriesLoader.loadFactoryNames
(this.getSpringFactoriesLoaderFactoryClass(), this.getBeanClassLoader());
        Assert.notEmpty(configurations, "No auto configuration classes found 
in META-INF/spring.factories. If you are using a custom packaging, make sure 
that file is correct.");
        return configurations;
    }

}
```

其中，getCandidateConfigurations() 方法会调用 SpringFactoriesLoader 类的 loadFactoryNames()获取所有的自动配置类的类名，源码如下所示：

```
public final class SpringFactoriesLoader {
    public static final String FACTORIES_RESOURCE_LOCATION = "META-INF/spring.
factories";
    private static final Log logger = LogFactory.getLog(SpringFactoriesLoader.
class);
    private static final Map<ClassLoader, MultiValueMap<String, String>> 
cache = new ConcurrentReferenceHashMap();

    private SpringFactoriesLoader() {
    }

    public static <T> List<T> loadFactories(Class<T> factoryType, @Nullable 
ClassLoader classLoader) {
        Assert.notNull(factoryType, "'factoryType' must not be null");
        ClassLoader classLoaderToUse = classLoader;
        if (classLoader == null) {
            classLoaderToUse = SpringFactoriesLoader.class.getClassLoader();
        }

        List<String> factoryImplementationNames = loadFactoryNames(factoryType, 
classLoaderToUse);
        if (logger.isTraceEnabled()) {
            logger.trace("Loaded [" + factoryType.getName() + "] names: " + 
factoryImplementationNames);
        }

        List<T> result = new ArrayList(factoryImplementationNames.size());
        Iterator var5 = factoryImplementationNames.iterator();
```

```java
            while(var5.hasNext()) {
                String factoryImplementationName = (String)var5.next();
                result.add(instantiateFactory(factoryImplementationName, factoryType, classLoaderToUse));
            }

            AnnotationAwareOrderComparator.sort(result);
            return result;
        }

    public static List<String> loadFactoryNames(Class<?> factoryType, @Nullable ClassLoader classLoader) {
        String factoryTypeName = factoryType.getName();
        return (List)loadSpringFactories(classLoader).getOrDefault(factoryTypeName, Collections.emptyList());
    }

    private static Map<String, List<String>> loadSpringFactories(@Nullable ClassLoader classLoader) {
        MultiValueMap<String, String> result = (MultiValueMap)cache.get(classLoader);
        if (result != null) {
            return result;
        } else {
            try {
                Enumeration<URL> urls = classLoader != null ? classLoader.getResources("META-INF/spring.factories") : ClassLoader.getSystemResources("META-INF/spring.factories");
                LinkedMultiValueMap result = new LinkedMultiValueMap();

                while(urls.hasMoreElements()) {
                    URL url = (URL)urls.nextElement();
                    UrlResource resource = new UrlResource(url);
                    Properties properties = PropertiesLoaderUtils.loadProperties(resource);
                    Iterator var6 = properties.entrySet().iterator();

                    while(var6.hasNext()) {
                        Entry<?, ?> entry = (Entry)var6.next();
                        String factoryTypeName = ((String)entry.getKey()).trim();
                        String[] var9 = StringUtils.commaDelimitedListToStringArray((String)entry.getValue());
                        int var10 = var9.length;
```

```java
                    for(int var11 = 0; var11 < var10; ++var11) {
                        String factoryImplementationName = var9[var11];
                        result.add(factoryTypeName, factoryImplementationName.trim());
                    }
                }
            }

            cache.put(classLoader, result);
            return result;
        } catch (IOException var13) {
            throw new IllegalArgumentException("Unable to load factories from location [META-INF/spring.factories]", var13);
        }
    }
}

    private static <T> T instantiateFactory(String factoryImplementationName, Class<T> factoryType, ClassLoader classLoader) {
        try {
            Class<?> factoryImplementationClass = ClassUtils.forName(factoryImplementationName, classLoader);
            if (!factoryType.isAssignableFrom(factoryImplementationClass)) {
                throw new IllegalArgumentException("Class [" + factoryImplementationName + "] is not assignable to factory type [" + factoryType.getName() + "]");
            } else {
                return ReflectionUtils.accessibleConstructor(factoryImplementationClass, new Class[0]).newInstance();
            }
        } catch (Throwable var4) {
            throw new IllegalArgumentException("Unable to instantiate factory class [" + factoryImplementationName + "] for factory type [" + factoryType.getName() + "]", var4);
        }
    }
}
```

AutoConfigurationImportSelector 会将所有需要导入的组件以全类名的方式返回，这些组件就会被注册到 IOC 容器中。

由以上源码可以得出结论：Spring Boot 在启动的时候从类路径 META-INF/spring.

factories 中获取 EnableAutoConfiguration 指定的配置项，在过滤后将这些值作为自动配置类导入容器中。

有哪些自动配置类呢？

可以在 spring-boot-autoconfigure-2.3.7.RELEASE.jar 依赖中查看。点击项目目录中的"External Libraries"可以看到所有依赖 jar 包，点开"Maven:org.springframework.boot:spring-boot-autoconfigure:2.3.7.RELEASE"，就能够看到"META-INF/spring.factories"文件，如图 5-5 所示。

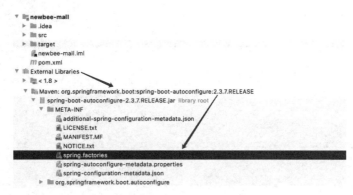

图 5-5　META-INF/spring.factories 文件

spring.factories 文件内容如下所示：

```
# Initializers
org.springframework.context.ApplicationContextInitializer=\
org.springframework.boot.autoconfigure.SharedMetadataReaderFactoryContex
tInitializer,\
org.springframework.boot.autoconfigure.logging.ConditionEvaluationReport
LoggingListener

# Application Listeners
org.springframework.context.ApplicationListener=\
org.springframework.boot.autoconfigure.BackgroundPreinitializer

# Auto Configuration Import Listeners
org.springframework.boot.autoconfigure.AutoConfigurationImportListener=\
org.springframework.boot.autoconfigure.condition.ConditionEvaluationRepo
rtAutoConfigurationImportListener

# Auto Configuration Import Filters
org.springframework.boot.autoconfigure.AutoConfigurationImportFilter=\
```

```
org.springframework.boot.autoconfigure.condition.OnBeanCondition,\
org.springframework.boot.autoconfigure.condition.OnClassCondition,\
org.springframework.boot.autoconfigure.condition.OnWebApplicationCondition

# Auto Configure
org.springframework.boot.autoconfigure.EnableAutoConfiguration=\
org.springframework.boot.autoconfigure.admin.SpringApplicationAdminJmxAutoConfiguration,\
org.springframework.boot.autoconfigure.aop.AopAutoConfiguration,\
……省略部分
org.springframework.boot.autoconfigure.web.servlet.DispatcherServletAutoConfiguration,\
org.springframework.boot.autoconfigure.web.servlet.ServletWebServerFactoryAutoConfiguration,\
org.springframework.boot.autoconfigure.web.servlet.error.ErrorMvcAutoConfiguration,\
org.springframework.boot.autoconfigure.web.servlet.HttpEncodingAutoConfiguration,\
org.springframework.boot.autoconfigure.web.servlet.MultipartAutoConfiguration,\
……省略部分

# Failure analyzers
org.springframework.boot.diagnostics.FailureAnalyzer=\
org.springframework.boot.autoconfigure.data.redis.RedisUrlSyntaxFailureAnalyzer,\
org.springframework.boot.autoconfigure.diagnostics.analyzer.NoSuchBeanDefinitionFailureAnalyzer,\
org.springframework.boot.autoconfigure.flyway.FlywayMigrationScriptMissingFailureAnalyzer,\
org.springframework.boot.autoconfigure.jdbc.DataSourceBeanCreationFailureAnalyzer,\
org.springframework.boot.autoconfigure.jdbc.HikariDriverConfigurationFailureAnalyzer,\
org.springframework.boot.autoconfigure.r2dbc.ConnectionFactoryBeanCreationFailureAnalyzer,\
org.springframework.boot.autoconfigure.session.NonUniqueSessionRepositoryFailureAnalyzer

# Template availability providers
org.springframework.boot.autoconfigure.template.TemplateAvailabilityProvider=\
org.springframework.boot.autoconfigure.freemarker.FreeMarkerTemplateAvai
```

```
labilityProvider,\
org.springframework.boot.autoconfigure.mustache.MustacheTemplateAvailabi
lityProvider,\
org.springframework.boot.autoconfigure.groovy.template.GroovyTemplateAva
ilabilityProvider,\
org.springframework.boot.autoconfigure.thymeleaf.ThymeleafTemplateAvaila
bilityProvider,\
org.springframework.boot.autoconfigure.web.servlet.JspTemplateAvailabili
tyProvider
```

spring.factories 文件的 org.springframework.boot.autoconfigure.EnableAutoConfiguration 配置项定义的就是 Spring Boot 默认加载的所有自动配置类。不过，这些自动配置类在项目启动时并不会全部被执行，因为每个自动配置类是否正常执行有对应的条件。以 org.springframework.boot.autoconfigure.web.servlet.DispatcherServletAutoConfiguration 自动配置类为例，如果 pom.xml 文件中没有引入 spring-boot-starter-web，那么该自动配置类就不会工作。这是因为缺少相应的 jar 包和依赖，自动配置无法生效，其他自动配置类同理。

除了 org.springframework.boot.autoconfigure.EnableAutoConfiguration 配置项，spring.factories 文件中还有 org.springframework.context.ApplicationContextInitializer、org.springframework.context.ApplicationListener 等配置项，这些都是 Spring Boot 重要的组成部分。

在自动配置类被执行时，Spring Boot 会启动相关组件和类的自动装配工作。以往需要开发人员在 XML 文件中手动配置的内容都由 Spring Boot 完成，从而免去了手动编写配置、注入相应功能组件等工作。

5.3.6 @ComponentScan 注解

使用 Spring 框架开发 Java Web 项目的读者一定都用过@Controller、@Service、@Repository 等注解。查看其源码会发现，在这些注解的定义上都会标注一个共同的注解 @Component。而在 Spring IOC 容器中@Controller、@Service、@Repository、@Component 等注解的默认装配标识是@ComponentScan 注解。

在普通的 Spring 项目开发中，一般会在 Spring 配置文件中编写如下配置，将对应包下的所有组件扫描并注册到容器中：

```
<!-- 自动扫描 -->
<context:component-scan base-package="com.ssm.demo.dao"/>
```

```xml
<context:component-scan base-package="com.ssm.demo.service"/>
<context:component-scan base-package="com.ssm.demo.controller"/>
```

使用注解的方式与这种 XML 配置文件的方式所实现的效果是相同的。@ComponentScan 注解的作用就是让 Spring 容器从对应包下获取需要注册的类。开发人员通过注解来定义哪些包需要被自动扫描并装配。一旦指定了相应的包名，Spring 将会在被指定的包及其子包中寻找标注了以上注解的 Bean 并注册到容器中。

5.4 SpringApplication启动流程解析

Spring Boot 项目通过运行启动类中的 run()方法就可以将整个应用启动。那么这个方法究竟做了哪些神奇的事情呢？SpringApplication 启动流程又做了哪些操作呢？接下来通过源码一探究竟。

点击启动类中的 run()方法进入 SpringApplication 类，源码及注释如下所示：

```java
public class SpringApplication {
    ...省略部分代码

    public static ConfigurableApplicationContext run(Class<?> primarySource,
String... args) {
        return run(new Class<?>[] { primarySource }, args);
    }

    public static ConfigurableApplicationContext run(Class<?>[]
primarySources, String[] args) {
        //1.实例化SpringApplication，之后执行run()方法
        return new SpringApplication(primarySources).run(args);
    }

    // 最终执行的run()方法
    public ConfigurableApplicationContext run(String... args) {
        StopWatch stopWatch = new StopWatch();
        // 2.代码执行时间监控开启
stopWatch.start();
        ConfigurableApplicationContext context = null;
        Collection<SpringBootExceptionReporter> exceptionReporters = new
ArrayList<>();
        // 3.配置headless属性，默认为true
configureHeadlessProperty();
        // 4.获取SpringApplicationRunListener 集合
```

```java
        SpringApplicationRunListeners listeners = getRunListeners(args);
            // 5.回调所有 SpringApplicationRunListener 对象的 starting()方法
    listeners.starting();
        try {
        // 6.创建 ApplicationArguments 对象
            ApplicationArguments applicationArguments = new
DefaultApplicationArguments(args);
            // 7.创建 Environment 对象,加载属性配置。
    ConfigurableEnvironment environment = prepareEnvironment(listeners,
applicationArguments);
            // 8.设置系统参数
    configureIgnoreBeanInfo(environment);
            // 9.获取打印的 Spring Boot banner
    Banner printedBanner = printBanner(environment);
    // 10.创建 Spring 容器 ApplicationContext
            context = createApplicationContext();
            exceptionReporters = getSpringFactoriesInstances(SpringBoot
ExceptionReporter.class,
                    new Class[] { ConfigurableApplicationContext.class },
context);
            // 11.准备容器
    prepareContext(context, environment, listeners, applicationArguments,
printedBanner);
            // 12.刷新 Spring 容器
    refreshContext(context);
            // 13.执行 Spring 容器初始化的后置逻辑
    afterRefresh(context, applicationArguments);
            // 14.代码执行时间监控结果
    stopWatch.stop();
            if (this.logStartupInfo) {
    // 打印 Spring Boot 的启动时长日志
                new StartupInfoLogger(this.mainApplicationClass).logStarted
(getApplicationLog(), stopWatch);
            }
            // 15.发布容器启动事件
    listeners.started(context);
            // 16.调用 ApplicationRunner 或者 CommandLineRunner 的运行方法
    callRunners(context, applicationArguments);
        }
        catch (Throwable ex) {
            handleRunFailure(context, ex, exceptionReporters, listeners);
            throw new IllegalStateException(ex);
        }
```

```java
    try {
        listeners.running(context);
    }
    catch (Throwable ex) {
        handleRunFailure(context, ex, exceptionReporters, null);
        throw new IllegalStateException(ex);
    }
    return context;
}
```

Spring Boot 项目启动步骤分析如下所示。

（1）实例化 SpringApplication 对象。

在执行 run()方法前，使用 new SpringApplication()构造 SpringApplication 对象。SpringApplication 类的构造方法如下所示：

```java
public SpringApplication(ResourceLoader resourceLoader, Class<?>... primarySources) {
    this.resourceLoader = resourceLoader;
    Assert.notNull(primarySources, "PrimarySources must not be null");
    this.primarySources = new LinkedHashSet<>(Arrays.asList(primarySources));
    // 设置当前应用类型 NONE、SERVLET、REACTIVE
    this.webApplicationType = WebApplicationType.deduceFromClasspath();
    // 加载 ApplicationContextInitializer,配置在 META-INF/spring.factories 文件中
    setInitializers((Collection) getSpringFactoriesInstances(ApplicationContextInitializer.class));
    // 加载 ApplicationListener, 配置在 META-INF/spring.factories 文件中
    setListeners((Collection) getSpringFactoriesInstances(ApplicationListener.class));
    this.mainApplicationClass = deduceMainApplicationClass();
}
```

这一步主要是构造 SpringApplication 对象，并为 SpringApplication 的属性赋值，在构造完成后，开始执行 run()方法。

比较重要的一个知识点是 webApplicationType 值的设置，其目的是获取当前应用的类型，对后续步骤构造容器环境和 Spring 容器的初始化起到作用。该值的获取是通过调用 WebApplicationType.deduceFromClasspath()方法得到的，该方法源码及注释如下所示：

```java
public enum WebApplicationType {
```

```java
    /**
     * The application should not run as a web application and should not start an
     * embedded web server.
     * 当前应用不是Web应用，不需要启动一个内嵌的Web服务器
     */
    NONE,

    /**
     * The application should run as a servlet-based web application and should start an
     * embedded servlet web server.
     * 当前应用需要启动一个内嵌的基于Servlet的服务器
     */
    SERVLET,

    /**
     * The application should run as a reactive web application and should start an
     * embedded reactive web server.
     * 当前应用需要启动一个内嵌的基于Reactive的服务器
     */
    REACTIVE;

    private static final String[] SERVLET_INDICATOR_CLASSES = { "javax.servlet.Servlet",
            "org.springframework.web.context.ConfigurableWebApplicationContext" };

    private static final String WEBMVC_INDICATOR_CLASS = "org.springframework.web.servlet.DispatcherServlet";

    private static final String WEBFLUX_INDICATOR_CLASS = "org.springframework.web.reactive.DispatcherHandler";

    private static final String JERSEY_INDICATOR_CLASS = "org.glassfish.jersey.servlet.ServletContainer";

    static WebApplicationType deduceFromClasspath() {
        // 1.加载 org.springframework.web.reactive.DispatcherHandler 类，如果该类存在并且 org.springframework.web.servlet.DispatcherServlet 类和 org.glassfish.jersey.servlet.ServletContainer 类不存在，则当前应用为 REACTIVE 类型
```

```
            if (ClassUtils.isPresent(WEBFLUX_INDICATOR_CLASS, null)
    && !ClassUtils.isPresent(WEBMVC_INDICATOR_CLASS, null)
                && !ClassUtils.isPresent(JERSEY_INDICATOR_CLASS, null)) {
            return WebApplicationType.REACTIVE;
        }
        for (String className : SERVLET_INDICATOR_CLASSES) {
       // 2.如果javax.servlet.Servlet类和 org.springframework.web.context.
ConfigurableWebApplicationContext类不存在，则当前应用为NONE类型
            if (!ClassUtils.isPresent(className, null)) {
                return WebApplicationType.NONE;
            }
        }
        // 3.与第2项相反，当前应用为SERVLET类型
        return WebApplicationType.SERVLET;
    }
}
```

WebApplicationType 的值有 3 个，分别如下所示。

①SERVLET：Servlet 环境。

②REACTIVE：Reactive 环境。

③NONE：非 Web 环境。

在 deduceFromClasspath()方法中代码多次调用 ClassUtils.isPresent()方法，以此判断在常量中的类是否存在。该方法的最终实现原理通过 Class.forName 加载某个类，如果成功加载，则证明这个类存在，反之则代表该类不存在。

deduceFromClasspath()方法的实现逻辑如下：先判断 webflux 相关的类是否存在，存在则认为当前应用为 REACTIVE 类型；不存在则继续判断 SERVLET 相关的类是否存在，都不存在则为 NONE 类型；否则，当前应用为 SERVLET 类型。具体的类加载判断方法可以直接查看源码，相关的代码注释笔者也已经标注在代码中。

以 newbee-mall 项目举例，由于项目中引用了 spring-boot-starter-web 且并未引用 webflux 相关的类，所以 newbee-mall 项目类型为 SERVLET 类型。

（2）开始执行 run()方法，代码执行时间的监控开启，在 Spring Boot 应用启动成功后会打印启动时间。

（3）配置 headless 属性，java.awt.headles 是 J2SE 的一种模式，用于在缺失显示屏、鼠标或者键盘时的系统配置，默认为 true。通俗而言，该行代码的作用是 Spring Boot 应用在启动时，没有检测到显示器也能够继续执行后面的步骤。

（4）获取 SpringApplicationRunListeners，getRunListeners()方法的源码如下所示：

```
private SpringApplicationRunListeners getRunListeners(String[] args) {
    Class<?>[] types = new Class[]{SpringApplication.class, String[].class};
    return new SpringApplicationRunListeners(logger, this.getSpringFactoriesInstances(SpringApplicationRunListener.class, types, this, args));
}

private <T> Collection<T> getSpringFactoriesInstances(Class<T> type, Class<?>[] parameterTypes, Object... args) {
    ClassLoader classLoader = this.getClassLoader();
    Set<String> names = new LinkedHashSet(SpringFactoriesLoader.loadFactoryNames(type, classLoader));
    List<T> instances = this.createSpringFactoriesInstances(type, parameterTypes, classLoader, args, names);
    AnnotationAwareOrderComparator.sort(instances);
    return instances;
}
```

这里会调用 SpringFactoriesLoader 类中的 loadFactoryNames()方法。该方法在介绍自动配置时已经讲解过，与获取自动配置类的类名相同。也就是在 getRunListeners()方法中调用该方法是从类路径 META-INF/spring.factories 中获取 SpringApplicationRunListener 指定类的。在 spring-boot-2.3.7.RELEASE.jar 包中的 META-INF 目录下找到了 spring.factories 文件，当前文件中只有一个 RunListener，即 org.springframework.boot.context.event.EventPublishingRunListener，如图 5-6 所示。

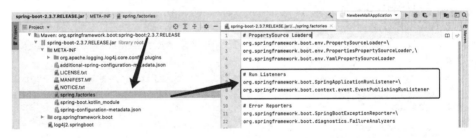

图 5-6　META-INF/spring.factories 文件

通过 debug 模式也可以得出该类为 org.springframework.boot.context.event.EventPublishingRunListener。在 "listeners.starting();" 代码前输入一个断点，之后通过 debug 模式启动项目，可以看出此时加载的 listener 为 EventPublishingRunListener，如图 5-7 所示。

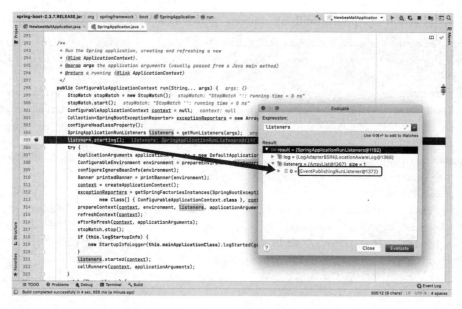

图 5-7　EventPublishingRunListener 类

（5）回调 SpringApplicationRunListener 对象的 starting() 方法。

（6）解析 run() 方法的 args 参数并封装为 DefaultApplicationArguments 类。

（7）prepareEnvironment() 方法的作用与它的方法名的含义相同，就是为当前应用准备一个 Environment 对象，也就是运行环境。它主要完成对 ConfigurableEnvironment 的初始化工作。该方法的源码及解析如下所示：

```java
    private ConfigurableEnvironment prepareEnvironment(SpringApplication
RunListeners listeners,ApplicationArguments applicationArguments) {
    // 创建环境对象
        ConfigurableEnvironment environment = getOrCreateEnvironment();
    // 配置环境
        configureEnvironment(environment, applicationArguments.getSourceArgs());
        ConfigurationPropertySources.attach(environment);
    // 触发监听器
        listeners.environmentPrepared(environment);
    // 为 SpringApplication 绑定 environment
        bindToSpringApplication(environment);
        if (!this.isCustomEnvironment) {
            environment = new EnvironmentConverter(getClassLoader()).
convertEnvironmentIfNecessary(environment,
                    deduceEnvironmentClass());
        }
```

```
    // 为 environment 其他附件配置信息
        ConfigurationPropertySources.attach(environment);
        return environment;
    }
```

由于项目中存在 spring-boot-starter-web 依赖,webApplicationType 的值为 WebApplicationType.SERVLET,所以 getOrCreateEnvironment() 方法返回的是 StandardServletEnvironment 对象,是一个标准的 Servlet 环境。StandardServletEnvironment 是整个 Spring Boot 项目运行环境的实现类,后续关于环境的设置都基于此类。

在创建环境完成后,接下来是配置环境,configureEnvironment()方法的源码如下所示:

```
    protected void configureEnvironment(ConfigurableEnvironment environment, String[] args) {
        if (this.addConversionService) {
            ConversionService conversionService = ApplicationConversionService.getSharedInstance();
            environment.setConversionService((ConfigurableConversionService) conversionService);
        }
        // 加载启动命令行配置属性
        configurePropertySources(environment, args);
        // 设置active属性
        configureProfiles(environment, args);
    }
```

该方法主要加载一些默认配置,在执行完这一步骤后,会触发监听器(主要触发 ConfigFileApplicationListener),将会加载 application.properties 或者 application.yml 配置文件。

(8)设置系统参数,configureIgnoreBeanInfo()方法的源码如下所示:

```
    private void configureIgnoreBeanInfo(ConfigurableEnvironment environment) {
        if (System.getProperty(CachedIntrospectionResults.IGNORE_BEANINFO_PROPERTY_NAME) == null) {
            Boolean ignore = environment.getProperty("spring.beaninfo.ignore", Boolean.class, Boolean.TRUE);
            System.setProperty(CachedIntrospectionResults.IGNORE_BEANINFO_PROPERTY_NAME, ignore.toString());
        }
    }
```

查看源码可知，该方法会获取 spring.beaninfo.ignore 配置项的值，即使未获取也没有关系。代码的最后还是给该配置项输入了一个默认值 true，表示跳过对 BeanInfo 类的搜索，它无特别含义，不用深究该步骤。

（9）获取需要打印的 Spring Boot 启动 Banner 对象，源码如下所示：

```java
private Banner printBanner(ConfigurableEnvironment environment) {
    if (this.bannerMode == Banner.Mode.OFF) {
        return null;
    }
    ResourceLoader resourceLoader = (this.resourceLoader != null) ? this.resourceLoader
            : new DefaultResourceLoader(null);
    SpringApplicationBannerPrinter bannerPrinter = new SpringApplicationBannerPrinter(resourceLoader, this.banner);
    if (this.bannerMode == Mode.LOG) {
        return bannerPrinter.print(environment, this.mainApplicationClass, logger);
    }
    return bannerPrinter.print(environment, this.mainApplicationClass, System.out);
}
```

首先判断当前是否允许打印 Banner，默认会打印到控制台上，之后获取 Banner 对象。而 Spring Boot 目前支持图片 Banner 和文字 Banner，如果开发人员做了 Banner 配置则会在控制台打印开发人员配置的 Banner，否则打印默认 Banner。默认 Banner 的实现类为 org.springframework.boot.SpringBootBanner，源码如下所示：

```java
class SpringBootBanner implements Banner {

    private static final String[] BANNER = { "", "  .   ____          _            __ _ _",
            " /\\\\ / ___'_ __ _ _(_)_ __  __ _ \\ \\ \\ \\", "( ( )\\___ | '_ | '_| | '_ \\/ _` | \\ \\ \\ \\",
            " \\\\/  ___)| |_)| | | | | || (_| |  ) ) ) )", "  '  |____| .__|_| |_|_| |_\\__, | / / / /",
            " =========|_|==============|___/=/_/_/_/" };

    private static final String SPRING_BOOT = " :: Spring Boot :: ";

    private static final int STRAP_LINE_SIZE = 42;

    @Override
    public void printBanner(Environment environment, Class<?> sourceClass, PrintStream printStream) {
```

```java
        for (String line : BANNER) {
            printStream.println(line);
        }
        String version = SpringBootVersion.getVersion();
        version = (version != null) ? " (v" + version + ")" : "";
        StringBuilder padding = new StringBuilder();
        while (padding.length() < STRAP_LINE_SIZE - (version.length() + SPRING_BOOT.length())) {
            padding.append(" ");
        }

        printStream.println(AnsiOutput.toString(AnsiColor.GREEN, SPRING_BOOT, AnsiColor.DEFAULT, padding.toString(),
                AnsiStyle.FAINT, version));
        printStream.println();
    }
}
```

BANNER 变量就是在默认情况下打印在控制台上的 Banner。而 printBanner()方法，就是把定义好的 Banner 和 Spring Boot 的版本号打印出来。

其实在 Banner 打印流程中也能够看出 Spring Boot 框架约定优于配置的特性。开发人员配置 Banner 就使用开发人员配置的，如果没有，就使用 Spring Boot 默认的。

Spring Boot 框架的约定优于配置理念正是"你配置就用你配置的，你不配置就用约定好的"。

（10）创建 Spring 容器 ApplicationContext，createApplicationContext()方法的源码如下所示：

```java
    /**
     * The class name of application context that will be used by default for non-web
     * environments.
     */
    public static final String DEFAULT_CONTEXT_CLASS = "org.springframework.context."
            + "annotation.AnnotationConfigApplicationContext";

    /**
     * The class name of application context that will be used by default for web
     * environments.
     */
```

```java
     */
    public static final String DEFAULT_SERVLET_WEB_CONTEXT_CLASS = "org.springframework.boot."
            + "web.servlet.context.AnnotationConfigServletWebServerApplicationContext";

    /**
     * The class name of application context that will be used by default for reactive web
     * environments.
     */
    public static final String DEFAULT_REACTIVE_WEB_CONTEXT_CLASS = "org.springframework."
            + "boot.web.reactive.context.AnnotationConfigReactiveWebServerApplicationContext";

    protected ConfigurableApplicationContext createApplicationContext() {
        Class<?> contextClass = this.applicationContextClass;
        if (contextClass == null) {
            try {
    // 根据webApplicationType 决定创建哪种容器
                switch (this.webApplicationType) {
                case SERVLET:
                    contextClass = Class.forName(DEFAULT_SERVLET_WEB_CONTEXT_CLASS);
                    break;
                case REACTIVE:
                    contextClass = Class.forName(DEFAULT_REACTIVE_WEB_CONTEXT_CLASS);
                    break;
                default:
                    contextClass = Class.forName(DEFAULT_CONTEXT_CLASS);
                }
            }
            catch (ClassNotFoundException ex) {
                throw new IllegalStateException(
                        "Unable create a default ApplicationContext, please specify an ApplicationContextClass", ex);
            }
        }
        return (ConfigurableApplicationContext) BeanUtils.instantiateClass(contextClass);
    }
```

通过源码可以看出 createApplicationContext() 方法的执行逻辑：根据 webApplicationType 决定创建哪种 contextClass。webApplicationType 变量赋值的过程在前文中已经介绍过。因为该类型为 WebApplicationType.SERVLET 类型，所以会通过反射装载对应的字节码 DEFAULF_SERVLET_WEB_CONTEXT_CLASS 创建。创建的容器类型为 AnnotationConfigServletWebServerApplicationContext，在后续步骤中的操作都会基于该容器。

（11）准备 ApplicationContext 实例，prepareContext()方法的源码如下所示：

```java
    private void prepareContext(ConfigurableApplicationContext context,
ConfigurableEnvironment environment,
        SpringApplicationRunListeners listeners, ApplicationArguments
applicationArguments, Banner printedBanner) {
    // 将应用环境设置到容器中，包括各种变量
        context.setEnvironment(environment);
    // 执行容器后置处理
        postProcessApplicationContext(context);
    // 执行容器中的 ApplicationContextInitializer
        applyInitializers(context);
    // 回调 SpringApplicationRunListener 的 contextPrepared()方法
        listeners.contextPrepared(context);
        if (this.logStartupInfo) {
            logStartupInfo(context.getParent() == null);
            logStartupProfileInfo(context);
        }
    // 添加特定的 Bean
        ConfigurableListableBeanFactory beanFactory = context.getBeanFactory();
        beanFactory.registerSingleton("springApplicationArguments",
applicationArguments);
        if (printedBanner != null) {
            beanFactory.registerSingleton("springBootBanner", printedBanner);
        }
        if (beanFactory instanceof DefaultListableBeanFactory) {
            ((DefaultListableBeanFactory) beanFactory)
                .setAllowBeanDefinitionOverriding(this.allowBeanDefinitionOverriding);
        }
        if (this.lazyInitialization) {
            context.addBeanFactoryPostProcessor(new LazyInitializationBeanFactoryPostProcessor());
        }
    // 加载资源
        Set<Object> sources = getAllSources();
```

```
    Assert.notEmpty(sources, "Sources must not be empty");
// 加载启动类，将启动类注入容器中
    load(context, sources.toArray(new Object[0]));
// 回调 SpringApplicationRunListener 的 contextLoaded()方法
    listeners.contextLoaded(context);
}
```

在创建对应的 Spring 容器后，程序会进行初始化、加载主启动类等预处理工作。至此，主启动类加载完成，容器准备好。

（12）刷新容器，refreshContext()方法的源码如下所示：

```
private void refreshContext(ConfigurableApplicationContext context) {
    if (this.registerShutdownHook) {
        try {
//注册一个 Hook 函数，Hook 的作用是监听 JVM 在关闭时销毁的 IOC 容器
            context.registerShutdownHook();
        }
        catch (AccessControlException ex) {
            // 在某些环境中不被允许
        }
    }
    refresh((ApplicationContext) context);
}

protected void refresh(ApplicationContext applicationContext) {
    Assert.isInstanceOf(ConfigurableApplicationContext.class, application
Context);
    refresh((ConfigurableApplicationContext) applicationContext);
}

protected void refresh(ConfigurableApplicationContext applicationContext) {
    applicationContext.refresh();
}
```

程序首先注册一个 Hook 函数，然后调用 refresh()方法，经过层层调用，程序执行 ServletWebServerApplicationContext 类中的 refresh()方法，源码如下所示：

```
public final void refresh() throws BeansException, IllegalStateException {
    try {
        super.refresh();
    }
    catch (RuntimeException ex) {
        WebServer webServer = this.webServer;
        if (webServer != null) {
```

```
            webServer.stop();
        }
        throw ex;
    }
}
```

ServletWebServerApplicationContext 会调用父类 AbstractApplicationContext 的 refresh()方法，因此最终执行的 refresh()方法源码如下所示：

```
public void refresh() throws BeansException, IllegalStateException {
    synchronized(this.startupShutdownMonitor) {
        // 初始化上下文环境，对环境变量或者系统属性进行准备和校验
        this.prepareRefresh();
        // 初始化 BeanFactory
        ConfigurableListableBeanFactory beanFactory = this.obtainFreshBeanFactory();
        // 设置 BeanFactory 的基本属性：类加载器、添加 BeanPostProcesser
        this.prepareBeanFactory(beanFactory);

        try {
            // 空实现，允许子类上下文中对 BeanFactory 进行后置处理
            this.postProcessBeanFactory(beanFactory);
            // 执行在 BeanFactory 创建后的后置处理器
            this.invokeBeanFactoryPostProcessors(beanFactory);
            // 注册 Bean 的后置处理器
            this.registerBeanPostProcessors(beanFactory);
            // 初始化 MessageSource 组件，如国际化文件、消息解析、绑定等
            this.initMessageSource();
            // 初始化 ApplicationContext 事件广播器
            this.initApplicationEventMulticaster();
            // 无具体实现，子类重写该方法并在容器刷新时自定义处理逻辑
            this.onRefresh();
            // 注册监听器，BeanFactory 创建完成
            this.registerListeners();
            // 初始化剩余的单例 Bean
            this.finishBeanFactoryInitialization(beanFactory);
            // 完成容器的创建工作，通知生命周期处理器 LifecycleProcessor 刷新，发布 ContextRefreshEvent 通知
            this.finishRefresh();
        } catch (BeansException var9) {
            if (this.logger.isWarnEnabled()) {
                this.logger.warn("Exception encountered during context initialization - cancelling refresh attempt: " + var9);
            }
```

```
            this.destroyBeans();
            this.cancelRefresh(var9);
            throw var9;
        } finally {
            this.resetCommonCaches();
        }

    }
}
```

该方法是 Spring Bean 加载的核心，用于刷新整个 Spring 上下文信息，定义整个 Spring 上下文加载的流程。其包括实例的初始化和属性设置、自动配置类的加载和执行、内置 Tomcat 服务器的启动等步骤。在后续章节中笔者也会结合源码对这些过程进行介绍。

（13）调用 afterRefresh()方法，执行 Spring 容器初始化的后置逻辑，默认实现是一个空的方法：

```
protected void afterRefresh(ConfigurableApplicationContext context,
ApplicationArguments args) {
    }
```

（14）代码执行时间的监控停止，即知道了启动应用所花费的时间。

（15）发布容器启动事件。

（16）在 ApplicationContext 完成启动后，程序会对 ApplicationRunner 和 CommandLineRunner 进行回调处理，查找当前 ApplicationContex 中是否注册有 CommandLineRunner，如果有，则遍历执行它们。

另外，在 SpringApplication 启动过程中，如果出现问题会由异常处理器接管，并对异常进行统一处理，源码如下所示：

```
private void handleRunFailure(ConfigurableApplicationContext context,
Throwable exception,
        Collection<SpringBootExceptionReporter> exceptionReporters,
SpringApplicationRunListeners listeners) {
    try {
        try {
            handleExitCode(context, exception);
            if (listeners != null) {
// 回调 SpringApplicationRunListener 的 failed()方法
                listeners.failed(context, exception);
            }
```

```
            }
            finally {
                reportFailure(exceptionReporters, exception);
                if (context != null) {
                    context.close();
                }
            }
        }
        catch (Exception ex) {
            logger.warn("Unable to close ApplicationContext", ex);
        }
        ReflectionUtils.rethrowRuntimeException(exception);
    }
```

  本章讲解的源码都来自 Spring Boot2.3.7.RELEASE 版本，它与其他版本的代码可能有些不同。读者想更好地理解 Spring Boot 及其启动过程的原理，可以参考本章给出的提示并自行通过 debug 模式进行调试。理论结合实践才能更好地理解 Spring Boot 在启动过程中的操作。

  通过源码解读和启动流程的介绍，相信读者对于 Spring Boot 框架有了进一步的认识。Spring Boot 的核心依然是 Spring。它只是在 Spring 框架的基础之上，针对 Spring 应用启动流程进行了规范和封装。Spring 的核心启动方法是 refresh()，Spring Boot 在启动时依然会调用该核心方法。在平时的 Spring 项目开发中，这些组件通常是通过 XML 配置文件进行定义和装载的，而 Spring Boot 将该过程简化并通过自动配置的方式实现该过程，减少了开发人员需要做的配置工作量。它更像是基于 Spring 框架的一个增强版的应用启动器。

第 6 章

Spring Boot 之 DispatcherServlet 自动配置及源码解读

如果没有使用 Spring Boot 开发 Web 项目，那么为了使 Spring MVC 中的组件生效，开发人员需要对 Spring MVC 的核心分发器 DispatcherServlet 做一系列的配置工作。而当使用了 Spring Boot 开发 Web 项目，开发人员就只需要导入 spring-boot-starter-web 场景启动器即可，无须再进行任何配置就能够使得 Spring MVC 的核心分发器 DispatcherServlet 正常加载并使用。

本章将结合源码介绍 Spring MVC 的核心分发器 DispatcherServlet 自动配置的流程，主要包括三部分内容。

（1）DispatcherServletAutoConfiguration 自动配置类具体做什么事情。

（2）DispatcherServletAutoConfiguration 自动配置类具体执行过程。

（3）DispatcherServlet 被装载到 Servlet 容器（Tomcat 服务器）中并生效的具体过程。

6.1　Spring MVC的核心分发器DispatcherServlet

6.1.1　核心分发器 DispatcherServlet 介绍

在未使用 Spring Boot 开发 Web 应用时，如果想要使 Spring MVC 生效，则需要增加

第 6 章 Spring Boot 之 DispatcherServlet 自动配置及源码解读

Spring MVC 相关的配置文件，并且在 web.xml 中配置 DispatcherServlet 以及对它的 servlet-mapping 节点进行请求地址的映射，配置内容如下所示：

```xml
<!--Start spring mvc servlet-->
<Servlet>
    <Servlet-name>springMVC</Servlet-name>
    <Servlet-class>org.springframework.web.Servlet.DispatcherServlet</Servlet-class>
    <init-param>
        <param-name>contextConfigLocation</param-name>
        <param-value>classpath:spring-mvc.xml</param-value>
    </init-param>
    <load-on-startup>1</load-on-startup>
</Servlet>
<!--End spring mvc servlet-->
<!--Start Servlet-mapping -->
<Servlet-mapping>
    <Servlet-name>springMVC</Servlet-name>
    <url-pattern>/</url-pattern>
</Servlet-mapping>
<!--End Servlet-mapping -->
```

在项目开发完成后启动 Servlet 容器，比如 Tomcat 服务器或者 Jetty 服务器装载 DispatcherServlet，就可以进行基于 Spring MVC 的 Web 项目开发及测试了。

DispatcherServlet 是 Spring MVC 核心的前端控制器，负责接收用户的请求，并根据用户的请求返回相应的视图给用户，其执行流程如图 6-1 所示。

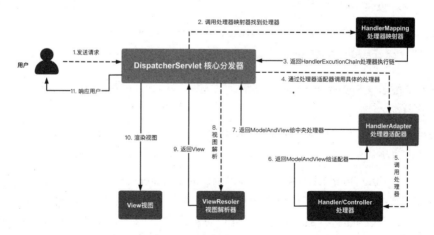

图 6-1　DispatcherServlet 执行流程

首先，用户发起请求，请求到达前端控制器 DispatcherServlet，前端控制器再根据用户的 URL 请求调用处理器映射器查找匹配该 URL 的 Handler，并返回一个执行链。然后，处理器适配器调用相应的 Handler 进行业务处理并返回给前端控制器一个 ModelAndView 对象。前端控制器再请求视图解析器对返回的逻辑视图进行解析。最后前端控制器将返回的视图进行渲染，并把数据封装到 Request 域中返回给用户。

6.1.2 DispatcherServlet 自动配置的日志输出

在第 4 章中笔者创建了第一个 Spring Boot 项目。在初始化时选择了其中的 Web 模块，即在 pom 文件中引入了 spring-boot-starter-web。该模块包含了 Spring MVC 相关依赖，开发了第一个 Web 功能，即在控制器类中实现了一个简单的字符串返回方法，并对该方法进行了请求地址映射设置，没有再做其他配置操作。该项目启动后就能够进行正常的 Web 请求并正确返回数据。

结合 DispatcherServlet 的原理和工作流程可知，Spring Boot 项目在启动后，Spring MVC 的核心分发器 DispatcherServlet 正常配置并装载，才使得/hello 请求能够正常返回数据。

但是，在使用 Spring Boot 开发 Web 项目时，开发人员只导入了 spring-boot-starter-web 场景启动器，根本未进行任何配置，DispatcherServlet 就可以正常加载并使用。这是怎么一回事呢？DispatcherServlet 是如何被装载到 Servlet 容器中的呢？Spring Boot 究竟做了哪些配置，让 Spring MVC 的相关配置都注入 IOC 容器中并生效，从而使得开发人员可以零配置进行 Web 开发呢？

在 Spring Boot 2.1.0 版本的项目启动之后，启动日志中包含一条记录：

```
Servlet dispatcherServlet mapped to [/]
```

如图 6-2 所示，Spring Boot 项目在启动过程中 DispatcherServlet 就已经被装载并且映射到 [/]路径上了。

启动日志是通过 ServletRegistrationBean 类的 addRegistration() 方法打印出来的，源码如图 6-3 所示。

不过，在 Spring Boot 2.1.1 版本之后，"Servlet dispatcherServlet mapped to [/]" 这行日志输出代码就被 Spring 官方开发人员删除了。本书所使用的项目源码的 Spring Boot 版本是 2.3.7。在该版本的代码中也没有这行日志输出语句。2.3.7 版本的 addRegistration() 方法源码如图 6-4 所示。

第 6 章　Spring Boot 之 DispatcherServlet 自动配置及源码解读

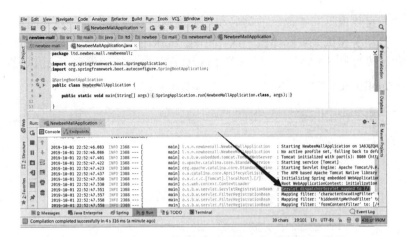

图 6-2　Spring Boot 2.1.0 版本项目的启动日志

```
protected Dynamic addRegistration(String description, ServletContext servletContext) {
    String name = this.getServletName();
    logger.info( o: "Servlet " + name + " mapped to " + this.urlMappings);
    return servletContext.addServlet(name, this.servlet);
}
```

图 6-3　Spring Boot 2.1.0 版本的 addRegistration()方法源码

```
protected Dynamic addRegistration(String description, ServletContext servletContext) {
    String name = this.getServletName();
    return servletContext.addServlet(name, this.servlet);
}
```

图 6-4　Spring Boot 2.3.7 版本的 addRegistration()方法源码

虽然没有了这行日志，但是对于 DispatchServlet 的注册和自动配置没有影响。流程和源码并没有太多改动。接下来笔者就根据这行消失的日志代码来回答前文提出的疑问。

笔者将结合源码来介绍 Spring MVC 的核心分发器 DispatcherServlet 在 Spring Boot 项目中自动配置的知识，让读者明白 Spring Boot 针对 DispatcherServlet 做了哪些自动配置以及如何使这些配置生效的。

6.2　DispatcherServlet 自动配置的源码调试记录

对部分读者来说，阅读源码是一件枯燥且困难的事情，如果毫无头绪或者看不懂源码的含义是很难理解这部分内容的。为了增加可读性，笔者在本章内容中加入了在整理 DispatcherServlet 自动配置原理时调试代码的记录，分别在重要的类中添加了调试。

首先读者可以根据书中提示的类名和行数进行断点设置，并使用 debug 模式启动项目，然后自己动手调试并实际体验一次 Spring Boot 的启动过程。除了 DispatchServlet 类的注册和装载的源码，本章也包括内置 Tomcat 服务器的初始化和启动流程的演示源码。

SpringApplication 类的第 315 行，如图 6-5 所示。

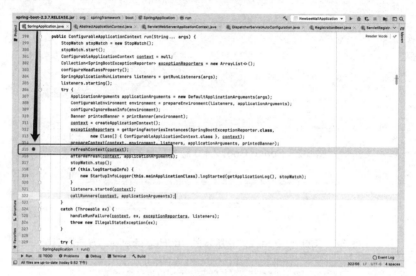

图 6-5　SpringApplication 类中增加调试断点

AbstractApplicationContext 类的第 545 行，如图 6-6 所示。

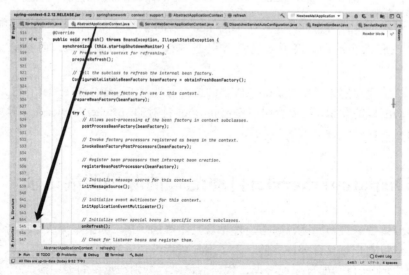

图 6-6　AbstractApplicationContext 类中增加调试断点

第 6 章　Spring Boot 之 DispatcherServlet 自动配置及源码解读

ServletWebServerApplicationContext 类的第 158 行、第 177 行、第 178 行、第 230 行，如图 6-7 所示。

图 6-7　ServletWebServerApplicationContext 类中增加调试断点

DispatcherServletAutoConfiguration 类的第 96 行，如图 6-8 所示。

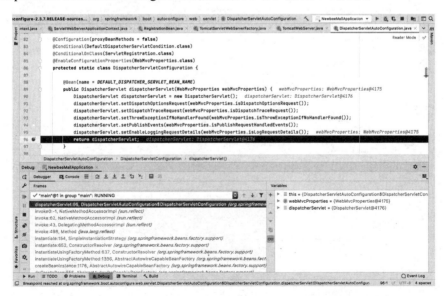

图 6-8　DispatcherServletAutoConfiguration 类中增加调试断点

TomcatServletWebServerFactory 类的第 193 行,如图 6-9 所示。

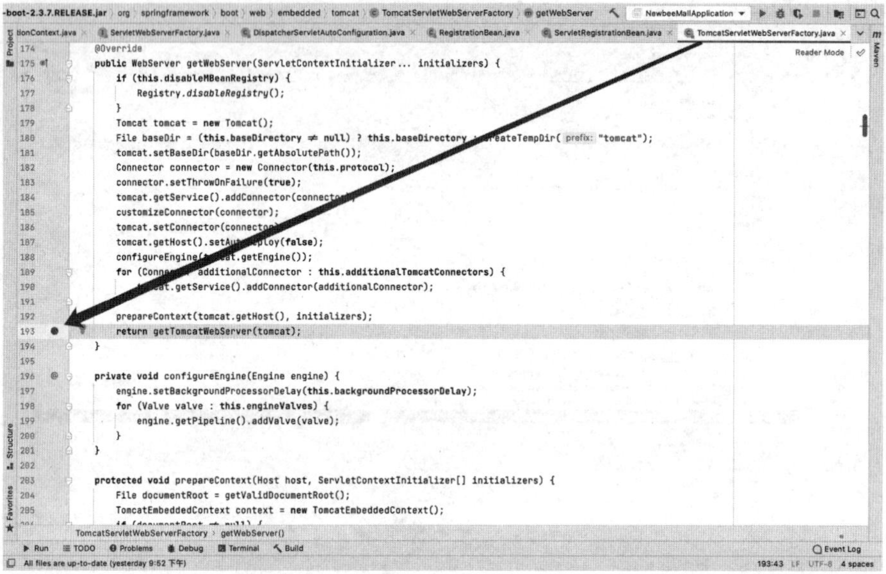

图 6-9　TomcatServletWebServerFactory 类中增加调试断点

TomcatWebServer 类的第 123 行,如图 6-10 所示。

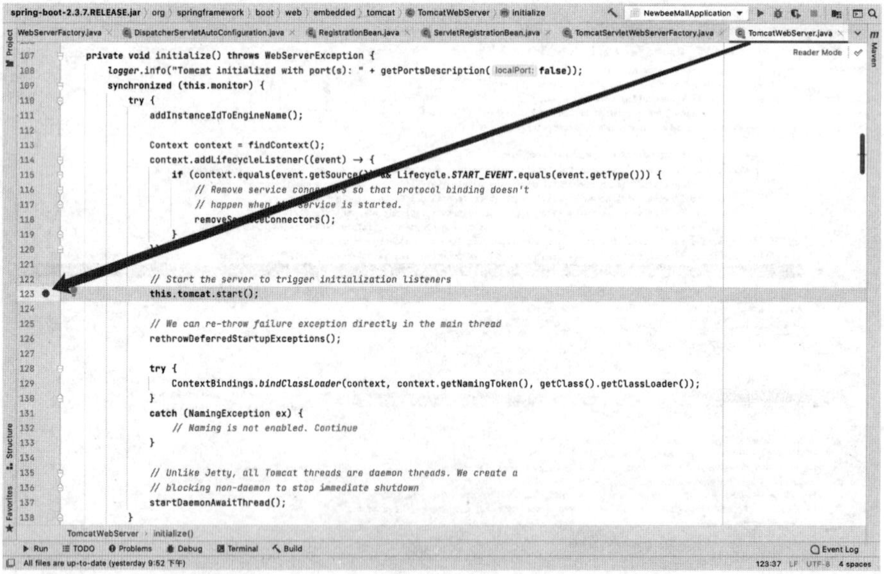

图 6-10　TomcatWebServer 类中增加调试断点

RegistrationBean 类的第 53 行,如图 6-11 所示。

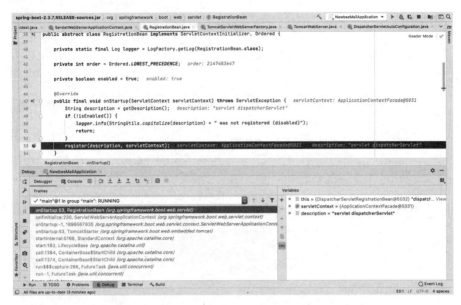

图 6-11　RegistrationBean 类中增加调试断点

ServletRegistrationBean 类的第 178 行,如图 6-12 所示。

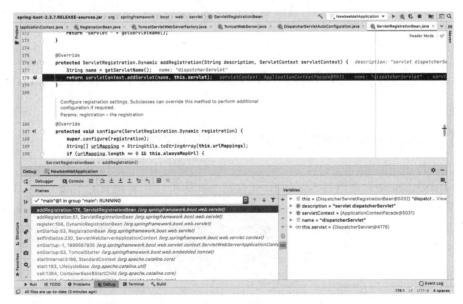

图 6-12　ServletRegistrationBean 类中增加调试断点

ApplicationContext 类的第 853 行，如图 6-13 所示。

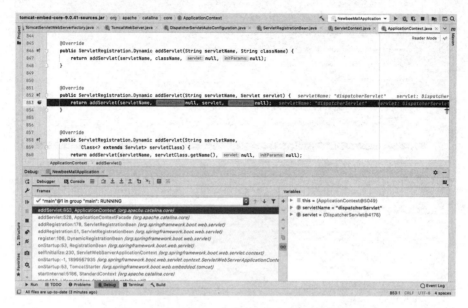

图 6-13　ApplicationContext 类中增加调试断点

在设置断点之后，一定要使用 debug 模式启动项目，并且手动调试几遍，完整地走完几次启动流程，再结合源码来理解 DispatcherServlet 自动配置过程。

6.3　自动配置类DispatcherServletAutoConfiguration

6.3.1　DispatcherServletAutoConfiguration 类的讲解

DispatcherServlet 是通过 Spring Boot 自动配置的机制注册到 IOC 容器中的。自动配置类名称为 DispatcherServletAutoConfiguration，该自动配置类定义在 spring-boot-autoconfigure-2.3.7.RELEASE.jar 包的 org.springframework.boot.autoconfigure.web 包中，如图 6-14 所示。

第 6 章 Spring Boot 之 DispatcherServlet 自动配置及源码解读

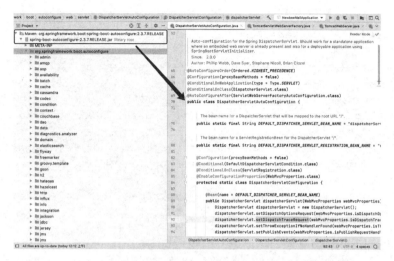

图 6-14 DispatcherServletAutoConfiguration 自动配置类

DispatcherServletAutoConfiguration 类的源码及注释如下所示：

```
@AutoConfigureOrder(Ordered.HIGHEST_PRECEDENCE)
@Configuration(proxyBeanMethods = false)
@ConditionalOnWebApplication(type = Type.SERVLET)
@ConditionalOnClass(DispatcherServlet.class)
@AutoConfigureAfter(ServletWebServerFactoryAutoConfiguration.class)
public class DispatcherServletAutoConfiguration {

    /**
     * The bean name for a DispatcherServlet that will be mapped to the root
URL "/".
     */
    public static final String DEFAULT_DISPATCHER_SERVLET_BEAN_NAME =
"dispatcherServlet";

    /**
     * The bean name for a ServletRegistrationBean for the DispatcherServlet
"/".
     */
    public static final String DEFAULT_DISPATCHER_SERVLET_REGISTRATION_
BEAN_NAME = "dispatcherServletRegistration";

    @Configuration(proxyBeanMethods = false)
    @Conditional(DefaultDispatcherServletCondition.class)
    @ConditionalOnClass(ServletRegistration.class)
```

```java
    @EnableConfigurationProperties(WebMvcProperties.class)
    protected static class DispatcherServletConfiguration {

        // 注册一个名称为"dispatcherServlet"的 Bean 到 Spring IOC 容器中，类型为 DispatcherServlet
        @Bean(name = DEFAULT_DISPATCHER_SERVLET_BEAN_NAME)
        public DispatcherServlet dispatcherServlet(WebMvcProperties webMvcProperties) {
            DispatcherServlet dispatcherServlet = new DispatcherServlet();
            dispatcherServlet.setDispatchOptionsRequest(webMvcProperties.isDispatchOptionsRequest());
            dispatcherServlet.setDispatchTraceRequest(webMvcProperties.isDispatchTraceRequest());
            dispatcherServlet.setThrowExceptionIfNoHandlerFound(webMvcProperties.isThrowExceptionIfNoHandlerFound());
            dispatcherServlet.setPublishEvents(webMvcProperties.isPublishRequestHandledEvents());
            dispatcherServlet.setEnableLoggingRequestDetails(webMvcProperties.isLogRequestDetails());
            return dispatcherServlet;
        }

        @Bean
        @ConditionalOnBean(MultipartResolver.class)
        @ConditionalOnMissingBean(name = DispatcherServlet.MULTIPART_RESOLVER_BEAN_NAME)
        public MultipartResolver multipartResolver(MultipartResolver resolver) {
            // 检测用户是否创建 MultipartResolver，但命名错误
            return resolver;
        }

    }

    @Configuration(proxyBeanMethods = false)
    @Conditional(DispatcherServletRegistrationCondition.class)
    @ConditionalOnClass(ServletRegistration.class)
    @EnableConfigurationProperties(WebMvcProperties.class)
    @Import(DispatcherServletConfiguration.class)
    protected static class DispatcherServletRegistrationConfiguration {

        // 注册一个名称为"dispatcherServletRegistration"的 Bean 到 Spring IOC 容器中，类型为 DispatcherServletRegistrationBean
```

```java
        @Bean(name = DEFAULT_DISPATCHER_SERVLET_REGISTRATION_BEAN_NAME)
        @ConditionalOnBean(value = DispatcherServlet.class, name = 
DEFAULT_DISPATCHER_SERVLET_BEAN_NAME)
        public DispatcherServletRegistrationBean dispatcherServlet
Registration(DispatcherServlet dispatcherServlet,
                WebMvcProperties webMvcProperties, ObjectProvider
<MultipartConfigElement> multipartConfig) {
            DispatcherServletRegistrationBean registration = new 
DispatcherServletRegistrationBean(dispatcherServlet,
                    webMvcProperties.getServlet().getPath());
            registration.setName(DEFAULT_DISPATCHER_SERVLET_BEAN_NAME);
            registration.setLoadOnStartup(webMvcProperties.getServlet().
getLoadOnStartup());
            multipartConfig.ifAvailable(registration::setMultipartConfig);
            return registration;
        }

    }

    ...省略部分代码

}
```

DispatcherServletAutoConfiguration 类中有两个静态内部类，分别为 DispatcherServletConfiguration 类和 DispatcherServletRegistrationConfiguration 类，这三个类的定义都有一些条件注解，表示类的生效条件。

DispatcherServletAutoConfiguration 类的注解释义如下所示。

@AutoConfigureOrder(Ordered.HIGHEST_PRECEDENCE)：类的加载顺序，数值越小越优先加载。

@Configuration(proxyBeanMethods = false)：指定该类为配置类。

@ConditionalOnWebApplication(type = Type.SERVLET)：当前应用是一个 Servlet Web 应用，这个配置类才生效。

@ConditionalOnClass(DispatcherServlet.class)：判断当前 classpath 是否存在 DispatcherServlet 类，存在则生效。

@AutoConfigureAfter(ServletWebServerFactoryAutoConfiguration.class)：自动配置的生效时间在 ServletWebServerFactoryAutoConfiguration 类之后。

通过源码可知，DispatcherServletAutoConfiguration 自动配置类的自动配置触发条件：当前项目类型必须为 SERVLET 且当前 classpath 存在 DispatcherServlet 类（因为当

前项目引入了 spring-boot-starter-web 场景启动器，该启动器包含了相关依赖，因此两个条件都需要满足，自动配置类才会生效）。而@AutoConfigureAfter 注解又定义了自动配置类生效时间在 ServletWebServerFactory 自动配置之后。

　　DispatcherServletConfiguration 类的注解释义如下所示。

　　@Configuration(proxyBeanMethods = false)：指定该类为配置类。

　　@Conditional(DefaultDispatcherServletCondition.class)：在 DefaultDispatcherServletCondition 条件类被满足的情况下该配置类才会生效。

　　@ConditionalOnClass(ServletRegistration.class)：判断当前 classpath 是否存在指定类，若存在则实例化当前类。

　　@EnableConfigurationProperties(WebMvcProperties.class)：实例化 WebMvcProperties 配置属性类。

　　通过源码可知，DispatcherServletConfiguration 自动配置类的自动配置触发条件：在 DefaultDispatcherServletCondition 条件类被满足且当前 classpath 存在 ServletRegistration 类时才会注册 DispatcherServle 至 IOC 容器中。DefaultDispatcherServletCondition 条件类的主要逻辑是判断在 BeanFactory 中是否已经存在 DispatcherServlet 实例，不存在才会触发。在触发后程序将会读取 application.properties 配置文件或者在 application.yml 配置文件中读取 spring.http 和 spring.mvc 前缀的配置项并完成配置。

　　DispatcherServletRegistrationConfiguration 类的注解释义如下所示。

　　@Configuration(proxyBeanMethods = false)：指定该类为配置类。

　　@Conditional(DispatcherServletRegistrationCondition.class)：在 DispatcherServletRegistrationCondition 条件类被满足的情况下该配置类才会生效。

　　@ConditionalOnClass(ServletRegistration.class)：判断当前 classpath 是否存在指定类，若存在则实例化当前类。

　　@EnableConfigurationProperties(WebMvcProperties.class)：实例化 WebMvcProperties 配置属性类。

　　通过源码可知，DispatcherServletConfiguration 自动配置类的自动配置触发条件：在 DispatcherServletRegistrationCondition 条件类被满足且当前 classpath 存在 ServletRegistration 类时才会注册 DispatcherServletRegistrationBean 至 IOC 容器中。DispatcherServletRegistrationCondition 条件类的主要逻辑是判断在当前 BeanFactory 中是否已经存在 DispatcherServlet 实例和 DispatcherServletRegistration 实例，DispatcherServlet 存在且 DispatcherServletRegistration 不存在才会触发。

6.3.2　DispatcherServletAutoConfiguration 在 IOC 容器中的注册

通过前文的条件注解分析可知，当自动配置流程开始且条件被满足后，自动配置类 DispatcherServletAutoConfiguration 在进行自动配置时会注册一个名称为 "dispatcherServlet" 的 Bean 至 IOC 容器中。这就是文章中一直提到的 Spring MVC 的核心分发器 DispatcherServlet。在 Spring Boot 项目中之所以可以不进行配置，是因为自动配置类会将其注册到 IOC 容器中。

在 DispatcherServlet 注册成功之后会继续注册一个名称为 "dispatcherServlet Registration" 的 Bean 至 IOC 容器中。这个 DispatcherServletRegistrationBean 有点特殊，它并不是一个普通意义上的 Bean。在类定义中可以看到它实现了 ServletContextInitializer 接口。DispatcherServletRegistrationBean 继承路径如图 6-15 所示。

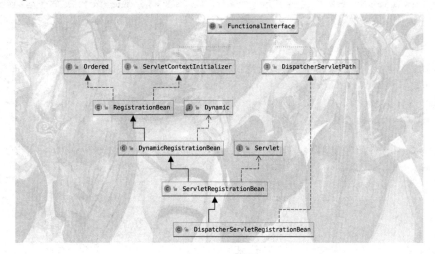

图 6-15　DispatcherServletRegistrationBean 继承路径

ServletContextInitializer 接口用于以编程的方式配置 Servlet 3.0+的上下文，它在 DispatcherServlet 装载中起到了重要作用。后文也会提到这个知识点。该接口的详细内容可以参照 Spring 官方文档：ServletContextInitializer-doc.html。

6.4　DispatcherServlet 自动配置流程

DispatcherServletAutoConfiguration 自动配置类会在 IOC 容器中注册

DispatcherServlet 和 DispatcherServletRegistrationConfiguration 两个 Bean，此为自动配置类的配置结果。那么这个自动配置类何时生效？配置流程又是如何的呢？

这里需要结合 Spring Boot 的启动流程来进行讲解和介绍。

6.4.1 注册至 IOC 容器

DispatcherServletAutoConfiguration 类的条件注解 @AutoConfigureAfter(ServletWebServerFactoryAutoConfiguration.class)定义了自动配置类生效的时间是在 ServletWebServerFactory 自动配置之后，那么首先要找到这个自动配置流程的生效时间点。

结合前文中 SpringApplication.run()方法调用的过程，Spring Boot 项目在启动过程中调用了 AbstractApplicationContext.refresh()方法，源码如下所示：

```java
public abstract class AbstractApplicationContext extends DefaultResourceLoader
        implements ConfigurableApplicationContext {
    ...省略部分代码
    public void refresh() throws BeansException, IllegalStateException {
        synchronized (this.startupShutdownMonitor) {
            try {
                // Allows post-processing of the bean factory in context subclasses.
                postProcessBeanFactory(beanFactory);
                // Invoke factory processors registered as beans in the context.
                invokeBeanFactoryPostProcessors(beanFactory);
                // Register bean processors that intercept bean creation.
                registerBeanPostProcessors(beanFactory);
                // Initialize message source for this context.
                initMessageSource();
                // Initialize event multicaster for this context.
                initApplicationEventMulticaster();
                // 在特殊的 context 子类中初始化其他的特殊 Bean
                onRefresh();
                // Check for listener beans and register them.
                registerListeners();
                // Instantiate all remaining (non-lazy-init) singletons.
                finishBeanFactoryInitialization(beanFactory);
                // Last step: publish corresponding event.
                finishRefresh();
```

```
            }
            catch (BeansException ex) {
                if (logger.isWarnEnabled()) {
                    logger.warn("Exception encountered during context
initialization - " +
                            "cancelling refresh attempt: " + ex);
                }
                throw ex;
            }
            finally {
                resetCommonCaches();
            }
        }
    }
```

点击进入 onfresh() 方法可以看到,onfresh() 方法最终会调用 ServletWebServerApplicationContext 类的 createWebServer() 方法。在该方法中程序会进行 ServletWebServerFactory 对象的获取。在 ServletWebServerFactory 对象初始化完成后,程序就会进行 DispatcherServlet 的自动配置,也就是在 ServletWebServerApplicationContext 类的 177 行会完成 ServletWebServerFactory 对象的自动配置。

如果没有发生异常,在 ServletWebServerFactory 对象配置完成后会触发 DispatcherServletAutoConfiguration 类进行自动配置工作。也就是在这个时间点自动配置类会向 IOC 容器注入两个 DispatcherServlet 相关的 Bean,createWebServer() 方法的源码如下所示:

```
public class ServletWebServerApplicationContext extends GenericWeb
ApplicationContext
        implements ConfigurableWebServerApplicationContext {

    @Override
    protected void onRefresh() {
        super.onRefresh();
        try {
            // 创建 WebServer
            createWebServer();
        }
        catch (Throwable ex) {
            throw new ApplicationContextException("Unable to start web
server", ex);
        }
    }
```

```
    ...省略部分代码

    private void createWebServer() {
        WebServer webServer = this.webServer;
        ServletContext ServletContext = getServletContext();
        if (webServer == null && ServletContext == null) {
    // 获取ServletWebServerFactory对象，在对象初始化完成后，会进行
DispatcherServlet的自动配置
            ServletWebServerFactory factory = getWebServerFactory();
            this.webServer = factory.getWebServer(getSelfInitializer());
        }
        else if (ServletContext != null) {
            try {
                getSelfInitializer().onStartup(servletContext);
            }
            catch (ServletException ex) {
                throw new ApplicationContextException("Cannot initialize
Servlet context",
                    ex);
            }
        }
        initPropertySources();
    }
```

获取 ServletWebServerFactory 对象是为了获取 WebServer 对象（在本次实例中 WebServer 对象为 Tomcat）。接下来就是创建内嵌的 Tomcat 实例并进行配置，在配置完成后服务器就启动了。这些步骤在 178 行的 getWebServer()方法中可以查看。在启动内嵌的 Tomcat 服务器成功后程序才可以装载 DispatcherServlet。

6.4.2 创建并启动嵌入式的 Tomcat 对象

DispatchServlet 在上一个步骤中已经完成了注册。此时，在 IOC 容器中已经含有名称为 "dispatcherServlet" 的 Bean。不过此时 DispatcherServlet 并没有生效，只是完成了在 IOC 容器中的注册。而 Servlet 的运行环境在 Servlet 容器中，如果没有 Servlet 容器或者说 Servlet 容器没有启动的话，Servlet 是没有任何作用的。因此还需要一个创建并启动嵌入式的 Tomcat 对象的流程。

查看 createWebServer()方法，源码如下所示：

```
    private void createWebServer() {
        WebServer webServer = this.webServer;
```

```
        ServletContext servletContext = getServletContext();
        if (webServer == null && servletContext == null) {
// 获取嵌入式的Servlet容器创建工厂（完成后开始DispatcherServlet的自动配置）
            ServletWebServerFactory factory = getWebServerFactory();
// 通过创建工厂进行嵌入式的Servlet容器对象创建，并启动该Servlet容器
            this.webServer = factory.getWebServer(getSelfInitializer());
            getBeanFactory().registerSingleton("webServerGracefulShutdown",
                    new WebServerGracefulShutdownLifecycle(this.webServer));
            getBeanFactory().registerSingleton("webServerStartStop",
                    new WebServerStartStopLifecycle(this, this.webServer));
        }
        else if (servletContext != null) {
            try {
                getSelfInitializer().onStartup(servletContext);
            }
            catch (ServletException ex) {
                throw new ApplicationContextException("Cannot initialize servlet context", ex);
            }
        }
        initPropertySources();
    }
```

通过源码可以发现，在 ServletWebServerApplicationContext 类的第 177 行可以获取嵌入式的 Servlet 容器创建工厂对象 ServletWebServerFactory，在创建成功之后调用 getWebServer()方法再创建 Servlet 容器对象。本案例中所创建的容器为 Tomcat。

getWebServer()方法的作用是创建并启动 Tomcat Server。最终调用的实例方法是 TomcatServletWebServerFactory 类的 getWebServer()方法，源码如下所示：

```
public class TomcatServletWebServerFactory extends AbstractServletWeb
ServerFactory
        implements ConfigurableTomcatWebServerFactory, ResourceLoaderAware {
... 省略部分代码

    @Override
    public WebServer getWebServer(ServletContextInitializer... initializers) {
        if (this.disableMBeanRegistry) {
            Registry.disableRegistry();
        }
        Tomcat tomcat = new Tomcat();
// 给嵌入式的Tomcat创建一个临时文件夹
        File baseDir = (this.baseDirectory != null) ? this.baseDirectory :
```

```java
createTempDir("tomcat");
    tomcat.setBaseDir(baseDir.getAbsolutePath());
//Tomcat 的核心概念 Connector,默认的协议为 Http11NioProtocol
    Connector connector = new Connector(this.protocol);
    connector.setThrowOnFailure(true);
    tomcat.getService().addConnector(connector);
    customizeConnector(connector);
    tomcat.setConnector(connector);
// 关闭热部署
    tomcat.getHost().setAutoDeploy(false);
    configureEngine(tomcat.getEngine());
    for (Connector additionalConnector : this.additionalTomcatConnectors) {
        tomcat.getService().addConnector(additionalConnector);
    }
// 生成TomcatEmbeddedContext,感兴趣的读者可以继续研究
    prepareContext(tomcat.getHost(), initializers);
// 创建 TomcatWebServer
    return getTomcatWebServer(tomcat);
}

... 省略部分代码

protected TomcatWebServer getTomcatWebServer(Tomcat tomcat) {
    return new TomcatWebServer(tomcat, getPort() >= 0, getShutdown());
}
}
```

在 getTomcatWebServer()方法中,调用了 TomcatWebServer 类的构造方法,TomcatWebServer 类源码如下所示:

```java
public class TomcatWebServer implements WebServer {

    private static final Log logger = LogFactory.getLog(TomcatWebServer.class);

    private static final AtomicInteger containerCounter = new AtomicInteger(-1);

    private final Object monitor = new Object();

    private final Map<Service, Connector[]> serviceConnectors = new HashMap<>();

    private final Tomcat tomcat;
```

```java
    private final boolean autoStart;

    private final GracefulShutdown gracefulShutdown;

    private volatile boolean started;

    /**
     * Create a new {@link TomcatWebServer} instance.
     * @param tomcat the underlying Tomcat server
     */
    public TomcatWebServer(Tomcat tomcat) {
        this(tomcat, true);
    }

    /**
     * Create a new {@link TomcatWebServer} instance.
     * @param tomcat the underlying Tomcat server
     * @param autoStart if the server should be started
     */
    public TomcatWebServer(Tomcat tomcat, boolean autoStart) {
        this(tomcat, autoStart, Shutdown.IMMEDIATE);
    }

    /**
     * Create a new {@link TomcatWebServer} instance.
     * @param tomcat the underlying Tomcat server
     * @param autoStart if the server should be started
     * @param shutdown type of shutdown supported by the server
     * @since 2.3.0
     */
    public TomcatWebServer(Tomcat tomcat, boolean autoStart, Shutdown shutdown) {
        Assert.notNull(tomcat, "Tomcat Server must not be null");
        this.tomcat = tomcat;
        this.autoStart = autoStart;
        this.gracefulShutdown = (shutdown == Shutdown.GRACEFUL) ? new GracefulShutdown(tomcat) : null;
        initialize();
    }

    private void initialize() throws WebServerException {
        logger.info("Tomcat initialized with port(s): " + getPortsDescription(false));
```

```
            synchronized (this.monitor) {
                try {
                    addInstanceIdToEngineName();

                    Context context = findContext();
                    context.addLifecycleListener((event) -> {
                        if (context.equals(event.getSource()) && Lifecycle.START_EVENT.equals(event.getType())) {
                            removeServiceConnectors();
                        }
                    });

                    // 启动 Tomcat
                    this.tomcat.start();

                    rethrowDeferredStartupExceptions();

                    try {
                        ContextBindings.bindClassLoader(context, context.getNamingToken(), getClass().getClassLoader());
                    }
                    catch (NamingException ex) {
                    }

                    startDaemonAwaitThread();
                }
                catch (Exception ex) {
                    stopSilently();
                    destroySilently();
                    throw new WebServerException("Unable to start embedded Tomcat", ex);
                }
            }
        }
... 省略部分代码

}
```

在 TomcatWebServer 构造方法中最后执行了 initialize()方法，在该方法中启动了嵌入式的 Tomcat 服务器。

在 Spring Boot 项目中嵌入式的 Tomcat 与平时使用的 Tomcat 在核心组件上是一样

的，都包含 Service、Connector、Engine、Host、Context，只是嵌入式的 Tomcat 服务器其初始化和启动流程都由 Spring Boot 来完成，开发人员无须操作。

Tomcat 的启动过程分为初始化和启动两个步骤，这一小节都已经介绍完毕，接下来介绍装载 DispatcherServlet 的步骤。

6.4.3 装载至 Servlet 容器

ServletWebServerApplicationContext 类第 178 行执行了 getWebServer()方法，创建并启动了 Tomcat，代码如下所示：

```
getWebServer(getSelfInitializer());
```

重点来看一下该方法的传参：getSelfInitializer()，点击查看该方法的实现，如下所示：

```
private org.springframework.boot.web.servlet.ServletContextInitializer
getSelfInitializer() {
    return this::selfInitialize;
}
```

由源码可知,该方法直接返回了一个 lambda 表达式作为 getWebServer()方法的传参。为什么可以这样呢？点开查看 getWebServer()方法的定义：

```
@FunctionalInterface
public interface ServletWebServerFactory {
    WebServer getWebServer(ServletContextInitializer... initializers);
}
```

getWebServer()方法的参数被定义为 ServletContextInitializer 类型的可变参数。ServletContextInitializer 接口在定义中标注了@FunctionalInterface 注解，是一个函数式接口。因此它可以直接将 lambda 表达式作为传参给 getWebServer()方法。而 selfInitialize()方法暂时不会被执行，而是在 Tomcat 服务器启动后被回调。selfInitialize()方法的代码定义在 ServletWebServerApplicationContext 类的第 225 行，源码如下所示：

```
private void selfInitialize(ServletContext servletContext) throws Servlet
Exception {
    prepareWebApplicationContext(servletContext);
    ConfigurableListableBeanFactory beanFactory = getBeanFactory();
    ExistingWebApplicationScopes existingScopes = new ExistingWeb
ApplicationScopes(
            beanFactory);
```

```
        WebApplicationContextUtils.registerWebApplicationScopes(beanFactory,
                getServletContext());
        existingScopes.restore();
        WebApplicationContextUtils.registerEnvironmentBeans(beanFactory,
                getServletContext());
        // 获取所有的 ServletContextInitializer 实例
        for (ServletContextInitializer beans : getServletContextInitializerBeans()) {
            // 循环并执行每个实例的 onStartup()方法
            beans.onStartup(servletContext);
        }
    }
```

首先来了解一下 Servlet 3.0 的规范。Servlet 3.0 提供了可以动态注册 Servlet、Filter、Listener 的 ServletContext 相关 API,开发人员可以将 web.xml 相关配置通过编码的方式实现,并由 javax.servlet.ServletContainerInitializer 的实现类负责在 Servlet 容器启动后进行加载。Spring 提供了一个实现类 org.springframework.web.SpringServletContainerInitializer,该类会调用所有实现类的 onStartup()方法将相关的组件装载到 Servlet 容器中。

selfInitialize() 会调用 getServletContextInitializerBeans() 方法获取所有 ServletContextInitializer 接口的实现类,DispatcherServletRegistrationBean 就是该特殊接口的实现类。在 DispatcherServletAutoConfiguration 执行过程中就已经在 IOC 容器中注册了名称为"dispatcherServletRegistrationBean"的 Bean,它会在执行 getServletContextInitializerBeans()方法时被获取,最后由执行它的 onStartup()方法来装载 DispatcherServlet 至 Tomcat 服务器中。

装载 Servlet 具体方法的调用链路如下所示。

(1) onStartup()方法的实现在 RegistrationBean 类中,该方法调用了 register()方法注册和配置 Bean,源码如下所示:

```
public abstract class RegistrationBean implements ServletContextInitializer, Ordered {

    ...省略部分代码

    @Override
    public final void onStartup(ServletContext servletContext) throws ServletException {
        String description = getDescription();
        if (!isEnabled()) {
            logger.info(StringUtils.capitalize(description)
                    + " was not registered (disabled)");
```

```
                return;
            }
// 注册 Bean
        register(description, servletContext);
    }
```

（2）register()方法的实现在 DynamicRegistrationBean 类中，该方法被执行时会调用 addRegistration()方法。这里所讲的 DispatcherServlet 就是在该方法内进行装载的，源码如下所示：

```
public abstract class DynamicRegistrationBean<D extends Registration.Dynamic>
        extends RegistrationBean {

    ...省略部分代码

    @Override
    protected final void register(String description, ServletContext servletContext) {
        // 添加过滤器或者Servlet
        D registration = addRegistration(description, servletContext);
        if (registration == null) {
            logger.info(StringUtils.capitalize(description) + " was not registered "
                    + "(possibly already registered?)");
            return;
        }
        configure(registration);
    }

    ...省略部分代码

}
```

（3）addRegistration()方法的实现在 ServletRegistrationBean 类中，该方法被执行时会调用 addServlet()方法装载 Servlet，源码如下所示：

```
public class ServletRegistrationBean<T extends Servlet>
        extends DynamicRegistrationBean<ServletRegistration.Dynamic> {

    ...省略部分代码

    @Override
    protected ServletRegistration.Dynamic addRegistration(String description,
```

```
        ServletContext servletContext) {
    String name = getServletName();
    // 装载 Servlet
    return servletContext.addServlet(name, this.servlet);
}

...省略部分代码

}
```

（5）addServlet()方法具体实现如下所示：

```
public class ApplicationContext implements ServletContext {

    ...省略部分代码

    @Override
    public ServletRegistration.Dynamic addServlet(String servletName,
Servlet servlet) {
        return addServlet(servletName, null, servlet, null);
    }

private ServletRegistration.Dynamic addServlet(String servletName, String servletClass,
        Servlet servlet, Map<String,String> initParams) throws
IllegalStateException {

    if (servletName == null || servletName.equals("")) {
        throw new IllegalArgumentException(sm.getString(
              "applicationContext.invalidServletName", servletName));
    }

    if (!context.getState().equals(LifecycleState.STARTING_PREP)) {
        //TODO Spec breaking enhancement to ignore this restriction
        throw new IllegalStateException(
              sm.getString("applicationContext.addServlet.ise",
                    getContextPath()));
    }

    Wrapper wrapper = (Wrapper) context.findChild(servletName);

    if (wrapper == null) {
        wrapper = context.createWrapper();
        wrapper.setName(servletName);
```

```java
            context.addChild(wrapper);
        } else {
            if (wrapper.getName() != null &&
                    wrapper.getServletClass() != null) {
                if (wrapper.isOverridable()) {
                    wrapper.setOverridable(false);
                } else {
                    return null;
                }
            }
        }

        ServletSecurity annotation = null;
        if (servlet == null) {
            wrapper.setServletClass(servletClass);
            Class<?> clazz = Introspection.loadClass(context, servletClass);
            if (clazz != null) {
                annotation = clazz.getAnnotation(ServletSecurity.class);
            }
        } else {
            wrapper.setServletClass(servlet.getClass().getName());
            wrapper.setServlet(servlet);
            if (context.wasCreatedDynamicServlet(servlet)) {
                annotation = servlet.getClass().getAnnotation(ServletSecurity.class);
            }
        }

        if (initParams != null) {
            for (Map.Entry<String, String> initParam: initParams.entrySet()) {
                wrapper.addInitParameter(initParam.getKey(), initParam.getValue());
            }
        }

        ServletRegistration.Dynamic registration =
                new ApplicationServletRegistration(wrapper, context);
        if (annotation != null) {
            registration.setServletSecurity(new ServletSecurityElement(annotation));
        }
        return registration;
```

```
    }
        ...省略部分代码
}
```

最终，通过 onStartup()方法将 DispatchServlet 装载到 Servlet 容器中（即 Tomcat 服务器），在 Tomcat 服务器启动后就能够使用 DispatchServlet 进行请求映射和拦截处理了。

综上所述，由于 Spring Boot 的自动配置，开发人员在 Spring Boot 项目中引入 spring-boot-starter-web 场景启动器之后无须进行任何设置也可以进行 Web 开发。

本章的主要内容就是介绍 Spring Boot 中 DispatcherServlet 自动配置的全部流程和知识点，主要围绕以下三个问题展开了讨论。

（1）DispatcherServletAutoConfiguration 自动配置类做了哪些事？

主要是向 IOC 容器中注册了两个 Bean，名称分别为"DispatcherServlet"和"DispatcherServletRegistration"，即 org.springframework.web.servlet.DispatcherServlet 和 org.springframework.boot.autoconfigure.web.servlet.DispatcherServletRegistrationBean。

（2）DispatcherServletAutoConfiguration 自动配置类是何时执行的？

结合 Spring Boot 项目启动过程可以得出，自动配置类的执行是在 ServletWebServerFactory 对象获取之后触发的，方法调用链如图 6-16 所示。

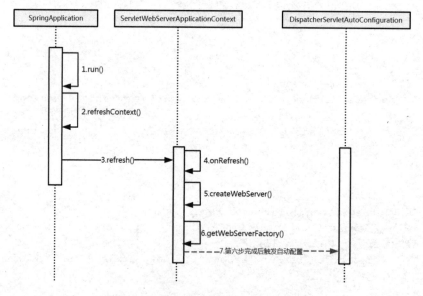

图 6-16 Spring Boot 项目启动方法调用链

（3）DispatcherServlet 是如何被装载到 Servlet 容器中并生效的？

Servlet 容器在启动之后会回调 selfInitialize()方法，在该方法完成了 DispatcherServlet 的装载过程，该方法调用链如图 6-17 所示。

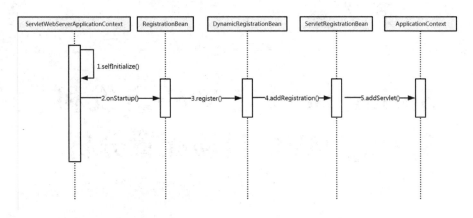

图 6-17　DispatcherServlet 装载方法的调用链

以上所涉及的类和方法读者可以按照书中的提示，自行查看 Spring Boot 的源码并手动调试，这样才能更好地理解整个 DispatcherServlet 自动配置和装载的过程。

另外，文章中涉及的源码都来自 Spring Boot 2.3.7-RELEASE 版本，与其他版本的代码可能些许不同。如果是 Spring Boot 2.0 之前的版本差异会更大，这一点需要注意。

第 7 章

Spring Boot 之 Web 开发和 Spring MVC 自动配置分析

Spring Boot 为 Spring MVC 的相关组件提供了自动配置，使得开发人员能够非常方便地进行 Web 项目开发。本章将继续结合实践案例和 Spring Boot 源码进行研究，讲解 Spring Boot 针对 Web 开发增加的功能。

7.1 Spring MVC 自动配置内容

如图 7-1 所示，Spring Boot 2.3.7 版本的官方解释文档介绍了 Spring MVC 的自动配置内容，文档地址为 boot-features-spring-mvc-auto-configuration。

通过官方文档的介绍可以发现，除了装载 DispatcherServlet 之外，Spring Boot 还做了如下默认配置。

① 自动配置了视图解析器
② 静态资源文件处理
③ 自动注册了大量的转换器和格式化器
④ 提供 HttpMessageConverter 对请求参数和返回结果进行处理

第 7 章 Spring Boot 之 Web 开发和 Spring MVC 自动配置分析

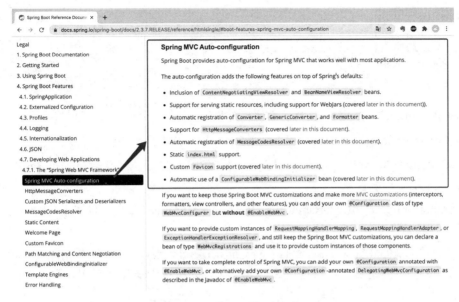

图 7-1 Spring MVC 自动配置内容

⑤自动注册了 MessageCodesResolver

⑥默认欢迎页配置

⑦favicon 自动配置

⑧可配置的 Web 初始化绑定器

以上自动配置都是在 WebMvcAutoConfiguration 自动配置类中操作的。

7.2 WebMvcAutoConfiguration源码分析

WebMvcAutoConfiguration 自动配置类的定义同样也在 spring-boot-autoconfigure-2.3.7. RELEASE.jar 包的 org.springframework.boot.autoconfigure.web 包中。WebMvcAutoConfiguration 类的源码如下所示：

```
@Configuration(proxyBeanMethods = false)
@ConditionalOnWebApplication(type = Type.SERVLET)
@ConditionalOnClass({ Servlet.class, DispatcherServlet.class, WebMvcConfigurer.class })
@ConditionalOnMissingBean(WebMvcConfigurationSupport.class)
@AutoConfigureOrder(Ordered.HIGHEST_PRECEDENCE + 10)
@AutoConfigureAfter({ DispatcherServletAutoConfiguration.class, TaskExecution
```

```
AutoConfiguration.class,
        ValidationAutoConfiguration.class })
public class WebMvcAutoConfiguration {
    ... 省略部分代码
}
```

WebMvcAutoConfiguration 类的注解释义如下所示。

@Configuration(proxyBeanMethods=false)：指定该类为配置类。

@ConditionalOnWebApplication(type=Type.SERVLET)：当前应用是一个 Servlet Web 应用，这个配置类才会生效。

@AutoConfigureOrder(Ordered.HIGHEST_PRECEDENCE+10)：类的加载顺序，数值越小越优先加载。

@ConditionalOnClass({Servlet.class, DispatcherServlet.class, WebMvcConfigurer.class})：判断当前 classpath 是否存在指定类 Servlet 类、DispatcherServle 类和 WebMvcConfigurer 类，存在则生效。

@ConditionalOnMissingBean(WebMvcConfigurationSupport.class)：判断 IOC 容器中是否存在 WebMvcConfigurationSupport 类型的 Bean，不存在则生效。

@AutoConfigureAfter({ DispatcherServletAutoConfiguration.class, TaskExecutionAutoConfiguration.class, ValidationAutoConfiguration.class })：自动配置的生效时间在 DispatcherServletAutoConfiguration 等三个自动配置类之后。

通过源码可知，WebMvcAutoConfiguration 自动配置类的自动配置触发条件：当前项目类型必须为 SERVLET，当前 classpath 存在 Servlet 类、DispatcherServle 类和 WebMvcConfigurer 类，未向 IOC 容器中注册 WebMvcConfigurationSupport 类型的 Bean，并且@AutoConfigureAfter 注解定义了自动配置类生效时间在 DispatcherServletAutoConfiguration、TaskExecutionAutoConfiguratio、ValidationAutoConfiguration 自动配置之后。

WebMvcAutoConfiguration 中有 3 个主要的内部类，如图 7-2 所示。具体的自动配置逻辑实现都是在这 3 个内部类中实现的。

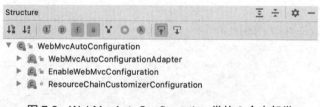

图 7-2 WebMvcAutoConfiguration 类的 3 个内部类

7.3 ViewResolver视图解析器的自动配置

Spring MVC 中的 Controller 控制器可以返回各种各样的视图，比如 JSP、JSON、Velocity、FreeMarker、Thymeleaf、HTML 字符流等。这些视图的解析就涉及各种视图（即 View）对应的各种视图解析器（即 ViewResolver）。视图解析器的作用是将逻辑视图转为物理视图，所有的视图解析器都必须实现 ViewResolver 接口。

Spring MVC 提供了不同的方式在 Spring Web 上下文中配置一种或多种解析策略，并可以指定它们之间的先后顺序，每一种映射策略对应一个具体的视图解析器实现类。开发人员可以设置一个视图解析器或混用多个视图解析器并指定解析器的优先顺序。Spring MVC 会按视图解析器的优先顺序对逻辑视图名进行解析，直到解析成功并返回视图对象，否则抛出异常。

在 WebMvcAutoConfigurationAdapter 内部类中，在前置条件满足的情况下自动配置类会向 IOC 容器中注册三个视图解析器，分别是 InternalResourceViewResolver、BeanNameViewResolver 和 ContentNegotiatingViewResolver。

源码和注释如下所示：

```java
public static class WebMvcAutoConfigurationAdapter implements WebMvcConfigurer {
    ...省略部分代码


    @Bean
    @ConditionalOnMissingBean
    //在 IOC 容器中没有 InternalResourceViewResolver 类的 Bean 时，向容器中注册一个 InternalResourceViewResolver 类型的 Bean
    public InternalResourceViewResolver defaultViewResolver() {
        InternalResourceViewResolver resolver = new InternalResourceViewResolver();
        resolver.setPrefix(this.mvcProperties.getView().getPrefix());
        resolver.setSuffix(this.mvcProperties.getView().getSuffix());
        return resolver;
    }

    @Bean
    @ConditionalOnBean(View.class)
     //IOC 容器中存在 View.class
    @ConditionalOnMissingBean
    // 在满足上面的一个条件时，如果 IOC 容器中没有 BeanNameViewResolver 类的
```

```java
Bean, 向容器中注册一个 BeanNameViewResolver 类型的 Bean
    public BeanNameViewResolver beanNameViewResolver() {
        BeanNameViewResolver resolver = new BeanNameViewResolver();
        resolver.setOrder(Ordered.LOWEST_PRECEDENCE - 10);
        return resolver;
    }

    @Bean
    @ConditionalOnBean(ViewResolver.class)
     //IOC 容器中存在 ViewResolver.class
    @ConditionalOnMissingBean(name = "viewResolver", value = ContentNegotiatingViewResolver.class)
// 在满足上面的一个条件时,如果 IOC 容器中没有名称为 viewResolver 且类型为 Content
NegotiatingViewResolver 类的 Bean, 向容器中注册一个 ContentNegotiatingView
Resolver 类型的 Bean
    public ContentNegotiatingViewResolver viewResolver(BeanFactory beanFactory) {
        ContentNegotiatingViewResolver resolver = new ContentNegotiatingViewResolver();
        resolver.setContentNegotiationManager(beanFactory.getBean(ContentNegotiationManager.class));
        // ContentNegotiatingViewResolver 使用其他所有视图解析器定位
        // 一个应该具有较高优先级的视图
        resolver.setOrder(Ordered.HIGHEST_PRECEDENCE);
        return resolver;
    }
```

BeanNameViewResolver：在控制器中，一个方法的返回值的字符串会根据 BeanNameViewResolver 查找 Bean 的名称并为返回字符串的 View 渲染视图。

InternalResourceViewResolver：常用的 ViewResolver，主要通过设置在前缀、后缀和控制器中的方法来返回视图名的字符串，从而得到实际视图内容。

ContentNegotiatingViewResolver：特殊的视图解析器，它并不会自己处理各种视图，而是委派给其他不同的 ViewResolver 来处理不同的 View，级别最高。

在普通的 Web 项目中，开发人员需要自己手动配置视图解析器，配置代码如下所示：

```xml
<!-- 视图解析器 -->
<bean id="viewResolver"
    class="org.springframework.web.servlet.view.InternalResourceViewResolver">
    <property name="prefix" value="/"/>
```

```
<property name="suffix" value=".jsp"></property>
</bean>
```

与之相对比，Spring Boot 的自动配置机制则会直接在项目启动过程中将视图解析器注册到 IOC 容器中，而不需要开发人员再做多余的配置。当然，如果不想使用默认的配置策略，也可以自行添加视图解析器到 IOC 容器中。

7.4 自动注册Converter、Formatter

在 WebMvcAutoConfigurationAdapter 内部类中，含有 addFormatters()方法，该方法会向 FormatterRegistry 添加在 IOC 容器中所拥有的 Converter、GenericConverter、Formatter 类型的 Bean。

addFormatters()方法的源码如下所示：

```
@Override
public void addFormatters(FormatterRegistry registry) {
    ApplicationConversionService.addBeans(registry, this.beanFactory);
}
```

实际调用的逻辑代码为 ApplicationConversionService 类的 addBeans()方法，该方法源码如下所示：

```
public static void addBeans(FormatterRegistry registry,
ListableBeanFactory beanFactory) {
    Set<Object> beans = new LinkedHashSet<>();
    beans.addAll(beanFactory.getBeansOfType(GenericConverter.class).values());
    beans.addAll(beanFactory.getBeansOfType(Converter.class).values());
    beans.addAll(beanFactory.getBeansOfType(Printer.class).values());
    beans.addAll(beanFactory.getBeansOfType(Parser.class).values());
    for (Object bean : beans) {
        if (bean instanceof GenericConverter) {
            registry.addConverter((GenericConverter) bean);
        }
        else if (bean instanceof Converter) {
            registry.addConverter((Converter<?, ?>) bean);
        }
        else if (bean instanceof Formatter) {
            registry.addFormatter((Formatter<?>) bean);
```

```
        }
        else if (bean instanceof Printer) {
            registry.addPrinter((Printer<?>) bean);
        }
        else if (bean instanceof Parser) {
            registry.addParser((Parser<?>) bean);
        }
    }
}
```

为了方便读者理解，这里简单举一个案例。

在 Controller 包中新建 TestController 类并新增 typeConversionTest()方法，参数分别如下所示。

goodsName：参数类型为 String。

weight：参数类型为 float。

type：参数类型为 int。

onSale：参数类型为 Boolean。

typeConversionTest()方法的代码如下所示：

```
@RestController
public class TestController {

    @RequestMapping("/test/type/conversion")
    public void typeConversionTest(String goodsName, float weight, int type, Boolean onSale) {
        System.out.println("goodsName:" + goodsName);
        System.out.println("weight:" + weight);
        System.out.println("type:" + type);
        System.out.println("onSale:" + onSale);
    }
}
```

在编码完成后重启 Spring Boot 项目，项目启动成功后在浏览器中输入地址进行请求，查看控制台的打印结果。

第一次请求：

```
http://localhost:8080/test/type/conversion?goodsName=iPhoneX&weight=174.5&type=1&onSale=true
```

打印结果如下:

```
goodsName:iPhoneX
weight:174.5
type:1
onSale:true
```

第二次请求:

```
http://localhost:8080/test/type/conversion?goodsName=iPhone8&weight=174.5&type=2&onSale=0
```

打印结果如下:

```
goodsName:iPhone8
weight:174.5
type:2
onSale:false
```

其实这就是 Spring MVC 中的类型转换，HTTP 请求传递的数据都是字符串 String 类型的。这个方法在 Controller 中被定义。该方法确保对应的地址接收到浏览器的请求，并且请求中 goodsName（String 类型）、weight（float 类型）、type（int 类型）、onSale（Boolean 类型）参数的类型转换都已经被正确执行。读者可以在本地自行测试几次。

以上是简单的类型转换。如果业务需要的话，也可以进行自定义类型转换器并添加到项目中。

7.5 消息转换器HttpMessageConverter

HttpMessageConverter 的设置也是通过 WebMvcAutoConfigurationAdapter 完成的，源码如下所示:

```
@Override
public void configureMessageConverters(List<HttpMessageConverter<?>> converters) {
    this.messageConvertersProvider
            .ifAvailable((customConverters) -> converters.addAll(customConverters.getConverters()));
}
```

在使用 Spring MVC 开发 Web 项目时，使用@RequestBody、@ResponseBody 注解

进行请求实体的转换和响应结果的格式化输出非常普遍。以 JSON 数据为例,这两个注解的作用分别可以将请求中的数据解析成 JSON 并绑定为实体对象以及将响应结果以 JSON 格式返回给请求发起者,但 HTTP 请求和响应是基于文本的。也就是说在 Spring MVC 内部维护了一套转换机制,也就是开发人员通常所说的"将 JSON 格式的请求信息转换为一个对象,将对象转换为 JSON 格式并输出为响应信息。"这些就是 HttpMessageConverter 的作用。

举一个简单的例子,首先在项目中新建 entity 包并定义一个实体类 SaleGoods,然后通过@RequestBody、@ResponseBody 注解进行参数的读取和响应,代码如下所示:

```java
// 实体类
public class SaleGoods {
    private Integer id;
    private String goodsName;
    private float weight;
    private int type;
    private Boolean onSale;
    public Integer getId() {
        return id;
    }
    public void setId(Integer id) {
        this.id = id;
    }
    public String getGoodsName() {
        return goodsName;
    }
    public void setGoodsName(String goodsName) {
        this.goodsName = goodsName;
    }
    public float getWeight() {
        return weight;
    }
    public void setWeight(float weight) {
        this.weight = weight;
    }
    public Boolean getOnSale() {
        return onSale;
    }
    public void setOnSale(Boolean onSale) {
        this.onSale = onSale;
    }
    public int getType() {
        return type;
```

```java
    }
    public void setType(int type) {
        this.type = type;
    }
    @Override
    public String toString() {
        return "SaleGoods{" +
                "id=" + id +
                ", goodsName='" + goodsName + '\'' +
                ", weight=" + weight +
                ", type=" + type +
                ", onSale=" + onSale +
                '}';
    }
}
```

在 TestController 控制器中新增 httpMessageConverterTest() 方法,代码如下所示:

```java
@RestController
public class TestController {

    @RequestMapping(value = "/test/httpmessageconverter", method = RequestMethod.POST)
    public SaleGoods httpMessageConverterTest(@RequestBody SaleGoods saleGoods) {
        System.out.println(saleGoods.toString());
        saleGoods.setType(saleGoods.getType() + 1);
        saleGoods.setGoodsName("商品名: " + saleGoods.getGoodsName());
        return saleGoods;
    }

}
```

上述代码的作用就是拿到封装好的 SaleGoods 对象,在进行简单的属性修改后,最后将对象数据返回。

在编码完成后重启项目,并发送请求数据进行测试,请求数据如下所示:

```
{
    "id":1,
    "goodsName":"Spring Boot 2 教程",
    "weight":10.5,
    "type":2,
    "onSale":true
}
```

由于这里是 POST 请求，因此没有直接使用浏览器访问，而使用 postman 软件进行模拟请求，最终获得的结果如图 7-3 所示。

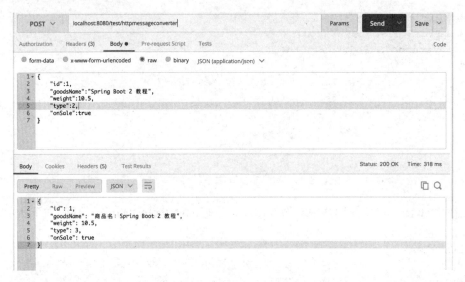

图 7-3　使用 postman 发送请求

由于消息转换器的存在，对象数据的读取不仅简单而且完全正确，响应时也不用自行封装工具类，使得开发过程变得更加灵活和高效。开发人员使用 Spring Boot 开发项目完全不用再做额外的配置工作，只需关心业务编码即可。

7.6　Spring Boot对静态资源的映射规则

与普通 Spring Web 项目相比，Spring Boot 项目的目录结构中仅有 java 和 resources 两个目录。用于存放资源文件的 webapp 目录在 Spring Boot 项目的目录结构中根本不存在。那么 Spring Boot 是如何处理静态资源的呢？ WebMVC 在自动配置时针对资源文件的访问又做了哪些配置呢？

由源码可知，这部分配置依然是通过 WebMvcAutoConfigurationAdapter 内部类完成的，源码如下所示：

```
@Override
public void addResourceHandlers(ResourceHandlerRegistry registry) {
    if (!this.resourceProperties.isAddMappings()) {
```

```
            logger.debug("Default resource handling disabled");
            return;
        }
        Duration cachePeriod = this.resourceProperties.getCache().
getPeriod();
        CacheControl cacheControl = this.resourceProperties.getCache().
getCachecontrol().toHttpCacheControl();
        // webjars 文件访问配置
        if (!registry.hasMappingForPattern("/webjars/**")) {
            customizeResourceHandlerRegistration(registry.
addResourceHandler("/webjars/**")
     .addResourceLocations("classpath:/META-INF/resources/webjars/")
                    .setCachePeriod(getSeconds(cachePeriod)).
setCacheControl(cacheControl));
        }
        // 静态资源映射配置
        String staticPathPattern = this.mvcProperties.getStatic
PathPattern();
        if (!registry.hasMappingForPattern(staticPathPattern)) {
            customizeResourceHandlerRegistration(registry.addResource
Handler(staticPathPattern)
                    .addResourceLocations(getResourceLocations(this.
resourceProperties.getStaticLocations()))
                    .setCachePeriod(getSeconds(cachePeriod)).setCache
Control(cacheControl));
        }
    }
```

如以上源码所示，静态资源的映射是在 addResourceHandlers()方法中进行映射配置的，它类似于在 Spring MVC 配置文件中的如下配置代码：

```
<mvc:resources mapping="/images/**" location="/images/" />
```

回到 addResourceHandlers()源码中来，staticPathPattern 的变量值为"/**"，其默认值在 WebMvcProperties 类中。实际的静态资源存放目录通过 getResourceLocations()方法获取，该方法源码如下所示：

```
@ConfigurationProperties(prefix = "spring.resources", ignoreUnknownFields
= false)
public class ResourceProperties {
```

```
    private static final String[] CLASSPATH_RESOURCE_LOCATIONS = {
            "classpath:/META-INF/resources/", "classpath:/resources/",
            "classpath:/static/", "classpath:/public/" };

    private String[] staticLocations = CLASSPATH_RESOURCE_LOCATIONS;

    public String[] getStaticLocations() {
        return this.staticLocations;
    }
}
```

由此可知，Spring Boot 默认的静态资源处理目录为："classpath:/META-INF/resources/"、"classpath:/resources/"、"classpath:/static/"、"classpath:/public/"。

由于访问当前项目的任何资源都能在静态资源的文件夹中查找，而不存在的资源则会显示相应的错误页面，因此在开发 Web 项目时只需要包含这几个目录中的任意一个或者多个，并将静态资源文件放入其中即可。

为了验证该配置，可以在类路径下分别创建 public 目录（PNG 格式文件）、resources 目录（CSS 格式文件）、stati 目录（HTML 格式文件和 JS 格式文件），并分别在三个文件夹中放入静态文件，如图 7-4 所示。

图 7-4　静态资源文件

重启 Spring Boot，在启动成功后打开浏览器并输入以下请求地址分别进行请求：

```
http://localhost:8080/logo.png
http://localhost:8080/main.css
http://localhost:8080/test.html
http://localhost:8080/test.js
```

访问结果如图 7-5～7-8 所示。

第 7 章　Spring Boot 之 Web 开发和 Spring MVC 自动配置分析

图 7-5　logo.png 请求结果

图 7-6　main.css 请求结果

图 7-7　test.html 请求结果

图 7-8　test.js 请求结果

通过以上请求结果可以发现，静态资源虽然在不同的目录中，但都能被正确返回。这就是 Spring Boot 对静态资源的拦截处理。

当然，开发时也可以在 Spring Boot 项目配置文件中修改这些属性。比如将拦截路径改为"/static/"，并将静态资源目录修改为"/file-test"，那么默认配置就会失效并使用开发人员自定义的配置。修改 application.properties 文件，添加如下配置：

```
spring.mvc.static-path-pattern=/static/**
spring.resources.static-locations=classpath:/file-test/
```

在修改后重启 Spring Boot 项目，再次使用原来的 URL 访问以上三个资源文件将会报 404 的错误，如图 7-9 所示。

图 7-9　404 错误

如果想要正常访问文件，则需要新建 static-test 目录并将静态资源文件移至 file-test 目录下，且修改访问路径为：

```
http://localhost:8080/static/logo.jpg
http://localhost:8080/static/main.css
http://localhost:8080/static/test.js
```

此时页面就不会出现 404 错误。

7.7　welcomePage 和 favicon 配置

7.7.1　welcomePage 配置

除了静态资源映射之外，Spring Boot 也默认配置了 welcomePage 和 favicon，这两个配置都和静态资源映射相关联。welcomePage 即默认欢迎页面，其配置源码如下所示：

```
@Bean
public WelcomePageHandlerMapping welcomePageHandlerMapping(Application
Context applicationContext,
            FormattingConversionService mvcConversionService, Resource
UrlProvider mvcResourceUrlProvider) {
        WelcomePageHandlerMapping welcomePageHandlerMapping = new
WelcomePageHandlerMapping(
```

```java
                new TemplateAvailabilityProviders(applicationContext),
applicationContext, getWelcomePage(),
                this.mvcProperties.getStaticPathPattern());
        welcomePageHandlerMapping.setInterceptors(getInterceptors
(mvcConversionService, mvcResourceUrlProvider));
        welcomePageHandlerMapping.setCorsConfigurations(getCors
Configurations());
        return welcomePageHandlerMapping;
    }

    private Optional<Resource> getWelcomePage() {
        String[] locations = getResourceLocations(this.resource
Properties.getStaticLocations());
        return Arrays.stream(locations).map(this::getIndexHtml).
filter(this::isReadable).findFirst();
    }

    private Resource getIndexHtml(String location) {
        // 在静态资源目录下的 index.html 文件
        return this.resourceLoader.getResource(location + "index.html");
    }
```

通过源码可以看出，在进行 WebMVC 自动配置时程序会向 IOC 容器注册一个 WelcomePageHandlerMapping 类型的 Bean，即默认欢迎页。其路径为静态资源目录下的 index.html。

在实际进行该功能测试时可以先访问一下当前项目根路径，比如在启动项目后访问 http://localhost:8080 地址，结果如图 7-10 所示。

图 7-10　访问项目根路径的页面结果

此时，服务器返回的是 404 错误页面。

但是，如果开发人员在静态资源目录下增加 index.html 文件就能够看到欢迎页面效果，比如选择默认的/static/目录，如图 7-11 所示。

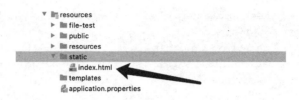

图 7-11 增加 index.html 文件

index.html 文件的代码如下所示：

```html
<!DOCTYPE html>
<html lang="en">
<head>
    <meta charset="UTF-8">
    <title>welcome page</title>
</head>
<body>
这里是默认欢迎页
</body>
</html>
```

编码完成，在重启项目成功后，再访问 http://localhost:8080 地址，结果如图 7-12 所示。

图 7-12 访问项目根路径的页面结果

此时，可以看到默认欢迎页面已经不再是错误页面了。

7.7.2 favicon 配置

favicon 是 favorites icon 的缩写，亦被称为 website icon（网页图标）、page icon（页面图标）或 urlicon（URL 图标）。favicon 是与某个网站或网页相关联的图标。

不同的网站会放置自身特有的 favicon 图标，图 7-13 分别是 Spring、百度、掘金、

GitHub 官网的 favicon 图标.

图 7-13　不同网站的 favicon 图标

Spring Boot 支持开发人员对 favicon 图标进行配置并显示。不过由于版本的迭代，对于 favicon 图标的支持官方做了一些调整。在 Spring Boot 2.2.x 版本之前，Spring Boot 会默认提供一个 favicon 图标，比如图 7-14 左侧类似叶子一样的图标。而 Spring Boot 2.2.x 版本之后不再提供默认的 favicon 图标。本书所讲解的案例和源码选择的 Spring Boot 版本都是 2.3.7。对于该版本，网页已经不显示 favicon 图标，右侧的浏览器标签栏也不存在 favicon 图标。但是，开发人员可以自定义配置 favicon 图标。

图 7-14　Spring Boot 项目的 favicon 图标

Spring 官方并没有对 favicon 图标做出特别的说明。不过，官方开发人员在 Spring Boot 开源仓库的 issue 中有提及此事，删除默认图标的原因是担心网站信息泄露。如果 Spring Boot 继续提供默认的 favicon 图标，这个绿色叶子的小图标很容易被看出是用 Spring Boot 开发的。

在 Spring Boot 2.2.x 之前的版本中，favicon 图标进行了默认设置，源码如下所示：

```
@Configuration //配置类
@ConditionalOnProperty(value = "spring.mvc.favicon.enabled",
matchIfMissing = true) //通过 spring.mvc.favicon.enabled 配置来确定是否进行设置，
默认为 true
    public static class FaviconConfiguration implements ResourceLoaderAware {
```

```java
            private final ResourceProperties resourceProperties;

            private ResourceLoader resourceLoader;

            public FaviconConfiguration(ResourceProperties resourceProperties) {
                this.resourceProperties = resourceProperties;
            }
            @Override
            public void setResourceLoader(ResourceLoader resourceLoader) {
                this.resourceLoader = resourceLoader;
            }
            @Bean
            public SimpleUrlHandlerMapping faviconHandlerMapping() {
                SimpleUrlHandlerMapping mapping = new SimpleUrlHandlerMapping();
                mapping.setOrder(Ordered.HIGHEST_PRECEDENCE + 1);
                mapping.setUrlMap(Collections.singletonMap("**/favicon.ico",
                    faviconRequestHandler()));
                return mapping;
            }


            @Bean
            public ResourceHttpRequestHandler faviconRequestHandler() {
                ResourceHttpRequestHandler requestHandler = new ResourceHttpRequestHandler();
                requestHandler.setLocations(resolveFaviconLocations());
                return requestHandler;
            }

            private List<Resource> resolveFaviconLocations() {
                String[] staticLocations = getResourceLocations(
                    this.resourceProperties.getStaticLocations());
                List<Resource> locations = new ArrayList<>(staticLocations.length + 1);
                Arrays.stream(staticLocations).map(this.resourceLoader::getResource)
                    .forEach(locations::add);
                locations.add(new ClassPathResource("/"));
                return Collections.unmodifiableList(locations);
            }
```

而在 Spring Boot 2.2.x 之后的版本中，这部分源码已经被删除，spring.mvc.favicon.enabled 配置项也被标记为"过时"。在 Spring Boot 官方文档中也能够看出，其实 Spring Boot 依然支持 favicon 图标的显示，只是该图标文件需要开发人员自行配置。

接下来就通过一个实际案例来讲解如何在 Spring Boot 项目中配置开发人员自定义的 favicon 图标。

首先需要制作一个 favicon 文件，并将其放入 static 目录或者其他静态资源目录。然后重启项目访问，可以看到页面已经替换为自定义设置的 favicon 图标了，如图 7-15 所示。由于浏览器缓存的原因，可能会出现"自定义 favicon 图标未生效"的错觉，读者可以尝试刷新几次页面。

图 7-15　自定义 favicon 图标

通过源码学习和实例讲解，可以发现 Spring Boot 在进行 Web 项目开发时为开发人员提供了全面而便利的默认配置，以往需要在 web.xml 或者 Spring MVC 配置文件中设置的内容，都改为以编码的方式进行自动注入和实现。开发人员在使用 Spring Boot 进行项目开发时，甚至可以零配置直接上手开发。不用做任何配置就已经有了视图解析器，不用自行添加消息转换器，Spring MVC 需要的一些功能也已经默认加载完成。对于开发人员来说，在开发 Web 项目时 Spring Boot 算得上是一件"神兵利器"。当然，如果这些默认配置不符合实际的业务需求，开发人员也可以自行配置，Spring Boot 提供了对应的配置参数和辅助类进行实现，非常灵活。

第 8 章

Thymeleaf 模板引擎使用详解

本章讲解当下较为流行的 Thymeleaf 模板引擎技术。首先会介绍模板引擎的工作原理和选择原因，然后讲解通过实际编码进行 Spring Boot 和 Thymeleaf 的整合操作。同时，也会结合具体的案例来介绍 Thymeleaf 模板引擎的常用语法。

8.1 模板引擎技术介绍

随着技术的发展，JSP+Servlet 的开发方式渐渐无法完全满足业务需求。一个中小型项目全部使用 JSP+Servlet 进行开发和维护，其成本会很高。至于大型项目，以此开发简直就是一场灾难，其开发成本、迭代成本、维护成本将会更高。

真正的企业应用开发有几个比较重要的关注点：代码复用、标准化、可维护性、开发成本。而 JSP+Servlet 的封装和抽象程度与框架技术相比毫无优势可言。业务的增长需求也反向驱动着技术的进步，间接催生出很多 JSP+Servlet 的替代方案和封装度更好的技术框架。以 JSP 为例，在前端渲染技术选型上，出现了 Thymeleaf、FreeMarker、Velocity 等模板引擎技术作为替代方案。

模板引擎（这里特指用于 Web 开发的模板引擎）是为了使用户看到的页面与业务数据分离而产生的一种模板技术。它可以生成特定格式的文档，用于网站的模板引擎就会生产出标准的 HTML 静态页面内容。在 Java Web 开发技术栈中，常见的模板引擎有 FreeMarker、Velocity、Thymeleaf 等。JSP 也可以理解为一种模板引擎技术。

模板引擎的工作原理如图 8-1 所示。

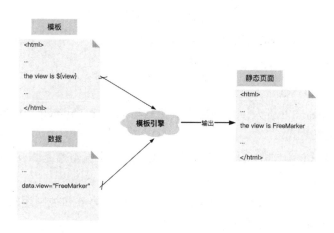

图 8-1　模板引擎的工作原理

使用模板引擎技术可以动态加载数据。在开发过程中，开发人员首先需要制作出模板引擎文件（不同的模板引擎技术的规范不同），并在控制器中将模板需要的数据组装好，然后将二者都交给模板引擎，模板引擎会根据数据和模板表达式语法解析并填充其到指定的位置进行页面渲染，最终生成 HTML 内容响应给客户端。

8.2　Thymeleaf模板引擎

8.2.1　Thymeleaf 模板引擎介绍

Thymeleaf 是目前比较受欢迎的模板引擎技术，主要因为其"原型即页面"的理念与 Spring Boot 倡导的快速开发非常契合。同时，Thymeleaf 模板引擎技术也确实拥有其他技术所不具备的优点。Thymeleaf 模板引擎官方图标如图 8-2 所示。

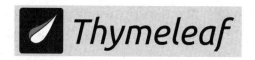

图 8-2　Thymeleaf 模板引擎官方图标

以 Thymeleaf 3.0 为例，其核心特性如下所示：
①完整的 HTML5 标记支持，全新的解析器

②自带多种模板模式，也可扩展支持其他模板格式

③在 Web 和非 Web 环境（离线）下都可以正常工作

④对 Spring Web 开发的支持非常完善

⑤独立的 Servlet API

另外，Thymeleaf 3.0 引入了一种新型表达式作为一般 Thymeleaf 标准表达系统的一部分，即 "*片段表达式;"。

在 Thymeleaf 3.0 中 Thymeleaf 标准表达式的另一个新特性是 NO-OP（无操作）令牌，由符号（_）表示。

Thymeleaf 3.0 允许在 HTML 和 XML 模式下的模板内容和控制逻辑完全解耦，从而实现 100%-Thymeleaf-free 无逻辑模板。

Thymeleaf 3.0 采用全新的方言系统。

Thymeleaf 3.0 完成了核心 API 的重构。

Thymeleaf 是高级语言的模板引擎，语法更简单，功能也更强大。

8.2.2　Thymeleaf 并非 Spring Boot 默认的模板引擎

坊间一直有一种说法：Thymeleaf 是 Spring Boot 默认的模板引擎。

这种说法的出处笔者无法详加考证，不过这种说法是不准确的。笔者看过多个版本的 Spring Boot 官方文档，并没有出现 "Thymeleaf 是 Spring Boot 默认的模板引擎" 之类的表述。

Spring Boot 官方文档对于模板引擎的介绍并没有太多的篇幅，提到 Thymeleaf 模板引擎的地方主要包括四个方面。

（1）Spring Boot 支持多种模板引擎技术，其中包括 Thymeleaf、FreeMarker、JSP、Mustache 等。

（2）针对不同的模板引擎，Spring Boot 也提供对应的视图解析器，包括 ThymeleafViewResolver、FreeMarkerViewResolver、GroovyMarkupViewResolver 和 MustacheViewResolver。

（3）Thymeleaf 模板引擎的配置项包括 spring.thymeleaf.cache、spring.thymeleaf.enabled、spring.thymeleaf.prefix 等配置项。

（4）在 Spring Boot 官方提供的代码样例中，大部分都使用 Thymeleaf 模板引擎。

除此之外，Spring Boot 官方文档对于 Thymeleaf 模板引擎技术并没有额外的记录。形成坊间传闻的原因可能是 Spring Boot 官方文档在介绍这些模板引擎时，Thymeleaf 模板引擎总是在第一位，并且 Spring Boot 提供的代码样例大部分使用了 Thymeleaf 模板引擎技术。这也许说明 Thymeleaf 模板引擎本身足够优秀，Spring Boot 官方更为推荐开发人员使用它。

本书的代码案例，使用的模板引擎技术全部都是 Thymeleaf。笔者选择它的原因如下所示。

（1）语法简单且功能强大。

（2）相关的教程比较多，学习成本低。

（3）Spring MVC 集成 Thymeleaf 模板引擎，非常方便。

以往在开发 Java Web 项目时通常会使用 JSP 技术，而 Spring 官方明确建议不在 Spring Boot 项目中使用 JSP 作为模板引擎。

Spring 官方表示，如果可能的话，尽量不要使用 JSP，因为 JSP 在内嵌容器中有许多局限性。Spring 官方文档也说明了相关局限性，如图 8-3 所示。

JSP Limitations

When running a Spring Boot application that uses an embedded servlet container (and is packaged as an executable archive), there are some limitations in the JSP support.

- With Jetty and Tomcat, it should work if you use war packaging. An executable war will work when launched with `java -jar`, and will also be deployable to any standard container. JSPs are not supported when using an executable jar.
- Undertow does not support JSPs.
- Creating a custom `error.jsp` page does not override the default view for error handling. Custom error pages should be used instead.

图 8-3　JSP 在内嵌容器中的局限性

（1）Jetty 和 Tomcat 服务器支持 JSP 模板引擎，但是打包方式有限制，使用 war 包可以正常运行，使用 jar 包则不支持。

（2）Undertow 服务器不支持 JSP。

（3）对 JSP 的自定义错误页面支持不够友好。

因此，笔者最终选择 Thymeleaf 模板引擎技术。

接下来将结合代码演示 Thymeleaf 与 Spring Boot 的整合过程，以及 Thymeleaf 模板引擎的语法介绍。

8.3　Spring Boot 之 Thymeleaf 整合

这一节主要介绍 Thymeleaf 的整合过程和注意事项，并通过一个实践案例进行讲解。

8.3.1 引入 Thymeleaf 依赖

因为 Spring Boot 官方提供了 Thymeleaf 的场景启动器 spring-boot-starter-thymeleaf，因此可以直接在 pom.xml 文件中添加该场景启动器，最终的 pom.xml 文件代码如下所示：

```xml
<?xml version="1.0" encoding="UTF-8"?>
<project xmlns="http://maven.apache.org/POM/4.0.0" xmlns:xsi="http://www.w3.org/2001/XMLSchema-instance"
    xsi:schemaLocation="http://maven.apache.org/POM/4.0.0 https://maven.apache.org/xsd/maven-4.0.0.xsd">
    <modelVersion>4.0.0</modelVersion>
    <parent>
        <groupId>org.springframework.boot</groupId>
        <artifactId>spring-boot-starter-parent</artifactId>
        <version>2.3.7.RELEASE</version>
        <relativePath/> <!-- lookup parent from repository -->
    </parent>
    <groupId>ltd.newbee.mall</groupId>
    <artifactId>newbee-mall</artifactId>
    <version>0.0.1-SNAPSHOT</version>
    <name>newbee-mall</name>
    <description>thymeleaf-demo</description>

    <properties>
        <project.build.sourceEncoding>UTF-8</project.build.sourceEncoding>
        <project.reporting.outputEncoding>UTF-8</project.reporting.outputEncoding>
        <java.version>1.8</java.version>
    </properties>

    <dependencies>
        <dependency>
            <groupId>org.springframework.boot</groupId>
            <artifactId>spring-boot-starter-web</artifactId>
        </dependency>

        <!--Thymeleaf 模板引擎依赖 -->
        <dependency>
            <groupId>org.springframework.boot</groupId>
            <artifactId>spring-boot-starter-thymeleaf</artifactId>
```

```xml
        </dependency>

        <dependency>
            <groupId>org.springframework.boot</groupId>
            <artifactId>spring-boot-starter-test</artifactId>
            <scope>test</scope>
            <exclusions>
                <exclusion>
                    <groupId>org.junit.vintage</groupId>
                    <artifactId>junit-vintage-engine</artifactId>
                </exclusion>
            </exclusions>
        </dependency>
    </dependencies>

    <build>
        <plugins>
            <plugin>
                <groupId>org.springframework.boot</groupId>
                <artifactId>spring-boot-maven-plugin</artifactId>
            </plugin>
        </plugins>
    </build>

    <repositories>
        <repository>
            <id>alimaven</id>
            <name>aliyun maven</name>
            <url>http://maven.aliyun.com/nexus/content/repositories/central/</url>
            <releases>
                <enabled>true</enabled>
            </releases>
            <snapshots>
                <enabled>false</enabled>
            </snapshots>
        </repository>
    </repositories>

</project>
```

8.3.2 创建模板文件

在 resources/templates 目录下新建模板文件 thymeleaf.html。Thymeleaf 模板引擎的默认后缀名即为.html。在新增文件后,首先在模板文件的标签中导入 Thymeleaf 的名称空间:

```
<html lang="en" xmlns:th="http://www.thymeleaf.org">
```

导入该名称空间主要是为了 Thymeleaf 语法的提示和 Thymeleaf 标签的使用。接下来在模板中增加与 JSP 中相同的显示内容,最终的模板文件代码如下所示:

```html
<!DOCTYPE html>
<html lang="en" xmlns:th="http://www.thymeleaf.org">
<head>
    <title>Thymeleaf demo</title>
</head>
<body>
<p>description 字段值为: </p>
<p th:text="${description}">这里显示的是 description 字段内容</p>
</body>
</html>
```

9.3.3 编辑 Controller 代码

在 controller 包下新增 ThymeleafController.java 文件,将模板文件所需的 description 字段赋值并转发至模板文件,代码如下所示:

```java
package com.lou.springboot.controller;
import org.springframework.stereotype.Controller;
import org.springframework.web.bind.annotation.GetMapping;
import org.springframework.web.bind.annotation.RequestParam;
import javax.servlet.http.HttpServletRequest;

@Controller
public class ThymeleafController {

    @GetMapping("/thymeleaf")
    public String hello(HttpServletRequest request, @RequestParam(value =
```

```
"description", required = false, defaultValue = "springboot-thymeleaf")
String description) {
    request.setAttribute("description", description);
    return "thymeleaf";
    }
}
```

最终的代码目录结构如图 8-4 所示。

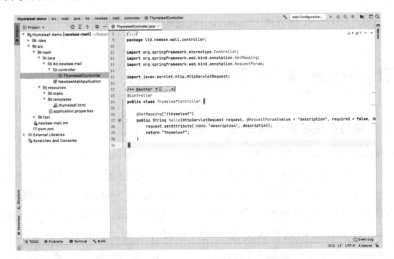

图 8-4　Thymeleaf 整合完成后的目录结构

在项目启动成功后，打开浏览器并输入本地域名和端口号，访问如下地址：

`http://localhost:8080/thymeleaf`

首先可以看到在原来静态 html 标签中的内容已经替换为 "springboot-thymeleaf" 字符串，而不再是默认内容，页面效果如图 8-5 所示。

图 8-5　Thymeleaf 模板引擎渲染结果 1

然后把 description 参数值传给后端处理方法，将模板中的字段改为 "我是十三"，页面效果如图 8-6 所示。

图 8-6　Thymeleaf 模板引擎渲染结果 2

8.3.4　Thymeleaf 模板引擎使用注意事项

1. 模板引擎的后缀名称注意事项

虽然 Thymeleaf 模板引擎文件的后缀名称是 .html，但是这种文件严格来说并不属于静态资源文件，而是模板文件，存放的目录也是在模板目录中，与前文中演示的 HTML 静态页面有一些区别。第 7 章用到的 test.html 页面是放在 static 目录中的，部署后可以直接访问其路径。但是模板引擎文件一般不允许直接访问，而是要经过 Controller 控制器的处理，将动态数据返回到模板文件中进行读取并渲染。比如本章演示的几个请求，都在控制器中实现了对应的方法并最终返回一个 Thymeleaf 视图对象在浏览器中呈现。

2. 必须引入名称空间

名称空间引入方法如下所示：

```
<html lang="en" xmlns:th="http://www.thymeleaf.org">
```

虽然不引入以上名称空间，静态资源访问和模板动态访问也不会报错，但是建议在开发过程中最好引入该名称空间。因为在引入之后会有 Thymeleaf 代码的语法提示，能够提升开发效率，也能减少人为造成的低级错误。有些开发人员可能会忽略这个事情。

3. 禁用模板缓存

Thymeleaf 的默认缓存设置是通过配置文件的 spring.thymeleaf.cache 配置属性决定的。通过图 8-7 中展示的 Thymeleaf 模板配置属性类 ThymeleafProperties 可以发现该属性默认为 true，因此 Thymeleaf 默认是使用模板缓存的。该设置有助于改善应用程序的性能，因此模板只需编译一次即可。

第 8 章 Thymeleaf 模板引擎使用详解

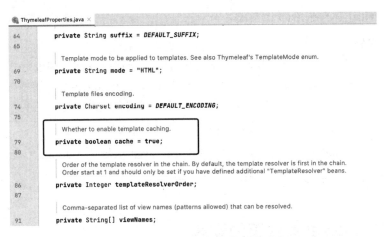

图 8-7 Thymeleaf 的默认缓存设置

除非重启应用程序，否则在开发过程中不能实时看到页面变更的效果，因此建议将该属性设置为 false，在配置文件中修改代码如下所示：

```
spring.thymeleaf.cache=false
```

8.4 Thymeleaf属性值讲解

8.4.1 Thymeleaf 模板解读

在第一个 Thymeleaf 模板引擎整合案例中，读取动态数据并渲染的语句如下所示：

```
<p th:text="${description}">这里显示的是 description 字段内容</p>
```

这是 Thymeleaf 模板引擎技术常见的语法规则，接下来详细分析这个知识点。

如图 8-8 所示，该模板文件语句中包含三部分内容。

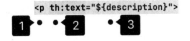

图 8-8 Thymeleaf 模板语句

①html 标签

②Thymeleaf 模板引擎的 th 标签

③Thymeleaf 表达式

前半部分 p 为 HTML 的标签，后半部分是 Thymeleaf 属性标签和表达式语法。th:text 表示文本替换、${description}表示读取后台设置的 description 字段。该模板文件语句的作用就是动态地将 p 标签的内容替换为 description 字段的内容。

同时该语句也展示了 Thymeleaf 模板文件的编写规则。

①任意的 Thymeleaf 属性标签 th:* 需要写在 html 标签中（th:block 除外）

②Thymeleaf 表达式写在 Thymeleaf 属性标签中

8.4.2 Thymeleaf 属性值的设置

这部分将介绍一些关于 Thymeleaf 设置属性值的内容。

th:text 对应 HTML5 语法中的 text 属性。除 th:text 属性以外，Thymeleaf 模板引擎也提供了其他的标签属性来替换 HTML5 原生属性的值。标签属性节选如图 8-9 所示。

th:alt	th:archive	th:audio
th:autocomplete	th:axis	th:background
th:bgcolor	th:border	th:cellpadding
th:cellspacing	th:challenge	th:charset
th:cite	th:class	th:classid
th:codebase	th:codetype	th:cols
th:colspan	th:compact	th:content
th:contenteditable	th:contextmenu	th:data
th:datetime	th:dir	th:draggable
th:dropzone	th:enctype	th:for
th:form	th:formaction	th:formenctype
th:formmethod	th:formtarget	th:fragment
th:frame	th:frameborder	th:headers
th:height	th:high	th:href
th:hreflang	th:hspace	th:http-equiv
th:icon	th:id	th:inline
th:keytype	th:kind	th:label
th:lang	th:list	th:longdesc
th:low	th:manifest	th:marginheight
th:marginwidth	th:max	th:maxlength
th:media	th:method	th:min
th:name	th:onabort	th:onafterprint
th:onbeforeprint	th:onbeforeunload	th:onblur

图 8-9 Thymeleaf 标签属性节选

其中 th:background 对应 HTML5 中的背景属性；th:class 对应 HTML5 中的 class 属性；th:href 对应 HTML5 中的链接地址属性；th:id 和 th:name 分别对应 HTML5 中的 id 和 name 属性。

th:block 比较特殊，它是 Thymeleaf 提供的唯一的一个 Thymeleaf 块级元素。其特殊性在于 Thymeleaf 模板引擎在处理 <th:block> 的时候会删掉其本身，而保留标签中的内容。

这里只列举了部分属性，完整内容可以查看 thymeleaf-attributes。

8.4.3 修改属性值实践

在 templates 目录下新建 attributes.html 文件，笔者将演示使用 Thymeleaf 语法来修改其属性值。

attributes.html 代码如下所示：

```html
<!DOCTYPE html>
<html lang="en" xmlns:th="http://www.thymeleaf.org">
<head>
    <title>Thymeleaf setting-value-to-specific-attributes</title>
    <meta charset="UTF-8">
</head>
<!-- background 标签-->
<body th:background="${th_background}" background="#D0D0D0">
<!-- text 标签-->
<h1 th:text="${title}">html 标签演示</h1>
<div>
    <h5>id、name、value 标签:</h5>
    <!-- id、name、value 标签-->
    <input id="input1" name="input1" value="1" th:id="${th_id}" th:name="${th_name}" th:value="${th_value}"/>
</div>
<br/>
<div class="div1" th:class="${th_class}">
    <h5>class、href 标签:</h5>
    <!-- class、href 标签-->
    <a th:href="${th_href}" href="##/">链接地址</a>
</div>
</body>
</html>
```

包含 background、id、name、class 等的 html 标签都设置了默认值，并在每个标签体中都添加了对应的 th 标签来读取动态数据。这里直接选择该文件，右键点击并选择 Open in Browser 查看页面效果，如图 8-10 所示。

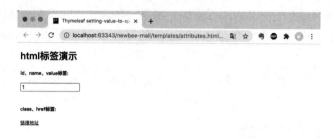

图 8-10　attributes.html 页面默认显示效果

需要注意的是，当前页面是直接打开的，并没有通过 Web 服务器。此时，attributes.html 页面能够直接正常访问且页面中的内容和元素的属性值都是默认值。

修改控制类代码，在 ThymeleafController 类中新增 attributes() 方法并将请求转发至该模板页面，代码如下所示：

```java
@GetMapping("/attributes")
public String attributes(ModelMap map) {
    // 更改 html 内容
    map.put("title", "Thymeleaf 标签演示");
    // 更改 id、name、value
    map.put("th_id", "thymeleaf-input");
    map.put("th_name", "thymeleaf-input");
    map.put("th_value", "13");
    // 更改 class、href
    map.put("th_class", "thymeleaf-class");
    map.put("th_href", "http://13blog.site");
    return "attributes";
}
```

在编码完成后，重启 Spring Boot 项目。在项目启动成功后，可以打开浏览器并访问如下地址：

```
http://localhost:8080/attributes
```

得到的页面结果如图 8-11 所示。

第 8 章 Thymeleaf 模板引擎使用详解

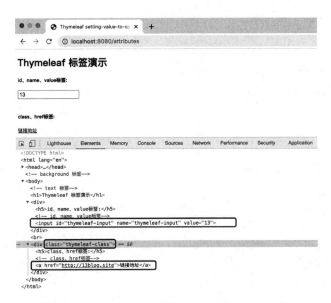

图 8-11 attributes.html 模板引擎渲染后的显示效果

打开浏览器控制台查看该页面的节点，即图 8-11 所框选的内容。

由于 th 标签的存在，页面在通过 Thymeleaf 渲染后，与静态页面相比较，内容和元素属性已经动态切换了，原来的默认值都变成了 attributes() 方法中设置的值。

这部分内容可以结合笔者提供的源码进行理解，也可以使用其他常用的 th 标签来练习。

8.5 Thymeleaf 语法讲解

8.5.1 Thymeleaf 语法

Thymeleaf 官方对于标准表达式特性的总结如下所示。

（1）表达式语法

①变量表达式：${...}

②选择变量表达式：*{...}

③信息表达式：#{...}

④链接 URL 表达式：@{...}

⑤分段表达式：~{...}

（2）字面量

①字符串：'one text'、'Another one!' ......

②数字：0、34、3.0、12.3 ......

③布尔值：true、false

④Null 值：null

⑤字面量标记：one、sometext、main ......

（3）文本运算

①字符串拼接：+

②字面量置换：|The name is ${name}|

（4）算术运算

①二元运算符：+、-、*、/、%

②负号（一元运算符）：(unary operator): -

（5）布尔运算

①二元运算符：and、or

②布尔非（一元运算符）：!、not

（6）比较运算

①比较：>、<、>=、<= (gt、lt、ge、le)

②相等运算符：==、!= (eq、ne)

比较运算符也可以使用转义字符，比如大于号，可以使用 Thymeleaf 语法 gt，也可以使用转义字符">"。

（7）条件运算符

①If-then: (if) ? (then)

②If-then-else: (if) ? (then) : (else)

③Default: (value) ?: (defaultvalue)

（8）特殊语法

无操作：_

接下来通过编码的方式实践这些知识点，并将知识点进行串联以接近实际开发情况。

8.5.2 Thymeleaf 简单语法实践

在 templates 目录下新建 simple.html 模板页面。该案例主要介绍字面量及简单的运算操作，包括字符串、数字、布尔值等常用的字面量及常用的运算和拼接操作，代码如下所示：

```html
<!DOCTYPE html>
<html lang="en" xmlns:th="http://www.thymeleaf.org">
<head>
    <title>Thymeleaf simple syntax</title>
    <meta charset="UTF-8">
</head>
<body>
<h1>Thymeleaf 简单语法</h1>
<div>
    <h5>基本类型操作(字符串):</h5>
    <p>一个简单的字符串：<span th:text="'thymeleaf text'">default text</span>.</p>
    <p>字符串连接：<span th:text="'thymeleaf text concat,'+${thymeleafText}">default text</span>.</p>
    <p>字符串连接：<span th:text="|thymeleaf text concat,${thymeleafText}|">default text</span>.</p>
</div>
<div>
    <h5>基本类型操作(数字):</h5>
    <p>一个简单的神奇的数字：<span th:text="2022">1000</span>.</p>
    <p>算术运算：2021+1=<span th:text="${number1}+1">0</span>.</p>
    <p>算术运算：14-1=<span th:text="14-1">0</span>.</p>
    <p>算术运算：1011 * 2=<span th:text="1011*${number2}">0</span>.</p>
    <p>算术运算：39 ÷ 3=<span th:text="39/3">0</span>.</p>
</div>
<div>
    <h5>基本类型操作(布尔值):</h5>
    <p>一个简单的数字比较：2022 > 2021=<span th:text="2022&gt;2021"> </span>.</p>
    <p>字符串比较：thymeleafText == 'spring-boot', 结果为<span th:text="${thymeleafText} == 'spring-boot'">0</span>.</p>
    <p>数字比较：13 == 39/3 结果为：<span th:text="13 == 39/3">0</span>.</p>
</div>
</body>
</html>
```

在浏览器中打开 simple.html 模板页面，结果如图 8-12 所示。

图 9-12　simple.html 页面默认显示效果

模板文件包含的部分变量为后台设置的值，并与字面量结合进行了计算和显示。在 ThymeleafController 类中新增对应的 simple() 方法并将请求转发至 simple.html 模板页面，代码如下所示：

```
@GetMapping("/simple")
public String simple(ModelMap map) {
    map.put("thymeleafText", "spring-boot");
    map.put("number1", 2021);
    map.put("number2", 2);
    return "simple";
}
```

在编码完成后，重启 Spring Boot 项目。在项目启动成功后，可以打开浏览器并访问如下地址：

```
http://localhost:8080/simple
```

得到的页面结果如图 8-13 所示。

第 8 章 Thymeleaf 模板引擎使用详解

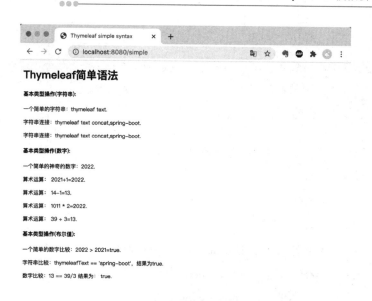

图 8-13 simple.html 模板引擎渲染后的显示效果

图 8-12 为静态 html 结果，图 8-13 为 Thymeleaf 模板引擎渲染的结果，可以看到字面量的展示及运算结果。以上为 Thymeleaf 语法中变量的使用方法和简单的运算操作，读者可以参考以上代码进行学习并适当修改数值以尽快掌握该知识点。

8.5.3 Thymeleaf 表达式

Thymeleaf 表达式包括：变量表达式${...}、选择变量表达式*{...}、信息表达式#{...}、链接 URL 表达式@{...}、分段表达式~{...}。这些表达式一般只写在 Thymeleaf 模板文件的 th 标签中，否则不会生效。表达式语法的主要作用就是获取变量值、获取绑定对象的变量值、国际化变量取值、URL 拼接与生成、Thymeleaf 模板布局。接下来笔者会选择一些常用的表达式进行介绍和实践。

1. 变量表达式

变量表达式即 OGNL 表达式或 Spring EL 表达式，其作用是获取模板中与后端返回数据所绑定对象的值，写法为${...}。这是最常见的一个表达式，在取值赋值、逻辑判断、循环语句中都可以使用该表达式，示例如下所示：

```
<!-- 读取参数 -->
<p>算术运算： 2021+1=<span th:text="${number1}+1">0</span>.</p>
<!-- 读取参数并进行三元运算 -->
```

```html
<div th:class="${path}=='users'?'nav-link active':'nav-link'"></div>
<!-- 读取对象中的属性 -->
<p>读取 goodsDetail 对象中 goodsName 字段：<span th:text="${goodsDetail.goodsName}">default text</span>.</p>
<!-- 循环遍历 -->
<li th:each="goods : ${goodsList}">
```

变量表达式也可以使用内置的基本对象，如下所示。

①ctx：上下文对象

②vars：上下文变量

③locale：上下文语言环境

④request：在 Web 环境下的 HttpServletRequest 对象

⑤response：在 Web 环境下的 HttpServletResponse 对象

⑥session：在 Web 环境下的 HttpSession 对象

⑦servletContext：在 Web 环境下的 ServletContext 对象

代码示例如下所示：

```html
<p>读取内置对象中 request 中的内容：<span th:text="${#request.getAttribute('requestObject')}">default text</span>.</p>
<p>读取内置对象中 session 中的内容：<span th:text="${#session.getAttribute('sessionObject')}">default text</span>.</p>
```

同时，Thymeleaf 还提供了一系列 Utility 工具对象（内置于 Context 中），可以通过 #直接访问，工具对象如下所示。

①dates：java.util.Date 的功能方法类

②calendars：类似#dates，面向 java.util.Calendar

③numbers：格式化数字的工具方法类

④strings：字符串对象的工具方法类，contains、startWiths、prepending/appending 等

⑤bools：求布尔值的工具方法

⑥arrays：数组的工具方法

⑦lists：java.util.List 的工具方法

⑧sets：java.util.Set 的工具方法

⑨maps：java.util.Map 的工具方法

在项目开发中，可以将这些方法视为工具类，通过这些方法可以使得 Thymeleaf 模

板引擎在操作变量时更加方便。

2. 选择（星号）表达式

选择表达式与变量表达式类似，不过它会用一个预先选择的对象代替上下文变量容器来执行。语法为*{goodsName}。被指定的对象由 th:object 标签属性在外层进行定义。前文读取 goodsDetail 对象的 goodsName 字段可以替换为：

```
<p th:object="${goodsDetail}">读取 goodsDetail 对象中 goodsName 字段：<span th:text="*{goodsName}">text</span>.</p>
```

如果在不考虑上下文的情况下，两者没有区别，使用${...}读取的内容也完全可以替换为使用*{...}进行读取。唯一的区别是使用*{...}前可以预先在父标签中通过 th:object 定义一个对象并进行操作，代码如下所示：

```
<p>读取 goodsDetail 对象中 goodsName 字段<span th:text="*{goodsDetail.goodsName}">default text</span></p>
<p>读取 text 字段：<span th:text="*{text}">default text</span>.</p>
```

3. URL 表达式

th:href 对应的是 html 中的 href 标签，它将计算并替换 href 标签中的 URL 地址，th:href 可以直接设置为静态地址，也可以使用表达式语法对读取的变量值动态拼接为 URL 地址。

比如一个详情页的 URL 地址：

```
http://localhost:8080/goods/detail/1
```

当使用 URL 表达式时，可以写成这样：

```
<a th:href="@{'http://localhost:goods/detail/1'}">商品详情页</a>
```

也可以根据 id 值进行替换，写法为：

```
<a th:href="@{'/goods/detail/'+${goods.goodsId}}">商品详情页</a>
```

或者也可以写成这样：

```
<a th:href="@{/goods/{goodsId}(goodsId=${goods.goodsId})}">商品详情页</a>
```

以上三种表达式生成的 URL 结果都是相同的。开发人员可以自己使用字符串拼接的方法组装 URL（第二种写法），也可以使用 URL 表达式提供的语法进行 URL 组装（第三种写法）。如果有多个参数可以自行拼装字符串，或者使用逗号进行分隔，写法如下：

```html
<a th:href="@{/goods/{goodsId}(goodsId=${goods.goodsId},title=${goods.
goodsName},tag='手机')}">商品详情页</a>
```

最终生成的 URL 为"http://localhost:8080/goods/1?title=iPhone13&tag=手机"。另外，URL 中以 "/" 开头的路径（比如 /goods/1），默认生成的 URL 会加上该项目的当前地址形成完整的 URL 。

4. Thymeleaf 复杂语法实践

这里将结合前文中的知识点进行更加复杂的语法实践，主要涉及后续实战章节会出现的一些方法，比如判断语句、循环语句、工具类使用等。

在 templates 目录下新建 complex.html 模板文件，代码如下所示：

```html
<!DOCTYPE html>
<html lang="en" xmlns:th="http://www.thymeleaf.org">
<head>
    <meta charset="UTF-8">
    <title th:text="${title}">语法测试</title>
</head>
<body>
<h3>#strings 工具类测试 </h3>
<div th:if="${not #strings.isEmpty(testString)}" >
    <p>testString初始值 : <span th:text="${testString}"/></p>
    <p>toUpperCase : <span th:text="${#strings.toUpperCase(testString)}"/>
</p>
    <p>toLowerCase : <span th:text="${#strings.toLowerCase(testString)}"/>
</p>
    <p>equalsIgnoreCase : <span th:text="${#strings.equalsIgnoreCase
(testString, '13')}"/></p>
    <p>indexOf : <span th:text="${#strings.indexOf(testString, 'r')}"/></p>
    <p>substring : <span th:text="${#strings.substring(testString, 1, 4)}"/>
</p>
    <p>startsWith : <span th:text="${#strings.startsWith(testString,
'Spring')}"/></p>
    <p>contains : <span th:text="${#strings.contains(testString, 'Boot')}"/>
</p>
</div>
<h3>#bools 工具类测试</h3>
<!-- 如果bool的值为false的话，该div将不会显示-->
<div th:if="${#bools.isTrue(bool)}">
    <p th:text="${bool}"></p>
</div>
```

```html
<h3>#arrays 工具类测试</h3>
<div th:if="${not #arrays.isEmpty(testArray)}">
    <p>length : <span th:text="${#arrays.length(testArray)}"/></p>
    <p>contains : <span th:text="${#arrays.contains(testArray, 5)}"/></p>
    <p>containsAll : <span th:text="${#arrays.containsAll(testArray,testArray)}"/></p>
    <p>循环读取 : <span th:each="i:${testArray}" th:text="${i+' '}"/></p>
</div>
<h3>#lists 工具类测试</h3>
<div th:unless="${#lists.isEmpty(testList)}">
    <p>size : <span th:text="${#lists.size(testList)}"/></p>
    <p>contains : <span th:text="${#lists.contains(testList, 0)}"/></p>
    <p>sort : <span th:text="${#lists.sort(testList)}"/></p>
    <p>循环读取 : <span th:each="i:${testList}" th:text="${i+' '}"/></p>
</div>
<h3>#maps 工具类测试</h3>
<div th:if="${not #maps.isEmpty(testMap)}">
    <p>size : <span th:text="${#maps.size(testMap)}"/></p>
    <p>containsKey : <span th:text="${#maps.containsKey(testMap, 'platform')}"/></p>
    <p>containsValue : <span th:text="${#maps.containsValue(testMap, '13')}"/></p>
    <p>读取 map 中键为 title 的值 : <span th:if="${#maps.containsKey(testMap, 'title')}" th:text="${testMap.get('title')}"/></p>
</div>
<h3>#dates 工具类测试</h3>
<div>
    <p>year : <span th:text="${#dates.year(testDate)}"/></p>
    <p>month : <span th:text="${#dates.month(testDate)}"/></p>
    <p>day : <span th:text="${#dates.day(testDate)}"/></p>
    <p>hour : <span th:text="${#dates.hour(testDate)}"/></p>
    <p>minute : <span th:text="${#dates.minute(testDate)}"/></p>
    <p>second : <span th:text="${#dates.second(testDate)}"/></p>
    <p>格式化: <span th:text="${#dates.format(testDate)}"/></p>
    <p>yyyy-MM-dd HH:mm:ss 格式化: <span th:text="${#dates.format(testDate, 'yyyy-MM-dd HH:mm:ss')}"/></p>
</div>
</body>
</html>
```

在 ThymeleafController 类中新增对应的 complex()方法并将请求转发至 complex.html 模板页面，代码如下所示：

```java
@GetMapping("/complex")
public String complex(ModelMap map) {
    map.put("title", "Thymeleaf 语法测试");
    map.put("testString", "Spring Boot 商城");
    map.put("bool", true);
    map.put("testArray", new Integer[]{2021, 2022, 2023, 2024});
    map.put("testList", Arrays.asList("Spring", "Spring Boot", "Thymeleaf", "MyBatis", "Java"));
    Map testMap = new HashMap();
    testMap.put("platform", "book");
    testMap.put("title", "Spring Boot 商城项目实战");
    testMap.put("author", "十三");
    map.put("testMap", testMap);
    map.put("testDate", new Date());
    return "complex";
}
```

在编码完成后，重启 Spring Boot 项目。在项目启动成功后，可以打开浏览器并访问如下地址：

```
http://localhost:8080/complex
```

得到的页面结果如图 8-14 所示。

在 strings 工具类测试中，首先使用了 th:if 标签进行逻辑判断。th:if="${not #strings.isEmpty(testString)}"即为一条判断语句。${...}表达式会返回一个布尔值结果，如果为 true 则该 div 中的内容会继续显示，否则将不会显示 th:if 所在的主标签。#strings.isEmpty 的作用为字符串判空，如果 testString 为空则会返回 true，而表达式前面的 not 则表示逻辑非运算，即如果 testString 不为空则继续展示该 div 中的内容。与 th:if 类似的判断标签为 th:unless，它与 th:if 标签刚好相反，即当表达式中返回的结果为 false 时，它所在标签中的内容才会继续显示。这里在#lists 工具类测试使用了 th:unless 标签，读者在调试代码时可以比较二者的区别。

Thymeleaf 模板引擎的循环语句语法为 th:each="i:${testList}"，类似于 JSP 中的 c:foreach 表达式，主要表示循环的逻辑，很多页面逻辑在生成时会使用到该语法。还有读取 Map 对象的方式为${testMap.get('title')} ，与 Java 语言也很类似。

第 8 章　Thymeleaf 模板引擎使用详解

图 8-14　complex.html 模板引擎渲染后的显示效果

逻辑判断、循环语句这两个知识点是系统开发中比较常用也是比较重要的内容，希望读者能够结合代码练习并牢牢掌握这些知识点。

本章讲解内容较多，且多为实践内容，完整的代码目录如图 8-15 所示。

读者可以参考本章提供的源码和案例进行练习。Thymeleaf 模板引擎的知识和语法实践这里讲解完成，后续在项目开发的章节中不再进行单独介绍。为了后续学习能够更顺畅，建议读者一定要多练习 Thymeleaf 模板引擎的基础语法。

图 8-15　Thymeleaf 模板引擎实例代码的完整目录结构

第 9 章

Spring Boot 操作 MySQL 数据库

本章介绍 Spring Boot 如何进行数据库相关的功能开发,包括基础的整合操作,JDBC 的整合和 MyBatis 框架的整合。另外,也会介绍相关组件的自动配置知识,比如数据库连接池的自动配置、JdbcTemplate 的自动配置、MyBatis 的自动配置。本章将通过代码实践及 Spring Boot 源码的讲解,让读者既能掌握在 Spring Boot 项目中的 MySQL 数据库连接和数据操作,也能掌握 Spring Boot 底层自动化配置的原理,达到简单、高效操作数据库的目的。

9.1　Spring Boot连接MySQL实践

9.1.1　Spring Boot 对数据库连接的支持

无论关系型数据库(如 PostgreSQL、MySQL、Oracle 等)还是非关系型数据库(如 Elastic Search、Redis、Cassandra、MangoDB 等),都是软件系统不可或缺的一部分。因为 Spring Boot 底层针对于这些数据库都提供了良好的支持,所以这些技术方案都可以很方便地整合到 Spring Boot 项目中来。通过 Spring Boot 提供的数据相关的场景启动器也能知晓一二,如图 9-1 所示。

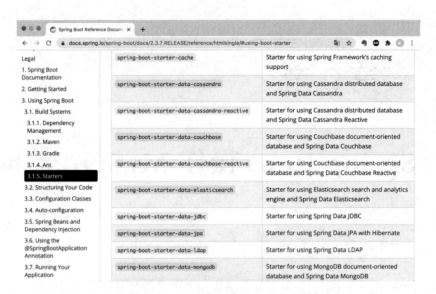

图 9-1　Spring Boot 提供的数据相关的场景启动器

　　Spring Boot 默认提供的关于数据操作的场景启动器有很多，如 Redis 场景启动器、JDBC 场景启动器、Elastic Search 场景启动器、MongoDB 场景启动器等。在企业级项目开发中的大部分数据库选型，Spring Boot 都已经提供了对应的解决方案，可以很方便地实现这些数据库的整合。

　　由于本书的实践教程使用 MySQL 数据库作为底层的数据存储工具，因此本章将会讲解 MySQL 的相关知识及实践内容。

9.1.2　Spring Boot 整合 spring-boot-starter-jdbc

　　MySQL 是最流行的关系型数据库管理系统之一。在 Web 应用开发中，MySQL 关系型数据库是一个十分不错的选择，接下来就会讲解在 Spring Boot 项目中如何连接 MySQL 数据库。

　　Java 程序在进行与 MySQL 的连接时需要通过 JDBC 来实现。JDBC 全称为 Java Data Base Connectivity（Java 数据库连接），主要由接口组成，是一种用于执行 SQL 语句的 Java API。各个数据库厂家基于它都各自实现了自己的驱动程序（Driver），如图 9-2 所示。

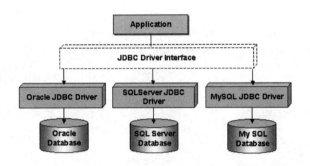

图 9-2　不同厂商的数据库驱动程序

Java 程序在获取数据库连接时，需要以 URL 方式指定不同类型数据库的 Driver，在获得特定的 Connection 连接后，可按照 JDBC 规范对不同类型的数据库进行数据操作，代码如下所示：

```
// 第一步，注册驱动程序
Class.forName("数据库驱动的完整类名");
// 第二步，获取一个数据库的连接
Connection conn = DriverManager.getConnection("数据库地址","用户名","密码");
// 第三步，创建一个会话
Statement stmt=conn.createStatement();
// 第四步，执行 SQL 语句
stmt.executeUpdate("SQL 语句");
// 或者查询记录
ResultSet rs = stmt.executeQuery("查询记录的 SQL 语句");
// 第五步，对查询的结果进行处理
while(rs.next()){
// 操作
}
// 第六步，关闭连接
rs.close();
stmt.close();
conn.close();
```

上面几行代码读者并不陌生，这是在初学 JDBC 连接时的代码。虽然现在可能用了一些数据层 ORM 框架（比如 MyBatis 或者 Hibernate），但是底层实现依然和这里的代码一样。通过使用 JDBC，开发人员可以直接使用 Java 程序来对关系型数据库进行操作。接下来将对 Spring Boot 如何使用 JDBC 进行实例演示。

首先，需要在项目中引入相应的场景启动器，也就是引入依赖的 jar 包，添加的内容如下所示：

```xml
<dependency>
    <groupId>org.springframework.boot</groupId>
    <artifactId>spring-boot-starter-jdbc</artifactId>
</dependency>

<dependency>
    <groupId>mysql</groupId>
    <artifactId>mysql-connector-java</artifactId>
    <scope>runtime</scope>
</dependency>
```

或者在 Spring Boot Initializr 创建时选择 MySQL 场景依赖和 JDBC 场景依赖，如图 9-3 所示。

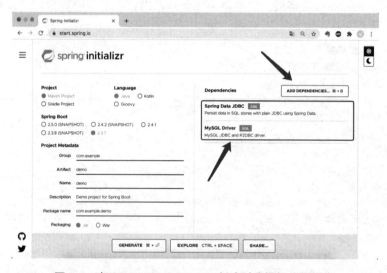

图 9-3 在 Spring Boot Initializr 创建时选择相关数据库

在添加完相关依赖之后，需要启动 MySQL 数据库并在新建的 Spring Boot 项目中配置数据库连接的地址和账号密码，这样才能正确连接到数据库。在 application.properties 配置文件中添加如下配置代码：

```
spring.datasource.name=newbee-mall-datasource
spring.datasource.driverClassName=com.mysql.cj.jdbc.Driver
spring.datasource.url=jdbc:mysql://localhost:3306/test_db?useUnicode=true&serverTimezone=Asia/Shanghai&characterEncoding=utf8&autoReconnect=true&useSSL=false&allowMultiQueries=true&useAffectedRows=true
spring.datasource.username=root
spring.datasource.password=123456
```

9.1.3 Spring Boot 连接 MySQL 数据库验证

这里编写一个测试类来测试能否连接数据库。在测试类 NewbeeMallApplicationTests 中添加 datasourceTest()单元测试方法，源码及注释如下所示：

```java
package ltd.newbee.mall;

import org.junit.jupiter.api.Test;
import org.springframework.beans.factory.annotation.Autowired;
import org.springframework.boot.test.context.SpringBootTest;

import javax.sql.DataSource;
import java.sql.Connection;
import java.sql.SQLException;

@SpringBootTest
class NewbeeMallApplicationTests {

    // 注入数据源对象
    @Autowired
    private DataSource defaultDataSource;

    @Test
    public void datasourceTest() throws SQLException {
        // 获取数据库连接对象
        Connection connection = defaultDataSource.getConnection();
        System.out.print("获取连接：");
        // 判断连接对象是否为空
        System.out.println(connection != null);
        connection.close();
    }

    @Test
    void contextLoads() {
    }

}
```

在编码完成后，点击运行该单元测试方法，操作步骤如下所示。

（1）点击 datasourceTest()方法左侧工具栏的启动按钮。

（2）弹出操作栏后，点击"Run 'datasourceTest()'"就可以运行该单元测试方法了，如图9-4所示。

图9-4　运行单元测试方法datasourceTest()

单元测试方法datasourceTest()运行结果如图9-5所示。

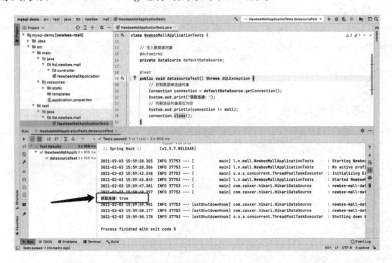

图9-5　单元测试方法datasourceTest()运行结果

在控制台中可以查看打印结果，如果connection对象不为空，则证明数据库连接成功。在对connection对象进行判空操作时，得到的结果是connection非空。如果数据库连接对象connection没有正常获取，则需要检查数据库是否正确启动或者数据库信息是否配置正确。

9.2 Spring Boot数据源自动配置源码详解

9.2.1 Spring Boot 默认数据源 HikariDataSource

通过前一节的演示读者也能够观察到,开发人员在配置文件中只需要加上数据库的相关信息即可获取数据库连接对象。那么 Spring Boot 究竟做了哪些自动配置操作,使得开发人员如此简单就可以直接获取数据库的连接呢?接下来,结合源码解释一下这个问题。

首先,来分析一下注入的 DataSource 数据源对象。新增单元测试方法 datasourceClassTest(),用来获取数据源类型,代码如下所示:

```
@Test
public void datasourceClassTest() throws SQLException {
    // 获取数据源类型
    System.out.println("默认数据源为:" + defaultDataSource.getClass());
}
```

点击运行该单元测试方法,在控制台中输出的内容如图 9-6 所示。

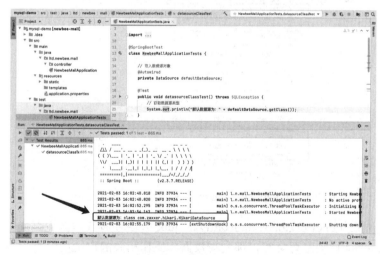

图 9-6 单元测试方法 datasourceClassTest()运行结果

最终得到的结果为：

默认数据源为：class com.zaxxer.hikari.HikariDataSource

基于以上结果可以得出结论，在 Spring Boot 项目启动时，Spring Boot 已经默认向 IOC 容器中注册了一个类型为 HikariDataSource 的数据源对象，不然在使用@Autowired 进行数据源对象的引入时肯定会报错。那么接下来的内容将围绕两个问题来分析数据源自动配置的相关源码。

（1）Spring Boot 如何将 DataSource 对象注册到 IOC 容器中的？

（2）为什么默认注册的数据源是 class com.zaxxer.hikari.HikariDataSource 类型？

9.2.2 数据源自动配置类 DataSourceAutoConfiguration

在 Spring Boot 中数据源的自动配置类名称为 DataSourceAutoConfiguration，该自动配置类定义在 spring-boot-autoconfigure-2.3.7.RELEASE.jar 的 org.springframework.boot.autoconfigure.jdbc 包中，如图 9-7 所示。

图 9-7　org.springframework.boot.autoconfigure.jdbc 包

在 org.springframework.boot.autoconfigure.jdbc 包中还有数据源配置类、数据源属性类、JdbcTemplate 自动配置类等，都是与数据库操作相关的自动配置类。

这里分析的是数据源自动配置机制，因此点开该目录下的 DataSourceAutoConfiguration 类进行分析，源码及注释如下所示：

```java
@Configuration(proxyBeanMethods = false) // 配置类
@ConditionalOnClass({ DataSource.class, EmbeddedDatabaseType.class }) // 自
动配置生效条件
@ConditionalOnMissingBean(type = "io.r2dbc.spi.ConnectionFactory") // 自动
配置生效条件
@EnableConfigurationProperties(DataSourceProperties.class) //属性值配置
@Import({ DataSourcePoolMetadataProvidersConfiguration.class, DataSource
InitializationConfiguration.class })
public class DataSourceAutoConfiguration {

    @Configuration(proxyBeanMethods = false)
    @Conditional(EmbeddedDatabaseCondition.class)
    @ConditionalOnMissingBean({ DataSource.class, XADataSource.class })
    @Import(EmbeddedDataSourceConfiguration.class)
    protected static class EmbeddedDatabaseConfiguration {

    }

    @Configuration(proxyBeanMethods = false)
    @Conditional(PooledDataSourceCondition.class)
    @ConditionalOnMissingBean({ DataSource.class, XADataSource.class })
    @Import({DataSourceConfiguration.Hikari.class, DataSourceConfiguration.Tomcat.class,
            DataSourceConfiguration.Dbcp2.class, DataSourceConfiguration.Generic.class,
            DataSourceJmxConfiguration.class })// 导入数据源配置类
    protected static class PooledDataSourceConfiguration {

    }

    ...省略部分代码

}
```

DataSourceAutoConfiguration 类的注解释义如下所示。

@Configuration(proxyBeanMethods = false)：指定该类为配置类。

@ConditionalOnClass({ DataSource.class, EmbeddedDatabaseType.class })：判断当前 classpath 是否存在指定类 DataSource 类和 EmbeddedDatabaseType 类，存在则生效。

@ConditionalOnMissingBean(type = "io.r2dbc.spi.ConnectionFactory")：判断 IOC 容器中是否存在 io.r2dbc.spi.ConnectionFactory 类型的 Bean，不存在则生效。

通过源码可知，DataSourceAutoConfiguration 自动配置类的自动配置触发条件：当前 classpath 存在 DataSource 类和 EmbeddedDatabaseType 类，且在 IOC 容器中不存在 io.r2dbc.spi.ConnectionFactory 类型的 Bean 时才会生效。因为当前项目引入了 spring-boot-starter-jdbc 场景启动器，该启动器包含了相关依赖。ConnectionFactory 相关依赖并未引入，也就不可能被注册到 IOC 容器中，因此两个条件都会满足，即自动配置类会生效。

9.2.3 属性绑定

如果在类声明中包含@EnableConfigurationProperties(DataSourceProperties.class)，那么该自动配置类就通过 DataSourceProperties 类来进行相关属性值的配置，DataSourceProperties 源码如下所示：

```
@ConfigurationProperties(prefix = "spring.datasource")
public class DataSourceProperties implements BeanClassLoaderAware,
InitializingBean {
    ...省略部分代码
}
```

@ConfigurationProperties(prefix = "spring.datasource")表示通过绑定配置文件中以 spring.datasource 开头的属性到配置类中。在 application.properties 文件中设置的数据库信息会在这里被读取绑定，并应用到后续的数据源对象的初始化步骤中。

9.2.4 DataSourceConfiguration 源码解析

在 DataSourceAutoConfiguration 自动配置类中定义了 PooledDataSourceConfiguration 内部类，但是该内部类并没有实现代码，而是通过 @Import 注解引入了 DataSourceConfiguration 类，点进该类查看代码，源码及注释如下所示：

```
abstract class DataSourceConfiguration {

    @SuppressWarnings("unchecked")
    protected static <T> T createDataSource(DataSourceProperties properties,
Class<? extends DataSource> type) {
        return (T)
properties.initializeDataSourceBuilder().type(type).build();
    }
```

```java
    /**
     * Tomcat Pool DataSource configuration.
     */
    @Configuration(proxyBeanMethods = false)
    @ConditionalOnClass(org.apache.tomcat.jdbc.pool.DataSource.class)
    @ConditionalOnMissingBean(DataSource.class)
    @ConditionalOnProperty(name = "spring.datasource.type", havingValue = "org.apache.tomcat.jdbc.pool.DataSource",
            matchIfMissing = true)
    static class Tomcat {

        @Bean
        @ConfigurationProperties(prefix = "spring.datasource.tomcat")
        org.apache.tomcat.jdbc.pool.DataSource dataSource(DataSourceProperties properties) {
            org.apache.tomcat.jdbc.pool.DataSource dataSource = createDataSource(properties,
                    org.apache.tomcat.jdbc.pool.DataSource.class);
            DatabaseDriver databaseDriver = DatabaseDriver.fromJdbcUrl(properties.determineUrl());
            String validationQuery = databaseDriver.getValidationQuery();
            if (validationQuery != null) {
                dataSource.setTestOnBorrow(true);
                dataSource.setValidationQuery(validationQuery);
            }
            return dataSource;
        }

    }

    /**
     * Hikari DataSource configuration.
     */
    @Configuration(proxyBeanMethods = false)
    @ConditionalOnClass(HikariDataSource.class)//判断当前classpath下是否存在指定类
    @ConditionalOnMissingBean(DataSource.class) // beanFactory中不存在DataSource 类型的 Bean
    @ConditionalOnProperty(name = "spring.datasource.type", havingValue = "com.zaxxer.hikari.HikariDataSource", matchIfMissing = true) // spring.datasource.type 属性值默认为 com.zaxxer.hikari.HikariDataSource
    static class Hikari {
```

```java
        @Bean
        @ConfigurationProperties(prefix = "spring.datasource.hikari")
        HikariDataSource dataSource(DataSourceProperties properties) {
            HikariDataSource dataSource = createDataSource(properties, HikariDataSource.class);
            if (StringUtils.hasText(properties.getName())) {
                dataSource.setPoolName(properties.getName());
            }
            return dataSource;
        }

    }

    /**
     * DBCP DataSource configuration.
     */
    @Configuration(proxyBeanMethods = false)
    @ConditionalOnClass(org.apache.commons.dbcp2.BasicDataSource.class)
    @ConditionalOnMissingBean(DataSource.class)
    @ConditionalOnProperty(name = "spring.datasource.type", havingValue = "org.apache.commons.dbcp2.BasicDataSource",
            matchIfMissing = true)
    static class Dbcp2 {

        @Bean
        @ConfigurationProperties(prefix = "spring.datasource.dbcp2")
        org.apache.commons.dbcp2.BasicDataSource dataSource(DataSourceProperties properties) {
            return createDataSource(properties, org.apache.commons.dbcp2.BasicDataSource.class);
        }

    }

    /**
     * Generic DataSource configuration.
     */
    @Configuration(proxyBeanMethods = false)
    @ConditionalOnMissingBean(DataSource.class)
    @ConditionalOnProperty(name = "spring.datasource.type")
    static class Generic {
```

```
    @Bean
    DataSource dataSource(DataSourceProperties properties) {
        return properties.initializeDataSourceBuilder().build();
    }

}
```

该类中含有 Tomcat.class、Hikari.class、Dbcp2.class、Generic.class 四个内部类，但是由于每个内部类上都包含有 @Conditional 条件注解，即只有满足条件才会创建一个特定的 Bean。四个内部类并不会全部生效，在默认情况下只有 Hikari 内部类满足了所有的限定条件。

接下来来分析一下在 Hikari 内部类上的所有限定条件。

（1）判断当前 classpath 是否存在指定类 HikariDataSource.class。

在 pom 文件中引入 spring-boot-starter-jdbc 依赖，如图 9-8 所示。

图 9-8　spring-boot-starter-jdbc 依赖

点击 spring-boot-starter-jdbc 并进入 spring-boot-starter-jdbc-2.3.7.RELEASE.pom 文件，源码如图 9-9 所示。

这里在 spring-boot-starter-jdbc 的 pom 文件中配置了 HikariCP 的依赖，已经默认引入了 Hikari 的相关依赖，因此第一个条件满足。

（2）IOC 容器中不存在 DataSource 类型的 Bean。

由于没有在配置文件中进行数据源指定，也没有进行自定义数据源的注入，因此在 IOC 容器中肯定不存在 DataSource 类型的对象，即该条件满足。

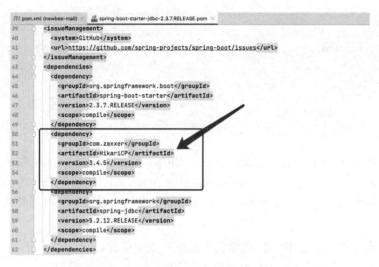

图 9-9　spring-boot-starter-jdbc-2.3.7.RELEASE.pom 源码

（3）判断当前绑定属性中 spring.datasource.type 的值。

在配置文件中并没有进行该配置项的设置，因此该配置项为空，不能满足 havingValue = "com.zaxxer.hikari.HikariDataSource"条件。但是 matchIfMissing = true，表示如果没有指定 spring.datasource.type 的值，则默认为 HikariDataSource，即该条件也成立。

基于以上条件判断，最终 DataSourceAutoConfiguration 数据源自动配置类会将 DataSource 对象注册到 IOC 容器中，对象类型即为默认的数据源类型 com.zaxxer.hikari.HikariDataSource。Spring Boot 2 版本中已经将默认数据源更改为 Hikari 数据源，而在更早的 Spring Boot 版本中的默认数据源为 org.apache.tomcat.jdbc.pool.DataSource。

开发人员如果想更换数据源，可以通过引入相关数据源的依赖，并修改配置文件中的 spring.datasource.type 配置项。

9.3　使用JdbcTemplate进行数据库的增、删、改、查

在完成数据源的配置和数据库的正确连接后，接下来将演示如何在 Spring Boot 项目中对 MySQL 数据库进行常规的 SQL 操作。

9.3.1 JdbcTemplate 介绍

在平时的项目开发中，开发人员对于数据库的操作通常是基于 ORM 框架实现的，如 MyBatis 框架、Hibernate 框架等。当然，也可以直接使用 JDBC 原生 API 来进行数据库操作。不过，使用 JDBC 原生 API 进行数据库操作非常烦琐，Spring Boot 也并没有默认集成相关的 ORM 框架，只是提供了 JdbcTemplate 对象来简化开发人员对数据库的操作流程。

JdbcTemplate 是 Spring 对 JDBC 的封装，目的是让 JDBC 更加易于使用。更为关键的一点是，JdbcTemplate 对象也是通过自动配置机制注册到 IOC 容器中的。JdbcTemplate 的自动配置类是 JdbcTemplateAutoConfiguration，该自动配置类也在 spring-boot-autoconfigure-2.3.7.RELEASE.jar 中的 org.springframework.boot.autoconfigure.jdbc 包中。感兴趣的读者可以自行查看和分析其源码。这里给出 JdbcTemplateAutoConfiguration 自动配置类的结果：在 DataSourceAutoConfiguration 自动配置后，程序会使用 IOC 容器中的 dataSource 对象作为构造参数创建一个 JdbcTemplate 对象并注册到 IOC 容器中。

在正确配置数据源后，开发人员可以直接在代码中使用 JdbcTemplate 对象进行数据库操作。

9.3.2 详解 JdbcTemplate 对数据库的增、删、改、查

接下来通过编码操作来完成使用 JdbcTemplate 进行数据库的增、删、改、查操作。

首先在数据库中创建一张测试表，建表代码如下所示：

```sql
DROP TABLE IF EXISTS 'jdbc_test';

CREATE TABLE 'jdbc_test' (
  'ds_id' int(11) NOT NULL AUTO_INCREMENT COMMENT '主键id',
  'ds_type' varchar(100) DEFAULT NULL COMMENT '数据源类型',
  'ds_name' varchar(100) DEFAULT NULL COMMENT '数据源名称',
  PRIMARY KEY ('ds_id') USING BTREE
) ENGINE = InnoDB CHARACTER SET = utf8;

/*Data for the table 'jdbc_test' */

insert into 'jdbc_test' ('ds_id', 'ds_type', 'ds_name') values (1,'com.
```

```
zaxxer.hikari.HikariDataSource','hikari 数据源
'),(2,'org.apache.commons.dbcp2.BasicDataSource','dbcp2 数据源');
```

为了演示方便，首先在 controller 包中新建 JdbcController 类并直接注入 JdbcTemplate 对象。然后创建四个方法，分别实现根据传入的参数向 jdbc_test 表中新增数据、修改 jdbctest 表中的数据、删除 jdbc_test 表中的数据、查询 jdbc_test 表中的数据，实现代码如下所示：

```java
package ltd.newbee.mall.controller;

import org.springframework.beans.factory.annotation.Autowired;
import org.springframework.jdbc.core.JdbcTemplate;
import org.springframework.util.CollectionUtils;
import org.springframework.util.StringUtils;
import org.springframework.web.bind.annotation.GetMapping;
import org.springframework.web.bind.annotation.RestController;

import java.util.List;
import java.util.Map;

@RestController
public class JdbcController {

    //已经自动配置，因此可以直接通过@Autowired注入进来
    @Autowired
    JdbcTemplate jdbcTemplate;

    // 新增一条记录
    @GetMapping("/insert")
    public String insert(String type, String name) {
        if (StringUtils.isEmpty(type) || StringUtils.isEmpty(name)) {
            return "参数异常";
        }
        jdbcTemplate.execute("insert into jdbc_test('ds_type', 'ds_name') value (\"" + type + "\",\"" + name + "\")");
        return "SQL 执行完毕";
    }

    // 删除一条记录
    @GetMapping("/delete")
    public String delete(int id) {
        if (id < 0) {
```

```java
        return "参数异常";
    }
    List<Map<String, Object>> result = jdbcTemplate.queryForList("select * from jdbc_test where ds_id = \"" + id + "\"");
    if (CollectionUtils.isEmpty(result)) {
        return "不存在该记录,删除失败";
    }
    jdbcTemplate.execute("delete from jdbc_test where ds_id=\"" + id + "\"");
    return "SQL 执行完毕";
}

// 修改一条记录
@GetMapping("/update")
public String update(int id, String type, String name) {
    if (id < 0 || StringUtils.isEmpty(type) || StringUtils.isEmpty(name)) {
        return "参数异常";
    }
    List<Map<String, Object>> result = jdbcTemplate.queryForList("select * from jdbc_test where ds_id = \"" + id + "\"");
    if (CollectionUtils.isEmpty(result)) {
        return "不存在该记录,无法修改";
    }
    jdbcTemplate.execute("update jdbc_test set ds_type=\"" + type + "\", ds_name= \"" + name + "\" where ds_id=\"" + id + "\"");
    return "SQL 执行完毕";
}

// 查询所有记录
@GetMapping("/queryAll")
public List<Map<String, Object>> queryAll() {
    List<Map<String, Object>> list = jdbcTemplate.queryForList("select * from jdbc_test");
    return list;
}
}
```

在编码完成后启动 Spring Boot 项目。在项目启动成功后,在浏览器中打开并对以上四个功能进行验证。

（1）使用 JdbcTemplate 向数据库中新增记录。

在地址栏输入如下地址：

```
http://localhost:8080/insert?type=test&name=测试类
```

传参分别为 test 和测试类，表示向数据库中新增一条记录，其中 dstype 字段值为 test，dsname 字段值为测试类，页面返回结果如图 9-10 所示。

图 9-10　JdbcTemplate 新增记录测试

此时查看数据库中的记录，可以看到已经新增成功，如图 9-11 所示。

ds_id	ds_type	ds_name
1	com.zaxxer.hikari.HikariDataSource	hikari数据源
2	org.apache.commons.dbcp2.BasicDataSource	dbcp2数据源
3	test	测试类

图 9-11　JdbcTemplate 新增记录测试结果

（2）使用 JdbcTemplate 删除数据库中的记录。

在地址栏输入如下地址：

```
http://localhost:8080/delete?id=3
```

传参为 3，表示从数据库表中删除一条 ds_id 为 3 的记录。在 JdbcController 中 delete() 方法的实现逻辑是先查询是否存在对应的记录，如果不存在则不执行删除的 SQL 语句，如果存在才会执行删除的 SQL 语句。页面返回结果如图 9-12 所示。

图 9-12　JdbcTemplate 删除记录测试

此时查看数据库中的记录，ds_id 为 3 的记录已经被删除成功，如图 9-13 所示。

	ds_id	ds_type	ds_name
1	1	com.zaxxer.hikari.HikariDataSource	hikari数据源
2	2	org.apache.commons.dbcp2.BasicDataSource	dbcp2数据源

图 9-13　JdbcTemplate 删除记录测试结果

（3）使用 JdbcTemplate 修改数据库中的记录。

在地址栏输入如下地址：

```
http://localhost:8080/update?id=2&type=BasicDataSource&name=数据源修改测试
```

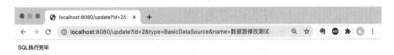

传参分别为 2、BasicDataSource 和数据源修改测试，表示修改数据库表中 ds_id 为 2 的记录。在 JdbcController 中 update()方法的实现逻辑是先查询是否存在对应的记录，如果不存在则不执行修改的 SQL 语句，如果存在才会执行修改的 SQL 语句。页面返回结果如图 9-14 所示。

图 9-14　JdbcTemplate 修改记录测试

此时查看数据库中的记录，ds_id 为 2 的记录已经修改成功，如图 9-15 所示。

	ds_id	ds_type	ds_name
1	1	com.zaxxer.hikari.HikariDataSource	hikari数据源
2	2	BasicDataSource	数据源修改测试

图 9-15　JdbcTemplate 修改记录测试结果

（4）使用 JdbcTemplate 查询数据库中的记录。

在地址栏输入如下地址：

```
http://localhost:8080/queryAll
```

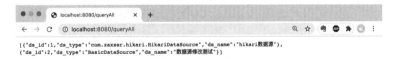

该请求会查询出数据库中的所有记录，页面返回结果如图 9-16 所示。

图 9-16　JdbcTemplate 查询记录测试

以上为笔者在进行功能测试的步骤。读者在测试时也可以尝试多添加几条记录。如果能够正常获取记录并正确操作 jdbc_test 表中的记录就表示功能整合成功！

在 Spring Boot 项目中，仅需要几行配置代码即可完成数据库的连接操作，并不需要多余的设置。另外，Spring Boot 自动配置了 jdbcTemplate 对象，开发人员可以直接上手进行数据库的相关开发工作。

作为知识的补充和拓展，接下来会讲解 Spring Boot 和 ORM 框架（MyBatis）的整合实践。

9.4　Spring Boot项目中MyBatis相关组件的自动配置讲解

9.4.1　MyBatis 简介

MyBatis 的前身是 Apache 社区的一个开源项目 iBatis，于 2010 年更名为 MyBatis。

MyBatis 是支持定制化 SQL、存储过程和高级映射的优秀持久层框架。它避免了几乎所有的 JDBC 代码、手动设置参数和获取结果集的操作，使得开发人员更加关注 SQL 本身和业务逻辑，不用再花费时间关注整个复杂的 JDBC 操作过程。

图 9-17 为 MyBatis 的结构图。

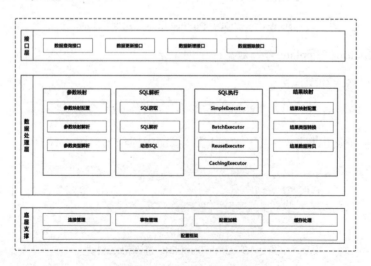

图 9-17　MyBatis 的结构图

MyBatis 的优点如下所示。

①封装了 JDBC 大部分操作，减少了开发人员的工作量

②半自动化的操对于编写 SQL 语句灵活度更高

③Java 代码与 SQL 语句分离，降低维护难度

④自动映射结果集，减少重复的编码工作

⑤开源社区十分活跃，文档齐全，学习成本低

9.4.2 mybatis-springboot-starter 介绍

Spring 官方并没有提供 MyBatis 的场景启动器，但是 MyBatis 官方却提供了 MyBatis 整合 Spring Boot 项目的场景启动器，也就是 mybatis-springboot-starter。通过命名方式也能够发现，Spring 官方提供的启动器的命名方式都为 spring-boot-starter-*，与 MyBatis 官方提供的 starter 组件不同。接下来就介绍 mybatis-springboot-starter 场景启动器。

读者可以去官网查看更多内容。

mybatis-spring-boot-starter 可以帮助开发人员快速创建基于 Spring Boot 的 MyBatis 应用程序。那么使用 mybatis-spring-boot-starter 具体可以做什么呢？

①构建独立的 MyBatis 应用程序

②零模板搭建

③更少甚至无 XML 配置代码

9.4.3 MyBatis 自动配置详解

在使用 MyBatis 开发功能之前，先来结合其自动配置源码了解一下 mybatis-spring-boot-starter 场景启动器的具体操作。

1. mybatis-spring-boot-starter 依赖

在 pom 文件中引入 spring-boot-starter-jdbc 依赖，如图 9-18 所示。

点击 mybatis-spring-boot-starter 并进入 mybatis-spring-boot-starter-2.1.3.pom 文件，其源码如图 9-19 所示。

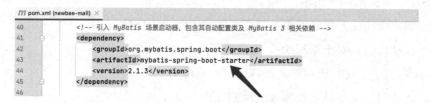

图 9-18　引入 spring-boot-starter-jdbc 依赖

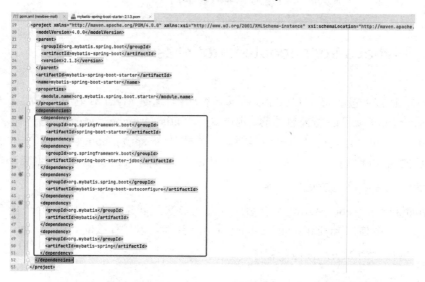

图 9-19　mybatis-spring-boot-starter-2.1.3.pom 源码

通过源码可以看到，在该 pom 文件中引入了 MyBatis 相关依赖及其自动配置依赖包，同时也包含 JDBC 场景启动器。如果开发人员没有在主 pom 配置文件中引入的话，这里也会引入并使用默认的 HikariCP 数据源。

2. mybatis-spring-boot-starter 自动配置类

通过 Maven Libraries 查找 MyBatis 的依赖文件可以看到其自动配置类，名称为 MybatisAutoConfiguration。该自动配置类与官方的自动配置类不同，它并不是定义在 spring-boot-autoconfigure-2.3.7.RELEASE.jar 的 org.springframework.boot.autoconfigure 包中，而是定义在 mybatis-spring-boot-autoconfigure-2.1.3.jar 的 org.mybatis.spring.boot.autoconfigure 包中，目录结构如图 9-20 所示。

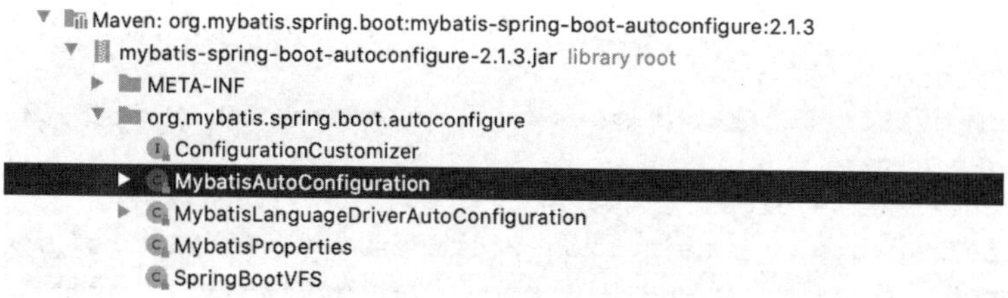

图 9-20 org.mybatis.spring.boot.autoconfigure 包

查看 MybatisAutoConfiguration 类，其源码及注释如下所示：

```
@org.springframework.context.annotation.Configuration // 配置类
@ConditionalOnClass({ SqlSessionFactory.class,
SqlSessionFactoryBean.class }) // 判断当前classpath是否存在指定类，若存在则将当
前的配置装载入Spring容器
@ConditionalOnSingleCandidate(DataSource.class) // 在当前IOC容器中存在
DataSource 数据源对象
@EnableConfigurationProperties(MybatisProperties.class) //MyBatis 的配置项
@AutoConfigureAfter({ DataSourceAutoConfiguration.class,
MybatisLanguageDriverAutoConfiguration.class })// 自动配置时机
public class MybatisAutoConfiguration implements InitializingBean {

  private static final Logger logger =
LoggerFactory.getLogger(MybatisAutoConfiguration.class);

  private final MybatisProperties properties;

  private final Interceptor[] interceptors;

  private final TypeHandler[] typeHandlers;

  private final LanguageDriver[] languageDrivers;

  private final ResourceLoader resourceLoader;

  private final DatabaseIdProvider databaseIdProvider;

  private final List<ConfigurationCustomizer> configurationCustomizers;

  public MybatisAutoConfiguration(MybatisProperties properties, ObjectProvider
<Interceptor[]> interceptorsProvider,
      ObjectProvider<TypeHandler[]> typeHandlersProvider,
ObjectProvider<LanguageDriver[]> languageDriversProvider,
```

```java
        ResourceLoader resourceLoader, ObjectProvider<DatabaseIdProvider> databaseIdProvider,
        ObjectProvider<List<ConfigurationCustomizer>> configurationCustomizersProvider) {
    this.properties = properties;
    this.interceptors = interceptorsProvider.getIfAvailable();
    this.typeHandlers = typeHandlersProvider.getIfAvailable();
    this.languageDrivers = languageDriversProvider.getIfAvailable();
    this.resourceLoader = resourceLoader;
    this.databaseIdProvider = databaseIdProvider.getIfAvailable();
    this.configurationCustomizers = configurationCustomizersProvider.getIfAvailable();
}

@Bean // 注册 SqlSessionFactory 到 IOC 容器中
@ConditionalOnMissingBean // 在当前 IOC 容器中不存在 SqlSessionFactory 类型的 Bean 时则注册
public SqlSessionFactory sqlSessionFactory(DataSource dataSource) throws Exception {
    SqlSessionFactoryBean factory = new SqlSessionFactoryBean();
    factory.setDataSource(dataSource);
    factory.setVfs(SpringBootVFS.class);
    if (StringUtils.hasText(this.properties.getConfigLocation())) {
        factory.setConfigLocation(this.resourceLoader.getResource(this.properties.getConfigLocation()));
    }
    applyConfiguration(factory);
    if (this.properties.getConfigurationProperties() != null) {
        factory.setConfigurationProperties(this.properties.getConfigurationProperties());
    }
    if (!ObjectUtils.isEmpty(this.interceptors)) {
        factory.setPlugins(this.interceptors);
    }
    if (this.databaseIdProvider != null) {
        factory.setDatabaseIdProvider(this.databaseIdProvider);
    }
    if (StringUtils.hasLength(this.properties.getTypeAliasesPackage())) {
        factory.setTypeAliasesPackage(this.properties.getTypeAliasesPackage());
    }
    if (this.properties.getTypeAliasesSuperType() != null) {
        factory.setTypeAliasesSuperType(this.properties.getTypeAliasesSuperType());
    }
```

```java
    }
    if (StringUtils.hasLength(this.properties.getTypeHandlersPackage())) {
      factory.setTypeHandlersPackage(this.properties.getTypeHandlersPackage());
    }
    if (!ObjectUtils.isEmpty(this.typeHandlers)) {
      factory.setTypeHandlers(this.typeHandlers);
    }
    if (!ObjectUtils.isEmpty(this.properties.resolveMapperLocations())) {
      factory.setMapperLocations(this.properties.resolveMapperLocations());
    }
    Set<String> factoryPropertyNames = Stream
        .of(new BeanWrapperImpl(SqlSessionFactoryBean.class).getPropertyDescriptors()).map(PropertyDescriptor::getName)
        .collect(Collectors.toSet());
    Class<? extends LanguageDriver> defaultLanguageDriver = this.properties.getDefaultScriptingLanguageDriver();
    if (factoryPropertyNames.contains("scriptingLanguageDrivers") && !ObjectUtils.isEmpty(this.languageDrivers)) {
      // Need to mybatis-spring 2.0.2+
      factory.setScriptingLanguageDrivers(this.languageDrivers);
      if (defaultLanguageDriver == null && this.languageDrivers.length == 1) {
        defaultLanguageDriver = this.languageDrivers[0].getClass();
      }
    }
    if (factoryPropertyNames.contains("defaultScriptingLanguageDriver")) {
      // Need to mybatis-spring 2.0.2+
      factory.setDefaultScriptingLanguageDriver(defaultLanguageDriver);
    }

    return factory.getObject();
}

@Bean // 注册 SqlSessionTemplate 到 IOC 容器中
@ConditionalOnMissingBean // 在当前 IOC 容器中不存在 SqlSessionTemplate 类型的 Bean 时则注册
public SqlSessionTemplate sqlSessionTemplate(SqlSessionFactory sqlSessionFactory) {
    ExecutorType executorType = this.properties.getExecutorType();
    if (executorType != null) {
      return new SqlSessionTemplate(sqlSessionFactory, executorType);
    } else {
      return new SqlSessionTemplate(sqlSessionFactory);
    }
}
```

```java
    }

    public static class AutoConfiguredMapperScannerRegistrar implements
BeanFactoryAware, ImportBeanDefinitionRegistrar {

      private BeanFactory beanFactory;

      @Override
      public void registerBeanDefinitions(AnnotationMetadata importingClass
Metadata, BeanDefinitionRegistry registry) {

        if (!AutoConfigurationPackages.has(this.beanFactory)) {
          logger.debug("Could not determine auto-configuration package,
automatic mapper scanning disabled.");
          return;
        }

        logger.debug("Searching for mappers annotated with @Mapper");

        List<String> packages = AutoConfigurationPackages.get(this.beanFactory);
        if (logger.isDebugEnabled()) {
          packages.forEach(pkg -> logger.debug("Using auto-configuration base
package '{}'", pkg));
        }

        BeanDefinitionBuilder builder = BeanDefinitionBuilder.genericBean
Definition(MapperScannerConfigurer.class);
        builder.addPropertyValue("processPropertyPlaceHolders", true);
        builder.addPropertyValue("annotationClass", Mapper.class);
        builder.addPropertyValue("basePackage", StringUtils.collectionTo
CommaDelimitedString(packages));
        BeanWrapper beanWrapper = new BeanWrapperImpl(MapperScannerConfigurer.class);
        Stream.of(beanWrapper.getPropertyDescriptors())
            // Need to mybatis-spring 2.0.2+
            .filter(x -> x.getName().equals("lazyInitialization")).findAny()
            .ifPresent(x -> builder.addPropertyValue("lazyInitialization",
"${mybatis.lazy-initialization:false}"));
        registry.registerBeanDefinition(MapperScannerConfigurer.class.getName(),
builder.getBeanDefinition());
      }

      @Override
      public void setBeanFactory(BeanFactory beanFactory) {
```

```
    this.beanFactory = beanFactory;
  }
}

@org.springframework.context.annotation.Configuration // 配置类
@Import(AutoConfiguredMapperScannerRegistrar.class)// 引入
AutoConfiguredMapperScannerRegistrar 类
@ConditionalOnMissingBean({ MapperFactoryBean.class, MapperScanner
Configurer.class }) // 在当前 IOC 容器中不存在 MapperFactoryBean 和 MapperScanner
Configurer 类型的 Bean 时则注册
public static class MapperScannerRegistrarNotFoundConfiguration implements
InitializingBean {

  @Override
  public void afterPropertiesSet() {
    logger.debug(
        "Not found configuration for registering mapper bean using @Mapper
Scan, MapperFactoryBean and MapperScannerConfigurer.");
  }

 }
}
```

DataSourceAutoConfiguration 类的注解释义如下所示。

@Configuration(proxyBeanMethods = false)：指定该类为配置类。

@ConditionalOnClass({{ SqlSessionFactory.class, SqlSessionFactoryBean.class })：判断当前 classpath 是否存在指定类 SqlSessionFactory 类和 SqlSessionFactoryBean 类，存在则生效。

@ConditionalOnSingleCandidate(DataSource.class)：判断 IOC 容器中是否存在类型为 DataSource 的 Bean，存在则生效。

@AutoConfigureAfter({DataSourceAutoConfiguration.class, MybatisLanguageDriverAutoConfiguration.class})：自动配置的生效时间。

结合该自动配置类的源码及源码注释可知，Mybatis 自动配置类的自动配置触发机制：当前项目的类路径中存在 SqlSessionFactory 类和 SqlSessionFactoryBean 类且 IOC 容器中存在 DataSource 类型的 Bean。因为当前项目引入了 spring-boot-starter-jdbc 和 mybatis-spring-boot-starter 两个场景启动器，启动器中包含了相关依赖，且在引入 spring-boot-starter-jdbc 后数据源已自动配置，因此 IOC 容器中存在 DataSource 类型的

Bean,即 MyBatis 自动配置类会生效。

3. DataSourceAutoConfiguration 自动配置类的作用

另外一个问题,MyBatis 自动配置类的作用是什么?

通过前文的条件注解分析可知,在自动配置流程开始且条件满足后,自动配置类 DataSourceAutoConfiguration 在进行自动配置时会判断当前 IOC 容器中是否存在 SqlSessionFactory 和 SqlSessionTemplate 类型的 Bean,MyBatis 框架对于数据库的访问和相关操作都是基于这两个对象展开的。

除此之外,DataSourceAutoConfiguration 自动配置类还有一个静态内部类 MapperScannerRegistrarNotFoundConfiguration,当项目中不存在@MapperScan 注解时它则生效。它是用于扫描在项目中标注了@Mapper 注解的接口。程序将这些接口与 xxxMapper.xml 文件一一映射后注入到 IOC 容器中以供业务代码调用并操作数据库。

查看@Mapper 注解的源码可知,它的定义中并没有@Component 注解。因此 Spring 框架本身并不支持@Mapper 注解的扫描,所以需要由 MyBatis 框架来完成这个操作。这一步的操作就由 MapperScannerRegistrarNotFoundConfiguration 类引入的 AutoConfigured MapperScannerRegistrar 类来完成。当项目中存在@MapperScan 注解时,Mapper 文件的扫描和配置工作就由 MapperScannerRegistrar 类来完成。感兴趣的读者可以自行查看这部分源码,这里就不再展开叙述。

在未使用 Spring Boot 配置 MyBatis 组件时,需要在 Spring 配置文件中增加 MyBatis 相关配置后才能够正常使用 MyBatis 框架,配置文件的代码如下所示:

```xml
<bean id="dataSource"
    class="org.springframework.jdbc.datasource.DriverManagerDataSource">
 <property name="driverClassName" value="com.mysql.jdbc.Driver"/>
 <property name="url"
        value="jdbc:mysql://localhost:3306/db"/>
 <property name="username" value="root"/>
 <property name="password" value="123456"/>
</bean>

<!-- 配置 MyBatis 的 sqlSessionFactory -->
<bean id="sqlSessionFactory" class="org.mybatis.spring.SqlSessionFactoryBean">
 <property name="dataSource" ref="dataSource"/>
 <!-- 自动扫描 mappers.xml 文件 -->
 <property name="mapperLocations" value="classpath:com/**/mappers/*.xml">
</property>
 <!-- MyBatis 配置文件 -->
 <property name="configLocation" value="classpath:mybatis-config.xml">
```

```xml
    </property>
</bean>

<!-- DAO接口所在的包名，Spring会自动查找其下的类 -->
<bean class="org.mybatis.spring.mapper.MapperScannerConfigurer">
    <property name="basePackage" value="com.core.dao"/>
    <property name="sqlSessionFactoryBeanName" value="sqlSessionFactory">
    </property>
</bean>
```

其实，在以上源码分析中，DataSourceAutoConfiguration 自动配置类所做的就是之前开发人员需要手动完成的 MyBatis 配置文件中的内容。只是在 Spring Boot 项目中，这个步骤由 Spring Boot 接管了。与 Spring 项目中需要手动配置 MyBatis 相比，Spring Boot 自动配置了使用 MyBatis 框架时所需的相关组件。这个过程对于开发人员来说是无感知的。虽然其源码看起来比较吃力，但是实际使用的时候并不会有这种感觉。书中介绍这部分源码只是为了读者能够更好地理解 Spring Boot 项目对于 MyBatis 框架的自动配置。

在实际开发过程中，开发人员只需引入 starter 依赖即可零配置使用 MyBatis 框架进行编码。这也是 Spring Boot 框架约定优于配置特性的体现。开发人员也可以自己配置 MyBatis 框架相关的组件，有了自定义的配置，MyBatis 自动配置流程的部分步骤就不会被执行。

9.5 Spring Boot整合MyBatis的过程

接下来将结合实际的代码案例讲解 Spring Boot 整合 MyBatis 的过程。

9.5.1 添加依赖

想要把 MyBatis 框架整合到 Spring Boot 项目中，首先需要将其依赖配置增加到 pom.xml 文件中。在本书代码案例中选择的 mybatis-springboot-starter 版本为 2.1.3，需要 Spring Boot 版本达到 2.0 或者以上版本才行。同时需要将数据源依赖和 JDBC 依赖也添加到配置文件中。由于前文中已经将这些依赖放入 pom.xml 配置文件中，因此只需要将 mybatis-springboot-starter 依赖放入 pom.xml 文件中即可。更新后的 pom.xml 配置文件如下所示：

```xml
<?xml version="1.0" encoding="UTF-8"?>
<project xmlns="http://maven.apache.org/POM/4.0.0" xmlns:xsi="http://www.
```

```xml
w3.org/2001/XMLSchema-instance"
    xsi:schemaLocation="http://maven.apache.org/POM/4.0.0 https://maven.apache.org/xsd/maven-4.0.0.xsd">
    <modelVersion>4.0.0</modelVersion>
    <parent>
        <groupId>org.springframework.boot</groupId>
        <artifactId>spring-boot-starter-parent</artifactId>
        <version>2.3.7.RELEASE</version>
        <relativePath/> <!-- lookup parent from repository -->
    </parent>
    <groupId>ltd.newbee.mall</groupId>
    <artifactId>newbee-mall</artifactId>
    <version>0.0.1-SNAPSHOT</version>
    <name>newbee-mall</name>
    <description>mysql-demo</description>

    <properties>
        <project.build.sourceEncoding>UTF-8</project.build.sourceEncoding>
        <project.reporting.outputEncoding>UTF-8</project.reporting.outputEncoding>
        <java.version>1.8</java.version>
    </properties>

    <dependencies>
        <dependency>
            <groupId>org.springframework.boot</groupId>
            <artifactId>spring-boot-starter-web</artifactId>
        </dependency>

        <dependency>
            <groupId>org.springframework.boot</groupId>
            <artifactId>spring-boot-starter-jdbc</artifactId>
        </dependency>

        <dependency>
            <groupId>mysql</groupId>
            <artifactId>mysql-connector-java</artifactId>
            <scope>runtime</scope>
        </dependency>

        <!-- 引入MyBatis场景启动器，包含其自动配置类及MyBatis3相关依赖 -->
        <dependency>
```

```xml
            <groupId>org.mybatis.spring.boot</groupId>
            <artifactId>mybatis-spring-boot-starter</artifactId>
            <version>2.1.3</version>
        </dependency>

        <dependency>
            <groupId>org.springframework.boot</groupId>
            <artifactId>spring-boot-starter-test</artifactId>
            <scope>test</scope>
            <exclusions>
                <exclusion>
                    <groupId>org.junit.vintage</groupId>
                    <artifactId>junit-vintage-engine</artifactId>
                </exclusion>
            </exclusions>
        </dependency>
    </dependencies>

    <build>
        <plugins>
            <plugin>
                <groupId>org.springframework.boot</groupId>
                <artifactId>spring-boot-maven-plugin</artifactId>
            </plugin>
        </plugins>
    </build>

    <repositories>
        <repository>
            <id>alimaven</id>
            <name>aliyun maven</name>
            <url>http://maven.aliyun.com/nexus/content/repositories/central/</url>
            <releases>
                <enabled>true</enabled>
            </releases>
            <snapshots>
                <enabled>false</enabled>
            </snapshots>
        </repository>
    </repositories>

</project>
```

这样，MyBatis 的场景启动器和相关依赖就整合进 Spring Boot 项目中了。

9.5.2 application.properties 的配置

在 Spring Boot 整合 MyBatis 时，有几个需要注意的配置参数，如下所示：

```
mybatis.config-location=classpath:mybatis-config.xml
mybatis.mapper-locations=classpath:mapper/*Dao.xml
mybatis.type-aliases-package=ltd.newbee.mall.entity
```

mybatis.config-location：配置 mybatis-config.xml 路径，在 mybatis-config.xml 中配置 MyBatis 基础属性，如果项目中配置了 mybatis-config.xml 文件就需要设置该参数。

mybatis.mapper-locations：配置 Mapper 文件对应的 XML 文件路径。

mybatis.type-aliases-package：配置项目中实体类包的路径。

在开发时只配置 mapper-locations 即可，最终的 application.properties 文件代码如下所示：

```
spring.datasource.name=newbee-mall-datasource
spring.datasource.driverClassName=com.mysql.cj.jdbc.Driver
spring.datasource.url=jdbc:mysql://localhost:3306/test_db?useUnicode=true&serverTimezone=Asia/Shanghai&characterEncoding=utf8&autoReconnect=true&useSSL=false&allowMultiQueries=true&useAffectedRows=true
spring.datasource.username=root
spring.datasource.password=123456

mybatis.mapper-locations=classpath:mapper/*Mapper.xml
```

9.5.3 启动类增加 Mapper 扫描

在启动类中添加对 Mapper 包的扫描@MapperScan，Spring Boot 在启动的时候会自动加载包路径下的 Mapper 接口。代码如下所示：

```
@SpringBootApplication
@MapperScan("ltd.newbee.mall.dao") //添加 @Mapper 注解
public class NewbeeMallApplication {
    public static void main(String[] args) {
        SpringApplication.run(Application.class, args);
```

```
        }
}
```

当然也可以直接在每个 Mapper 接口上面添加@Mapper 注解。但是如果 Mapper 接口数量较多，在每个 Mapper 接口上加注解是比较烦琐的，建议使用扫描注解。

9.6 Spring Boot整合MyBatis进行数据库的增、删、改、查

在开发项目之前需要在 MySQL 中先创建一张表，SQL 语句如下所示：

```sql
DROP TABLE IF EXISTS 'tb_user';

CREATE TABLE 'tb_user' (
  'id' INT(11) NOT NULL AUTO_INCREMENT COMMENT '主键',
  'name' VARCHAR(100) NOT NULL DEFAULT '' COMMENT '登录名',
  'password' VARCHAR(100) NOT NULL DEFAULT '' COMMENT '密码',
  PRIMARY KEY ('id')
) ENGINE=INNODB AUTO_INCREMENT=1 DEFAULT CHARSET=utf8;

/*Data for the table 'jdbc_test' */

insert into 'tb_user'('id','name','password') values (1,'Spring Boot',
'123456'),(2,'MyBatis','123456'),(3,'Thymeleaf','123456'),(4,'Java',
'123456'),(5,'MySQL','123456'),(6,'IDEA','123456');
```

这里在数据库中新建了一个名称为 tb_user 的数据表，表中有 id、name、password 三个字段，在本机测试时可以直接将以上 SQL 拷贝到 MySQL 中执行即可。

接下来是功能实现步骤，使用 MyBatis 进行数据的增、删、改、查操作。

9.6.1 新建实体类和 Mapper 接口

新建 entity 包并在 entity 包下新建 User 类，将 tb_user 中的字段映射到该实体类中，代码如下所示：

```java
package ltd.newbee.mall.entity;

public class User {
```

```java
    private Integer id;
    private String name;
    private String password;

    public Integer getId() {
        return id;
    }

    public void setId(Integer id) {
        this.id = id;
    }

    public String getName() {
        return name;
    }

    public void setName(String name) {
        this.name = name;
    }

    public String getPassword() {
        return password;
    }

    public void setPassword(String password) {
        this.password = password;
    }
}
```

新建 dao 包并在 dao 包中新建 UserDao 接口，并定义增、删、改、查四个方法，代码如下所示：

```java
package ltd.newbee.mall.dao;
import ltd.newbee.mall.entity.User;
import java.util.List;

/**
 * @author 十三
 *MyBatis 功能测试
 */
public interface UserDao {
    /**
     * 返回数据列表
     *
```

```java
     * @return
     */
    List<User> findAllUsers();

    /**
     * 添加
     *
     * @param User
     * @return
     */
    int insertUser(User User);

    /**
     * 修改
     *
     * @param User
     * @return
     */
    int updUser(User User);

    /**
     * 删除
     *
     * @param id
     * @return
     */
    int delUser(Integer id);
}
```

9.6.2 创建 Mapper 接口的映射文件

在 resources 目录下新建 mapper 目录,并在 mapper 目录下新建 Mapper 接口的映射文件 UserMapper.xml,再进行映射文件的编写。

首先,定义映射文件与 Mapper 接口的对应关系。比如在该示例中,需要将 UserMapper.xml 文件与对应的 UserDao 接口类之间的关系定义出来:

```xml
<mapper namespace="ltd.newbee.mall.dao.UserDao">
```

然后,配置表结构和实体类的对应关系:

```xml
<resultMap type="ltd.newbee.mall.entity.User" id="UserResult">
    <result property="id" column="id"/>
```

```xml
    <result property="name" column="name"/>
    <result property="password" column="password"/>
</resultMap>
```

最后，按照对应的接口方法，编写增、删、改、查方法的具体 SQL 语句，最终的 UserMapper.xml 文件如下所示：

```xml
<?xml version="1.0" encoding="UTF-8"?>
<!DOCTYPE mapper PUBLIC "-//mybatis.org//DTD Mapper 3.0//EN" "http://mybatis.org/dtd/mybatis-3-mapper.dtd">
<mapper namespace="ltd.newbee.mall.dao.UserDao">
  <resultMap type="ltd.newbee.mall.entity.User" id="UserResult">
    <result property="id" column="id"/>
    <result property="name" column="name"/>
    <result property="password" column="password"/>
  </resultMap>

  <select id="findAllUsers" resultMap="UserResult">
    select id,name,password from tb_user
    order by id desc
  </select>

  <insert id="insertUser" parameterType="ltd.newbee.mall.entity.User">
    insert into tb_user(name,password)
    values(#{name},#{password})
  </insert>

  <update id="updUser" parameterType="ltd.newbee.mall.entity.User">
    update tb_user
    set
    name=#{name},password=#{password}
    where id=#{id}
  </update>

  <delete id="delUser" parameterType="int">
    delete from tb_user where id=#{id}
  </delete>
</mapper>
```

9.6.3　新建 MyBatisController

为了对 MyBatis 进行功能测试，在 controller 包下新建 MyBatisController 类，并新增 4 个方法分别接收对于 tb_user 表的增、删、改、查请求，代码如下所示：

```java
package ltd.newbee.mall.controller;

import ltd.newbee.mall.dao.UserDao;
import ltd.newbee.mall.entity.User;
import org.springframework.util.StringUtils;
import org.springframework.web.bind.annotation.GetMapping;
import org.springframework.web.bind.annotation.RestController;

import javax.annotation.Resource;
import java.util.List;

@RestController
public class MyBatisController {

    @Resource
    UserDao userDao;

    // 查询所有记录
    @GetMapping("/users/mybatis/queryAll")
    public List<User> queryAll() {
        return userDao.findAllUsers();
    }

    // 新增一条记录
    @GetMapping("/users/mybatis/insert")
    public Boolean insert(String name, String password) {
        if (StringUtils.isEmpty(name) || StringUtils.isEmpty(password)) {
            return false;
        }
        User user = new User();
        user.setName(name);
        user.setPassword(password);
        return userDao.insertUser(user) > 0;
    }
```

```java
// 修改一条记录
@GetMapping("/users/mybatis/update")
public Boolean update(Integer id, String name, String password) {
    if (id == null || id < 1 || StringUtils.isEmpty(name) || StringUtils.isEmpty(password)) {
        return false;
    }
    User user = new User();
    user.setId(id);
    user.setName(name);
    user.setPassword(password);
    return userDao.updUser(user) > 0;
}

// 删除一条记录
@GetMapping("/users/mybatis/delete")
public Boolean delete(Integer id) {
    if (id == null || id < 1) {
        return false;
    }
    return userDao.delUser(id) > 0;
}
```

在上述步骤完成后，项目的代码目录就如图 9-21 所示。

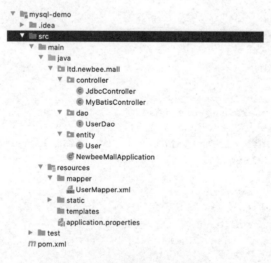

图 9-21　整合 MyBatis 后的目录结构

第 9 章　Spring Boot 操作 MySQL 数据库

在编码完成后启动 Spring Boot 项目。在启动成功后打开浏览器，对以上四个功能进行验证。

（1）Spring Boot 整合 MyBatis 向数据库中新增记录。

在地址栏输入如下地址：

`http://localhost:8080/users/mybatis/insert?name=十三&password=1234567`

传参分别为十三和 1234567，表示向数据库中新增一条记录。其中 name 字段值为十三，password 字段值为 1234567。页面返回结果如图 9-22 所示。

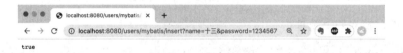

图 9-22　整合 MyBatis 向数据库中新增记录测试

此时查看数据库中的记录，可以看到已经新增成功，新的记录已经生成，如图 9-23 所示。

id	name	password
1	Spring Boot	123456
2	MyBatis	123456
3	Thymeleaf	123456
4	Java	123456
5	MySQL	123456
6	IDEA	123456
7	十三	1234567

图 9-23　整合 MyBatis 向数据库中新增记录测试结果

（2）Spring Boot 整合 MyBatis 删除数据库中的记录。

在地址栏输入如下地址：

`http://localhost:8080/delete?id=3`

传参为 5，表示从数据库表中删除一条 id 为 5 的记录，页面返回结果如图 9-24 所示。

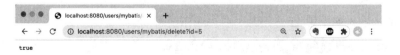

图 9-24　整合 MyBatis 删除记录测试

此时查看数据库中的记录，id 为 5 的记录已经被删除成功，如图 9-25 所示。

图 9-25　整合 MyBatis 删除记录测试结果

（3）Spring Boot 整合 MyBatis 修改数据库中的记录。

在地址栏输入如下地址：

```
http://localhost:8080/users/mybatis/update?id=1&name=book01&password=12345678
```

传参分别为 1、book01 和 12345678，表示修改数据库表中 id 为 1 的记录，页面返回结果如图 9-26 所示。

图 9-26　整合 MyBatis 修改记录测试

此时查看数据库中的记录，id 为 1 的记录已经修改成功，如图 9-27 所示。

图 9-27　整合 MyBatis 修改记录测试结果

（4）Spring Boot 整合 MyBatis 查询数据库中的记录。

在地址栏输入如下地址：

```
http://localhost:8080/users/mybatis/queryAll
```

该请求会查询出数据库中的所有记录，页面返回结果如图 9-28 所示。

第 9 章 Spring Boot 操作 MySQL 数据库

图 9-28 整合 MyBatis 查询记录

以上为笔者进行功能测试的步骤。读者在测试时也可以尝试多添加几条记录，如果能够正常获取记录并正确操作 tb_user 表中的记录就表示功能整合成功！

第 10 章 分页功能的讲解和编码实现

商城端和商城后台管理系统都会用到分页功能。在讲解具体的功能模块开发之前，先介绍一下分页功能的知识点。

分页是网站系统非常重要也十分常用的功能。在 MVC 开发模式下开发人员通常把它放在 ModelAndView 对象中，在页面代码中循环遍历列表数据并渲染到网站页面上。它也可以通过接口返回实现，即在前端通过 Ajax 调用接口数据并通过插件来实现分页数据的渲染和翻页功能。这两种方式的实现代码本书都会讲解。在商城项目中有多个页面会用到分页功能，不同的页面会使用不同的交互方式，但它们的分页原理都是一样的。

10.1 分页功能介绍

分页功能存在于各类电商网站、新闻网站、音乐网站，以及各类后台管理系统中。这里笔者整理了一些常用网站的分页展示效果。

10.1.1 百度分页功能演示

在百度首页搜索框中输入"Spring Boot"查询并跳转到搜索结果页面，页面大致会有 10 条左右的数据列表。此时展示的是第 1 页的数据，如图 10-1 所示。

第 10 章 分页功能的讲解和编码实现

图 10-1 百度分页功能示例

如果想浏览更多的搜索内容点击页面下方的数字或翻页按钮即可，如图 10-2 所示。

图 10-2 百度分页功能的翻页展示

10.1.2 GitHub 分页功能演示

在 GitHub 页面的搜索框中输入"Spring Boot"查询相关代码仓库，搜索结果页面如图 10-3 所示。页面中有 10 条左右的仓库数据列表，当前展示的是第 1 页的数据。

图 10-3 GitHub 分页功能示例

如果想浏览更多的 Spring Boot 仓库，点击页面下方的数字或翻页按钮即可，如图 10-4 所示。

第 10 章　分页功能的讲解和编码实现

图 10-4　GitHub 分页功能的翻页展示

10.1.3　商城后台管理系统分页功能演示

商城后台管理系统的大部分功能模块也加入了分页功能。数据列表页面没有把所有的内容都展示出来，而是通过分页功能分批次展示出来的。比如商品列表数据和订单列表数据的展示，如图 10-5 所示。

图 10-5　商品后台管理系统分页功能示例

10.1.4　商品搜索页分页功能演示

通常商城系统的商品数量非常大，不可能在一个页面中把商品数据全部展示出来，所以在搜索结果页面也需要加入分页功能的设计。

以新蜂商城为例，首先在搜索框输入关键字"华为"，查询出与华为相关的商品，然后可以在搜索结果页面看到对应的商品。该页面也用分页功能展示了部分商品，每一页展示 10 条商品数据，可以通过页面底部的数字或翻页按钮来查看更多的商品信息，如图 10-6 所示。

第 10 章 分页功能的讲解和编码实现

图 10-6 商品搜索页分页功能示例

当然，商城系统还有其他功能模块也用到了分页功能，在后续开发过程中都会逐一讲解。

10.2 分页功能的作用

分页功能作为各类网站和系统不可或缺的部分，其作用有 5 个。

（1）减少系统资源的消耗

数据查询结果是放在系统内存中的，如果在数据量很大的情况下一次性将所有内容都显示出来，会占用过多内存，而通过分页可以减少这种系统资源的消耗。

（2）提高数据库的查询性能

后端服务与数据库间通过网络传输数据，一次传输 10 条数据结果集与一次传输 2 万条数据结果集相比，肯定是前者消耗的网络资源更少。

（3）提升页面的访问速度

浏览器与后端服务间的传输也是通过网络进行的。因为数据包的大小有差别，返回 10 条数据明显要比返回 2 万条数据速度更快。

（4）符合用户的浏览习惯

以搜索结果或者商品展示为例，用户通常只查看展示在前面的部分数据，而将所有数据都展示出来不符合用户浏览习惯。

（5）适配页面的排版

由于设备屏幕的大小固定，一个屏幕能够展示的信息有限，如果一次展示太多的数据，无论排版还是页面美观度都会被影响。分页的作用也是为了适配页面的排版。

10.3 分页功能的设计

10.3.1 前端分页功能设计

GitHub 分页按钮如图 10-7 所示。

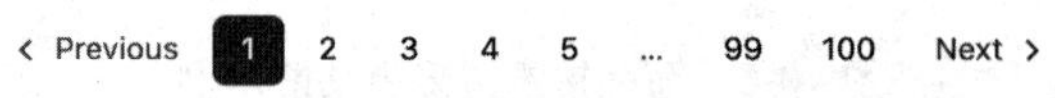

图 10-7　GitHub 分页按钮

百度分页按钮如图 10-8 所示。

图 10-8　百度分页按钮

通过这种前端页面分页展示区的设计可以看出，分页展示区比较重要的 3 个参数：页码展示、当前页码、每页条数。

此外，有些页面也会加上首页、尾页、跳转页码等功能。开发人员根据功能需要和页面设计可以自行增加或删减。

10.3.2 后端分页功能设计

分页功能在前端页面的执行是渲染数据和分页信息展示，在后端页面则需要按照前端传输而来的请求，将分页所需的数据正确地查询出来并返回给前端。两端的侧重点并不相同。比如，前端需要展示所有页码，而后端则只需要提供总页数即可，不需要对总页数进行其他操作。再比如前端需要根据用户的操作记录当前页码的参数，以便对页码信息进行调整和限制，而后端只需要接收前端传输过来的页码并进行相应判断和查询即可。

后端分页必不可少的两个参数：页码、条数（每页）。

在实现分页功能时，使用不同的数据库实现方式也不同。因为不同数据库实现分页功能的关键字有些差别，比如 SQL Server 的 top 关键字、Oracle 的 rownum 关键字、MySQL 的 limit 关键字。limit 关键字如下所示：

```
//MySQL 数据库中的查询语句：

select * from tb_xxxx limit 10,20;
```

分页功能的最终实现就是通过页码和条数，确定数据库需要查询的数据。比如查询第 1 页且每页 20 条的数据，就是查询数据库中从第 1 到 20 条的数据；查询第 4 页且每页 10 条的数据就是查询数据库中第 30 到 40 条的数据。因此，对于后端代码的实现来说，页码和条数两个参数就显得特别重要，缺少这两个参数查询逻辑就不成立，分页数据也就无从查起。

此外，为了前端分页区的展示，还要将数据总量或者总页数返回给前端。数据总量是必不可少的，而总页数可以计算出来（即数据总量除以每页条数）。数据总量的获取方式：

```
select count(*) from tb_xxxx;
```

至此，将分页数据进行数据封装，并返回给前端即可。

10.4 分页功能的编码实现

这里实现的分页功能是一个通用的分页接口。笔者将分页数据封装到一个返回结果对象中，并通过 JSON 格式返回，目的是让读者理解分页功能的主要逻辑和简单的代码实现。

接下来将结合 tb_user 表进行简单的查询与分页功能的实现，即在前端请求对应的页数时，返回那一页的所有数据。

10.4.1 新增分页测试数据

为了演示分页功能，需要在表中多增加一些数据，如果数量太少就无法充分模拟分页功能。

执行 SQL 向数据库中新增一些测试数据，代码如下所示：

```sql
-- ----------------------------
-- Table structure for tb_user
-- ----------------------------
DROP TABLE IF EXISTS 'tb_user';

CREATE TABLE 'tb_user' (
  'id' INT(11) NOT NULL AUTO_INCREMENT COMMENT '主键',
  'name' VARCHAR(100) NOT NULL DEFAULT '' COMMENT '登录名',
  'password' VARCHAR(100) NOT NULL DEFAULT '' COMMENT '密码',
  PRIMARY KEY ('id')
) ENGINE=INNODB AUTO_INCREMENT=1 DEFAULT CHARSET=utf8;

-- ----------------------------
-- Records of tb_user
-- ----------------------------
INSERT INTO 'tb_user' VALUES (1, 'admin', 'e10adc3949ba59abbe56e057f20f883e');
INSERT INTO 'tb_user' VALUES (2, 'test2', '098f6bcd4621d373cade4e832627b4f6');
INSERT INTO 'tb_user' VALUES (3, 'test3', '098f6bcd4621d373cade4e832627b4f6');
INSERT INTO 'tb_user' VALUES (4, 'test4', '098f6bcd4621d373cade4e832627b4f6');
INSERT INTO 'tb_user' VALUES (5, 'test5', '098f6bcd4621d373cade4e832627b4f6');
INSERT INTO 'tb_user' VALUES (6, 'test6', '098f6bcd4621d373cade4e832627b4f6');
INSERT INTO 'tb_user' VALUES (7, 'test7', '098f6bcd4621d373cade4e832627b4f6');
INSERT INTO 'tb_user' VALUES (8, 'test8', '098f6bcd4621d373cade4e832627b4f6');
INSERT INTO 'tb_user' VALUES (9, 'test9', '098f6bcd4621d373cade4e832627b4f6');
INSERT INTO 'tb_user' VALUES (10, 'test10', '098f6bcd4621d373cade4e832627b4f6');
INSERT INTO 'tb_user' VALUES (11, 'test11', '098f6bcd4621d373cade4e832627b4f6');
INSERT INTO 'tb_user' VALUES (12, 'test12', '098f6bcd4621d373cade4e832627b4f6');
INSERT INTO 'tb_user' VALUES (13, 'test13', '098f6bcd4621d373cade4e832627b4f6');
INSERT INTO 'tb_user' VALUES (14, 'test14', '098f6bcd4621d373cade4e832627b4f6');
INSERT INTO 'tb_user' VALUES (15, 'test15', '098f6bcd4621d373cade4e832627b4f6');
INSERT INTO 'tb_user' VALUES (16, 'test16', '098f6bcd4621d373cade4e832627b4f6');
```

```sql
INSERT INTO 'tb_user' VALUES (17, 'test17', '098f6bcd4621d373cade4e832627b4f6');
INSERT INTO 'tb_user' VALUES (18, 'test18', '098f6bcd4621d373cade4e832627b4f6');
INSERT INTO 'tb_user' VALUES (19, 'test19', '098f6bcd4621d373cade4e832627b4f6');
INSERT INTO 'tb_user' VALUES (20, 'admin2', '098f6bcd4621d373cade4e832627b4f6');
INSERT INTO 'tb_user' VALUES (21, 'admin3', '098f6bcd4621d373cade4e832627b4f6');
INSERT INTO 'tb_user' VALUES (22, 'admin4', '098f6bcd4621d373cade4e832627b4f6');
INSERT INTO 'tb_user' VALUES (23, 'admin5', '098f6bcd4621d373cade4e832627b4f6');
INSERT INTO 'tb_user' VALUES (24, 'admin6', '098f6bcd4621d373cade4e832627b4f6');
INSERT INTO 'tb_user' VALUES (25, 'admin7', '098f6bcd4621d373cade4e832627b4f6');
INSERT INTO 'tb_user' VALUES (26, 'admin8', '098f6bcd4621d373cade4e832627b4f6');
INSERT INTO 'tb_user' VALUES (27, 'admin9', '098f6bcd4621d373cade4e832627b4f6');
INSERT INTO 'tb_user' VALUES (28, 'admin10', '098f6bcd4621d373cade4e832627b4f6');
INSERT INTO 'tb_user' VALUES (29, 'admin11', '098f6bcd4621d373cade4e832627b4f6');
INSERT INTO 'tb_user' VALUES (30, 'admin12', '098f6bcd4621d373cade4e832627b4f6');
INSERT INTO 'tb_user' VALUES (31, 'admin13', '098f6bcd4621d373cade4e832627b4f6');
INSERT INTO 'tb_user' VALUES (32, 'admin14', '098f6bcd4621d373cade4e832627b4f6');
INSERT INTO 'tb_user' VALUES (33, 'admin15', '098f6bcd4621d373cade4e832627b4f6');
INSERT INTO 'tb_user' VALUES (34, 'admin16', '098f6bcd4621d373cade4e832627b4f6');
INSERT INTO 'tb_user' VALUES (35, 'admin17', '098f6bcd4621d373cade4e832627b4f6');
INSERT INTO 'tb_user' VALUES (36, 'admin18', '098f6bcd4621d373cade4e832627b4f6');
INSERT INTO 'tb_user' VALUES (37, 'admin19', '098f6bcd4621d373cade4e832627b4f6');
INSERT INTO 'tb_user' VALUES (38, 'admin011', '098f6bcd4621d373cade4e832627b4f6');
INSERT INTO 'tb_user' VALUES (39, 'admin02', '098f6bcd4621d373cade4e832627b4f6');
INSERT INTO 'tb_user' VALUES (40, 'admin03', '098f6bcd4621d373cade4e832627b4f6');
INSERT INTO 'tb_user' VALUES (41, 'admin04', '098f6bcd4621d373cade4e832627b4f6');
INSERT INTO 'tb_user' VALUES (42, 'admin05', '098f6bcd4621d373cade4e832627b4f6');
INSERT INTO 'tb_user' VALUES (43, 'admin06', '098f6bcd4621d373cade4e832627b4f6');
INSERT INTO 'tb_user' VALUES (44, 'admin07', '098f6bcd4621d373cade4e832627b4f6');
INSERT INTO 'tb_user' VALUES (45, 'admin08', '098f6bcd4621d373cade4e832627b4f6');
INSERT INTO 'tb_user' VALUES (46, 'admin09', '098f6bcd4621d373cade4e832627b4f6');
INSERT INTO 'tb_user' VALUES (47, 'admin010', '098f6bcd4621d373cade4e832627b4f6');
INSERT INTO 'tb_user' VALUES (48, 'admin011', '098f6bcd4621d373cade4e832627b4f6');
INSERT INTO 'tb_user' VALUES (49, 'admin012', '098f6bcd4621d373cade4e832627b4f6');
INSERT INTO 'tb_user' VALUES (50, 'admin013', '098f6bcd4621d373cade4e832627b4f6');
INSERT INTO 'tb_user' VALUES (51, 'admin014', '098f6bcd4621d373cade4e832627b4f6');
INSERT INTO 'tb_user' VALUES (52, 'admin015', '098f6bcd4621d373cade4e832627b4f6');
INSERT INTO 'tb_user' VALUES (53, 'admin016', '098f6bcd4621d373cade4e832627b4f6');
INSERT INTO 'tb_user' VALUES (54, 'admin017', '098f6bcd4621d373cade4e832627b4f6');
INSERT INTO 'tb_user' VALUES (55, 'admin018', '098f6bcd4621d373cade4e832627b4f6');
INSERT INTO 'tb_user' VALUES (56, 'admin019', '098f6bcd4621d373cade4e832627b4f6');
INSERT INTO 'tb_user' VALUES (57, 'ZHENFENG13', '77c9749b451ab8c713c48037ddfbb2c4');
INSERT INTO 'tb_user' VALUES (58, '213312', 'eqwfasdfa');
INSERT INTO 'tb_user' VALUES (59, '14415143', '51435135');
```

```sql
INSERT INTO 'tb_user' VALUES (60, 'shisan', 'e10adc3949ba59abbe56e057f20f883e');
INSERT INTO 'tb_user' VALUES (61, 'zhangsan', 'fcea920f7412b5da7be0cf42b8c93759');
INSERT INTO 'tb_user' VALUES (62, 'test-user1', '3d0faa930d336ba748607ab7076ebce2');
INSERT INTO 'tb_user' VALUES (63, '3123213213', '6fdce2f14f4baf2d666fa13dfd8d1945');
```

10.4.2 分页功能返回结果的封装

在之前的文章中，返回结果的数据格式都比较简单，如果遇到复杂的数据类型则需要修改代码。比如这里的分页数据结果，就需要返回多个字段。在实际的项目开发工作中开发人员通常会将返回结果进行抽象并封装成一个常用的 Java 类。

新建 util 包并在 util 包中新建 Result 通用结果类，代码如下所示：

```java
package ltd.newbee.mall.util;

import java.io.Serializable;

/**
 * @author 13
 * @qq 交流群 796794009
 * @email 2449207463@qq.com
 * @link https://github.com/newbee-ltd
 */
public class Result<T> implements Serializable {
    private static final long serialVersionUID = 1L;
    private int resultCode;
    private String message;
    private T data;

    public Result() {
    }

    public Result(int resultCode, String message) {
        this.resultCode = resultCode;
        this.message = message;
    }

    public int getResultCode() {
        return resultCode;
    }

    public void setResultCode(int resultCode) {
```

```java
        this.resultCode = resultCode;
    }

    public String getMessage() {
        return message;
    }

    public void setMessage(String message) {
        this.message = message;
    }

    public T getData() {
        return data;
    }

    public void setData(T data) {
        this.data = data;
    }

    @Override
    public String toString() {
        return "Result{" +
                "resultCode=" + resultCode +
                ", message='" + message + '\'' +
                ", data=" + data +
                '}';
    }
}
```

后端接口返回的数据会根据以上格式进行数据封装,包括业务码、返回信息、实际的数据结果。这个格式是开发人员自行设置的,如果有其他更好的方案也可以进行适当的调整。如要为了返回结果的数据统一,前端接收该结果后需要对数据进行解析,并通过业务码进行相应的逻辑操作,再获取 data 属性中的数据并进行页面渲染或者进行信息提示。

实际返回的数据格式示例如下所示。

对象列表数据:

```
{
    "resultCode": 200,
    "message": "SUCCESS",
    "data": [{
        "id": 2,
```

```
        "name": "user1",
        "password": "123456"
    }, {
        "id": 1,
        "name": "13",
        "password": "12345"
    }]
}
```

单条对象数据：

```
{
    "resultCode": 200,
    "message": "SUCCESS",
    "data": {
        "id": 2,
        "name": "user1",
        "password": "123456"
    }
}
```

简单对象数据：

```
{
    "resultCode": 200,
    "message": "SUCCESS",
    "data": true
}
```

以上代码分别是不同格式的数据返回，在后端进行业务处理后程序将会返回给前端一串 JSON 格式的数据。resultCode 等于 200 表示数据请求成功，该字段也可以自行定义，比如 0、1001、500 等。message 值为 SUCCESS，也可以自行定义返回信息，比如获取成功、列表数据查询成功等字符串。这些都需要与前端约定好，一个 resultCode 码只能表示一种含义。而 data 属性使用泛型定义，因此在 data 字段中的数据可以是一个对象数组，也可以是一个字符串、数字等类型，它根据不同的业务返回不同的结果，在接下来的实践内容中都会以这种方式返回数据。

分页结果数据也做了统一的封装，根据前文的分析，分页结果一般需要以下几个字段：总记录数、每页记录数、总页数、当前页数和实际的列表数据。

在 util 包中新建 PageResult 类，PageResult 分页结果集的数据格式定义如下所示：

```
package ltd.newbee.mall.util;
```

```java
import java.io.Serializable;
import java.util.List;

/**
 * 分页工具类
 *
 * @author 13
 * @qq 交流群 796794009
 * @email 2449207463@qq.com
 * @link https://github.com/newbee-ltd
 */
public class PageResult implements Serializable {

    //总记录数
    private int totalCount;
    //每页记录数
    private int pageSize;
    //总页数
    private int totalPage;
    //当前页数
    private int currPage;
    //列表数据
    private List<?> list;

    /**
     * 分页
     *
     * @param list       列表数据
     * @param totalCount 总记录数
     * @param pageSize   每页记录数
     * @param currPage   当前页数
     */
    public PageResult(List<?> list, int totalCount, int pageSize, int currPage) {
        this.list = list;
        this.totalCount = totalCount;
        this.pageSize = pageSize;
        this.currPage = currPage;
        this.totalPage = (int) Math.ceil((double) totalCount / pageSize);
    }
```

```java
    public int getTotalCount() {
        return totalCount;
    }

    public void setTotalCount(int totalCount) {
        this.totalCount = totalCount;
    }

    public int getPageSize() {
        return pageSize;
    }

    public void setPageSize(int pageSize) {
        this.pageSize = pageSize;
    }

    public int getTotalPage() {
        return totalPage;
    }

    public void setTotalPage(int totalPage) {
        this.totalPage = totalPage;
    }

    public int getCurrPage() {
        return currPage;
    }

    public void setCurrPage(int currPage) {
        this.currPage = currPage;
    }

    public List<?> getList() {
        return list;
    }

    public void setList(List<?> list) {
        this.list = list;
    }
}
```

10.4.3 分页功能代码的具体实现

1. 数据层分页代码实现

由于 tb_user 的实体类和 Mapper 文件已经创建，所以新增对应的方法即可。在 UserDao 接口中新增两个方法 findUsers()和 getTotalUser()，代码如下所示：

```java
/**
 * 返回分页数据列表
 *
 * @param pageUtil
 * @return
 */
List<User> findUsers(PageQueryUtil pageUtil);

/**
 * 返回数据总数
 *
 * @param pageUtil
 * @return
 */
int getTotalUser(PageQueryUtil pageUtil);
```

该段代码表示根据传入的分页参数获取分页数据列表并返回数据数量。

在 UserMapper.xml 文件中新增这两个方法的映射语句，代码如下所示：

```xml
<!-- 查询用户列表 -->
<select id="findUsers" parameterType="Map" resultMap="UserResult">
  select id,name,password from tb_user
  order by id desc
  <if test="start!=null and limit!=null">
    limit #{start},#{limit}
  </if>
</select>

<!-- 查询用户总数 -->
<select id="getTotalUser" parameterType="Map" resultType="int">
  select count(*) from tb_user
</select>
```

2. 业务层分页代码实现

新建 service 包，并新增业务类 UserService，代码如下所示：

```java
package ltd.newbee.mall.service;

import ltd.newbee.mall.dao.UserDao;
import ltd.newbee.mall.entity.User;
import ltd.newbee.mall.util.PageQueryUtil;
import ltd.newbee.mall.util.PageResult;
import org.springframework.beans.factory.annotation.Autowired;
import org.springframework.stereotype.Service;

import java.util.List;

@Service
public class UserService {

    @Autowired
    private UserDao userDao;

    public PageResult getUserPage(PageQueryUtil pageUtil) {
        // 当前页码中的数据列表
        List<User> users = userDao.findUsers(pageUtil);
        // 数据总条数，用于计算分页数据
        int total = userDao.getTotalUser(pageUtil);
        // 分页信息封装
        PageResult pageResult = new PageResult(users, total, pageUtil.getLimit(), pageUtil.getPage());
        return pageResult;
    }
}
```

首先根据当前页面和每页条数查询当前页的数据集合，然后调用 select count(*) 语句查询数据的总条数用于计算分页数据，最后将获取的数据封装到 PageResult 对象中并返回给控制层。

3. 控制层分页代码实现

在 controller 包中新建 PageTestController 类，用于实现分页请求的处理并返回查询结果，代码如下所示：

```java
package ltd.newbee.mall.controller;

import ltd.newbee.mall.service.UserService;
import ltd.newbee.mall.util.PageResult;
import ltd.newbee.mall.util.Result;
import ltd.newbee.mall.util.PageQueryUtil;
import org.springframework.beans.factory.annotation.Autowired;
import org.springframework.util.StringUtils;
import org.springframework.web.bind.annotation.RequestMapping;
import org.springframework.web.bind.annotation.RequestMethod;
import org.springframework.web.bind.annotation.RequestParam;
import org.springframework.web.bind.annotation.RestController;

import java.util.Map;

/**
 * @author 13
 * @qq 交流群 796794009
 * @email 2449207463@qq.com
 * @link https://github.com/newbee-ltd
 */
@RestController
@RequestMapping("/users")
public class PageTestController {

    @Autowired
    private UserService userService;

    /**
     * 分页功能测试
     */
    @RequestMapping(value = "/list", method = RequestMethod.GET)
    public Result list(@RequestParam Map<String, Object> params) {
        Result result = new Result();
        if (StringUtils.isEmpty(params.get("page")) || StringUtils.isEmpty(params.get("limit"))) {
            // 返回错误码
            result.setResultCode(500);
            // 错误信息
            result.setMessage("参数异常！");
```

```
            return result;
        }
        // 封装查询参数
        PageQueryUtil queryParamList = new PageQueryUtil(params);
        // 查询并封装分页结果集
        PageResult userPage = userService.getUserPage(queryParamList);
        // 返回成功码
        result.setResultCode(200);
        result.setMessage("查询成功");
        // 返回分页数据
        result.setData(userPage);
        return result;
    }
}
```

通过后端代码可以看出,分页功能的交互流程:前端将所需页码和条数参数传输给后端,后端在接收分页请求后对分页参数进行计算,并利用 MySQL 的 limit 关键字查询对应的记录,在查询结果被封装后返回给前端。在 TestUserControler 类上使用的是 @RestController 注解,该注解相当于@ResponseBody+@Controller 的组合注解。

10.4.4　分页功能测试

在编码完成后启动 Spring Boot 项目。在启动成功后打开浏览器,在地址栏输入如下地址:

```
http://localhost:8080/users/list?page=1&limit=10
```

请求参数分别为 1 和 10,表示以每页 10 条的数据查询第 1 页的数据。在结果页面中可以看到分页数据的结果返回。接着按住 F12 键或者右键进入浏览器调试模式,通过开发人员工具查看返回的分页数据结果。在 NetWork 面板中可以看到这次请求的信息,返回格式是标准的 Result 对象格式,其中包含 resultCode、message 和 data 三个字段。其中 data 对象中的数据格式为 PageResult 对象的数据格式。currPage 表示当前页,数值为 1。后端封装的第 1 页的 10 条数据都在 list 字段中。测试结果及返回的分页数据格式如图 10-9 所示。

接下来通过改变 page 参数和 limit 参数的值进行分页功能的测试。比如以每页 20 条的数据查询第 3 页的数据,则请求 URL 为:

```
http://localhost:8080/users/list?page=3&limit=20
```

得到的结果如图 10-10 所示。返回的数据格式依然没变,只是具体字段的值有些改动,比如 currPage、totalPage、list 字段。因为请求的参数不同,返回的分页数据肯定也会有所改动。

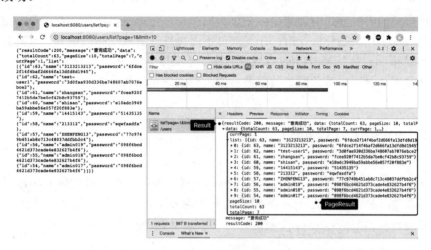

图 10-9 分页功能测试结果 1

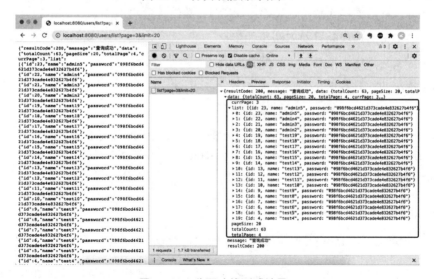

图 11-10 分页功能测试结果 2

读者可以结合该过程进行代码的学习和理解,并自行对比不同参数结果返回的不同。

10.5　jqGrid分页插件

接下来将会结合实际的页面和前端分页插件进行分页功能的效果展示。

这种分页的实现方式由后端提供分页数据接口，前端通过插件进行分页。由于分页插件已经实现了相关的分页逻辑，开发人员不用再进行前端代码的编写，只需要做一些配置工作即可。在接下来开发的商城项目中会把这种方式应用到后台系统的功能模块中。后台管理系统所有功能模块的分页功能采用的都是这种方式。

10.5.1　jqGrid 分页插件介绍

jqGrid 是一个用来显示网格数据的 jQuery 插件。开发人员通过使用 jqGrid 可以轻松实现前端页面与后台数据的 Ajax 异步通信并实现分页功能，其特点如下所示。

①兼容目前所有流行的 Web 浏览器

②拥有完善强大的分页功能

③支持多种数据格式解析，XML、JSON、数组等形式

④提供丰富的选项配置及方法事件接口

⑤支持表格排序、拖动列、隐藏列

⑥支持滚动加载数据

⑦开源免费

jqGrid 是一款代码开源的分页插件，源码也一直处于迭代更新的状态中。因此笔者选择 jqGrid 插件进行项目开发，jqGrid 在 GitHub 上的仓库主页为：

https://github.com/tonytomov/jqGrid。

jqGrid 的源文件下载地址如下：

https://github.com/tonytomov/jqGrid/releases

在浏览器中输入该地址就可以看到各个版本的 jqGrid 代码压缩包。如图 10-11 所示，选择其中一个版本的源码压缩包下载就可以在项目中使用了。本书所选择的 jqGrid 版本为 5.5.2。

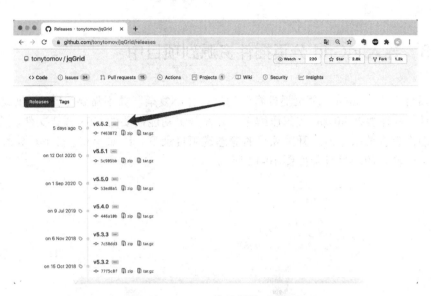

图 10-11　jqGrid 分页插件下载页面

将代码压缩包解压后可以看到 jqGrid 正式包的目录结构，其所有文件都在解压目录中。如果在项目中仅仅利用 jqGrid 实现分页这一个功能的话，并不需要把这些文件全部放到项目中，分页功能必要的文件如下所示：

```
## JS 文件
jquery.jqGrid.js
grid.locale-cn.js
jquery.jqGrid.min.js

## CSS 样式文件
ui.jqgrid-bootstrap-ui.css
ui.jqgrid-bootstrap.css
ui.jqgrid.css
```

以上列举的主要是 JS 文件和 CSS 样式文件，仅实现分页功能只需要笔者上述列举的几个文件即可。如果读者想使用 jqGrid 其他特性则需要引入其他对应的 JS 文件和 CSS 样式文件。

本书实战项目的所有模块的分页插件都是使用 jqGrid 插件实现的。它的分页功能十分强大，而且使用和学习起来都比较简单。jqGrid 还有其他优秀的特性，感兴趣的读者可以继续学习其相关知识并进行功能的优化升级。

10.5.3 导入 jqGrid 分页插件资源到项目中

在项目中导入 jqGrid 分页插件资源的过程并不复杂，并不需要配置 jar 包或者编写配置文件，只需要把 jqGrid 代码压缩包中开发分页功能需要的 CSS 样式文件、JS 文件、图片等静态资源放入当前所开发项目的静态资源目录下，比如 static 目录。导入 jqGrid 分页插件资源后的项目目录如图 10-12 所示。

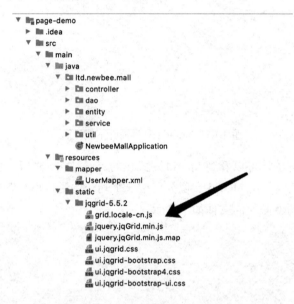

图 10-12 导入 jqGrid 分页插件后的代码目录

这样，就可以在各个需要分页的页面中引用这些静态资源文件并实现分页功能了。

10.5.3 使用 jqGrid 实现分页的步骤

首先，在前端页面代码中引入 jqGrid 分页插件所需的源文件，代码如下所示：

```
<link href="plugins/jqgrid-5.5.2/ui.jqgrid-bootstrap4.css" rel="stylesheet"/>
<!-- jqGrid 依赖 jQuery，因此需要先引入 jquery.min.js 文件，下方地址为字节跳动提供的 cdn 地址 -->
<script src="https://s3.pstatp.com/cdn/expire-1-M/jquery/3.3.1/jquery.min.
```

js"></script>

```
<!-- grid.locale-cn.js 为国际化所需的文件，-cn 表示中文 -->
<script src="plugins/jqgrid-5.5.2/grid.locale-cn.js"></script>
<script src="plugins/jqgrid-5.5.2/jquery.jqGrid.min.js"></script>
```

其次，在页面中需要展示分页数据的区域添加用于 jqGrid 初始化的代码：

```
<!-- jqGrid 必要 DOM,用于创建表格展示列表数据 -->
<table id="jqGrid" class="table table-bordered"></table>

<!-- jqGrid 必要 DOM,分页信息区域 -->
<div id="jqGridPager"></div>
```

最后，调用 jqGrid 分页插件的 jqGrid()方法渲染分页展示区域，代码如下所示：

```
$("#jqGrid").jqGrid({
    url: 'users/list',
    datatype: "json",
    colModel: [
        {label: 'id', name: 'id', index: 'id', width: 50, hidden: true, key: true},
        {label: '登录名', name: 'userName', index: 'userName', sortable: false, width: 80},
        {label: '添加时间', name: 'createTime', index: 'createTime', sortable: false, width: 80}
    ],
    height: 485,
    rowNum: 10,
    rowList: [10, 30, 50],
    styleUI: 'Bootstrap',
    loadtext: '信息读取中...',
    rownumbers: true,
    rownumWidth: 35,
    autowidth: true,
    multiselect: true,
    pager: "#jqGridPager",
    jsonReader: {
        root: "data.list",
        page: "data.currPage",
        total: "data.totalPage",
        records: "data.totalCount"
```

```
        },
        prmNames: {
            page: "page",
            rows: "limit",
            order: "order"
        },
        gridComplete: function () {
            //隐藏 grid 底部滚动条
            $("#jqGrid").closest(".ui-jqgrid-bdiv").css({"overflow-x": "hidden"});
        }
    });
```

jqGrid()方法中的参数及含义如表 10-1 所示。

表 10-1　jqGrid()方法中的参数及含义

参数名	类型	参数含义
url	string	请求的接口地址
datatype	string	从服务器端返回的数据类型，比如 JSON、XML 等
colModel	array	列表信息：表头、宽度、是否显示、渲染参数等属性
height	mixed	表格高度，可以是数字、像素值或者百分比，可自行调节
rowNum	int	默认一页显示多少条数据，可自行调节
rowList	array	在底部翻页控制条中，每页显示记录数可选集合
styleUI	string	在主题项目中选用的是 Bootstrap 主题
loadtext	string	数据在请求时显示的提示信息
rownumbers	boolean	是否显示行号，默认值是 false，不显示
rownumWidth	integer	行号列的宽度，如果 rownumbers 为 true，则可以设置 column 的宽度
autowidth	boolean	宽度是否能够自适应
multiselect	boolean	是否可以多选
pager	minxed	翻页导航栏对象，必须是一个有效的 HTML 元素
jsonReader	object	用于设置如何读取后端接口的返回数据
prmNames	object	在向后台请求时发送的参数

10.5.4 分页数据格式详解

在 jqGrid 整合中有如下代码：

```
jsonReader: {
  root: "data.list", //数据列表字段
  page: "data.currPage", //当前页码
  total: "data.totalPage", //数据总页码
  records: "data.totalCount" //数据总记录数
}
```

这里定义的是 jsonReader 对象如何对后端返回的 JSON 数据进行解析。比如数据列表为何读取 data.list，当前页码为何读取 data.currPage。这些都是由后端返回的数据格式所决定的，后端响应结果的数据格式定义在 ltd.newbee.mall.util 类中，如下所示：

```
public class Result<T> implements Serializable {
    //响应码 200 为成功
    private int resultCode;
    //响应 msg
    private String message;
    //返回数据
    private T data;
}
```

分页结果类的定义如下所示：

```
public class PageResult implements Serializable {

    //总记录数
    private int totalCount;
    //每页记录数
    private int pageSize;
    //总页数
    private int totalPage;
    //当前页数
    private int currPage;
    //列表数据
    private List<?> list;
}
```

由于jqGrid分页插件在实现分页功能时必须读取以下数据：当前页的所有数据列表、页码、总页码、总记录数量。所以在jqGrid读取时直接读取对应的参数即可。分页相关的数据都会被封装到PageResult中。因为后端统一的返回结果是将PageResult对象设置到Result结果类的data属性中的，所以在读取设置jsonReader时的数据列表root字段的值为data.list，当前页码的值为data.currPage，总页码的值为data.totalPage，数据总记录数的值为data.totalCount。

10.6 整合jqGrid实现分页功能

接下来将进行实际的编码，结合前文中实现的分页接口和jqGrid插件在一张前端页面中完成分页数据的展示和翻页功能。

10.6.1 前端页面制作

在resources/static目中新建jqgrid-page-test.html文件，代码如下所示：

```html
<!DOCTYPE html>
<html lang="en">
<head>
    <meta charset="UTF-8">
    <title>jqGrid分页测试</title>
    <!-- 引入bootstrap样式文件-->
    <link rel="stylesheet" href="bootstrap/css/bootstrap.css"/>
    <link href="jqgrid-5.5.2/ui.jqgrid-bootstrap4.css" rel="stylesheet"/>
</head>
<body>
<div style="margin: 24px;">
    <table id="jqGrid" class="table table-bordered">
    </table>
    <div id="jqGridPager"></div>
</div>
</body>
<!-- jqGrid依赖jQuery，因此需要先引入jquery.min.js文件，下方地址为字节跳动提供的cdn地址 -->
<script src="https://s3.pstatp.com/cdn/expire-1-M/jquery/3.3.1/jquery.min.js"></script>
<!-- grid.locale-cn.js为国际化所需的文件，-cn表示中文 -->
<script src="jqgrid-5.5.2/grid.locale-cn.js"></script>
```

```html
<script src="jqgrid-5.5.2/jquery.jqGrid.min.js"></script>
</html>
```

在页面中引入 jqGrid 分页插件的相关静态资源文件，并且在页面布局的分页数据展示区域增加如下代码：

```html
<!-- 数据展示列表，id 为 jqGrid-->
<table id="jqGrid" class="table table-bordered">
</table>

<!-- 分页按钮展示区 -->
<div id="jqGridPager"></div>
```

首先是一个 id 为 jqGrid 的 table 标签，它主要是为了显示分页数据，包括表头、数据、选择器等。然后是一个 id 为 jqGridPager 的 div 标签，它主要是用来显示 jqGrid 插件生成的分页按钮，包括前一页、后一页、跳页等操作按钮。

10.6.2　jqGrid 初始化

在 resources/static 目录下新建 jqgrid-page-test.js 文件，代码如下所示：

```javascript
$(function () {
    $("#jqGrid").jqGrid({
        url: 'users/list',
        datatype: "json",
        colModel: [
            {label: 'id', name: 'id', index: 'id', width: 50, hidden: true, key: true},
            {label: '登录名', name: 'name', index: 'name', sortable: false, width: 80},
            {label: '密码字段', name: 'password', index: 'password', sortable: false, width: 80}
        ],
        height: 485,
        rowNum: 10,
        rowList: [10, 30, 50],
        styleUI: 'Bootstrap',
        loadtext: '信息读取中...',
        rownumbers: true,
```

```
        rownumWidth: 35,
        autowidth: true,
        multiselect: true,
        pager: "#jqGridPager",
        jsonReader: {
            root: "data.list",
            page: "data.currPage",
            total: "data.totalPage",
            records: "data.totalCount"
        },
        prmNames: {
            page: "page",
            rows: "limit",
            order: "order"
        },
        gridComplete: function () {
            //隐藏 grid 底部滚动条
            $("#jqGrid").closest(".ui-jqgrid-bdiv").css({"overflow-x": "hidden"});
        }
    });

    $(window).resize(function () {
        $("#jqGrid").setGridWidth($(".card-body").width());
    });
});
```

该代码的含义：在页面加载时，首先调用 jqGrid 的初始化方法，将页面中 id 为 jqGrid 的 DOM 渲染为分页表格，并向后端发送请求，然后按照后端返回的 JSON 数据填充表格和表格下方的分页按钮。第一页、下一页、最后一页等逻辑都由 jqGrid 插件内部实现了，开发人员只需要将在初始化时所需要的几个数据设置好即可。最后修改 jqgrid-page-test.html 代码，引入 jqgrid-page-test.js 文件，代码如下所示：

```html
<script src="jqgrid-page-test.js"></script>
```

由于在 jqgrid-page-test.html 文件中引入了 jqgrid-page-test.js 文件，所以在页面加载完成后会进行数据列表的渲染及分页插件的渲染。用户可以直接使用翻页功能。本来这些功能需要开发人员自行实现，但是在使用 jqGrid 后这些都不需要再做逻辑实现，只需要调用其分页方法并将所需的参数设置好即可，十分方便。

10.6.3　整合 jqGrid 实现分页功能测试

在编码完成后启动 Spring Boot 项目，打开浏览器，在地址栏输入如下地址：
`http://localhost:8080/jqgrid-page-test.html`

jqgrid-page-test.js 代码中已经定义了在$(function () {})方法中执行 jqGrid 插件初始化的操作。在 jqgrid-page-test.html 页面的 DOM 文档加载完成后程序就会执行 jqGrid 的初始化操作，在打开页面就能够看到分页数据的展示效果，如图 10-13 所示。

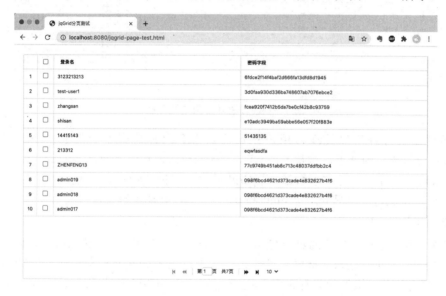

图 10-13　jqGrid 分页功能测试结果

在分页区域正确渲染出分页数据和翻页按钮之后，可以点击下方的翻页按钮进行测试。这里可以看到所有功能都正常，列表中展示的内容是登录名和密码两个字段。

接着笔者会结合页面上的按钮点击和接口请求的信息，对分页功能的实现进行具体的分析。

打开浏览器的开发人员控制台，并进入 Network 面板，之后点击页面下方的翻页按钮改变页码或者点击每页展示的数据条数，可以看到在 NetWork 面板中的多次请求数据，如图 10-14 所示。

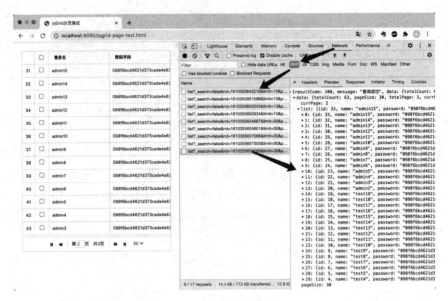

图 10-14　jqGrid 分页功能分析

通过调试过程可以发现，用户在每次点击分页按钮时都会发起一条向后端的 users/list 分页接口的请求。这是一个 GET 方式的请求，请求参数是_search、nd、limit、page、sidx 四个字段。这些参数都是 jqGrid 插件内部封装好的，默认会有这四个字段。后端接口在接受请求时只处理了 limit 和 page 参数，其他参数可以自行增加处理逻辑，也可以不作处理。

此外，翻页逻辑和参数的封装在 jqGrid 内部都已经实现，比如第一次请求以每页 10 条的数据向后端进行请求。此时总页数共有 7 页，当每页展示条数改为每页 30 条数据时，分页插件中的总页数就只有 3 页了。这个逻辑也由 jqGrid 插件实现。因此，Spring Boot 项目在整合该插件后，只需要处理后端逻辑即可，前端相关的分页逻辑都由该插件来完成，大大减少了开发人员的工作量。

第 11 章

Spring Boot 文件上传功能的实现

文件上传是被用户熟知的常见功能模块，其常用场景有头像设置、产品预览图、报表文件上传等。本章将结合实际案例讲解如何使用 Spring Boot 实现文件上传及其相关的注意事项，并结合源码对文件上传的流程及功能设计进行讲解。

11.1　Spring MVC处理文件上传的源码分析

笔者先通过源码分析具体讲解一下 Spring MVC 是如何进行文件上传处理的，包括源码调用过程和 Spring MVC 文件处理代码的解析。

11.1.1　文件上传功能源码调用链

利用 Spring MVC 实现文件上传功能，离不开对 MultipartResolver 接口的设置，该接口有两个实现类，分别是 CommonsMultipartResolver 和 StandardServletMultipartResolver。

MultipartResolver 接口的实现类，读者可以将其视为 Spring MVC 在实现文件上传功能时的工具类，该类只会在文件上传中发挥作用。接下来笔者会以 StandardServletMultipartResolver 实现类为例，讲解在文件上传流程中的源码调用。

一旦有请求被 DispatcherServlet 类处理，DispatcherServlet 类就会先调用在 MultipartResolver 实现类中的方法判断此请求是不是文件上传请求。如果是文件上传请求，DispatcherServlet 类会调用在 MultipartResolver 实现类中的 resolveMultipart(request)

方法对该请求对象进行装饰并返回一个新的 MultipartHttpServletRequest 对象供后续处理流程使用。

注意，此时的请求对象会由 HttpServletRequest 类型转换成 MultipartHttpServletRequest 类型或者 MultipartResolver 实现类。在 MultipartResolver 实现类中会包含所上传的文件对象，可供后续流程直接使用，无须开发人员在代码中自行实现对文件内容的读取逻辑。

由此，笔者绘制了 MultipartResolver 实现类的调用时序图，如图 11-1 所示。

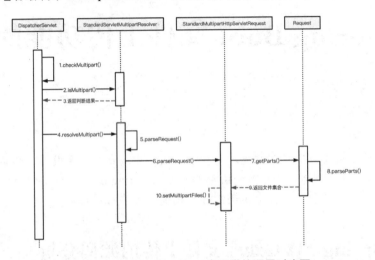

图 11-1　MultipartResolver 实现类的调用时序图

由图 11-1 可知，当收到请求时，在 DispatcherServlet 类中的 checkMultipart()方法会调用 MultipartResolver 实现类的 isMultipart()方法判断请求中是否包含文件。如果请求数据中包含文件，则首先调用 MultipartResolver 实现类的 resolveMultipart()方法对请求的数据进行解析，然后将文件数据解析成 MultipartFile 并封装在 MultipartHttpServletRequest（继承了 HttpServletRequest）对象中，最后传递给 Controller 控制器。

11.1.2　文件上传功能源码分析

DispatcherServlet 类的 checkMultipart()方法源码如下所示：

```
protected HttpServletRequest checkMultipart(HttpServletRequest request)
throws MultipartException {
    // 1.判断是否包含文件
    if (this.multipartResolver != null && this.multipartResolver.isMultipart
(request)) {
```

```java
        if (WebUtils.getNativeRequest(request, MultipartHttpServletRequest.class) != null) {
            if (request.getDispatcherType().equals(DispatcherType.REQUEST)) {
            logger.trace("Request already resolved to MultipartHttpServletRequest, e.g. by MultipartFilter");
            }
        }
        else if (hasMultipartException(request)) {
            logger.debug("Multipart resolution previously failed for current request - " +
                    "skipping re-resolution for undisturbed error rendering");
        }
        else {
            try {
                // 2.将文件对象封装到 Request 中
                return this.multipartResolver.resolveMultipart(request);
            }
            catch (MultipartException ex) {
                if (request.getAttribute(WebUtils.ERROR_EXCEPTION_ATTRIBUTE) != null) {
                    logger.debug("Multipart resolution failed for error dispatch", ex);
                }
                else {
                    throw ex;
                }
            }
        }
    }
    return request;
}
```

checkMultipart()方法的执行逻辑如下所示。

①先分析判断 HttpServletRequest 请求对象中是否包含文件信息。

②如果包含文件信息,则调用相应的方法将文件对象封装到 HttpServletRequest 对象中。

其中,this.multipartResolver.isMultipart(request)调用在 StandardServletMultipartResolver 类中的 isMultipart()方法来判断是否包含文件。isMultipart()方法的源码如下所示:

```java
public boolean isMultipart(HttpServletRequest request) {
return StringUtils.startsWithIgnoreCase(request.getContentType(),"multipart/");
}
```

isMultipart()方法的实现非常简单：对请求头中的 contentType 对象进行判断，如果请求头中的 contentType 不为空且 contentType 的值以 multipart/开头，此时会返回 true，否则不会将这次请求标示为文件上传请求。

在返回 true 后，表明在此次请求中含有文件。接下来 DispatcherServlet 将会调用 resolveMultipart(request)重新封装 Request 对象。实际调用的是 StandardServletMultipartResolver 类的 resolveMultipart()方法，源码如下所示：

```java
public MultipartHttpServletRequest resolveMultipart(HttpServletRequest request) throws MultipartException {
    return new StandardMultipartHttpServletRequest(request, this.resolveLazily);
}
```

跟踪源码调用链，可查出最终调用的源码如下所示：

```java
public StandardMultipartHttpServletRequest(HttpServletRequest request, boolean lazyParsing)
    throws MultipartException {

    super(request);
    if (!lazyParsing) {
        parseRequest(request);
    }
}


private void parseRequest(HttpServletRequest request) {
    try {
        Collection<Part> parts = request.getParts();
        this.multipartParameterNames = new LinkedHashSet<>(parts.size());
        MultiValueMap<String, MultipartFile> files = new LinkedMultiValueMap<>(parts.size());
        for (Part part : parts) {
            String headerValue = part.getHeader(HttpHeaders.CONTENT_DISPOSITION);
            ContentDisposition disposition = ContentDisposition.parse(headerValue);
            String filename = disposition.getFilename();
            if (filename != null) {
                if (filename.startsWith("=?") && filename.endsWith("?=")) {
                    filename = MimeDelegate.decode(filename);
                }
                files.add(part.getName(), new StandardMultipartFile(part, filename));
            }
```

```
        else {
            this.multipartParameterNames.add(part.getName());
        }
    }
    setMultipartFiles(files);
}
catch (Throwable ex) {
    handleParseFailure(ex);
}
}
```

由上面代码可以看出，该方法对请求对象中的文件参数进行解析和处理。文件解析最终的实现方法是 Request 类中的 getParts()方法。通过阅读源码可知，resolveMultipart() 方法最终会得到一个 StandardMultipartHttpServletRequest 对象。该对象是 MultipartHttpServletRequest 接口类的一个实现类，同时该类中含有已经解析的文件对象。通过以上步骤程序就可以在具体的 Controller 类中直接使用文件对象，而不用开发人员自行实现文件对象的解析代码。

11.1.3 Spring Boot 中 MultipartResolver 的自动配置

在 Spring Boot 项目中 MultipartResolver 的自动配置类为 MultipartAutoConfiguration，其源码如下所示：

```
@Configuration(proxyBeanMethods = false)
@ConditionalOnClass({ Servlet.class, StandardServletMultipartResolver.
class, MultipartConfigElement.class })
@ConditionalOnProperty(prefix = "spring.servlet.multipart", name = "enabled",
matchIfMissing = true)
@ConditionalOnWebApplication(type = Type.SERVLET)
@EnableConfigurationProperties(MultipartProperties.class)
public class MultipartAutoConfiguration {

    private final MultipartProperties multipartProperties;

    public MultipartAutoConfiguration(MultipartProperties multipart
Properties) {
        this.multipartProperties = multipartProperties;
    }
```

```java
    @Bean
    @ConditionalOnMissingBean({ MultipartConfigElement.class,
CommonsMultipartResolver.class })
    public MultipartConfigElement multipartConfigElement() {
        return this.multipartProperties.createMultipartConfig();
    }

    // 注册一个名称为multipartResolver的Bean到Spring IOC容器中,类型为
StandardServletMultipartResolver
    @Bean(name = DispatcherServlet.MULTIPART_RESOLVER_BEAN_NAME)
    @ConditionalOnMissingBean(MultipartResolver.class)
    public StandardServletMultipartResolver multipartResolver() {
        StandardServletMultipartResolver multipartResolver = new Standard
ServletMultipartResolver();
        multipartResolver.setResolveLazily(this.multipartProperties.is
ResolveLazily());
        return multipartResolver;
    }

}
```

MultipartAutoConfiguration 类的注解释义如下所示。

@Configuration(proxyBeanMethods = false)：指定该类为配置类。

@ConditionalOnWebApplication(type = Type.SERVLET)：当前应用是一个 Servlet Web 应用的时候，这个配置类才生效。

@ConditionalOnClass({ Servlet.class, StandardServletMultipartResolver.class, MultipartConfigElement.class })：判断当前 classpath 是否存在 Servlet 类、StandardServletMultipartResolver 类、MultipartConfigElement 类，存在则生效。

@ConditionalOnProperty(prefix = "spring.servlet.multipart", name = "enabled", matchIfMissing = true)：根据配置文件的 spring.servlet.multipart 配置项判断是否生效，该配置项默认为 true，即默认允许文件上传。

通过源码可知，MultipartAutoConfiguration 自动配置类的自动配置触发条件：当前项目类型必须为 SERVLET；当前 classpath 存在 Servlet 类、StandardServletMultipartResolver 类和 MultipartConfigElement 类（因为当前项目引入了 spring-boot-starter-web 场景启动器，以上条件都满足）；spring.servlet.multipart 配置项默认为 true，且在配置文件中并没有将该配置项设置为 false。

在以上三个条件都满足时，MultipartAutoConfiguration 类就会开始进行自动配置的

流程，并注册一个名称为 multipartResolver 的 Bean 到 Spring IOC 容器中，其类型为 StandardServletMultipartResolver。这样，在 Spring Boot 项目中就可以进行文件上传操作了

注意事项：

①在阅读源码时，可以结合书中提供的调用时序图查看。

②此次展示在文章中的源码，对应的版本号分别为 spring-boot-2.3.7.RELEASE 和 spring-5.2.12.RELEASE，其他版本的源码与书中列出的代码可能有微小的差别。

③在 Spring Boot 2.3.7.RELEASE 版本中默认的文件处理类为 StandardServletMultipartResolver，因此文中的源码都与该类有关。CommonsMultipartResolver 类也是 MultipartResolver 接口类的实现类，同样可以对文件进行处理，只是需要引入其他的依赖并进行 spring.servlet.multipart 的配置，感兴趣的读者可以尝试一下。

11.2　Spring Boot文件上传功能的实现案例

这一节将会讲解 Spring Boot 文件上传功能的一个实现案例。

11.2.1　Spring Boot 文件上传配置项

由于 Spring Boot 自动配置机制的存在，开发人员在 Spring Boot 项目中开发文件上传功能时并不需要进行多余的设置。只要在 pom.xml 文件中引入 spring-boot-starter-web 依赖即可直接调用文件上传功能。虽然不用配置也可以使用文件上传功能，但是开发人员在文件上传时也可能有一些特殊的需求。因此，这里需要对 Spring Boot 中 MultipartFile 的常用设置进行介绍，其配置项和默认值如图 11-2 所示。

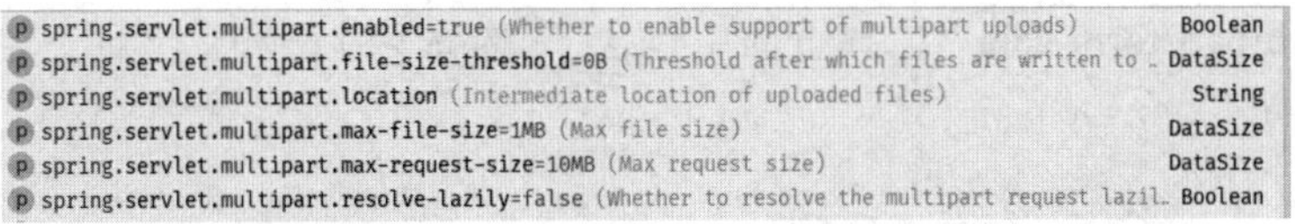

```
spring.servlet.multipart.enabled=true (Whether to enable support of multipart uploads)    Boolean
spring.servlet.multipart.file-size-threshold=0B (Threshold after which files are written to ... DataSize
spring.servlet.multipart.location (Intermediate location of uploaded files)                String
spring.servlet.multipart.max-file-size=1MB (Max file size)                                 DataSize
spring.servlet.multipart.max-request-size=10MB (Max request size)                          DataSize
spring.servlet.multipart.resolve-lazily=false (Whether to resolve the multipart request lazil. Boolean
```

图 11-2　文件上传的配置项和默认值

配置项目释义如下所示。

spring.servlet.multipart.enabled=true：是否支持 multipart 上传文件，默认支持。

spring.servlet.multipart.file-size-threshold=0B：文件大小阈值，当大于这个阈值时将

写入磁盘，否则存在内存中，（默认值 0，一般情况下不用特意修改）。

spring.servlet.multipart.location：上传文件的临时目录。

spring.servlet.multipart.max-file-size=1MB：最大支持的文件大小，默认 1MB，该值可适当调整。

spring.servlet.multipart.max-request-size=10MB：最大支持的请求大小，默认 10MB。

spring.servlet.multipart.resolve-lazily=false：判断是否要延迟解析文件（相当于懒加载，一般情况下不用特意修改）。

11.2.2 新建文件上传页面

在 static 目录中新建 upload-test.html，上传页面代码如下所示：

```html
<!DOCTYPE html>
<html lang="en">
<head>
    <meta charset="UTF-8">
    <title>Spring Boot 文件上传测试</title>
</head>
<body>
<form action="/uploadFile" method="post" enctype="multipart/form-data">
    <input type="file" name="file" />
    <input type="submit" value="文件上传" />
</form>
</body>
</html>
```

读者应该都很熟悉该文件上传页面的 demo，后端处理文件上传的请求地址为 /uploadFile，请求方法为 POST。需要注意的是在文件上传时需要设置 form 表单的 enctype 属性为 "multipart/form-data"。该页面中包含 "选择文件" 和 "文件上传" 的按钮，如图 11-3 所示。

图 11-3 upload-test.html 页面效果

11.2.3 新建文件上传处理 Controller 类

在 controller 包下新建 UploadController 类并编写实际的文件上传逻辑代码,如下所示:

```java
package ltd.newbee.mall.controller;

import org.springframework.stereotype.Controller;
import org.springframework.web.bind.annotation.RequestMapping;
import org.springframework.web.bind.annotation.RequestMethod;
import org.springframework.web.bind.annotation.RequestParam;
import org.springframework.web.bind.annotation.ResponseBody;
import org.springframework.web.multipart.MultipartFile;

import java.io.IOException;
import java.nio.file.Files;
import java.nio.file.Path;
import java.nio.file.Paths;
import java.text.SimpleDateFormat;
import java.util.Date;
import java.util.Random;

@Controller
public class UploadController {
    // 文件保存路径为在 D 盘下的 upload 文件夹,可以按照自己的习惯来修改
    private final static String FILE_UPLOAD_PATH = "D:\\upload\\";
    @RequestMapping(value = "/uploadFile", method = RequestMethod.POST)
    @ResponseBody
    public String uploadFile(@RequestParam("file") MultipartFile file) {
        if (file.isEmpty()) {
            return "上传失败";
        }
        String fileName = file.getOriginalFilename();
        String suffixName = fileName.substring(fileName.lastIndexOf("."));
        //生成文件名称通用方法
        SimpleDateFormat sdf = new SimpleDateFormat("yyyyMMdd_HHmmss");
        Random r = new Random();
        StringBuilder tempName = new StringBuilder();
        tempName.append(sdf.format(new Date())).append(r.nextInt(100)).append(suffixName);
```

```
        String newFileName = tempName.toString();
    try {
        // 保存文件
        byte[] bytes = file.getBytes();
        Path path = Paths.get(FILE_UPLOAD_PATH + newFileName);
        Files.write(path, bytes);
    } catch (IOException e) {
        e.printStackTrace();
    }
    return "上传成功";
    }
}
```

由于 Spring Boot 已经自动配置了 StandardServletMultipartResolver 类来处理文件上传请求，因此能够直接在控制器方法中使用 MultipartFile 读取文件信息。@RequestParam 中文件名称属性需要与前端页面中 input 文件输入框设置的 name 属性一致。如果文件为空则返回上传失败，如果不为空则根据日期生成一个新的文件名，读取文件流程并写入指定的路径中，最后返回上传成功的提示信息。

需要注意的是文件上传路径的设置。在上述代码中设置的文件保存路径为 D:\upload\，即在 D 盘下的 upload 文件夹。当然，开发人员可以按照自己的习惯修改为其他目录名称。D:\upload\这种写法是在 Windows 系统下的路径写法，如果项目部署在 Linux 系统中的话，写法与此不同。比如想要把文件上传到/opt/newbee/upload 目录下，就需要把路径设置的代码改为 private final static String FILE_UPLOAD_PATH = "/opt/newbee/upload/"。这一点需要读者注意，两种系统的写法存在一些差异。

回到本次文件上传测试，如果文件存储目录还没有创建的话，首先需要创建该目录，然后启动项目进行文件上传测试。

11.2.5 文件上传功能测试

在编码完成后，启动 Spring Boot 项目。在启动成功后，打开浏览器并输入测试页面地址：

```
http://localhost:8080/upload-test.html
```

首先在该页面点击"选择文件"按钮，然后选择需要上传的文件，这里选择了一个名称为 WX20210114-204745.png 的图片文件，如图 11-3 所示。

第 11 章 Spring Boot 文件上传功能的实现

图 11-3 选择文件

在文件选择完成后点击 "文件上传" 按钮，等待后端业务处理并返回结果，最终的页面效果如图 11-4 所示。

图 11-4 上传成功

首先该页面中可以看到上传成功的提示信息，然后确认文件是否已经上传到设定的文件目录中，即查看在 upload 目录下是否存在该文件，如果存在则功能正常实现，文件上传测试完成！

另外，在 Spring Boot 项目中支持单个文件的最大值默认为 1MB，支持单个请求最大值默认 10 MB。如果选择了大于默认值的文件进行上传，比如一个 1.2MB 的文件或者一个 11MB 的请求，会分别报出如下两种错误。

（1）org.apache.tomcat.util.http.fileupload.impl.FileSizeLimitExceededException：

```
2021-01-12 12:17:31.785 ERROR 62249 --- [nio-8080-exec-7] o.a.c.c.C.
[.[.[/].[dispatcherServlet]      : Servlet.service() for servlet [dispatcher
Servlet] in context with path [] threw exception [Request processing failed;
nested exception is org.springframework.web.multipart.MaxUploadSize
ExceededException: Maximum upload size exceeded; nested exception is
java.lang.IllegalStateException: org.apache.tomcat.util.http.fileupload.
impl.FileSizeLimitExceededException: The field file exceeds its maximum
permitted size of 1048576 bytes.] with root cause

... 省略部分日志
```

（2）org.apache.tomcat.util.http.fileupload.impl.SizeLimitExceededException：

```
org.apache.tomcat.util.http.fileupload.impl.SizeLimitExceededException:
the request was rejected because its size (13262412) exceeds the configured
maximum (10485760)
```

> …省略部分日志

以上异常分别表示表示单个文件大小超出设定值以及请求的大小超出设定值。此时就需要调整文件上传的配置项来避免这种异常信息的产生。文件上传的配置项可参考 11.2.1 节的内容。

11.3 Spring Boot文件上传路径回显

很多 Spring Boot 文件上传教程通常只讲如何实现文件上传功能,但是在具体的业务实现中,流程不止于文件上传成功。比如图片上传,在完成文件上传后,还需要知道访问该文件的访问路径,并且,最好能在页面中直接查看文件上传成功后的回显效果。

Spring Boot 项目与普通 Spring 项目的目录结构不同,并没有 webapp 目录,因此无法与普通的 Java Web 项目一样,上传文件到 webapp 目录中并直接根据目录进行访问。Spring Boot 项目中通常使用自定义静态资源映射目录,以此来实现文件上传整个流程的闭环。比如在前文文件上传案例中,在文件上传到 upload 目录后,会增加一个自定义静态资源映射配置,使得在 upload 下的静态资源可以通过该映射地址被访问到。

新建 config 包并在 config 包中新增 SpringBootWebMvcConfigurer 类,实现代码如下所示:

```java
package ltd.newbee.mall.config;

import org.springframework.context.annotation.Configuration;
import org.springframework.web.servlet.config.annotation.ResourceHandlerRegistry;
import org.springframework.web.servlet.config.annotation.WebMvcConfigurer;

@Configuration
public class NeeBeeMallWebMvcConfigurer implements WebMvcConfigurer {

    public void addResourceHandlers(ResourceHandlerRegistry registry) {
        registry.addResourceHandler("/upload/**").addResourceLocations("file:D:\\upload\\");
    }
}
```

通过以上代码配置,所有以 "/upload/" 开头的静态资源在请求时都会映射到 D 盘的 upload 目录下。路径的设置与前文中上传文件的设置目录类似,不同系统的文件路径的写法不同（比如 Linux 和 Windows）。同时需要注意在设置静态资源映射路径时，路径前需要添加"file:"前缀。

最后修改一下在文件上传时的返回信息，把路径拼装好并返回到页面上，以便于功能测试，UploadController 代码的修改如下所示：

```
return "上传成功，地址为：/upload/" + newFileName;
```

在编码完成后，启动 Spring Boot 项目测试一下上传的文件能否被访问到。在项目启动成功后进行文件上传，返回结果如图 11-5 所示。

图 11-5　upload-test.html 图片上传返回结果

那么，后端已经返回文件的访问地址就是/upload/20210113_14205625.jpg。接下来，根据后端返回的地址访问这张图片，请求地址为：

```
http://localhost:8080/upload/20210113_14205625.jpg
```

返回结果如图 11-6 所示。

图 11-6　图片上传成功后的回显效果

能够正常上传文件并且访问上传后的文件说明使用 Spring Boot 实现文件上传和回显的案例就完成了。

11.4 Spring Boot多文件上传功能的实现

前文讲解的是单文件上传功能，在某些业务中也会遇到多文件上传的情况。那么，在 Spring Boot 项目中如何处理多文件上传呢？

接下来，通过两个实际的案例来讲解，同时结合实际的编码，让读者也可以上手体验多文件上传功能。

11.4.1 文件名相同时的多文件上传处理

首先，在 static 目录中新建 upload-same-file-name.html，页面代码如下所示：

```html
<!DOCTYPE html>
<html lang="en">
<head>
    <meta charset="UTF-8">
    <title>Spring Boot 多文件上传测试（file name 相同）</title>
</head>
<body>
<form action="/uploadFilesBySameName" method="post" enctype="multipart/form-data">
    <input type="file" name="files"/><br><br>
    <input type="file" name="files"/><br><br>
    <input type="file" name="files"/><br><br>
    <input type="file" name="files"/><br><br>
    <input type="file" name="files"/><br><br>
    <input type="submit" value="文件上传"/>
</form>
</body>
</html>
```

多文件上传页面与单文件上传页面类似，不同点是新增了 4 个文件输入框，文件输入框的 name 属性统一命名为 files，文件名完全一致。后端处理文件上传的请求地址为 /uploadFilesBySameName，请求方法为 POST，页面如图 11-7 所示。

图 11-7 upload-same-file-name.html 页面显示效果

在 UploadController 类中新增 uploadFilesBySameName()方法，用于处理在文件名相同时的多文件上传问题，新增代码如下所示：

```java
@RequestMapping(value = "/uploadFilesBySameName", method = RequestMethod.POST)
@ResponseBody
public String uploadFilesBySameName(@RequestPart MultipartFile[] files) {
  if (files == null || files.length == 0) {
    return "参数异常";
  }
  if (files.length > 5) {
    return "最多上传5个文件";
  }
  String uploadResult = "上传成功，地址为：<br>";
  for (MultipartFile file : files) {
    String fileName = file.getOriginalFilename();
    if (StringUtils.isEmpty(fileName)) {
      //表示无文件信息，跳出当前循环
      continue;
    }
    String suffixName = fileName.substring(fileName.lastIndexOf("."));
    //生成文件名称的通用方法
    SimpleDateFormat sdf = new SimpleDateFormat("yyyyMMdd_HHmmss");
    Random r = new Random();
    StringBuilder tempName = new StringBuilder();
    tempName.append(sdf.format(new Date())).append(r.nextInt(100)).append(suffixName);
    String newFileName = tempName.toString();
    try {
      // 保存文件
      byte[] bytes = file.getBytes();
      Path path = Paths.get(FILE_UPLOAD_PATH + newFileName);
      Files.write(path, bytes);
      uploadResult += "/upload/" + newFileName + "<br>";
    } catch (IOException e) {
```

```
        e.printStackTrace();
    }
}
return uploadResult;
}
```

与单文件在上传时的 uploadFile()方法相比，多文件上传有两处改动。

第一，文件参数在接收时的代码改动。在多文件上传并接收参数时使用的是 @RequestPart 注解，且接收的文件参数是一个数组 MultipartFile。而单文件在上传时使用的是@RequestParam 注解，接收的文件是单个对象。

第二，文件在保存时增加循环逻辑。多文件保存的处理方式与单文件在上传时比较类似，只是增加了循环逻辑，对接收的 MultipartFile 数组中每一个文件进行存储操作，最后拼接文件的地址信息并返回。

另外一个需要注意的知识点，多文件上传在接收参数时，参数名称 files 需要完全对应 input 框中的 name 属性。比如本次演示，在 upload-same-file-name.html 文件中所有文件输入框的 name 属性都是 files；在后端处理时，uploadFilesBySameName()方法的参数名称也定义为 files，两个名称是对应的。如果所有文件输入框的 name 属性都改为 uploadFiles，那么 uploadFilesBySameName()方法的参数名称也需要改为 uploadFiles，否则接收的文件对象数组为空。

在编码完成后，启动 Spring Boot 项目。在在启动成功后打开浏览器并输入多图上传的测试页面地址：

```
http://localhost:8080/upload-same-file-name.html
```

在该页面依次点击"选择文件"按钮，并选择多张需要上传的图片，如图 11-8 所示。

图 11-8　选择多张需要上传的图片

首先在文件选择完成后点击"文件上传"按钮，然后等待后端业务处理并返回结果，页面效果如图 11-9 所示。

第 11 章 Spring Boot 文件上传功能的实现

图 11-9　upload-same-file-name.html 多图上传成功的返回结果

在结果页面中可以看到上传成功的提示以及所有上传文件的访问地址。接下来需要确认文件是否已经上传到设定的文件目录中,并测试通过后端返回的地址是否可以访问已经上传的文件,这里以/upload/20210113_14225118.png 为例进行请求测试,请求地址如下:

```
http://localhost:8080/upload/20210113_14225118.png
```

得到的结果如图 11-10 所示。

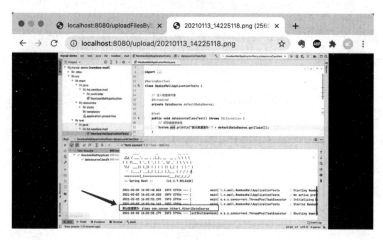

图 11-10　图片上传成功后的回显效果

至此,多文件上传功能测试完成!

11.4.2　文件名不同时的多文件上传处理

文件名相同时的处理逻辑已经实现完毕,如果这些文件的 name 属性不相同又该如何处理呢?

这种情况在企业项目开发中也会遇到，本书实战商城项目就有这个需求。在商城后台管理系统中集成了 wangEditor 富文本编辑器，该编辑器在进行多图上传时所使用的 input 框中的 name 属性就完全不同。比如上传 3 张图片，这 3 张图片对应 input 框分别为 files1、files2、files3，请求中的内容如图 11-11 所示。

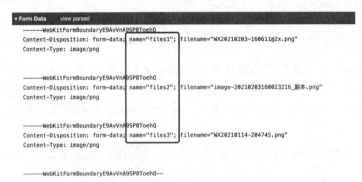

图 11-11　wangEditor 富文本编辑器在多图上传时的 name 属性

在这种情况下，因为无法正常读取参数，就无法使用 11.4.1 节中的代码进行多图文件上传的处理方式。接下来，就以实际的代码案例来讲解该如何处理文件名不同的多文件上传。

首先，在 static 目录中新建 upload-different-file-name.html，页面代码如下所示：

```html
<!DOCTYPE html>
<html lang="en">
<head>
    <meta charset="UTF-8">
    <title>Spring Boot 多文件上传测试（file name 不同）</title>
</head>
<body>
<form action="/uploadFilesByDifferentName" method="post" enctype="multipart/form-data">
    <input type="file" name="file1"/><br><br>
    <input type="file" name="file2"/><br><br>
    <input type="file" name="file3"/><br><br>
    <input type="file" name="file4"/><br><br>
    <input type="file" name="file5"/><br><br>
    <input type="submit" value="文件上传"/>
</form>
</body>
</html>
```

与upload-same-file-name.html页面代码相似,这里只是各个input文件输入框的name属性不同。后端处理文件上传的请求地址修改为/uploadFilesByDifferentName,请求方法为POST,页面如图11-12所示。

图11-12　upload-different-file-name.html页面效果

然后,在UploadController类中新增uploadFilesByDifferentName()方法,用于处理文件名在不相同时的多文件上传,新增代码如下所示:

```
@RequestMapping(value = "/uploadFilesByDifferentName", method = RequestMethod.POST)
@ResponseBody
public String uploadFilesByDifferentName(HttpServletRequest httpServletRequest) {
  List<MultipartFile> multipartFiles = new ArrayList<>(8);
  // 如果不是文件上传请求则不处理
  if (!standardServletMultipartResolver.isMultipart(httpServletRequest)) {
    return "请选择文件";
  }
  // 将HttpServletRequest对象转换为MultipartHttpServletRequest对象,并读取文件
  MultipartHttpServletRequest multiRequest = (MultipartHttpServletRequest) httpServletRequest;
  Iterator<String> iter = multiRequest.getFileNames();
  int total = 0;
  while (iter.hasNext()) {
    if (total > 5) {
      return "最多上传5个文件";
    }
    total += 1;
    MultipartFile file = multiRequest.getFile(iter.next());
    multipartFiles.add(file);
  }
  if (CollectionUtils.isEmpty(multipartFiles)) {
    return "请选择文件";
```

```
    }
    if (multipartFiles != null && multipartFiles.size() > 5) {
      return "最多上传5个文件";
    }
    String uploadResult = "上传成功,地址为: <br>";
    for (int i = 0; i < multipartFiles.size(); i++) {
      String fileName = multipartFiles.get(i).getOriginalFilename();
      if (StringUtils.isEmpty(fileName)) {
        //表示无文件信息,跳出当前循环
        continue;
      }
      String suffixName = fileName.substring(fileName.lastIndexOf("."));
      //生成文件名称的通用方法
      SimpleDateFormat sdf = new SimpleDateFormat("yyyyMMdd_HHmmss");
      Random r = new Random();
      StringBuilder tempName = new StringBuilder();
      tempName.append(sdf.format(new Date())).append(r.nextInt(100)).append(suffixName);
      String newFileName = tempName.toString();
      try {
        // 保存文件
        byte[] bytes = multipartFiles.get(i).getBytes();
        Path path = Paths.get(FILE_UPLOAD_PATH + newFileName);
        Files.write(path, bytes);
        uploadResult += "/upload/" + newFileName + "<br>";
      } catch (IOException e) {
        e.printStackTrace();
      }
    }
    return uploadResult;
}
```

　　与文件名相同时的多文件上传的 uploadFilesBySameName()方法相比,文件名不同时的改动只有一处,即文件参数在接收时的代码做了改动。在读取文件信息时的逻辑是自行实现的代码逻辑,首先调用 isMultipart()方法判断当前请求是否为文件上传请求,如果不是则不进行处理,如果是文件上传请求,则 HttpServletReques 对象转换为 MultipartHttpServletRequest 对象,并读取文件数据,在读取完成后再依次进行存储。存储文件的过程与之前的逻辑一致,最后拼接文件的地址信息并返回。

　　另外,需要判断当前请求是否为文件上传请求时要用到 StandardServletMultipartResolver 类 isMultipart()方法。前文中提到的 StandardServletMultipartResolver 类已经自动配置,所以可以直接在 UploadController 类中使用@Autowired 注解注入,代码如下

第 11 章　Spring Boot 文件上传功能的实现

所示：
```
@Autowired
private StandardServletMultipartResolver standardServletMultipartResolver;
```

在编码完成后，启动 Spring Boot 项目。在启动成功后打开浏览器并输入多图上传的测试页面地址：

```
http://localhost:8080/upload-different-file-name.html
```

在该页面依次点击"选择文件"按钮并选择多张需要上传的图片，如图 11-13 所示。

首先在文件选择完成后点击"文件上传"按钮，然后等待后端业务处理并返回结果，页面效果如图 11-14 所示。

图 11-13　选择多张需要上传的图片

图 11-14　upload-different-file-name.html 多图上传成功的返回结果

在结果页面中可以看到上传成功的提示以及所有上传文件的访问地址。接下来需要确认文件是否已经上传到设定的文件目录中并测试通过后端返回的地址是否可以访问已经上传的文件，这里以/upload/20210113_14261694.png 为例进行请求测试，请求地址如下所示：

```
http://localhost:8080/upload/20210113_14261694.png
```

得到的结果如图 11-15 所示。

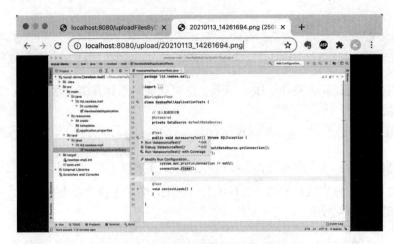

图 11-15 图片上传成功后的回显效果

至此,文件名不同的多文件上传功能测试完成!

由于图片文件看起来比较直观,所以本章中在上传时选择的都是图片文件,其他类型的文件也可以正常上传和访问,读者可以自行尝试。

第 12 章

Spring Boot 实现验证码生成及验证功能

本章将会介绍在网页开发中常用的验证码功能,并具体讲解如何使用 Spring Boot 生成验证码并进行后续的验证操作。

12.1 验证码介绍

12.1.1 什么是验证码

在日常上网的时候经常会看到验证码的设计,比如登录账号、论坛发帖、抢购商品等都会要求用户输入验证码。验证码的生成规则或者展现形式也各不相同。

验证码(CAPTCHA)是 Completely Automated Public Turing test to tell Computers and Humans Apart(全自动区分计算机和人类的图灵测试)的缩写,是一种区分用户是计算机还是人的公共全自动程序。

验证码设计的主要目的和作用就是防止不法分子在短时间内用机器进行批量的重复操作给网站带来破坏,从而保护系统的安全。通俗而言,验证码的存在就是让用户证明自己是人为操作,而非机器操作。

12.1.2 验证码的形式

随着技术的进步，验证码的展现形式也越来越多样化。

1. 传统输入式验证码

传统输入式验证码比较常见，用户识别验证字符并自行输入验证。

该实现方式比较简单，一般以图片形式呈现给用户，其中的字符通常由数字、字母混合组成，也有单纯使用数字或字母的。为了提高用户的识别难度系统可能提供数学运算表达式，或者在图片背景中添加干扰线等。

在新蜂商城项目中所采用的验证码就是这种类型，如图 12-1 所示。

2. 手机短信验证码

随着智能手机的普及，越来越多的移动端 App 和网页端系统都开始采用手机号码注册、登录的方式。图 12-2 是淘宝网站的登录页面，用户首先填写手机号获取验证码，然后输入收到的短信验证码进行验证，验证成功才能登录。

图 12-1　新蜂商城登录页面的验证码

图 12-2　淘宝网登录页面

相较于传统输入式验证码，该方式多了填写手机号的步骤。在系统开发过程中也需要对接短信服务系统并向运营商支付相应的短信费用，所以会更复杂一些。此外，如果短信接收不及时也会影响用户的体验。但是这种方式不仅包含了验证码的基础功能，可以防止恶意的注册和登录，同时也能够对手机号码的真实性进行校验，一举多得。因此在真实的企业开发中通常会使用这种方式。除了短信验证码之外，运营商也可以提供语音验证码。

与手机短信验证码类似的还有邮箱验证。

3. 图片识别与选择型验证码

这种方式与传统输入式验证码类似,都是由用户辨别验证码并进行验证,只是展现形式和用户的操作方式不同。图片识别与选择型验证码提供的并不是简单的字符,而是提供一系列的图片让用户辨认,辨认后进行点击选择并完成验证过程。整个过程并不需要用户进行键盘输入操作,只需要使用鼠标或者触屏点击就可以。

这种形式的验证码,最深入人心的莫过于 12306 网站上的验证码,如图 13-3 所示。

网页上出现多张图片和限定词,用户需要根据限定词去辨别图片并根据识别结果依次点击和选择图片,最终完成验证。

对于用户来说,辨别和选择的过程容易出错,耗时耗力,如果出现比较相似的图片或者难以辨别的图片很容易引起用户的反感。12306 网站为了防止黄牛行为或者其他刷票行为采用了这种验证方式。而一般的网站为了用户体验会简化验证方式,比如图 12-4 所示的点选型验证码。

图 12-3 12306 网站上的验证码

图 12-4 点选型验证码

4. 滑块类型验证码

滑块类型验证码是非常有创意的验证码形式,实现方式有滑块和轨迹线。用户需要识别并拖动滑块到指定位置或者根据验证码中的轨迹线沿着轨迹划线,最终完成验证操作,如图 12-5 所示。

图 12-5 滑块型验证码

这种形式的验证码趣味性强、操作简单，但是滑块类型的验证码开发起来比较复杂。

5．其他类型验证码

随着技术的发展和终端产品的升级，也可以通过用户的行为特征、设备的唯一标示码、大数据风控等技术，提供对用户来说近似于无感知的验证码方式。这种验证方式完全免去用户的常规验证步骤，但是一旦被判定为异常行为则会进行强制验证，比如指纹识别、人脸识别等。采用这种形式的验证需要强大的技术支撑能力，实现复杂，但是对于用户来说，这种方式是用户体验最好的。

验证码的展现形式非常多样化，在项目中需要选择哪种类型的验证码，需要综合考虑开发成本、用户体验等因素，一定要选择最合适的。

12.2　Spring Boot整合easy-captcha生成验证码

生成验证码的方式和案例有很多，在新蜂商城项目中验证码的形式为传统输入式验证码，所选择的实现方案是 easy-captcha 工具包。

easy-captcha 开源地址为：

https://github.com/whvcse/EasyCaptcha

它是一款国人开发的验证码工具，支持 GIF、中文、算术等类型，可用于 Java Web 等项目，生成的验证码形式如图 12-6 所示。

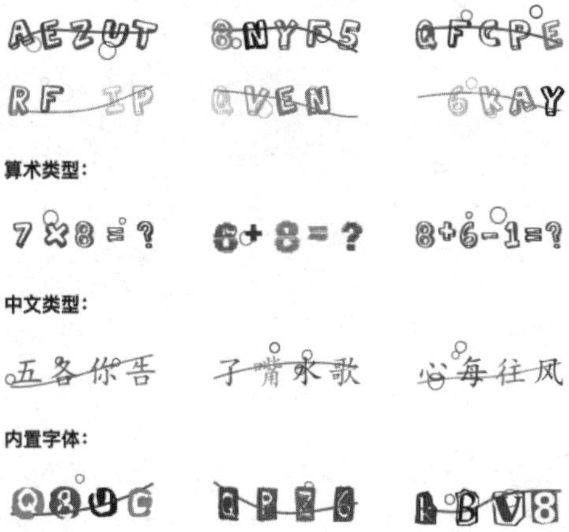

算术类型：

中文类型：

内置字体：

图 12-6　easy-captcha 生成的验证码示例

接下来将会通过一个代码案例来讲解如何使用 Spring Boot 生成验证码并在网页中显示验证码。

12.2.1　添加 easy-captcha 依赖

Spring Boot 整合 easy-captcha 的第一步就是增加依赖。首先需要将 easy-captcha 的依赖配置文件增加到 pom.xml 文件中，此时的 pom.xml 文件代码如下所示：

```xml
<?xml version="1.0" encoding="UTF-8"?>
<project xmlns="http://maven.apache.org/POM/4.0.0" xmlns:xsi="http://www.w3.org/2001/XMLSchema-instance"
    xsi:schemaLocation="http://maven.apache.org/POM/4.0.0 https://maven.apache.org/xsd/maven-4.0.0.xsd">
    <modelVersion>4.0.0</modelVersion>
    <parent>
        <groupId>org.springframework.boot</groupId>
        <artifactId>spring-boot-starter-parent</artifactId>
        <version>2.3.7.RELEASE</version>
        <relativePath/> <!-- lookup parent from repository -->
    </parent>
    <groupId>ltd.newbee.mall</groupId>
    <artifactId>newbee-mall</artifactId>
    <version>0.0.1-SNAPSHOT</version>
    <name>newbee-mall</name>
```

```xml
<description>captcha-demo</description>

<properties>
    <project.build.sourceEncoding>UTF-8</project.build.sourceEncoding>
    <project.reporting.outputEncoding>UTF-8</project.reporting.outputEncoding>
    <java.version>1.8</java.version>
</properties>

<dependencies>

    <dependency>
        <groupId>org.springframework.boot</groupId>
        <artifactId>spring-boot-starter-web</artifactId>
    </dependency>

    <!-- 验证码 -->
    <dependency>
        <groupId>com.github.whvcse</groupId>
        <artifactId>easy-captcha</artifactId>
        <version>1.6.2</version>
    </dependency>

    <dependency>
        <groupId>org.springframework.boot</groupId>
        <artifactId>spring-boot-starter-test</artifactId>
        <scope>test</scope>
        <exclusions>
            <exclusion>
                <groupId>org.junit.vintage</groupId>
                <artifactId>junit-vintage-engine</artifactId>
            </exclusion>
        </exclusions>
    </dependency>
</dependencies>

<build>
    <plugins>
        <plugin>
            <groupId>org.springframework.boot</groupId>
            <artifactId>spring-boot-maven-plugin</artifactId>
        </plugin>
    </plugins>
```

```xml
    </build>

    <repositories>
        <repository>
            <id>alimaven</id>
            <name>aliyun maven</name>
            <url>http://maven.aliyun.com/nexus/content/repositories/central/</url>
            <releases>
                <enabled>true</enabled>
            </releases>
            <snapshots>
                <enabled>false</enabled>
            </snapshots>
        </repository>
    </repositories>
</project>
```

如果是在本地开发的话，只需要等待 jar 包及相关依赖下载完成即可。

12.2.2 验证码格式

easy-captcha 验证码工具支持 GIF、中文、算术等类型，分别通过以下几个实例对象实现：SpecCaptcha、GifCaptcha、ChineseCaptcha、ChineseGifCaptcha 和 ArithmeticCaptcha。

他们依次为 PNG 类型的静态图片验证码、GIF 类型的图片验证码、中文类型的图片验证码、中文 GIF 类型的图片验证码和算数类型的验证码。读者可以针对自身项目的情况选择合适的验证码生成方式。一般常用 SpecCaptcha 和 GifCaptcha 两种方式来生成验证码。

12.2.3 验证码字符类型

对于验证码生成的规则，easy-captcha 提供了 6 种类型，如表 12-1 所示。

表 12-1 easy-captcha 验证码字符类型

类型	描述
TYPE_DEFAULT	数字和字母混合
TYPEONLYNUMBER	纯数字
TYPEONLYCHAR	纯字母
TYPEONLYUPPER	纯大写字母
TYPEONLYLOWER	纯小写字母
TYPENUMAND_UPPER	数字和大写字母

使用方法如下所示：

```
// 生成验证码对象
SpecCaptcha captcha = new SpecCaptcha(130, 48, 5);
// 设置验证码的字符类型
captcha.setCharType(Captcha.TYPE_ONLY_NUMBER);
```

12.2.4 字体设置

easy-captcha 默认提供了 10 种内置字体，如图 12-7 所示。

字体	效果
Captcha.FONT_1	ACTION
Captcha.FONT_2	Epilog
Captcha.FONT_3	FRESNEL
Captcha.FONT_4	Headache
Captcha.FONT_5	LEXOGRAPHER
Captcha.FONT_6	Prefix
Captcha.FONT_7	PROG.BOT
Captcha.FONT_8	beatae
Captcha.FONT_9	Robot Teacher
Captcha.FONT_10	SCANDAL

图 12-7 easy-captcha 内置字体

当然，如果不想用以上 10 种内置字体，也可以使用系统字体。默认字体和系统字体的设置方式如下所示：

```
// 生成验证码对象
SpecCaptcha captcha = new SpecCaptcha(130, 48, 5);

// 设置验证码字体为内置字体 1
captcha.setFont(Captcha.FONT_1);

// 设置设置验证码字体为系统字体
captcha.setFont(new Font("楷体", Font.PLAIN, 28));
```

12.2.5　验证码图片输出

这里可以选择输出为文件流，这是比较常见的处理方式。当然，也有一些 Web 项目会使用 base64 编码的图片。这两种方式 easy-captcha 都支持。

base64 编码的输出方法如下所示：

```
SpecCaptcha specCaptcha = new SpecCaptcha(130, 48, 5);
specCaptcha.toBase64();

// 如果不想要 base64 的头部 data:image/png;base64,加一个空的参数即可
// specCaptcha.toBase64("");
```

输出到磁盘上的方法如下所示：

```
// 输出到磁盘上
FileOutputStream outputStream = new FileOutputStream(new File("/home/project/captcha.png"));
SpecCaptcha specCaptcha = new SpecCaptcha(130, 48, 5);
specCaptcha.out(outputStream);
```

该段代码为生成一张图片并保存到磁盘目录中，这里可以使用 easy-captcha 工具自带的 out()方法输出。而在开发 Web 项目时，则会使用 Response 对象的输出流进行验证码的输出，接下来会结合代码进行详细讲解。

12.3 生成并显示验证码

12.3.1 后端逻辑实现：生成并输出验证码

在 controller 包中新建 KaptchaController 类，就可以新建一个方法。在方法里使用 GifCaptcha 可以生成一个 PNG 类型的验证码对象，并以图片流的方式输出到前端以供显示，代码如下所示：

```java
package ltd.newbee.mall.controller;

import com.wf.captcha.SpecCaptcha;
import com.wf.captcha.base.Captcha;
import org.springframework.stereotype.Controller;
import org.springframework.web.bind.annotation.GetMapping;

import javax.servlet.http.HttpServletRequest;
import javax.servlet.http.HttpServletResponse;

@Controller
public class KaptchaController {

    @GetMapping("/kaptcha")
    public void defaultKaptcha(HttpServletRequest httpServletRequest,
HttpServletResponse httpServletResponse) throws Exception {
        httpServletResponse.setHeader("Cache-Control", "no-store");
        httpServletResponse.setHeader("Pragma", "no-cache");
        httpServletResponse.setDateHeader("Expires", 0);
        httpServletResponse.setContentType("image/gif");

        // 三个参数分别为宽、高、位数
        SpecCaptcha captcha = new SpecCaptcha(75, 30, 4);

        // 设置类型为数字和字母混合
        captcha.setCharType(Captcha.TYPE_DEFAULT);

        // 设置字体
        captcha.setCharType(Captcha.FONT_9);

        // 验证码存入 session
```

```
            httpServletRequest.getSession().setAttribute("verifyCode", captcha.
text().toLowerCase());

            // 输出图片流
            captcha.out(httpServletResponse.getOutputStream());
    }
}
```

这里在控制器中新增了 defaultKaptcha()方法，该方法所拦截处理的路径为/kaptcha。在前端访问该路径后就可以接收一个图片流并显示在浏览器页面上。

12.3.2　前端逻辑实现：在页面中展示验证码

在 static 目录中新建 kaptcha.html 页面，在该页面中显示验证码，代码如下所示：

```html
<!DOCTYPE html>
<html lang="en">
  <head>
    <meta charset="UTF-8" />
    <title>验证码显示</title>
  </head>
  <body>
    <img src="/kaptcha" onclick="this.src='/kaptcha?d='+new Date()*1" />
  </body>
</html>
```

首先访问后端验证码路径/kaptcha，由于验证码是图片形式，所以将其显示在 img 标签中。然后定义 onclick 方法，在点击该 img 标签时可以动态切换显示一个新的验证码。点击时访问的路径为'/kaptcha?d='+new Date()*1，即原来的验证码路径后面带上一个时间戳参数 d。时间戳是会变化的，所以每次点击都会是一个与之前不同的请求。如果不这样处理的话，由于浏览器的缓存机制，在点击刷新验证码后可能不会重新发送请求，将导致在一段时间内一直显示同一张验证码图片。

在编码完成后，启动 Spring Boot 项目。在启动成功后打开浏览器并输入验证码显示的测试页面地址：

```
http://localhost:8080/kaptcha.html
```

页面显示效果如图 12-8 所示。

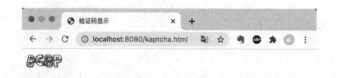

图 12-8　生成验证码示例

在页面中已经可以正确显示验证码图片，并且在每次点击后也可以动态切换。至此，验证码演示功能实现完成。

12.4　验证码的输入验证

在验证码的显示完成后，紧接着要做的就是对用户输入的验证码进行比对和验证。

一般的做法是在后端生成验证码后，首先对当前生成的验证码内容进行保存，可以选择保存在 session 对象中，或者保存在缓存中，或者保存在数据库中。然后，返回验证码图片并显示到前端页面。用户在识别验证码后，在页面对应的输入框中填写验证码并向后端发送请求，后端在接到请求后会对用户输入的验证码进行验证。如果用户输入的验证码与之前保存的验证码不相等的话，则返回"验证码错误"的提示消息且不会进行后续的流程，只有验证成功才会继续后续的流程。

12.4.1　后端逻辑实现

在 KaptchaController 类中新增 verify() 方法，代码如下所示：

```java
@GetMapping("/verify")
@ResponseBody
public String verify(@RequestParam("code") String code, HttpSession session) {
    if (!StringUtils.hasLength(code)) {
        return "验证码不能为空";
    }
    String kaptchaCode = session.getAttribute("verifyCode") + "";
    if (!StringUtils.hasLength(kaptchaCode) || !code.toLowerCase().equals(kaptchaCode)) {
        return "验证码错误";
    }
```

```
        return "验证成功";
    }
```

该方法所拦截处理的路径为/verify，请求参数为code，即用户输入的验证码。在进行基本的非空验证后，与之前保存在session中的verifyCode值进行比较，如果两个字符串不相等则返回"验证码错误"的提示，二者相同则返回"验证码成功"的提示。

12.4.2 前端逻辑实现

在static目录中新建verify.html，该页面会显示验证码，同时也包含供用户输入验证码的输入框和提交按钮，代码如下所示：

```html
<!DOCTYPE html>
<html lang="en">
<head>
    <meta charset="UTF-8" />
    <title>验证码测试</title>
</head>
<body>
<img src="/kaptcha" onclick="this.src='/kaptcha?d='+new Date()*1" />
<br>
<input type="text" maxlength="5" id="code" placeholder="请输入验证码" />
<button id="verify">验证</button>
<br>
<p id="verifyResult">
</p>
</body>
<!-- 下方地址为字节跳动提供的jQuery cdn地址 -->
<script src="https://s3.pstatp.com/cdn/expire-1-M/jquery/3.3.1/jquery.min.js"></script>
<script type="text/javascript">
    $(function () {
        // 验证按钮的点击事件
        $('#verify').click(function () {
            var code = $('#code').val();
            $.ajax({
                type: 'GET', // 方法类型
                url: '/verify?code=' + code,
                success: function (result) {
                    // 将验证结果显示在p标签中
```

```
        $('#verifyResult').html(result);
      },
      error: function () {
        alert('请求失败');
      },
    });
  });
});
</script>
</html>
```

用户识别显示在页面上的验证码后，就可以在 input 框中输入验证码并点击"验证"按钮。在 JS 代码中已经定义了"验证"按钮的点击事件，一旦点击，就会获取用户在输入框中输入的内容，并将其作为请求参数向后端发送请求，验证用户输入的验证码是否正确，后端在处理完成后会返回处理结果，拿到处理结果就显示在 id 为 verifyResult 的 p 标签中。

在编码完成后，启动 Spring Boot 项目。在启动成功后打开浏览器并输入验证的测试页面地址：

```
http://localhost:8080/verify.html
```

页面显示效果如图 12-9 所示。

图 12-9　verify.html 页面显示效果

可以看到该页面中显示的验证码内容为"CNUP"字符串，此时，在页面中输入"cnnb"字符串并点击"验证"按钮，会得到"验证码错误"的提示，如图 12-10 所示。

图 12-10　验证失败的显示效果

这证明这里并没有输入正确的验证码。如果未输入任何内容就点击验证，则会显示"验证码不能为空"的提示信息。

输入正确的字符串"cnup"并点击"验证"按钮，此时就会得到"验证成功"的提示信息，如图12-11所示。

图 12-11　验证成功的显示效果

验证结果与预期一致，功能测试完成。

另外，图片中的字母都是大写的形式，不过在后端代码处理时都使用了 toLowerCase() 方法对字符串进行了简单的处理，即不区分大小写，所以在输入框中输入"CNup"、"cnUP"、"CNUP"都能够验证成功。

网站验证码的功能逻辑主要包括验证码的生成、验证码的显示和验证码的比对。

不过对于不同形式的验证码，可能实现方式不同，实现难度也不一样。本章所讲解的是传统输入式验证码，并且已经将所有功能逐一实现，后续将会结合实际的项目开发对该功能逻辑进行完善。希望读者可以根据文中的步骤和提供的源码进行编码和测试。

第 13 章

商城项目需求分析与功能设计

前文介绍过 Spring Boot 的基本情况和学习方法,同时也介绍了如何统筹一个大型项目的功能设计和开发。本章将介绍使用 Spring Boot 为主要技术栈开发的新蜂商城项目,让读者了解本书最终的实战项目成品,包括新蜂商城的开发背景、新蜂商城项目的迭代记录、新蜂商城的功能模块设计和运行预览图等。

13.1 选择开发商城系统的原因

程序员是一个非常务实的群体,这种性格品质用最经典的一句话来概括就是"Talk is cheap,show me the code(废话少说,放码过来)",特别强调项目和实战经验。

大部分开发人员都特别注重项目实践。项目的种类很多,也会随着个人技术的不断提升而不同,最简单的可能是学生管理系统、工资管理系统等,难一点的可能是博客或者论坛项目,大型的可能是商城系统。本书最终的实战项目新蜂商城就属于商城系统。

13.1.1 什么是商城系统

商城系统就是功能完善的网上销售系统,与传统的市场一样,商城系统也会提供在交易时所必须的信息交换、支付结算和实物配送等基础服务,它可以让用户通过网络实现购物行为。

线上商城为个人用户和企业用户提供人性化的全方位服务,可以为用户创造亲切、轻松和愉悦的购物环境。京东、唯品会、天猫、拼多多等都属于线上商城。它们通常包含会员模块、商品模块、订单模块和支付模块等元素。

当然,由于系统的完善程度不同,可能还会有仓库模块、物流模块、营销模块等元素。图13-1是商城系统抽象图。

图 13-1　商城系统抽象图

13.1.2　为什么要做商城系统

1. 热度高

电子商务的高速发展加速了为电子商务服务的软件行业的发展,并随之诞生了很多与之密切相关的商城系统。无论开源的还是商业性质的商城系统,其实现技术都非常丰富,热度也非常高。因此,很多开发人员都会尝试开发一套商城系统。

2. 知识点复杂

关于此点,第1章已说明,此处不再赘述。

3. 产品流程完整

商城系统具有完整的产品开发流程,产品设计、原型设计、功能开发、功能测试、项目上线等环节都有涉及,如图13-2所示。开发人员一般不太关注完整的产品开发流程,但是能够掌握整个产品开发的流程对日后的职业提升有极大的帮助。

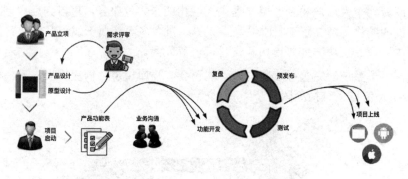

图 13-2 产品开发流程示意图

13.2 认识新蜂商城系统

13.2.1 新蜂商城系统介绍

newbee-mall 项目（新蜂商城）是笔者发布到开源平台的一套商城系统，包括 newbee-mall 商城系统及 newbee-mall-admin 商城后台管理系统，它基于 Spring Boot 2.x 及相关技术栈开发。前台商城系统包含首页门户、商品分类、新品上线、首页轮播、商品推荐、商品搜索、商品展示、购物车、订单结算、订单流程、个人订单管理、会员中心、帮助中心等模块。后台管理系统包含数据面板、轮播图管理、商品管理、订单管理、会员管理、分类管理、设置等模块。

该项目包括商城系统和商城后台管理系统。对应的用户体系包括商城会员和商城后台管理员。商城系统是所有用户都可以浏览使用的系统，商城会员在这里可以浏览、搜索、购买商品。管理员在商城后台管理系统中管理商品信息、订单信息、会员信息等，具体包括商城基本信息的录入和更改、商品信息的添加和编辑、处理订单的拣货和出库，还有商城会员信息的管理。

该项目具体特点如下所示。

（1）newbee-mall 对开发人员十分友好，无须复杂的操作步骤，仅需 2 秒就可以启动完整的商城项目。

（2）newbee-mall 是一个企业级别的 Spring Boot 大型项目，对于各个阶段的 Java 开发人员都是极佳的选择。

（3）开发人员可以把 newbee-mall 作为 Spring Boot 技术栈的综合实践项目，其在技术上符合要求，且代码开源、功能完备、流程完整、页面美观、交互顺畅。

（4）newbee-mall 涉及的技术栈新颖、知识点丰富，有助于读者理解和掌握相关知识，进一步提升开发人员的职业市场竞争力。

13.2.2 新蜂商城开发背景

图 13-3 是笔者的开源仓库主页。

图 13-3 笔者的开源仓库主页

2017 年 2 月 23 日，笔者在 GitHub 网站上发布了第一个开源项目，是 Spring+Spring MVC+MyBatis 框架的整合实践项目，仓库名称是 ssm-demo。后来，由于公司切换了开发框架，全面拥抱了 Spring Boot 的体系，笔者所做的开源项目也直接调转方向，转向了 Spring Boot 相关的仓库制作，包括 Spring Boot 框架基础整合、实践源码和在 Spring Boot 基础上的一些实战项目。笔者先后发布了很多基础代码，还有让开发人员能上手的实战项目，包括基础的后台管理系统、咨询发布系统、博客系统等。

笔者经过三年慢慢地整理和动手开发，从无到有、由小至大，最终制作并开源了这一系列的项目。读者可以看出这是一个循序渐进的过程。由于制作这些开源项目，笔者也创建了几个交流群以供使用这些开源项目的开发人员交流和答疑。在交流过程中笔者收到了不少反馈，其中大家对商城类的项目尤为感兴趣。

结合在交流群中的反馈和商城系统特点，开发一个开源商城的想法就浮现在笔者脑海中。

当时，网上已经有很多开源的商城项目，再做一个商城项目会显得很多余。于是笔者就实际调研了一些开源商城项目，发现它们有不少问题，会导致学习和使用的不便。其主要问题如下所示。

（1）项目不完整，没有完整的文档，要么缺少前端页面，要么缺少依赖，要么缺少数据库 SQL 文件。

（2）技术栈庞杂，Spring Cloud / Dubbo / Redis / Elastic Search / Docker 等同时存在，导致运行一个商城项目需要安装配置很多软件，对于新手来说是一个极大的挑战。

（3）部分开源商城项目存在技术老旧、页面不美观、交互体验差、更新迭代慢的问题。

考虑以上 3 个问题，笔者决定开发一个商城项目并发布到开源网站。当时的计划很明确，弥补某些开源商城项目存在的不足，开发一个能够轻松、顺利运行的商城项目，保证文件齐全、页面美观、交互体验良好。

以上就是开源项目新蜂商城的开发背景。

13.2.3　新蜂商城开源过程

2019 年 8 月 12 号笔者写下了新蜂商城项目的第一行代码，经过近两个月的开发和测试，新蜂商城项目于 2019 年 10 月 9 日正式开源在 GitHub 网站上，当时的提交记录如图 13-4 所示。

图 13-4　新蜂商城开源代码提交记录

由于弥补了其他开源商城项目的不足之处，并且学习和使用起来的成本不高，新蜂商城项目开源的第一年就取得不错的成绩，获得近 6000 个 Star 和 1500 个 Fork，成为一个比较受欢迎的开源项目。当时的开源仓库数据如图 13-5 所示。

最让笔者感到欣慰的一点是新蜂商城开源项目帮助了很多开发人员。在开源之后笔者经常会收到留言和邮件。这些开发人员说他们在学习和使用该开源商城项目后，对于 Spring Boot 技术栈有了更深刻的认识并且拥有了项目实战经验，让他们可以顺利地完成课程作业，甚至在找到心仪工作的过程中起到了关键作用。

第 13 章　商城项目需求分析与功能设计

图 13-5　newbee-mall 仓库开源第一年的数据

这些反馈不仅让人欣慰，也让笔者更加有动力不断地完善新蜂商城开源项目。为了让新蜂商城开源项目保持长久的生命力，并且帮助更多的开发人员，笔者也一直在优化和升级该项目。目前已经完成三个大版本的开发。

（1）新蜂商城 v1 版本，主要技术栈为 Spring Boot。

（2）新蜂商城 v2 前后端分离版本，主要技术栈为 Spring Boot+Vue。

（3）新蜂商城 Vue3 版本，主要技术栈为 Spring Boot+Vue 3。

新蜂商城 v1 版本只有 PC 端的页面，加入前后端分离 Vue 版本之后，新蜂商城的展现形式更加丰富，如图 13-6 所示。

图 13-6　新蜂商城的展示效果

从左到右依次为新蜂商城后台管理系统页面、新蜂商城 PC 端页面、新蜂商城 Vue 版本的三个页面。这里不仅是展现形式的增多和产品线的丰富，也加入了 Vue 技术栈，开发模式也变成了前后端分离模式。

由于篇幅原因，不可能将新蜂商城三个版本都写在同一本书中。本书主要讲解的是新蜂商城 v1 版本，技术栈为 Spring Boot+Thymeleaf+MyBatis，更适合后端开发人员学习。

关于新蜂商城的版本迭代记录，笔者也整理了重要版本的时间轴，如图 13-7 所示。今后也会一直完善和迭代新蜂商城项目。

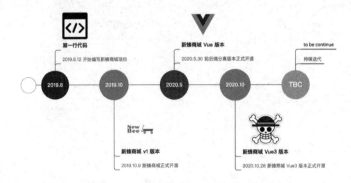

图 13-7　新蜂商城重要版本的时间轴

13.2.4　新蜂商城运行预览图

新蜂商城项目包含商城系统的一些共有属性和功能，页面设计和流程设计也完全按照商城系统来开发和处理。新蜂商城运行预览图包括商城端运行预览图和商城后台管理系统运行预览图，由于页面较多，本章只展示部分主要流程的预览图，更多运行效果图可以自行启动项目并查看。

1. 商城端运行预览图

商城首页预览图如图 13-8 所示。

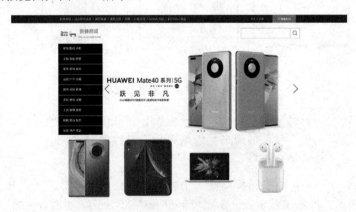

图 13-8　商城首页预览图

第 13 章 商城项目需求分析与功能设计

商城搜索页面预览图如图 13-9 所示。

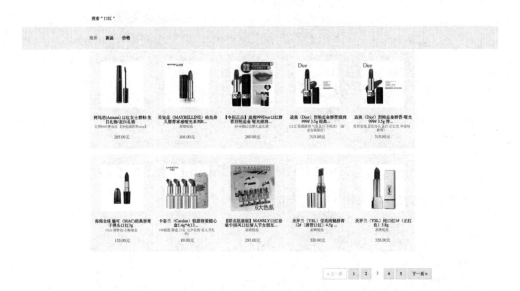

图 13-9 商城搜索页面预览图

商城购物车页面预览图如图 13-10 所示。

图 13-10 商城购物车页面预览图

商城结算页面预览图如图 13-11 所示。

图 13-11　商城结算页面预览图

支付选择页面预览图如图 13-12 所示。

图 13-12　支付选择页面预览图

订单列表页面预览图如图 13-13 所示。

图 13-13　订单列表页面预览图

2. 商城后台管理端运行预览图

后台管理系统登录页面预览图如图 13-14 所示。

图 13-14　后台管理系统登录页面预览图

后台管理系统轮播图管理页面预览图如图 13-15 所示。

图 13-15　后台管理系统轮播图管理页面预览图

后台管理系统新品上线管理页面预览图如图 13-16 所示。

图 13-16　后台管理系统新品上线管理页面预览图

后台管理系统分类管理页面预览图如图 13-17 所示。

图 13-17　后台管理系统分类管理页面预览图

后台管理系统商品管理页面预览图如图 13-18 所示。

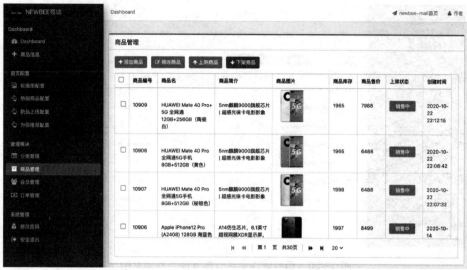

图 13-18　后台管理系统商品管理页面预览图

后台管理系统商品编辑页面预览图如图 13-19 所示。

图 13-19　后台管理系统商品编辑页面预览图

后台管理系统订单管理页面预览图如图 13-20 所示。

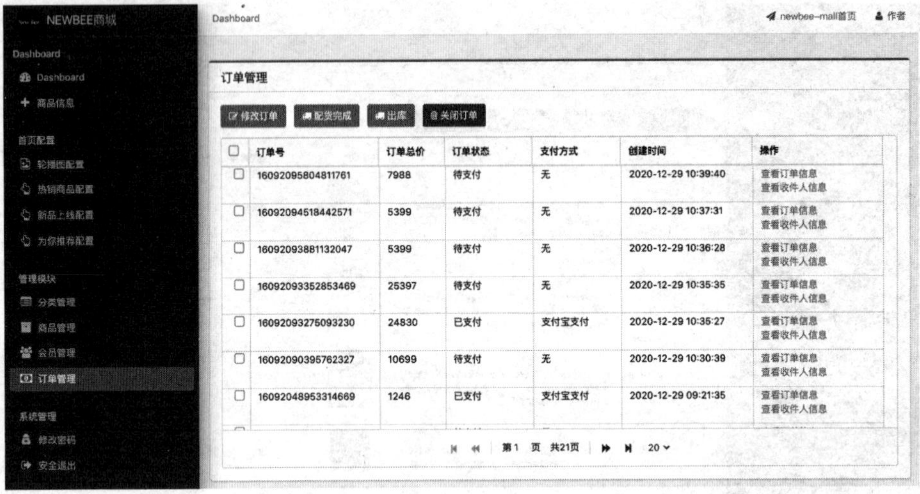

图 13-20　后台管理系统订单管理页面预览图

13.3　新蜂商城功能详解

商城系统属于大型项目，这种项目虽然复杂，但是也不是完全无法实现。只要计划合理，选用的解决方案有效就能够完成。而行业普遍的一个解决方案就是"拆"。

核心思路是化繁为简，首先将大项目拆解成若干个小项目，大系统拆分出若干个功能模块，大功能拆解成若干个小功能，然后再对各个环节或者各个功能做具体的实现和完善。比如做好功能、接口、表结构设计。具体到功能可能就有实现登录、文件上传、分页、分类的三级联动、搜索、订单流程等。当开发人员将这些功能模块各个击破并且全部完善的时候，这个完整的项目就逐渐建立起来。

为了加深读者的理解并且能够更好地学习该项目，笔者会将项目中所涉及的功能点全部列举出来。

13.3.1　商城端功能整理

新蜂商城的商城端功能点汇总如图 13-21 所示，主要包括商城首页、商品展示、商品搜索、会员模块、购物车模块、订单模块和支付模块。

第 13 章　商城项目需求分析与功能设计

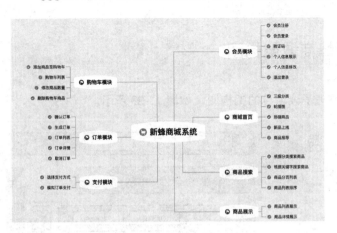

图 13-21　商城端功能点汇总

13.3.2　后台管理系统功能整理

新蜂商城后台管理系统功能点汇总如图 13-22 所示，主要包括系统管理员、轮播图管理、热销商品配置、新品上线配置、推荐商品配置、分类管理、商品管理、会员管理和订单管理。后台管理系统中的所有功能模块主要是为了让商城管理员操作运营数据以及管理用户交易数据。这里通常就是基本的增、删、改、查的功能。

图 13-22　后台管理系统功能点汇总

13.3.3 新蜂商城架构图

笔者简单整理了该项目的架构图，如图 13-23 所示。

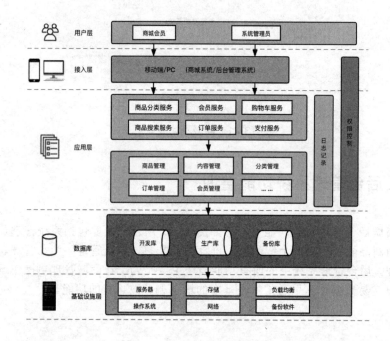

图 13-23 新蜂商城架构图

新蜂商城架构图主要包括系统的逻辑架构和实际的基础设施搭建。这里供读者参考和学习，希望通过图 13-23 能让读者在架构层面或者宏观角度有一定思考和提升，而不只是运行和使用新蜂商城，也不只是理解新蜂商城项目的源码。

第 14 章

项目初体验：启动和使用新蜂商城

本章主要是介绍商城项目的源码下载、目录结构、商城项目启动和注意事项，以便读者可以顺利运行源码并进行个性化修改。

14.1 下载商城项目的源码

在部署项目之前，首先需要把项目的源码下载到本地，newbee-mall 项目在 GitHub 和 Gitee 平台都创建了代码仓库。由于国内访问 GitHub 网站可能速度缓慢，所以笔者在 Gitee 上也创建了一个同名代码仓库，两个仓库会保持同步更新，它们的地址如下所示：

```
https://github.com/newbee-ltd/newbee-mall
https://gitee.com/newbee-ltd/newbee-mall
```

读者可以直接在浏览器中输入上述链接到对应的仓库中查看源码及相关文件。

14.1.1 使用 clone 命令下载源码

如果本地安装了 Git 环境的话，可以直接在命令行中使用 git clone 命令把仓库中的文件全部下载到本地。

通过 GitHub 下载源码，执行如下命令：

```
git clone https://github.com/newbee-ltd/newbee-mall.git
```

通过 Gitee 下载源码，执行如下命令：

```
git clone https://gitee.com/newbee-ltd/newbee-mall.git
```

首先打开 cmd 命令行，然后切换到对应的目录。比如下载到 D 盘的 java-dev 目录，那就先执行 cd 切换到该目录下，并执行 git clone 命令，过程如图 14-1 所示。

图 14-1 使用命令行下载代码

等待文件下载，在全部下载完成后就能够在 java-dev 目录下看到新蜂商城项目所有的源码了。如果使用 GitHub 的链接下载较慢的话，可以通过国内的 Gitee 链接执行 git clone 操作。

14.1.2 通过开源网站下载源码

除了通过命令行下载之外，读者也可以选择更直接的方式。GitHub 和 Gitee 两个开源平台都提供了对应的下载功能，读者可以在仓库中直接点击对应的下载按钮进行源码下载。

如果在 GitHub 网站上直接下载代码，就进入 newbee-mall 在 GitHub 网站中的仓库主页，如图 14-2 所示。

在 newbee-mall 代码仓库页面上有一个带着下载图标的绿色 "Code" 按钮，点击该按钮，如图 14-3 所示。再点击 "Download Zip" 就可以下载新蜂商城源码的压缩包文件，下载完成并解压，最后导入 IDEA 或者 Eclipse 编辑器中进行开发或者修改。

第 14 章 项目初体验：启动和使用新蜂商城

图 14-2　newbee-mall 仓库主页（GitHub）

图 14-3　点击"Code"按钮

Gitee 网站下载源码会更快一些。如果在 Gitee 网站上直接下载源码，就先进入 newbee-mall 在 Gitee 网站中的仓库主页，如图 14-4 所示。

· 277 ·

图 14-4　newbee-mall 仓库主页（Gitee）

在 newbee-mall 仓库主页上有一个"克隆/下载"按钮，点击该按钮，再点击"下载 Zip"，如图 14-5 所示。

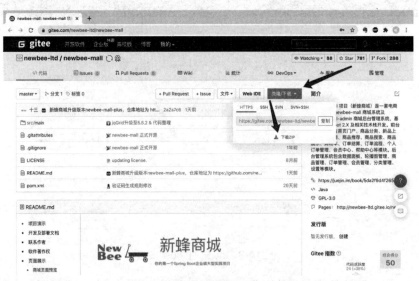

图 14-5　点击"克隆/下载"按钮

在 Gitee 网站上下载源码多了一步验证操作，点击"下载 Zip"后会跳转到验证页面，如图 14-6 所示。

图 14-6 下载源码前的验证

验证页面中的验证方式可能是滑块验证码，也可能是输入型验证码，在验证成功后就可以下载代码的压缩包文件，下载完成并解压，最后导入到 IDEA 或者 Eclipse 编辑器中进行开发或者修改。

14.2 新蜂商城目录结构讲解

在下载代码并解压后，在代码编辑器中打开项目，这是一个标准的 Maven 项目。笔者使用的开发工具是 IDEA，导入之后 newbee-mall 源码目录结构如图 14-7 所示。

笔者介绍一下目录的内容和作用，结果整理如下所示：

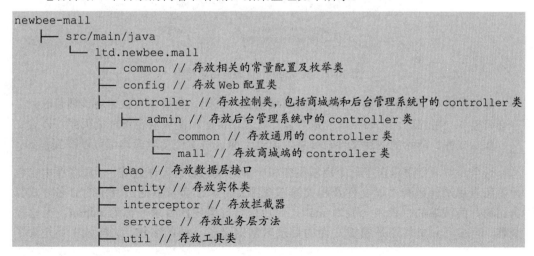

```
                └─ NewBeeMallApplication  // Spring Boot 项目的主类
        ├── src/main/resources
            ├── mapper // 存放 MyBatis 的通用 Mapper 文件
            ├── static // 默认的静态资源文件目录
                ├── admin // 存放后台管理系统端的静态资源文件目录
                └── mall // 存放商城端的静态资源文件目录
            ├── templates
                ├── admin // 存放后台管理系统端页面的模板引擎目录
                └── mall // 存放商城端页面的模板引擎目录
            ├── application.properties // 项目配置文件
            ├── newbee_mall_schema.sql // 项目所需的 SQL 文件
            └── upload.zip // 商品图片
    └── pom.xml // Maven 配置文件
```

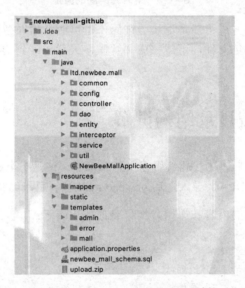

图 14-7　newbee-mall 源码目录结构

　　读者可以根据目录结构文档来理解商城项目的源码。有很多人在使用该项目时，经常会问笔者一些问题，比如源码中是否包含商城端页面和商城后台管理端页面？样式文件、脚本文件、商品的图片文件在哪里？通过上面这份文档读者应该都可以得到答案。

　　这个商城项目中包含了一个商城系统和一个商城的后台管理系统。开源仓库中也包含了所有后端源码和页面文件及相关静态资源文件。读者可以通过目录的命名方式得出结论，商城端的目录名一般为 mall，而后台管理系统的目录名称为 admin。无论控制器、静态资源文件还是模板文件的目录名都定义了不同的名称。这样做主要是为了方便使用者理解和区分不同功能，同时也方便后续项目的拆分和功能优化。

除了 Spring Boot 项目基础目录中的源码之外，笔者在 resources 目录中也上传了 newbee_mall_schema.sql 文件和 upload.zip 文件，这是项目启动时所需的两个文件。newbee_mall_schema.sql 是商城项目的 SQL 文件，包含了项目所需的所有表结构和初始化数据。upload.zip 文件则是商品图片文件，在商品表中存储了数百条记录。为了使用者可以获得更好的学习体验，这些数据所需的图片文件都在 upload.zip 压缩包中。如果没有这个压缩包，在启动项目后看到的所有页面的商品图片都会出现 404 错误。

14.3 启动商城项目

接下来会讲解新蜂商城项目的启动和启动前的准备工作。

14.3.1 导入数据库

打开 MySQL 管理软件，新建一个数据库，命名为 newbee_mall_db。在数据库创建完成后就可以将 newbee_mall_schema.sql 文件导入到该数据库中。在导入成功后可以看到数据库的表结构如图 14-8 所示。

图 14-8 newbee_mall_db 数据库的表结构

14.3.2 修改数据库连接配置

在导入成功后，进行第二步操作，打开 resources 目录下的 application.properties 配置文件，修改数据库连接的相关信息，默认的数据库配置如下所示：

```
spring.datasource.url=jdbc:mysql://localhost:3306/newbee_mall_db?useUnic
ode=true&serverTimezone=Asia/Shanghai&characterEncoding=utf8&autoReconne
ct=true&useSSL=false&allowMultiQueries=true
spring.datasource.username=root
spring.datasource.password=123456
```

需要修改的主要内容如下所示。

①数据库地址和数据库名称：localhost:3306/ newbee_mall_db

②数据库登录账户名称：root

③账户密码：123456

这里需要根据开发人员所安装的数据库地址和账号信息进行修改。数据库名称默认为 newbee_mal_db，如果更改为其他名称，则需要将配置文件中的数据库名称也修改了。application.properties 配置文件的其他配置项可以不进行修改，只有数据库连接的这三个配置项需要根据数据库配置的不同进行具体的修改。

14.3.3 静态资源目录设置

完成前面两个步骤其实就可以启动项目，但是在启动项目后，可能会出现图片无法显示的问题。这里就需要进行静态资源目录配置，tb_newbee_mall_goods_info 是商品表，该表在初始化时已经新增了数百条商品记录，这些记录中有商品主图的数据列且都有值，如图 14-9 所示。

图 14-9 商品表数据节选

所有的商品图片都在 upload.zip 压缩包中，在解压后就能够看到数百张商品图片文件。为了商品图片能够正确地显示，还需要进行两步操作。

第一步是解压 upload.zip 压缩包并将文件放到一个文件夹中，可根据个人习惯选择文件夹，比如 D 盘的 upload 文件夹，或者 E 盘的 mall\images 文件夹。第二步操作是配置静态资源目录，这个配置项在 ltd.newbee.mall.common 包的 Constants 类中，变量名称为 FILE_UPLOAD_DIC，其代码如图 14-10 所示。

这里默认的配置为 D:\upload\，即 D 盘的 upload 文件夹。如果将图片文件放到了 E 盘的 mall\images 文件夹中，则将该变量的值修改为 E:\mall\images\。如果是其他文件夹对应修改该变量值即可。

图 14-10　Constants 类的代码

一定要注意路径最后有一个斜杠，有不少使用者向笔者反馈过，因为没注意最后的斜杠导致无法访问图片。

以上都是在 Windows 系统下的写法。如果是 Linux 系统写法为"/opt/newbee/upload/"，即将 FILE_UPLOAD_DIC 变量的值修改为放置图片文件的目录路径即可。

14.3.4　启动并访问商城项目

在上面几个步骤做完后，就可以启动商城项目了，过程如图 14-11 所示。

点击启动按钮后，等待项目启动即可。在图 14-11 中笔者选择的是通过运行 main() 方法的方式启动 Spring Boot 项目，读者也可以通过其他方式启动。

如图 14-12 所示，在控制台中可以查看启动日志。通过日志信息也能够知道项目经过 3.2 秒就完成启动。监听的端口为 28089，这个端口号可以通过修改在 application.properties 中的 server.port 配置项来更改。

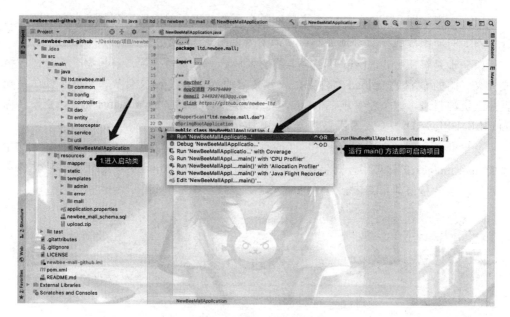

图 14-11　启动项目的过程

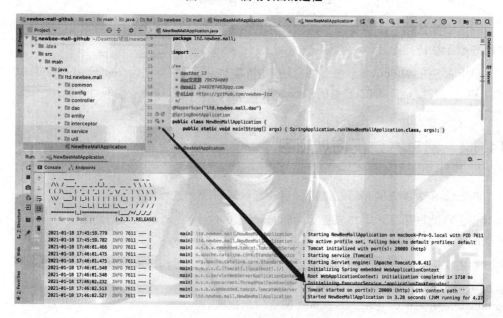

图 14-12　启动日志

启动 Spring Boot 商城项目，在启动成功后打开浏览器并输入首页地址：

```
http://localhost:28089
```

待页面加载完成，就能够看到商城项目的庐山真面目了！如图 14-13 所示。

图 14-13　新蜂商城首页

另外，如果修改了启动端口号，访问地址也需要对应修改。

14.4　注意事项

接下来讲一下项目的注意事项。这些注意事项是笔者经常会被读者问到的，这里做一个整理和总结。

14.4.1　关于项目地址

商城项目包括后台管理系统和商城系统，在启动成功后，两个系统都可以正常访问。商城系统的访问地址如下，键入任何一个地址都能够访问到商城首页。

①BASE_URL

②BASE_URL + '/'

③BASE_URL + /index

④BASE_URL + /index.html

后台管理系统的访问地址如下，键入任何一个地址都能够进入到商城后台管理系统。

①BASE_URL + '/admin'

②BASE_URL + '/admin/login'

③BASE_URL + '/admin/index'

其中 BASE_URL 为主机名+端口号，比如笔者的本地地址就是 localhost:28089。

14.4.2 关于账号及密码

项目启动成功了，但是由于权限问题无法完整地体验商城的所有功能。这就需要几个测试账号来解决。该项目在表中初始化了几条用户记录和管理员记录，数据分别存储在 tb_newbee_mall_user 表和 tb_newbee_mall_admin_user 表中，读者也可以自行添加和修改这些记录。

商城系统用户的测试账号及密码如下所示。

①账号：13700002703 密码：123456

②账号：13711113333 密码：shisan

③账号：13811113333 密码：shisan

后台管理系统管理员的测试账号及密码如下所示。

①账号：admin 密码：123456

②账号：newbee-admin1 密码：123456

③账号：newbee-admin2 密码：123456

对于两张表中存储的密码字段都是经过 MD5 加密后的字符串。如果想要直接通过数据库来修改密码的话，首先需要将密码进行 MD5 转换，然后将加密后的字符串放到数据表的密码这一列中。

14.4.3 商城登录和后台管理系统登录演示

商城系统的很多功能是需要登录才能够正常使用的，比如后台管理、下单流程。如果不是商城的用户或者管理员，肯定没有权限，大部分功能也无法操作。这一节就演示一下登陆过程。

首先是商城端的登录页面，打开浏览器并输入登录页面地址：
`http://localhost:28089`

如图 14-14 所示，在登录页面依次输入正确的账号密码和验证码，并点击"立即登录"按钮完成商城端的登录。这样就可以完成后续的商品下单等操作了。

图 14-14　商城端登录

然后是商城后台管理系统端的登录页面，打开浏览器并输入登录页面地址：
`http://localhost:28089/admin/login`

如图 14-15 所示，在登录页面依次输入正确的账号密码和验证码，并点击"登录"按钮就能够进入到后台管理系统中。

图 14-15 后台管理系统端登录

进入后台管理系统后，就可以对商品模块、轮播图模块等进行操作。

登录功能演示完成，希望读者按照本章介绍的内容，顺利运行和体验新蜂商城项目。

第 15 章

页面设计及商城后台管理系统页面布局的实现

本章会介绍前端的技术选型和一些页面设计的知识点，并结合实际的编码把后台管理系统的页面布局开发出来。

15.1 前端页面实现的技术选型

虽然前端页面的布局大致是固定的，但是由于设计风格、功能偏重不同，也就存在不同的页面风格。并且，不同前端开发人员开发的页面样式也风格迥异。笔者是一个后端开发人员，前端开发水平有限，所以在开发商城项目时选择了一些成熟的前端框架来进行二次开发。

商城项目选择的技术是 Bootstrap 和 AdminLTE3。接下来笔者会对二者进行介绍并列举选择的理由。

15.1.1 Bootstrap 产品介绍

Bootstrap 是美国 Twitter 公司的设计师 Mark Otto 和 Jacob Thornton 合作，并基于 HTML、CSS、JavaScript 开发的简洁、直观、强悍的前端开发框架。它可以使 Web 开发

更加快捷。

Bootstrap 提供了优雅的 HTML 和 CSS 规范，由动态 CSS 语言 Less 编写而成。Bootstrap 一经推出便受到欢迎，一直是 GitHub 上的热门开源项目，包括 NASA 旗下的 MSNBC Breaking News 都使用了该项目。国内一些前端开源框架也是基于 Bootstrap 源码进行性能优化升级而来的，比如 WeX5 等前端开源框架。

Bootstrap 的开源仓库地址为：

https://github.com/twbs/bootstrap

Bootstrap 火热和广泛流行的原因如下所示。
① 提供了一套完整的基础 CSS 插件
② 提供了一组基于 jQuery 的 JavaScript 插件集
③ 提供了非常丰富的组件与插件，组件包含小图标、按钮组、菜单导航、标签页等
④ 扩展性强，兼容各种脚本插件
⑤ 拥有现成 UI 组件，可以快速搭建网页页面
⑥ 可以让响应式网站的实现更加简单
⑦ 为用户提供了一套响应式移动设备优先的流式栅格系统，拥有完备的框架结构，整体和谐，支持谷歌、火狐、IE 等浏览器，项目开发方便快捷
⑧ 不断适应 Web 技术的发展，十分成熟，在大量的项目中已被充分使用和测试
⑨ 拥有完善的文档，使用起来十分方便

借助 Bootstap 提供的插件集、组件集可以很方便地搭建开发人员想要的网站。本书所实战的新蜂商城项目，商城端和后台管理系统端的前端页面都用到了 Bootstrap 中的部分组件。

15.1.2　为什么选择 Bootstrap

选择 Bootstrap 的主要原因可以归结为以下几点。

（1）符合规范。在网站的产品设计中，需要涉及各个元素和组件。Bootstrap 提供了从字体到页面布局再到交互体验所有的风格和设计，且提供了多种规范的风格和标准可替换主题。

（2）使用成本低。Bootstrap 官方提供了源码和工具包，只需要导入其 JS 文件和 CSS 样式表就可以轻松实现一个 Bootstrap 风格的页面，集成的过程十分简单、方便、快捷。

（3）代码复用性高。代码复用性是衡量一个工具或者框架优劣的重要指标。代码复用性高可以节省开发人员的开发时间且代码不会变得臃肿复杂。Bootstrap 包中包含了大量现成的代码片段可以直接拿来使用，只要找到符合产品的代码段即可直接使用，减少了开发人员重复编码的频率，提高了代码的复用性。

（4）学习成本低。Bootstrap 是 2011 年在 GitHub 上发布的开源产品，至今已有 10 个年头，但是依然更新频繁、不停迭代。由于其众多优异的特性使得其在开源社区十分活跃，且帮助文档十分齐全，在网站上的相关教程也比较多，学习起来很容易。

（5）可维护性高。开箱即用、上手快使得集成了 Bootstrap 的项目维护难度降低，同时 Bootstrap 具有现成的组件、CSS 样式和可以直接在代码中引用的插件，可以减少对代码的维护，以便高效地组织代码编写。

（6）时间成本低。详尽的学习文档使得 Bootstrap 学习成本降低，因此花费较少的时间就可以掌握并使用。Bootstrap 提供了非常丰富的组件与插件，可以显著地节省时间和精力，使得项目进展加快，实现快速开发，节省大量的时间成本。

（7）响应式。响应式网站最大的特点就是无论在电脑、平板，还是在手机上，都会根据屏幕尺寸自动调节大小、图片分辨率等。这已经渐渐成为一种技术趋势。而 Bootstrap 就是响应式框架，它能以超快的速度与极高的效率适应不同平台。

15.1.3 AdminLTE3 产品介绍

商城后台管理系统完全基于 AdminLTE3 模板来开发。AdminLTE3 是一个 WebApp 模板并且基于 Bootstrap 开发而来。所以，整个商城项目的页面布局和样式的实现可以说是基于 Bootstrap 开发的。

AdminLTE3 是一个完全响应式管理并基于 Bootstrap 开发的免费高级管理控制面板主题。它目前在 GitHub 开源，开源社区十分活跃且特别受欢迎。其开源协议是 MIT 许可证，拥有相对宽松的软件授权条款。开源、流行、高度可定制化、易于使用，这只是它众多优点的一部分。

AdminLTE3 的开源仓库地址为：

https://github.com/almasaeed2010/AdminLTE。

本书实战商城项目选用的是 AdminLTE3 版本，源文件的下载地址为：

https://github.com/ColorlibHQ/AdminLTE/releases/tag/v3.0.0

下载页面如图 15-1 所示，点击"Source code"即可下载 AdminLTE3 版本的源文件。

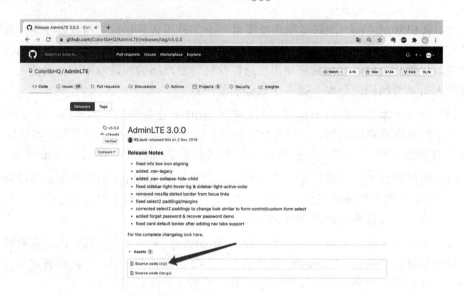

图 15-1　AdminLTE3 下载地址

解压后可以直接在浏览器中打开其中的 HTML 文件进行预览，AdminLTE3 的部分页面预览图如图 15-2 和图 15-3 所示。

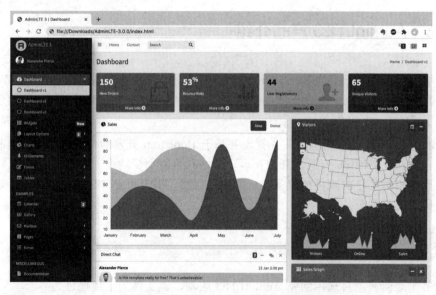

图 15-2　AdminLTE3 页面预览 1

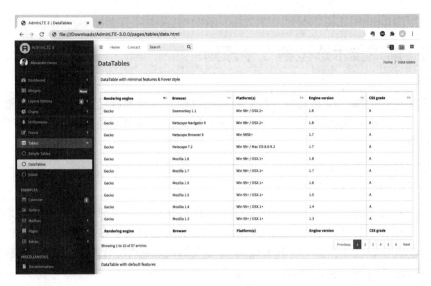

图 15-3　AdminLTE3 页面预览 2

读者看到这两张图片应该觉得非常熟悉，因为新蜂商城后台管理系统的布局和样式就是基于 AdminLTE3 进行开发的。由于 AdminLTE3 基于 Bootstrap 而开发，因此前文中提到的 Bootstrap 的优点 AdminLTE3 模板也都有。不仅仅如此，AdminLTE3 自适应多种屏幕分辨率，兼容 PC 端和手机移动端，内置了多个模板页面，包括仪表盘、邮箱、日历、锁屏、登录及注册、404 错误、500 错误等页面，拥有多种主题皮肤，支持多种浏览器，插件齐全。并且，AdminLTE3 是模块化设计的，非常容易进行二次开发。

15.1.4　为什么选择 AdminLTE3

选择 AdminLTE3 作为商城后台管理系统的模板原因如下所示。

（1）基于 Bootstrap。前文中谈到 Bootstrap 的优点 AdminLTE3 模板也都有。

（2）视觉效果好。笔者对比了当前比较流行的各种系统模板，AdminLTE3 是佼佼者，美观大气。

（3）社区活跃。关注度高、更新迭代快，一直在进步和优化。

（4）代码复用性高。工具齐全且内置了比较完善的后台管理系统页面，可以大大减少开发工作量。

（5）开源免费。一般成品的 WebApp 模板都是收费的，而 AdminLTE3 是完全免费的，代码全部开源在 GitHub 上。

15.1.5 前端技术选型的 5 个原则

关于如何做前端的技术选型，首先要明确目标，再根据自己的需求和技术能力进行选择。开发人员可以根据公司产品和技术栈有针对性地学习相关技术，在具体工作中才能事半功倍。

和后端技术选型类似，代码复用性和可维护性是十分重要的指标。工具性产品的使用就是为了减少代码臃肿和提升开发效率。

再者，开发人员需要从工具的成熟度、社区活跃度、学习成本来考虑。如果社区比较活跃且文档齐全的话，学习起来比较简单，遇到问题也可以很快解决。同时，社区活跃度高也会反哺该技术工具，使其 Bug 修复速度加快、迭代优化更频繁，形成良性循环。

总结下来，前端技术选型有 5 个原则。

①符合开发需求

②提升开发效率

③技术普及广泛

④社区活跃度高

⑤节约学习成本

不仅对前端技术选型如此，读者在工作中使用其他技术类型也可以按照这 5 个原则进行参照。至此，新蜂商城项目的基本技术选型都已经介绍完毕，后端框架使用 Spring Boot+Thymeleaf+MyBatis，前端页面及交互则使用 AdminLTE3+Bootstrap。

15.2 商城页面布局讲解

15.2.1 后台管理系统页面布局介绍

由于后台管理系统具有通用性，大部分网站都会有一个甚至多个后台管理系统。不仅如此，后台管理系统也能够简化管理员和运营人员的工作量、提升工作效率。相信大部分 Java Web 开发人员初次开发完成的项目应该也是与后台管理系统类似的项目，比如学生信息管理系统、图书管理系统、商城管理系统等。

通过观察这些通用的后台管理系统可以归纳出其布局的通用模板，如图 15-4 所示。

第 15 章　页面设计及商城后台管理系统页面布局的实现

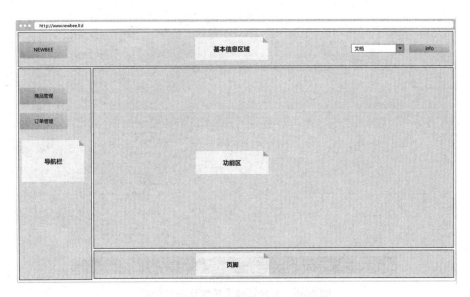

图 15-4　后台管理系统布局通用模板

整个设计版面被切分成四个部分。

（1）基本信息区域：可以放置 Logo 图片、管理系统名称，在本区域的右方还可能放置其他辅助信息。

（2）导航栏：一般会被设计在整个版面的左侧，也有可能放在页面上方基本信息区域之下。一般称之为菜单栏，用以实现页面跳转的管理。

（3）功能区：这个区域会占用整个版面的大部分面积，后台管理系统的重要信息都在这里展示，绝大部分的页面逻辑、功能实现都会在这个区域里，因此它是整个系统最重要的部分。

（4）页脚：这个区域占用的面积较小，通常会在整个版面的底部，用来展示辅助信息，如版权信息、公司信息、项目版本号等。

以上是对后台管理系统布局的大致切分，如果再细化讲解的话，一般的后台管理系统布局如图 15-5 所示。

导航栏中可能有多级菜单，菜单显示的方式有平铺模式或者树状模式。而功能区则又可以细分出按钮功能区域、搜索区域、列表显示区域、分页信息区域、信息编辑区域。

图 15-5　后台管理系统细化布局示例

15.2.2　商城端页面布局介绍

供用户浏览和使用的商城页面读者应该也都体验过。由于这些页面直接面向用户且功能并不通用，因此会有一些极具个性化的布局。商城端页面往往会出现页面不同则排版不同的情景，甚至每一张页面都会有自己的布局和设计。与后台管理系统相比，商城端页面与页面之间的差异性太大，只能抽取出三个功能区域，如图 15-6 所示。

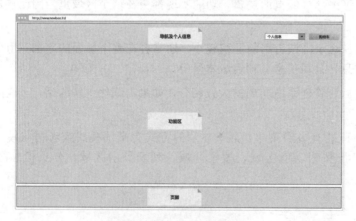

图 15-6　商城端页面布局

第 15 章 页面设计及商城后台管理系统页面布局的实现

另外，有的页面存在个人中心导航栏，一般会有侧边工具栏，而其他部分是三个功能区域，如图 15-7 所示。

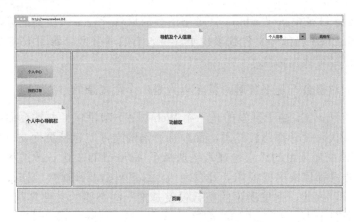

图 15-7　商城端个人中心页面布局

（1）导航及个人信息：该区域主要用于放置商城的导航链接、登录信息、购物车等内容。

（2）功能区：由于商城端页面并不具有通用性，这里只能统一将该区域称为功能区域，主要用于放置每个页面所独有的功能模块。

（3）页脚：该区域主要放置的是商城相关的链接、介绍信息、联系方式等。笔者参考了目前主要的几个商城网站的底部栏目。

（4）个人中心导航栏：与上面三个页面区域不同，这个区域并不是每个页面都有的，只有个人中心相关页面才存在。

15.3　后台管理系统页面制作

接下来，笔者会以商城后台管理系统的首页为例，讲解页面布局的实现。

15.3.1　AdminLTE3 整合到 Spring Boot 项目中

由于后台管理系统的样式和布局是基于 AdminLTE3 进行的二次修改，所以需要把 AdminLTE3 整合到 Spring Boot 项目中。整合过程就是把 AdminLTE3 代码压缩包中的部分样式文件、JS 文件、图片等静态资源，放入 Spring Boot 项目的静态资源目录中，比如默认的 static 目录。

首先在 static 目录中新建 admin 目录用于存放后台管理系统所需的静态资源文件。然后引入 AdminLTE3 的部分文件到项目中，引入后的项目目录如图 15-8 所示。其中几个重要的文件已经用线框标注了。

有些人可能对整合不是很理解，甚至以为是一个很复杂的过程。

这里笔者解释一下。由于开发的是一个 Web 项目，项目包括前端页面所需的文件和后端代码。后端代码读者都比较熟悉，而前端所需的文件就包括页面文件、样式文件、JS 文件等。新蜂商城项目的后台管理系统选择了 AdminLTE3 进行改造和开发。其中，大部分页面和样式都已经由模板作者开发完成。这里只需要针对性地修改一些页面供当前项目使用即可。因此，这里所说的整合仅仅是把一些必要的文件复制到 Spring Boot 项目的静态资源目录中。

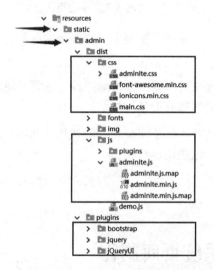

图 15-8　引入 AdminLTE3 的部分文件

15.3.2　后台管理系统页面制作

首先在 resources/templates 目录下新建 admin 目录用于存放后台管理系统页面的模板页面。然后新建 index-all.html 模板文件，打开 index-all.html 文件并在该模板文件的 <html> 标签中导入 Thymeleaf 的名称空间：

```html
<html lang="en" xmlns:th="http://www.thymeleaf.org">
```

导入该名称空间主要是为了 Thymeleaf 的语法提示和 Thymeleaf 标签的使用。接着在模板中使用 th 标签来修改静态资源的引用路径，最终代码如下所示：

```html
<!DOCTYPE html>
<html xmlns:th="http://www.thymeleaf.org">
<head>
    <meta charset="utf-8">
    <meta http-equiv="X-UA-Compatible" content="IE=edge">
    <title>newbee-mall | 后台管理系统</title>
    <meta name="viewport" content="width=device-width, initial-scale=1">
    <link rel="shortcut icon" th:href="@{/admin/dist/img/favicon.ico}"/>
    <link rel="stylesheet" th:href="@{/admin/dist/css/font-awesome.min.css}">
    <link rel="stylesheet" th:href="@{/admin/dist/css/ionicons.min.css}">
    <link rel="stylesheet" th:href="@{/admin/dist/css/main.css}">
    <link rel="stylesheet" th:href="@{/admin/plugins/bootstrap/css/bootstrap.css}"/>
    <link rel="stylesheet" th:href="@{/admin/dist/css/adminlte.min.css}">
</head>
<body class="hold-transition sidebar-mini">
<div class="wrapper">
    <nav class="main-header navbar navbar-expand bg-white navbar-light border-bottom">
        <ul class="navbar-nav">
            <li class="nav-item d-none d-sm-inline-block">
                <a th:href="@{/admin/index}" class="nav-link">Dashboard</a>
            </li>
        </ul>
        <ul class="navbar-nav ml-auto">
            <li class="nav-item dropdown">
                <a class="nav-link" th:href="@{/}">
                    <i class="fa fa-paper-plane">  newbee-mall 首页</i>
                </a>
            </li>
            <li class="nav-item dropdown">
                <a class="nav-link" data-toggle="dropdown" href="#">
                    <i class="fa fa-user">  作者</i>
                </a>
                <div class="dropdown-menu dropdown-menu-lg dropdown-menu-right">
                    <div class="dropdown-divider"></div>
                    <a href="#" class="dropdown-item">
```

```html
                <i class="fa fa-user-o mr-2"></i> 姓名
                <span class="float-right text-muted text-sm">十三 / 13</span>
              </a>
              <div class="dropdown-divider"></div>
              <a href="#" class="dropdown-item">
                <i class="fa fa-user-secret mr-2"></i> 身份
                <span class="float-right text-muted text-sm">Java 开发工程师</span>
              </a>
              <div class="dropdown-divider"></div>
              <a href="#" class="dropdown-item">
                <i class="fa fa-address-card mr-2"></i> 邮箱
                <span class="float-right text-muted text-sm">2449207463@qq.com</span>
              </a>
            </div>
          </li>
        </ul>
      </nav>
      <aside class="main-sidebar sidebar-dark-primary elevation-4">
        <!-- Brand Logo -->
        <a th:href="@{/admin/index}" class="brand-link">
          <img th:src="@{/admin/dist/img/new-bee-logo-1.png}" alt="ssm-cluster Logo"
               class="brand-image img-circle elevation-3"
               style="opacity: .8">
          <span class="brand-text font-weight-light">NEWBEE 商城</span>
        </a>
        <div class="sidebar">
          <nav class="mt-2">
            <ul class="nav nav-pills nav-sidebar flex-column" data-widget="treeview" role="menu"
                data-accordion="false">
              <li class="nav-header">Dashboard</li>
              <li class="nav-item">
                <a th:href="@{/admin/index}" th:class="${path}=='index'?'nav-link active':'nav-link'">
                  <i class="nav-icon fa fa-dashboard"></i>
                  <p>
                    Dashboard
                  </p>
                </a>
```

```html
                </li>
            </ul>
        </nav>
    </div>
</aside>
<div class="content-header">
    <div class="container-fluid">
    </div><!-- /.container-fluid -->
</div>
<div class="content-wrapper">
    <div class="content">
        <div class="card card-primary card-outline">
            <div class="card-header">
                <h3 class="card-title">首页</h3>
            </div> <!-- /.card-body -->
            <div class="card-body" style="min-height: 400px;">
                <div class="panel panel-default">
                    <div style="padding: 10px 0 20px 10px;">
                        页面布局制作及跳转逻辑实现
                    </div>
                </div>
            </div>
        </div>
    </div>
</div>
<footer class="main-footer">
    <strong>Copyright &copy; 2019-2029 <a href="https://gitee.com/newbee-ltd/newbee-mall"
                                                     target="_blank">newbee.ltd</a>.
</strong>
        All rights reserved.
    <div class="float-right d-none d-sm-inline-block">
        <b>newbee-mall #Version</b> 1.0.0
    </div>
</footer>
</div>
<!-- jQuery -->
<script th:src="@{/admin/plugins/jquery/jquery.min.js}"></script>
<!-- jQuery UI 1.11.4 -->
<script th:src="@{/admin/plugins/jQueryUI/jquery-ui.min.js}"></script>
<!-- Bootstrap 4 -->
<script th:src="@{/admin/plugins/bootstrap/js/bootstrap.bundle.min.js}"></script
```

```
>
<!-- AdminLTE App -->
<script th:src="@{/admin/dist/js/adminlte.min.js}"></script>
</body>
</html>
```

这样,前端文件制作完毕。制作过程就是对 AdminLTE3 中的 index.html 页面进行修改,删去大部分内容,保留页面的整体布局,引入 Thymeleaf 命名空间并使用 Thymeleaf 语法修改文件引入路径。

接下来新建 Controller 类来处理首页请求路径并跳转到对应的页面,将该页面与 15.1.3 节的 AdminLTE3 示例页面进行对比。

15.3.3 Controller 类处理页面跳转

在 controller 包下新建 admin 包并新建 AdminController.java 类,新增代码如下所示:

```
package ltd.newbee.mall.controller.admin;

import org.springframework.stereotype.Controller;
import org.springframework.web.bind.annotation.GetMapping;
import org.springframework.web.bind.annotation.RequestMapping;

@Controller
@RequestMapping("/admin")
public class AdminController {

    @GetMapping("/indexAll")
    public String indexAll() {
        return "admin/index-all";
    }
}
```

该方法用于处理 "/admin/indexAll" 请求,访问该方法的映射路径会跳转到 templates/admin 目录下的 index-all.html 模板文件中,至此,跳转逻辑实现完毕。

在编码完成后,启动 Spring Boot 项目。在启动成功后打开浏览器并输入测试页面地址:

```
http://localhost:8080/admin/indexAll
```

页面显示效果如图 15-9 所示。

第 15 章　页面设计及商城后台管理系统页面布局的实现

图 15-9　index-all.html 页面效果

与 15.1.3 节中 AdminLTE3 示例页面相比，此时的 index.html 页面只保留了整体的布局，左侧菜单栏仅保留了 Dashboard，右侧功能区的代码只保留了布局样式，其他内容全部删除并且增加了部分中文文案。

从图 15-9 可以看到页面已经按照前文讲解的布局切分为四个区域：基本信息区域、左侧导航栏区域、功能区、底部页脚区域。只是此时还没有填充具体的内容。后续章节在讲解具体的功能模块制作时会逐一添加。

15.3.4　公共页面抽取

前文提到了页面设计和后台管理系统中几个基本的页面区域。其中，页面顶部和页面底部两个区域在后台管理系统中都是相同的，因此需要对这两部分进行公共代码的抽取以减少重复编码。

在 resources/templates 目录下新建 header.html、footer.html 和 sidebar.html 三个模板文件并使用 th:fragment 属性来定义被包含的模版片段以供其他功能页面的引入。

header.html 的代码如下所示：

```
<!DOCTYPE html>
<html lang="en" xmlns:th="http://www.thymeleaf.org">
<head th:fragment="header-fragment">
```

```html
    <meta charset="utf-8">
    <meta http-equiv="X-UA-Compatible" content="IE=edge">
    <title>newbee-mall | 后台管理系统</title>
    <!-- Tell the browser to be responsive to screen width -->
    <meta name="viewport" content="width=device-width, initial-scale=1">
    <!-- Font Awesome -->
    <link rel="shortcut icon" th:href="@{/admin/dist/img/favicon.ico}"/>
    <link rel="stylesheet" th:href="@{/admin/dist/css/font-awesome.min.css}">
    <!-- Ionicons -->
    <link rel="stylesheet" th:href="@{/admin/dist/css/ionicons.min.css}">
    <link rel="stylesheet" th:href="@{/admin/dist/css/main.css}">
    <link rel="stylesheet" th:href="@{/admin/plugins/bootstrap/css/bootstrap.css}"/>
    <link rel="stylesheet" th:href="@{/admin/plugins/sweetalert/sweetalert.css}"/>
    <link rel="stylesheet" th:href="@{/admin/plugins/jqgrid-5.3.0/ui.jqgrid-bootstrap4.css}"/>
    <!-- Theme style -->
    <link rel="stylesheet" th:href="@{/admin/dist/css/adminlte.min.css}">
</head>
<!-- Navbar -->
<nav class="main-header navbar navbar-expand bg-white navbar-light border-bottom" th:fragment="header-nav">
    <!-- Left navbar links -->
    <ul class="navbar-nav">
        <li class="nav-item d-none d-sm-inline-block">
            <a th:href="@{/admin/index}" class="nav-link">Dashboard</a>
        </li>
    </ul>
    <!-- Right navbar links -->
    <ul class="navbar-nav ml-auto">
        <li class="nav-item dropdown">
            <a class="nav-link" th:href="@{/}">
                <i class="fa fa-paper-plane">  newbee-mall 首页</i>
            </a>
        </li>
        <li class="nav-item dropdown">
            <a class="nav-link" data-toggle="dropdown" href="#">
                <i class="fa fa-user">  作者</i>
            </a>
            <div class="dropdown-menu dropdown-menu-lg dropdown-menu-right">
```

```html
            <div class="dropdown-divider"></div>
            <a href="#" class="dropdown-item">
                <i class="fa fa-user-o mr-2"></i> 姓名
                <span class="float-right text-muted text-sm">十三 / 13</span>
            </a>
            <div class="dropdown-divider"></div>
            <a href="#" class="dropdown-item">
                <i class="fa fa-user-secret mr-2"></i> 身份
                <span class="float-right text-muted text-sm">Java 开发工程师</span>
            </a>
            <div class="dropdown-divider"></div>
            <a href="#" class="dropdown-item">
                <i class="fa fa-address-card mr-2"></i> 邮箱
                <span class="float-right text-muted text-sm">2449207463@qq.com</span>
            </a>
        </div>
    </li>
  </ul>
</nav>
<!-- /.navbar -->
</html>
```

footer.html 的代码如下所示：

```html
<!DOCTYPE html>
<html lang="en" xmlns:th="http://www.thymeleaf.org">
<footer class="main-footer" th:fragment="footer-fragment">
    <strong>Copyright &copy; 2019-2029 <a href="https://gitee.com/newbee-ltd/newbee-mall" target="_blank">newbee.ltd</a>.</strong>
    All rights reserved.
    <div class="float-right d-none d-sm-inline-block">
        <b>newbee-mall #Version</b> 1.0.0
    </div>
</footer>
</html>
```

sidebar.html 的代码如下所示：

```html
<!DOCTYPE html>
<html lang="en" xmlns:th="http://www.thymeleaf.org">
<aside th:fragment="sidebar-fragment(path)" class="main-sidebar sidebar-dark-primary elevation-4">
    <!-- Brand Logo -->
    <a th:href="@{/admin/index}" class="brand-link">
        <img th:src="@{/admin/dist/img/new-bee-logo-1.png}" alt="ssm-cluster Logo"
             class="brand-image img-circle elevation-3"
             style="opacity: .8">
        <span class="brand-text font-weight-light">NEWBEE商城</span>
    </a>
    <!-- Sidebar -->
    <div class="sidebar">
        <!-- Sidebar Menu -->
        <nav class="mt-2">
            <ul class="nav nav-pills nav-sidebar flex-column" data-widget="treeview" role="menu"
                data-accordion="false">
                <!-- Add icons to the links using the .nav-icon class
                     with font-awesome or any other icon font library -->
                <li class="nav-header">Dashboard</li>
                <li class="nav-item">
                    <a th:href="@{/admin/index}" th:class="${path}=='index'?'nav-link active':'nav-link'">
                        <i class="nav-icon fa fa-dashboard"></i>
                        <p>
                            Dashboard
                        </p>
                    </a>
                </li>
                <li class="nav-item">
                    <a th:href="@{/admin/category}" th:class="${path}=='category'?'nav-link active':'nav-link'">
                        <i class="nav-icon fa fa-bookmark"></i>
                        <p>
                            category 页面
                        </p>
                    </a>
                </li>
            </ul>
        </nav>
```

```html
        <!-- /.sidebar-menu -->
    </div>
    <!-- /.sidebar -->
</aside>
</html>
```

最后新建 index.html 并填充页面内容，将 header、footer、sidebar 三个公共页面中的代码通过分段表达式 th:replace 引入进来，分段表达式的代码如下所示：

```html
<!-- 引入页面头 header-fragment -->
<head th:replace="admin/header::head-fragment"></head>

<header th:replace="admin/header::header-fragment"></header>

<!-- 引入工具栏 sidebar-fragment -->
<div th:replace="admin/sidebar::sidebar-fragment(${path})"></div>

<!-- 引入页脚 footer-fragment -->
<footer th:replace="admin/footer::footer-fragment"></footer>
```

index.html 的最终代码如下：

```html
<!DOCTYPE html>
<html xmlns:th="http://www.thymeleaf.org">
<header th:replace="admin/header::header-fragment"></header>
<body class="hold-transition sidebar-mini">
<div class="wrapper">
    <!-- 引入页面头 header-fragment -->
    <div th:replace="admin/header::header-nav"></div>
    <!-- 引入工具栏 sidebar-fragment -->
    <div th:replace="admin/sidebar::sidebar-fragment(${path})"></div>
    <!-- Content Wrapper. Contains page content -->
    <div class="content-wrapper">
        <!-- Content Header (Page header) -->
        <div class="content-header">
            <div class="container-fluid">
            </div><!-- /.container-fluid -->
        </div>
        <!-- Main content -->
        <div class="content">
            <div class="card card-primary card-outline">
                <div class="card-header">
                    <h3 class="card-title">首页</h3>
                </div> <!-- /.card-body -->
```

```html
            <div class="card-body">
                <div style="padding: 10px 0 20px 10px;">
                    页面布局制作及跳转逻辑实现
                </div>
            </div>
        </div><!-- /.container-fluid -->
    </div>
    <!-- /.content -->
</div>
<!-- /.content-wrapper -->
<!-- 引入页脚 footer-fragment -->
<div th:replace="admin/footer::footer-fragment"></div>
</div>
<!-- jQuery -->
<script th:src="@{/admin/plugins/jquery/jquery.min.js}"></script>
<!-- jQuery UI 1.11.4 -->
<script th:src="@{/admin/plugins/jQueryUI/jquery-ui.min.js}"></script>
<!-- Bootstrap 4 -->
<script th:src="@{/admin/plugins/bootstrap/js/bootstrap.bundle.min.js}">
</script>
<!-- AdminLTE App -->
<script th:src="@{/admin/dist/js/adminlte.min.js}"></script>
</body>
</html>
```

该页面的显示效果与 index-all.html 模板文件相同，只是实现代码不同。读者可以简单理解成 index-all.html 模板文件被切分出 4 个模板文件，其中 footer.html 为底部公共区域的内容，header.html 为顶部区域的内容，而 sidebar.html 则是左侧导航栏的内容。index.html 最终引入公共的模板片段重新组合成一张完整的页面。

修改完成后在 AdminController 类中新建 index.html 模板页面的跳转处理方法，代码如下所示：

```java
@GetMapping("/index")
public String index() {
  return "admin/index";
}
```

在编码完成后，启动 Spring Boot 项目。在启动成功后打开浏览器并输入测试页面地址：

```
http://localhost:8080/admin/index
```

第 15 章 页面设计及商城后台管理系统页面布局的实现

如果显示效果与图 15-9 一致，则证明编码正常。

以上三个抽取出来的公共页面组件，会在之后的各个功能模块中引入。由于这部分公共代码抽取出来可以复用，在后续开发中也就不用在每个功能模块的页面中编写太多相似而又冗余的代码。

通过新蜂商城项目的 templates 源码结构能够看到，商城端的页面制作和公共页面抽取的步骤也是一样的，如图 15-10 所示。

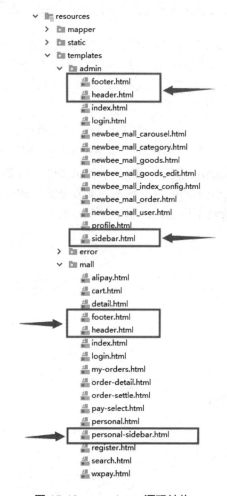

图 15-10 templates 源码结构

至此，后台管理系统和商城端的公共页面都已经抽取出来。在图 15-10 中被线框标注的即为公共页面，在其他功能页面中它们通过分段表达式引入。

15.3.5　分段表达式传参

在公共页面抽取和引入工具栏的 replace 和其他两个公共页面的 replace 有一些差异，代码如下所示：

```
<!-- 引入工具栏 sidebar-fragment -->
<div th:replace="admin/sidebar::sidebar-fragment(${path})"></div>
```

在引入 sidebar-fragment 片段时明显多了一个参数处理的写法，这是一个需要注意的知识点。Thymeleaf 模板引擎在声明片段时可以同时声明变量参数，并且在引入片段中使用变量参数，这与 Java 命名方法的声明变量类似。这里以商城后台管理系统的 sidebar.html 片段的引入为例讲一下。

如下图 15-11 和图 15-12 所示，访问首页时，左侧菜单栏中的"Dashboard"为选中状态；在访问分类页面时，左侧菜单栏中的"category 页面"为选中状态。

而侧边导航栏 sidebar.html 是单独的一个页面文件，如果不进行逻辑处理的话，在访问不同页面时导航栏呈现不同的选中状态的交互就无法实现。因此，这里需要增加一个变量并以此来判断在访问不同页面时哪个导航栏处于被选中状态。

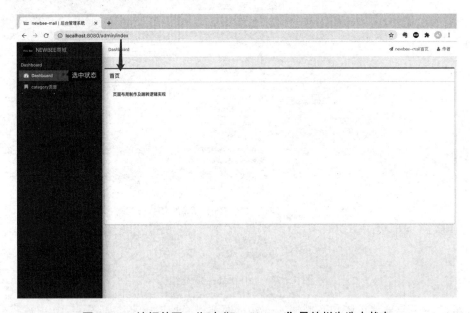

图 15-11　访问首页，此时"Dashboard"导航栏为选中状态

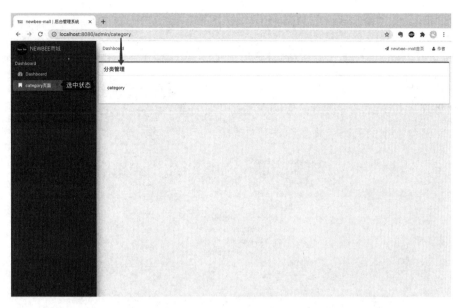

图 15-12 访问分类管理页面，此时"category 页面"导航栏为选中状态

比如在跳转到首页时，后端对应的 Controller 方法会在 request 域中设置一个 path 字段，值为 index；在跳转到分类管理页时，后端对应的 Controller 方法会在 request 域中设置一个 path 字段，值为 category。在 sidebar.html 片段中，使用简单的运算语法来判断当前的 path 变量值，并设置当前的导航栏 class 是否为 active，这样就可实现导航栏选中的交互效果。代码和注释如下所示：

```html
<li class="nav-item">
  <!-- 如果path变量的值为index,则当前a标签为选中状态-->
  <a th:href="@{/admin/index}" th:class="${path}=='index'?'nav-link active':'nav-link'">
    <i class="nav-icon fa fa-dashboard"></i>
    <p>
      Dashboard
    </p>
  </a>
</li>
<li class="nav-item">
  <!-- 如果path变量的值为category,则当前a标签为选中状态-->
  <a th:href="@{/admin/category}" th:class="${path}=='category'?'nav-link active':'nav-link'">
    <i class="nav-icon fa fa-bookmark"></i>
    <p>
      category 页面
```

```
        </p>
      </a>
    </li>
```

这样处理之后,导航栏选中的交互效果就完成了。商城端页面的分段引入也有类似的参数处理,具体编码实现会在后续商城端开发的章节中讲解。

本章最终实现的代码目录如图 15-13 所示。

图 15-13 本章实现的代码目录

后续实战章节中的代码将以本章的代码为基础进行功能模块的增加。

第 16 章

后台管理系统登录功能的实现

本章主要介绍后台管理系统的登录功能及其源码实现,包括页面制作和后端 Java 代码逻辑的开发。

16.1 登录流程设计

16.1.1 什么是登录

在互联网上,供多人使用的网站或程序应用系统会为每位用户配置一套独特的用户名和密码,用户可以使用各自的用户名和密码进入系统,以便系统识别该用户的身份,从而保存该用户的使用习惯或使用数据。用户使用这套用户名和密码进入系统,以及系统验证进入成功或失败的过程,被称为登录。登录页面的内容如图 16-1 所示。

在登录成功后,用户就可以合法地使用该账号具有的各项能力。例如,淘宝用户可以正常浏览商品和完成购买行为等;论坛用户可以查看或更改资料、发表和回复帖子等;OA 等系统管理员用户可以正常地处理各种数据和信息。从最简单的角度来说,登录就是输入用户名和密码进入一个系统进行访问和操作。

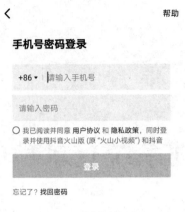

图 16-1 登录页面示例

16.1.2 用户登录状态

客户端（通常是浏览器）在连上 Web 服务器后，如果想获得 Web 服务器中的各种资源，就需要遵守一定的通讯格式。Web 项目通常使用的是 HTTP 协议，它用于定义客户端与 Web 服务器通讯的格式。而 HTTP 协议是无状态的协议，也就是说，这个协议是无法记录用户访问状态的，其每次请求都是独立的、没有任何关联的。一个请求就只是一个请求。

以新蜂商城的后台管理系统为例，它拥有多张页面。在页面跳转过程中和通过接口在进行数据交互时，系统需要知道用户的状态，尤其是用户登录的状态，以便服务器验证用户状态是否正常。这样系统才能判断是否可以让当前用户使用某些功能或者获取某些数据。

这时，就需要在每个页面对用户的身份进行验证和确认，但现实情况却不能如此。一个网站不可能让用户在每个页面上都输入用户名和密码，这是一个违反操作逻辑的设计，也不会有用户使用有这种设计的系统。

因此，在设计登录流程时，需要让用户只进行一次登录操作即可。为了实现这一功能就需要一些辅助技术，用得最多的技术就是浏览器的 Cookie。而在 Java Web 开发中，比较常见的是使用 Session 技术来实现。将用户登录的信息存放在 Cookie 或 Session 中，这样就可以通过读取在 Cookie 或 Session 中的用户登录信息，达到记录用户状态、验证用户状态的目的。

16.1.3 登录流程设计

登录的本质即身份验证和登录状态的保持，在实际编码中是如何实现的呢？

首先，在数据库中查询这条用户记录，伪代码如下所示：

```
select * from xxx_user where account_number = 'xxxx';
```

如果不存在这条记录则表示身份验证失败，登录流程终止；如果存在这条记录，则表示身份验证成功。

然后，进行登录状态的存储和验证，存储的伪代码如下所示：

```
//通过Cookie存储
Cookie cookie = new Cookie("userName",xxxxx);

//通过Session存储
session.setAttribute("userName",xxxxx);
```

验证逻辑的伪代码如下所示：

```
//通过Cookie获取需要验证的数据并进行比对校验
Cookie cookies[] = request.getCookies();
if (cookies != null){
    for (int i = 0; i < cookies.length; i++)
        {
            Cookie cookie = cookies[i];
            if (name.equals(cookie.getName()))
            {
                return cookie;
            }
        }
}

//通过Session获取需要验证的数据并进行比对校验
session.getAttribute("userName");
```

本次实践项目的登录状态是通过 Session 来保存的。在用户登录成功后，用户信息被放到 Session 对象中。在访问项目时，通过拦截器判断 Session 中是否有用户登录信息。用户登录信息存在则放行请求，没有就跳转到登录页面。

16.2 管理员登录功能实践

16.2.1 管理员登录页面的实现

由于选用了 AdminLTE3 作为模板，在制作登录页面时直接改造 AdminLTE3 模板中的登录页面代码即可。在 templates/admin 目录中新建 login.html 模板页面，模板引擎选择 Thymeleaf 技术，最终代码如下所示：

```html
<!-- Copyright (c) 2019-2020 十三 all rights reserved. -->
<!DOCTYPE html>
<html xmlns:th="http://www.thymeleaf.org">
<head>
    <meta charset="utf-8">
    <meta http-equiv="X-UA-Compatible" content="IE=edge">
    <title>newbee-mall 管理系统 | Log in</title>
    <!-- Tell the browser to be responsive to screen width -->
    <meta name="viewport" content="width=device-width, initial-scale=1">
    <link rel="shortcut icon" th:href="@{/admin/dist/img/favicon.ico}"/>
    <!-- Font Awesome -->
    <link rel="stylesheet" th:href="@{/admin/dist/css/font-awesome.min.css}">
    <!-- Ionicons -->
    <link rel="stylesheet" th:href="@{/admin/dist/css/ionicons.min.css}">
    <!-- Theme style -->
    <link rel="stylesheet" th:href="@{/admin/dist/css/adminlte.min.css}">
    <style>
        canvas {
            display: block;
            vertical-align: bottom;
        }
        #particles {
            background-color: #F7FAFC;
            position: absolute;
            top: 0;
            width: 100%;
            height: 100%;
            z-index: -1;
        }
    </style>
```

```html
</head>
<body class="hold-transition login-page">
<div id="particles">
</div>
<div class="login-box">
    <div class="login-logo" style="color: #1baeae;">
        <img th:src="@{/admin/dist/img/new-bee-logo.png}" style=" height: 58px;float: left;margin-left: 10px;">
        <h1>管理系统登录</h1>
    </div>
    <!-- /.login-logo -->
    <div class="card">
        <div class="card-body login-card-body">
            <p class="login-box-msg"> NewBee MALL , Let's Go !</p>
            <form th:action="@{/admin/login}" method="post">
                <div th:if="${not #strings.isEmpty(session.errorMsg)}" class="form-group">
                    <div class="alert alert-danger" th:text="${session.errorMsg}"></div>
                </div>
                <div class="form-group has-feedback">
                    <span class="fa fa-user form-control-feedback"></span>
                    <input type="text" id="userName" name="userName" class="form-control" placeholder="请输入账号"
                           required="true">
                </div>
                <div class="form-group has-feedback">
                    <span class="fa fa-lock form-control-feedback"></span>
                    <input type="password" id="password" name="password" class="form-control" placeholder="请输入密码"
                           required="true">
                </div>
                <div class="row">
                    <div class="col-6">
                        <input type="text" class="form-control" name="verifyCode" placeholder="请输入验证码" required="true">
                    </div>
                    <div class="col-6">
                        <img alt="点击图片刷新！" class="pointer" th:src="@{/common/kaptcha}"
                             onclick="this.src='/common/kaptcha?d='+new Date()*1">
                    </div>
```

```html
            </div>
            <div class="form-group has-feedback"></div>
            <div class="row">
                <div class="col-8">
                </div>
                <div class="col-4">
                    <button type="submit" class="btn btn-primary btn-block btn-flat">登录
                    </button>
                </div>
            </div>
        </form>

    </div>
    <!-- /.login-card-body -->
  </div>
</div>
<!-- /.login-box -->

<!-- jQuery -->
<script th:src="@{/admin/plugins/jquery/jquery.min.js}"></script>
<!-- Bootstrap 4 -->
<script th:src="@{/admin/plugins/bootstrap/js/bootstrap.bundle.min.js}">
</script>
<script th:src="@{/admin/dist/js/plugins/particles.js}"></script>
<script th:src="@{/admin/dist/js/plugins/login-bg-particles.js}"></script>
</body>
</html>
```

该页面代码直接修改自 AdminLTE3 模板的登录页。这里将页面中原本的文案修改为中文并微调了一下页面布局，同时增加了验证码的设计。

在 AdminController 类中新建登录模板页面的跳转处理方法，代码如下所示：

```
@GetMapping({"/login"})
public String login() {
  return "admin/login";
}
```

该方法拦截的路径为/admin/login，请求方式为 GET。在编码完成后，启动 Spring Boot 项目。在启动成功后打开浏览器并输入测试页面地址：

```
http://localhost:8080/admin/login
```

最终的页面效果如图 16-2 所示。

图 16-2 后台管理系统登录页面

用户在输入账号、密码和验证码后，点击"登录"按钮就会向后端发送登录请求。在 form 表单中已经定义了登录的请求路径：

```
<form th:action="@{/admin/login}" method="post">
```

请求登录的地址为 admin/login，请求类型为 POST。该请求与跳转到登录页面的请求路径都是/admin/login，但是请求方式不同。所以后端有两个方法分别处理这两个请求，后者只是简单地跳转到登录页面，前者则是处理登录请求。

通过 form 表单中的字段可以看出，该登录请求带着 3 个请求参数，分别是 userName、password 和 verifyCode。

在编写后端代码时需要接收这三个字段，并根据这三个字段进行相应的登录业务处理。

16.2.2 管理员表结构设计

前文有对 MySQL 数据库操作的教程，其中也有一些表结构的设计，不过那是用来讲解功能的测试表。商城系统正式的表都是以 tb_newbee_mall_为前缀的。这一点读者需要注意，如果表名不是以该前缀开头的则为测试表，是用来演示相关知识点的。后续在项目开发中会一直使用 newbee_mall_db 数据库。商城系统的后台管理员表结构设计如下所示：

```sql
CREATE DATABASE /*!32312 IF NOT EXISTS*/'newbee_mall_db ' /*!40100 DEFAULT CHARACTER SET utf8 */;

USE 'newbee_mall_db ';

DROP TABLE IF EXISTS 'tb_newbee_mall_admin_user';
CREATE TABLE 'tb_newbee_mall_admin_user' (
  'admin_user_id' int(11) NOT NULL AUTO_INCREMENT COMMENT '管理员id',
  'login_user_name' varchar(50) CHARACTER SET utf8 COLLATE utf8_general_ci NOT NULL COMMENT '管理员登录名称',
  'login_password' varchar(50) CHARACTER SET utf8 COLLATE utf8_general_ci NOT NULL COMMENT '管理员登录密码',
  'nick_name' varchar(50) CHARACTER SET utf8 COLLATE utf8_general_ci NOT NULL COMMENT '管理员显示昵称',
  'locked' tinyint(4) NULL DEFAULT 0 COMMENT '是否锁定 0未锁定 1已锁定无法登录',
  PRIMARY KEY ('admin_user_id') USING BTREE
) ENGINE = InnoDB AUTO_INCREMENT = 4 CHARACTER SET = utf8 COLLATE = utf8_general_ci ROW_FORMAT = Dynamic;

INSERT INTO 'tb_newbee_mall_admin_user' VALUES (1, 'admin', 'e10adc3949ba59abbe56e057f20f883e', '十三', 0);
INSERT INTO 'tb_newbee_mall_admin_user' VALUES (2, 'newbee-admin1', 'e10adc3949ba59abbe56e057f20f883e', '新蜂01', 0);
INSERT INTO 'tb_newbee_mall_admin_user' VALUES (3, 'newbee-admin2', 'e10adc3949ba59abbe56e057f20f883e', '新蜂02', 0);
```

该 SQL 文件的作用如下所示。

① 创建 newbee_mall_db 数据库

② 创建 tb_newbee_mall_admin_user 管理员表

③ 在 tb_newbee_mall_admin_user 管理员表中新增 3 条测试数据

新增的 3 条管理员数据用于测试，之后在演示登录功能时会用到。登录名分别为 admin、newbee-admin1 和 newbee-admin2，密码都是 123456，密码字段做了一次 MD5 的加密处理。

16.2.3 新建管理员实体类和 Mapper 接口

首先在 ltd.newbee.mall 包下新建 entity 包,再选中 entity 包并右键点击,在弹出的菜单中选择"New→Java Class",然后在弹出的窗口中输入"AdminUser",最后在 AdminUser 类中新增如下代码:

```java
package ltd.newbee.mall.entity;

public class AdminUser {
    private Integer adminUserId;

    private String loginUserName;

    private String loginPassword;

    private String nickName;

    private Byte locked;

    public Integer getAdminUserId() {
        return adminUserId;
    }

    public void setAdminUserId(Integer adminUserId) {
        this.adminUserId = adminUserId;
    }

    public String getLoginUserName() {
        return loginUserName;
    }

    public void setLoginUserName(String loginUserName) {
        this.loginUserName = loginUserName == null ? null : loginUserName.trim();
    }

    public String getLoginPassword() {
        return loginPassword;
    }

    public void setLoginPassword(String loginPassword) {
```

```
        this.loginPassword = loginPassword == null ? null : loginPassword.
trim();
    }

    public String getNickName() {
        return nickName;
    }

    public void setNickName(String nickName) {
        this.nickName = nickName == null ? null : nickName.trim();
    }

    public Byte getLocked() {
        return locked;
    }

    public void setLocked(Byte locked) {
        this.locked = locked;
    }
}
```

首先在 ltd.newbee.mall 包下新建 dao 包，再选中 dao 包并右键点击，在弹出的菜单中选择"New→Java Class"，然后在弹出的窗口中输入"AdminUserMapper"，并选中"Interface"选项，如图 16-3 所示。

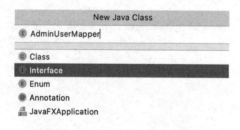

图 16-3 新建 AdminUserMapper

最后在 AdminUserMapper.java 文件中新增如下代码：

```
package ltd.newbee.mall.dao;

import ltd.newbee.mall.entity.AdminUser;
import org.apache.ibatis.annotations.Param;

public interface AdminUserMapper {
    /**
```

```
 * 登录方法
 *
 * @param userName
 * @param password
 * @return
 */
AdminUser login(@Param("userName") String userName, @Param("password")
String password);
}
```

这里定义了 login()方法，参数为 userName 和 password，返回对象的类型为 AdminUser。

16.2.4 创建 AdminUserMapper 接口的映射文件

首先在 resources 目录下新建 mapper 目录，并在 mapper 目录下新建 AdminUserMapper 接口的映射文件 AdminUserMapper.xml，然后进行映射文件的编写。

首先，定义映射文件与 Mapper 接口的对应关系。比如在该示例中，需要将 AdminUserMapper.xml 文件与对应的 AdminUserMapper 接口之间的关系定义出来：

```xml
<mapper namespace="ltd.newbee.mall.dao.AdminUserMapper">
```

然后，配置表结构和实体类的对应关系：

```xml
<resultMap id="BaseResultMap" type="ltd.newbee.mall.entity.AdminUser">
  <id column="admin_user_id" jdbcType="INTEGER" property="adminUserId" />
  <result column="login_user_name" jdbcType="VARCHAR" property="loginUserName" />
  <result column="login_password" jdbcType="VARCHAR" property="loginPassword" />
  <result column="nick_name" jdbcType="VARCHAR" property="nickName" />
  <result column="locked" jdbcType="TINYINT" property="locked" />
</resultMap>
```

最后，按照对应的接口方法，编写具体的 SQL 语句，最终的 AdminUserMapper.xml 文件如下所示：

```xml
<?xml version="1.0" encoding="UTF-8"?>
<!DOCTYPE mapper PUBLIC "-//mybatis.org//DTD Mapper 3.0//EN" "http://mybatis.org/dtd/mybatis-3-mapper.dtd">
<mapper namespace="ltd.newbee.mall.dao.AdminUserMapper">
```

```xml
<resultMap id="BaseResultMap" type="ltd.newbee.mall.entity.AdminUser">
    <id column="admin_user_id" jdbcType="INTEGER" property="adminUserId" />
    <result column="login_user_name" jdbcType="VARCHAR" property="loginUserName" />
    <result column="login_password" jdbcType="VARCHAR" property="loginPassword" />
    <result column="nick_name" jdbcType="VARCHAR" property="nickName" />
    <result column="locked" jdbcType="TINYINT" property="locked" />
</resultMap>
<sql id="Base_Column_List">
    admin_user_id, login_user_name, login_password, nick_name, locked
</sql>

<select id="login" resultMap="BaseResultMap">
    select
    <include refid="Base_Column_List" />
    from tb_newbee_mall_admin_user
    where login_user_name = #{userName,jdbcType=VARCHAR} AND login_password=#{password,jdbcType=VARCHAR} AND locked = 0
</select>
</mapper>
```

这里在 AdminUserMapper.xml 文件中定义了 login() 方法具体执行的 SQL 语句，并通过用户名和密码查询 tb_newbee_mall_admin_user 表中的记录。

16.2.5 业务层代码的实现

首先在 ltd.newbee.mall 包下新建 service 包，再选中 service 包并右键点击，在弹出的菜单中选择 "New→Java Class"，然后在弹出的窗口中输入 "AdminUserService"，并选中 "Interface" 选项，最后在 AdminUserService.java 文件中新增如下代码：

```java
package ltd.newbee.mall.service;

import ltd.newbee.mall.entity.AdminUser;

public interface AdminUserService {

    AdminUser login(String userName, String password);

}
```

首先在 ltd.newbee.mall.service 包下新建 impl 包，再选中 impl 包并右键点击，在弹出的菜单中选择"New→Java Class"，然后在弹出的窗口中输入"AdminUserServiceImpl"，最后在 AdminUserServiceImpl 类中新增如下代码：

```java
package ltd.newbee.mall.service.impl;

import ltd.newbee.mall.dao.AdminUserMapper;
import ltd.newbee.mall.entity.AdminUser;
import ltd.newbee.mall.service.AdminUserService;
import ltd.newbee.mall.util.MD5Util;
import org.springframework.stereotype.Service;

import javax.annotation.Resource;

@Service
public class AdminUserServiceImpl implements AdminUserService {

    @Resource
    private AdminUserMapper adminUserMapper;

    @Override
    public AdminUser login(String userName, String password) {
        String passwordMd5 = MD5Util.MD5Encode(password, "UTF-8");
        return adminUserMapper.login(userName, passwordMd5);
    }

}
```

该类为管理员业务层代码的实现类，用于编写具体的方法实现。login()方法的具体实现逻辑：首先将 password 参数转换为 MD5 格式，然后调用数据层的查询方法获取 AdminUser 对象。

MD5Util 类的代码此处省略，读者可以下载本章源码查看。

16.2.6 管理员登录控制层代码的实现

在 AdminController 类中新增 login()方法，代码如下所示：

```java
@PostMapping(value = "/login")
public String login(@RequestParam("userName") String userName,
                    @RequestParam("password") String password,
```

```java
                    @RequestParam("verifyCode") String verifyCode,
                    HttpSession session) {
    if (StringUtils.isEmpty(verifyCode)) {
        session.setAttribute("errorMsg", "验证码不能为空");
        return "admin/login";
    }
    if (StringUtils.isEmpty(userName) || StringUtils.isEmpty(password)) {
        session.setAttribute("errorMsg", "用户名或密码不能为空");
        return "admin/login";
    }
    String kaptchaCode = session.getAttribute("verifyCode") + "";
    if (StringUtils.isEmpty(kaptchaCode) || !verifyCode.equals(kaptchaCode)) {
        session.setAttribute("errorMsg", "验证码错误");
        return "admin/login";
    }
    AdminUser adminUser = adminUserService.login(userName, password);
    if (adminUser != null) {
        session.setAttribute("loginUser", adminUser.getNickName());
        session.setAttribute("loginUserId", adminUser.getAdminUserId());
        //session 过期时间设置为 7200 秒，即 2 小时
        return "redirect:/admin/index";
    } else {
        session.setAttribute("errorMsg", "登录失败");
        return "admin/login";
    }
}
```

这里首先对参数进行校验，参数中包括登录信息和验证码。验证码的显示和校验的逻辑前文已经介绍，这里就直接整合到了登录功能中。接下来使用参数 verifyCode 与存储在 Session 中的验证码值进行比较，然后调用 adminUserService 业务层代码查询用户对象，最后根据验证结果来跳转页面。如果登录成功则跳转到管理系统的首页；如果失败则带上错误信息返回到登录页，登录页中会显示出登录失败的错误信息。

16.2.7 管理员登录功能演示及注意事项

在编码完成后，启动 Spring Boot 项目。在启动成功后打开浏览器并输入后台管理系统登录页面地址：

```
http://localhost:8080/admin/login
```

第 16 章 后台管理系统登录功能的实现

页面显示效果如图 16-4 所示。

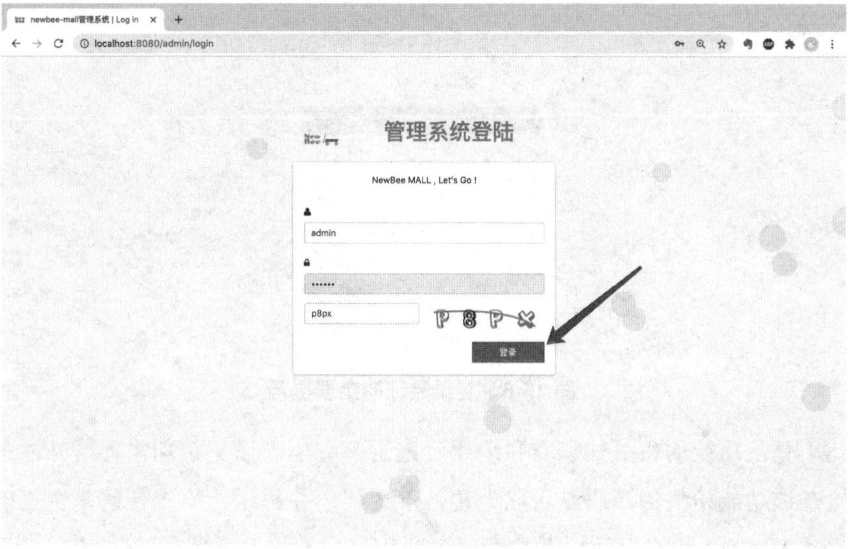

图 16-4 登录功能演示

在该页面依次输入用户名、密码和当前页面展示的验证码，点击"登录"按钮向后端发送请求并完成整个登录过程。如果一切信息都输入正确，则会跳转到 index 页面中，否则会在当前登录页面上提示对应的错误信息。如果登录信息输入错误，或者验证码填写错误，则会分别出现如下错误提示信息，如图 16-5 和图 16-6 所示。

图 16-5 登录信息错误提示

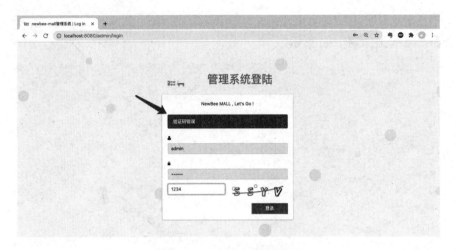

图 16-6 登录验证码错误提示

至此，登录功能的相关知识点和编码实践讲解完毕。以下事项需要格外注意。

①检查该功能相关的源码是否已上传，下载代码并解压，检查代码是否正确

②启动 MySQL 数据库并检查是否包含 tb_newbee_mall_admin_user 表

③检查数据库连接信息是否正确，并改为自己的数据库链接信息

④在项目成功启动后，默认的端口号是 8080，也可自行修改该配置项

希望各位读者都能够顺利进行测试和实践，后续步骤会对整个登录流程和用户身份验证进行完善。

16.3 后台管理系统登录拦截器的实现

接下来将继续对管理员功能模块进行完善。首先是在请求后台资源时进行身份认证，即请求拦截验证，然后完成管理员信息修改、密码修改、登录退出的功能。

16.3.1 登录拦截器

管理员的登录功能已经完成，不过身份认证的整个流程并没有完成。该流程应该包括登录功能、身份认证、访问拦截、退出功能。

登录功能的目的是能够访问后台页面，比如在 16.2 节中访问后台管理系统首页，路

径为/admin/index。但是在当前的代码中,即使不进行登录操作也可以访问该页面。

不仅是首页请求,后台管理系统中其他多个功能模块的页面和接口请求都与此时的 index 请求一样,如果不做权限处理的话,都可以在未登录的状态下访问相关页面。读者可以自行测试。也就是说此时的代码并没有在用户访问后台管理系统的资源时进行身份认证和访问拦截,所以这里需要访问拦截的功能实现。

这个功能比较常见的实现方法就是使用拦截器。

16.3.2 定义拦截器

定义一个拦截器(Interceptor)非常简单,方式也有几种。这里简单列举两种。

① 新建要实现 Spring 的 HandlerInterceptor 接口的类

② 新建继承了 HandlerInterceptor 接口的实现类,比如实现了 HandlerInterceptor 接口的抽象类 HandlerInterceptorAdapter

HandlerInterceptor 接口的主要实现方法如下所示:

```
boolean preHandle(HttpServletRequest request, HttpServletResponse response, Object handler)
        throws Exception;

void postHandle(
        HttpServletRequest request, HttpServletResponse response,
Object handler, ModelAndView modelAndView) throws Exception;

void afterCompletion(
        HttpServletRequest request, HttpServletResponse response,
Object handler, Exception ex) throws Exception;
```

preHandle:在业务处理器处理请求之前被调用,即预处理,可以进行编码、安全控制、权限校验等处理。

postHandle:在业务处理器处理请求执行完成后、生成视图之前执行。

afterCompletion:在 DispatcherServlet 完全处理完请求后被调用,用于清理资源、返回处理(已经渲染了页面)等。

在 ltd.newbee.mall 包下新建 interceptor 包,并在包中新建 AdminLoginInterceptor 类。该类需要实现 HandlerInterceptor 接口,代码如下所示:

```
package ltd.newbee.mall.interceptor;
```

```java
import org.springframework.stereotype.Component;
import org.springframework.web.servlet.HandlerInterceptor;
import org.springframework.web.servlet.ModelAndView;
import javax.servlet.http.HttpServletRequest;
import javax.servlet.http.HttpServletResponse;

/**
 * 后台系统身份验证拦截器
 */
@Component
public class AdminLoginInterceptor implements HandlerInterceptor {

    @Override
    public boolean preHandle(HttpServletRequest request, HttpServletResponse response, Object o) throws Exception {
        System.out.println("进入拦截器...");
        String uri = request.getRequestURI();
        if (uri.startsWith("/admin") && null == request.getSession().getAttribute("loginUser")) {
            request.getSession().setAttribute("errorMsg", "请登录");
            response.sendRedirect(request.getContextPath() + "/admin/login");
            System.out.println("未登录，拦截成功...");
            return false;
        } else {
            request.getSession().removeAttribute("errorMsg");
            System.out.println("已登录，放行...");
            return true;
        }
    }

    @Override
    public void postHandle(HttpServletRequest httpServletRequest, HttpServletResponse httpServletResponse, Object o, ModelAndView modelAndView) throws Exception {
    }

    @Override
    public void afterCompletion(HttpServletRequest httpServletRequest, HttpServletResponse httpServletResponse, Object o, Exception e) throws Exception {

    }
}
```

这里只需要完善 preHandle 方法即可。同时在类声明上方添加@Component 注解使其注册到 IOC 容器中。

通过上面代码可以看出，在请求的预处理过程中，程序会读取当前 Session 对象中是否存在 loginUser 对象。如果不存在则返回 false 并跳转至登录页面；如果已经存在则返回 true，继续后续处理流程。笔者在代码中添加了几条打印语句，可以在测试的时候判断拦截器是否真正产生作用。

16.3.3 配置拦截器

在实现拦截器的相关方法后，需要对该拦截器进行配置以使其生效。在开发时，Spring Boot 1.x 版本通常会继承 WebMvcConfigurerAdapter 类，但是在 Spring Boot 2.x 版本中，WebMvcConfigurerAdapter 被@deprecated 注解标注，说明已过时。而现在 WebMvcConfigurer 接口已经有了默认的空白方法，所以在 Spring Boot 2.x 版本下更好的做法还是实现 WebMvcConfigurer 接口。

首先在 ltd.newbee.mall 包下新建 config 包，然后新建 NeeBeeMallWebMvcConfigurer 类，该类需要实现 WebMvcConfigurer 接口，代码如下所示：

```java
package ltd.newbee.mall.config;

import ltd.newbee.mall.interceptor.AdminLoginInterceptor;
import org.springframework.beans.factory.annotation.Autowired;
import org.springframework.context.annotation.Configuration;
import org.springframework.web.servlet.config.annotation.InterceptorRegistry;
import org.springframework.web.servlet.config.annotation.WebMvcConfigurer;

@Configuration
public class NeeBeeMallWebMvcConfigurer implements WebMvcConfigurer {

    @Autowired
    private AdminLoginInterceptor adminLoginInterceptor;

    public void addInterceptors(InterceptorRegistry registry) {
        // 添加一个拦截器，拦截以/admin 为前缀的 URL 路径（后台登录拦截）
        registry.addInterceptor(adminLoginInterceptor)
                .addPathPatterns("/admin/**")
                .excludePathPatterns("/admin/login")
                .excludePathPatterns("/admin/dist/**")
```

```
                .excludePathPatterns("/admin/plugins/**");
    }
}
```

这里，在该配置类中的 addInterceptors()方法中注册 AdminLoginInterceptor 登录拦截器，并对该拦截器所拦截的路径进行配置。由于后端管理系统的所有请求路径都以/admin 开头，所以配置该拦截器需要拦截的路径为/admin/**。登录页面和部分静态资源文件也是以/admin 开头的，但是这些页面是不需要拦截的，因此需要将这些路径排除。读者注意 addPathPatterns()方法和 excludePathPatterns()两个方法，它们分别是添加路径拦截规则和排除路径拦截规则。

在编码完成后，启动 Spring Boot 项目。在启动成功后打开浏览器并输入后台管理系统首页的访问地址：

```
http://localhost:8080/admin/index
```

此时，用户处于未登录状态，如果访问 index 页面，会被 AdminLoginInterceptor 拦截器拦截并且跳回到登录页面。查看控制台记录也能看到拦截器生效的信息，如图 16-8 所示。

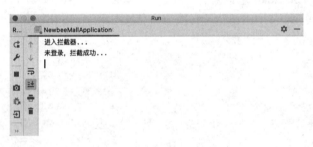

图 16-8　拦截成功的信息

如果管理员已经成功登录则不会出现上述情况。正常登录状态的用户进入拦截器后会被放行，能够访问 index 页面，且控制台会打印"已登录，放行…"的信息，如图 16-9 所示。

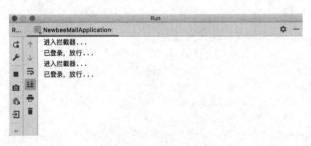

图 16-9　拦截器放行的信息

至此，后台管理系统的登录功能及登录权限的拦截配置就开发完成了。

16.4 管理员模块功能的完善

管理员模块不仅包括登录功能和身份状态验证，还包括用户信息修改、安全退出的功能。

修改 sidebar.html 页面，在侧边栏中加入"修改密码"和"退出登录"两个栏目，代码如下所示：

```html
<li class="nav-header">系统管理</li>
<li class="nav-item">
    <a th:href="@{/admin/profile}"
       th:class="${path}=='profile'?'nav-link active':'nav-link'">
        <i class="fa fa-user-secret nav-icon"></i>
        <p>修改密码</p>
    </a>
</li>
<li class="nav-item">
    <a th:href="@{/admin/logout}" class="nav-link">
        <i class="fa fa-sign-out nav-icon"></i>
        <p>安全退出</p>
    </a>
</li>
```

在 templates/admin 目录下新建 profile.html 模板文件，用于处理管理员信息的显示，前端模板代码如下所示：

```html
<!-- Copyright (c) 2019-2020 十三 all rights reserved. -->
<!DOCTYPE html>
<html xmlns:th="http://www.thymeleaf.org">
<header th:replace="admin/header::header-fragment"></header>
<body class="hold-transition sidebar-mini">
<div class="wrapper">
    <!-- 引入页面头 header-fragment -->
    <div th:replace="admin/header::header-nav"></div>
    <!-- 引入工具栏 sidebar-fragment -->
    <div th:replace="admin/sidebar::sidebar-fragment(${path})"></div>
    <!-- Content Wrapper. Contains page content -->
    <div class="content-wrapper">
        <!-- Content Header (Page header) -->
```

```html
                    <div class="content-header">
                        <div class="container-fluid">
                        </div><!-- /.container-fluid -->
                    </div>
                    <!-- Main content -->
                    <section class="content">
                        <div class="container-fluid">
                            <div class="row">
                                <div class="col-md-6">
                                    <div class="card card-primary card-outline">
                                        <div class="card-header">
                                            <h3 class="card-title">基本信息</h3>
                                        </div> <!-- /.card-body -->
                                        <div class="card-body">
                                            <form role="form" id="userNameForm">
                                                <div class="form-group col-sm-8">
                                                    <div class="alert alert-danger" id="updateUserName-info" style="display: none;"></div>
                                                </div>
                                                <!-- text input -->
                                                <div class="form-group">
                                                    <label>登陆名称</label>
                                                    <input type="text" class="form-control" id="loginUserName"
                                                           name="loginUserName"
                                                           placeholder="请输入登陆名称" required="true" th:value="${loginUserName}">
                                                </div>
                                                <div class="form-group">
                                                    <label>昵称</label>
                                                    <input type="text" class="form-control" id="nickName"
                                                           name="nickName"
                                                           placeholder="请输入昵称" required="true" th:value="${nickName}">
                                                </div>
                                                <div class="card-footer">
                                                    <button type="button" id="updateUserNameButton" onsubmit="return false;"
                                                            class="btn btn-danger float-right">确认修改
                                                    </button>
                                                </div>
```

```html
                </form>
            </div><!-- /.card-body -->
        </div>
    </div>
    <div class="col-md-6">
        <div class="card card-primary card-outline">
            <div class="card-header">
                <h3 class="card-title">修改密码</h3>
            </div> <!-- /.card-body -->
            <div class="card-body">
                <form role="form" id="userPasswordForm">
                    <div class="form-group col-sm-8">
                        <div class="alert alert-danger updatePassword-info" id="updatePassword-info" style="display: none;"></div>
                    </div>
                    <!-- input states -->
                    <div class="form-group">
                        <label class="control-label"><i class="fa fa-key"></i> 原密码</label>
                        <input type="text" class="form-control" id="originalPassword"
                            name="originalPassword"
                            placeholder="请输入原密码" required="true">
                    </div>
                    <div class="form-group">
                        <label class="control-label"><i class="fa fa-key"></i> 新密码</label>
                        <input type="text" class="form-control" id="newPassword" name="newPassword"
                            placeholder="请输入新密码" required="true">
                    </div>
                    <div class="card-footer">
                        <button type="button" id="updatePasswordButton" onsubmit="return false;"
                            class="btn btn-danger float-right">确认修改
                        </button>
                    </div>
                </form>
            </div><!-- /.card-body -->
        </div>
```

```html
                </div>
            </div>
        </div><!-- /.container-fluid -->
    </section>
    <!-- /.content -->
</div>
<!-- /.content-wrapper -->
<!-- 引入页脚 footer-fragment -->
<div th:replace="admin/footer::footer-fragment"></div>
</div>
<!-- jQuery -->
<script th:src="@{/admin/plugins/jquery/jquery.min.js}"></script>
<!-- Bootstrap 4 -->
<script th:src="@{/admin/plugins/bootstrap/js/bootstrap.bundle.min.js}">
</script>
<!-- AdminLTE App -->
<script th:src="@{/admin/dist/js/adminlte.min.js}"></script>
<!-- public.js -->
<script th:src="@{/admin/dist/js/public.js}"></script>
<!-- profile -->
<script th:src="@{/admin/dist/js/profile.js}"></script>
</body>
</html>
```

后端逻辑的实现也在 AdminController 类中，新增代码如下所示：

```java
@GetMapping("/profile")
public String profile(HttpServletRequest request) {
    Integer loginUserId = (int)
request.getSession().getAttribute("loginUserId");
    AdminUser adminUser = adminUserService.getUserDetailById(loginUserId);
    if (adminUser == null) {
        return "admin/login";
    }
    request.setAttribute("path", "profile");
    request.setAttribute("loginUserName", adminUser.getLoginUserName());
    request.setAttribute("nickName", adminUser.getNickName());
    return "admin/profile";
}

@PostMapping("/profile/password")
@ResponseBody
public String passwordUpdate(HttpServletRequest request, @RequestParam
("originalPassword") String originalPassword,
```

```java
                    @RequestParam("newPassword") String newPassword) {
    if (StringUtils.isEmpty(originalPassword) || StringUtils.isEmpty(newPassword)) {
        return "参数不能为空";
    }
    Integer loginUserId = (int) request.getSession().getAttribute("loginUserId");
    if (adminUserService.updatePassword(loginUserId, originalPassword, newPassword)) {
        //修改成功后清空 Session 中的数据，前端控制跳转至登录页
        request.getSession().removeAttribute("loginUserId");
        request.getSession().removeAttribute("loginUser");
        request.getSession().removeAttribute("errorMsg");
        return "success";
    } else {
        return "修改失败";
    }
}


@PostMapping("/profile/name")
@ResponseBody
public String nameUpdate(HttpServletRequest request, @RequestParam("loginUserName") String loginUserName,
                    @RequestParam("nickName") String nickName) {
    if (StringUtils.isEmpty(loginUserName) || StringUtils.isEmpty(nickName)) {
        return "参数不能为空";
    }
    Integer loginUserId = (int) request.getSession().getAttribute("loginUserId");
    if (adminUserService.updateName(loginUserId, loginUserName, nickName)) {
        return "success";
    } else {
        return "修改失败";
    }
}

@GetMapping("/logout")
public String logout(HttpServletRequest request) {
    request.getSession().removeAttribute("loginUserId");
    request.getSession().removeAttribute("loginUser");
    request.getSession().removeAttribute("errorMsg");
    return "admin/login";
}
```

在业务层 AdminUserServiceImpl 类中新增对应的方法，代码如下所示：

```java
@Override
public AdminUser getUserDetailById(Integer loginUserId) {
    return adminUserMapper.selectByPrimaryKey(loginUserId);
}

@Override
public Boolean updatePassword(Integer loginUserId, String originalPassword, String newPassword) {
    AdminUser adminUser = adminUserMapper.selectByPrimaryKey(loginUserId);
    //当前用户非空才可以进行更改
    if (adminUser != null) {
        String originalPasswordMd5 = MD5Util.MD5Encode(originalPassword, "UTF-8");
        String newPasswordMd5 = MD5Util.MD5Encode(newPassword, "UTF-8");
        //比较原密码是否正确
        if (originalPasswordMd5.equals(adminUser.getLoginPassword())) {
            //设置新密码并修改
            adminUser.setLoginPassword(newPasswordMd5);
            if (adminUserMapper.updateByPrimaryKeySelective(adminUser) > 0) {
                //修改成功则返回 true
                return true;
            }
        }
    }
    return false;
}

@Override
public Boolean updateName(Integer loginUserId, String loginUserName, String nickName) {
    AdminUser adminUser = adminUserMapper.selectByPrimaryKey(loginUserId);
    //当前用户非空才可以进行更改
    if (adminUser != null) {
        //设置新名称并修改
        adminUser.setLoginUserName(loginUserName);
        adminUser.setNickName(nickName);
        if (adminUserMapper.updateByPrimaryKeySelective(adminUser) > 0) {
            //修改成功则返回 true
            return true;
```

```
            }
        }
        return false;
}
```

对于需要重点关注的代码，笔者都已经加上了注释。读者可以结合注释来理解管理员昵称修改、管理员密码修改、注销登录这三个功能。

通过源码可以看出，Controller 层的代码首先会接收前端传输过来的参数并做基本的参数判断，然后调用业务层的方法完成功能实现。业务层代码则做具体的逻辑处理，最后调用 Dao 层的方法并执行对应的 SQL 语句完成实际的入库操作。

在管理员昵称修改、管理员密码修改两个功能中，前端请求使用的是 Ajax 方式，对应的 JS 代码在 static/admin/dist/js/profile.js 文件中，源码如下所示：

```
$(function () {
    //修改个人信息
    $('#updateUserNameButton').click(function () {
        $("#updateUserNameButton").attr("disabled",true);
        var userName = $('#loginUserName').val();
        var nickName = $('#nickName').val();
        if (validUserNameForUpdate(userName, nickName)) {
            //Ajax 提交数据
            var params = $("#userNameForm").serialize();
            $.ajax({
                type: "POST",
                url: "/admin/profile/name",
                data: params,
                success: function (r) {
                    console.log(r);
                    if (r == 'success') {
                        alert('修改成功');
                    } else {
                        alert('修改失败');
                    }
                }
            });
        }
    });
    //修改密码
    $('#updatePasswordButton').click(function () {
        $("#updatePasswordButton").attr("disabled",true);
        var originalPassword = $('#originalPassword').val();
```

```javascript
            var newPassword = $('#newPassword').val();
            if (validPasswordForUpdate(originalPassword, newPassword)) {
                var params = $("#userPasswordForm").serialize();
                $.ajax({
                    type: "POST",
                    url: "/admin/profile/password",
                    data: params,
                    success: function (r) {
                        console.log(r);
                        if (r == 'success') {
                            alert('修改成功');
                            window.location.href = '/admin/login';
                        } else {
                            alert('修改失败');
                        }
                    }
                });
            }

        });
    })

    /**
     * 名称验证
     */
    function validUserNameForUpdate(userName, nickName) {
        if (isNull(userName) || userName.trim().length < 1) {
            $('#updateUserName-info').css("display", "block");
            $('#updateUserName-info').html("请输入登陆名称！");
            return false;
        }
        if (isNull(nickName) || nickName.trim().length < 1) {
            $('#updateUserName-info').css("display", "block");
            $('#updateUserName-info').html("昵称不能为空！");
            return false;
        }
        if (!validUserName(userName)) {
            $('#updateUserName-info').css("display", "block");
            $('#updateUserName-info').html("请输入符合规范的登录名！");
            return false;
        }
        if (!validCN_ENString2_18(nickName)) {
            $('#updateUserName-info').css("display", "block");
            $('#updateUserName-info').html("请输入符合规范的昵称！");
```

```
            return false;
        }
        return true;
    }

    /**
     * 密码验证
     */
    function validPasswordForUpdate(originalPassword, newPassword) {
        if (isNull(originalPassword) || originalPassword.trim().length < 1) {
            $('#updatePassword-info').css("display", "block");
            $('#updatePassword-info').html("请输入原密码！");
            return false;
        }
        if (isNull(newPassword) || newPassword.trim().length < 1) {
            $('#updatePassword-info').css("display", "block");
            $('#updatePassword-info').html("新密码不能为空！");
            return false;
        }
        if (!validPassword(newPassword)) {
            $('#updatePassword-info').css("display", "block");
            $('#updatePassword-info').html("请输入符合规范的密码！");
            return false;
        }
        return true;
    }
```

该段代码的主要功能就是对输入的信息进行正则验证，封装数据并向后端发送请求来实现修改名称和密码的功能。

接下来实际操作一下用户模块的登录、登出及信息修改功能。

在编码完成后，启动 Spring Boot 项目。在启动成功后打开浏览器并输入后台管理系统登录页面地址：

```
http://localhost:8080/admin/login
```

在该页面依次输入用户名、密码和验证码登录到后台管理系统中，可以看到工具栏中增加了两个栏目，点击修改密码进入管理员基本信息页面，如图 16-10 所示。

这里笔者演示的是密码修改功能，依次输入原密码和新密码并点击"确认修改"按钮，如果输入的信息正确则会提示密码修改成功。在密码修改成功后，系统会清空 Session 中原来的登录信息并跳转到登录页面进行重新登录。修改密码的功能演示如图 16-11 所示。

图 16-10　管理员基本信息页面

图 16-11　修改管理员密码

名称修改和密码修改的交互过程类似，这里就不再演示了。

退出功能也比较简单，在点击"安全退出"菜单栏后，系统会请求后端的/admin/logout 地址。通过源码可以知道这个方法的作用就是清空 Session 中的登录信息，并直接跳转回登录页面。读者可以在本地自行测试一下。

用户的身份认证流程应该包括登录功能、身份认证、访问拦截、退出功能。本章已经把这些功能全部实现了，至此管理员模块的功能开发就完成了。接下来是商城后台管理系统中其他模块的功能开发。

第 17 章

轮播图管理模块的开发

大部分商城网站都会在商城首页放置轮播图模块。商城端的轮播图效果是呈现给用户看的，只需要展示功能就可以。而轮播图数据的添加和配置则需要在后台管理系统中实现。这样商城端才能够查询出数据并展现给用户。

17.1 轮播图模块介绍

横跨屏幕的轮播图是时下比较流行的网页设计手法。网站设计师会通过这种覆盖用户视线的图片，给用户营造一种身临其境的视觉感受，这非常符合人类视觉优先的信息获取方式。大部分网站也会在首屏选择这种设计，优质的首图能够让用户预先知道网站的内容。

购物网站在首屏轮播图中往往会有各种推荐商品、优惠活动等。在这个区域，商城管理者可以放置抓人眼球的商品图片，可以放置不久后即将上线的主力产品，也可以放置用户最关心的促销通知等。淘宝、京东、小米等商城也采取这种首屏轮播图的网页设计。

小米商城首屏轮播图如图 17-1 所示。

京东商城首屏轮播图如图 17-2 所示。

在新蜂商城中笔者也添加了首屏轮播图的效果，视觉效果如图 17-3 所示。

图 17-1　小米商城首屏轮播图

图 17-2　京东商城首屏轮播图

图 17-3　新蜂商城首屏轮播图

商城端首屏轮播图的实现会在商城首页开发中讲解，本章节主要讲解轮播图的后台管理模块。

不同的时间需要策划不同的促销活动，随着时间的推移商城管理员也会不断上线新品满足用户的需要，因此轮播区域的图片不可能是一成不变的，管理员用户需要根据需要对它进行配置和更改。

17.2 轮播图管理页面跳转逻辑的实现

17.2.1 导航栏中增加"轮播图配置"栏目

首先在左侧导航栏中新增轮播图管理页的导航按钮，在 sidebar.html 文件中新增如下代码：

```html
<li class="nav-header">首页配置</li>
<li class="nav-item">
  <a th:href="@{/admin/carousels}"
     th:class="${path}=='newbee_mall_carousel'?'nav-link active':'nav-link'">
    <i class="nav-icon fa fa-file-image-o"></i>
    <p>
      轮播图配置
    </p>
  </a>
</li>
```

这里，点击后的跳转路径为/admin/carousels，新建 Controller 类来处理该路径并跳转到对应的页面上。

17.2.2 控制类处理跳转逻辑

在 controller/admin 包下新建 NewBeeMallCarouselController.java 类，并新增如下代码：

```java
package ltd.newbee.mall.controller.admin;

import org.springframework.stereotype.Controller;
import org.springframework.web.bind.annotation.*;
```

```
import javax.servlet.http.HttpServletRequest;

@Controller
@RequestMapping("/admin")
public class NewBeeMallCarouselController {

    @GetMapping("/carousels")
    public String carouselPage(HttpServletRequest request) {
        request.setAttribute("path", "newbee_mall_carousel");
        return "admin/newbee_mall_carousel";
    }
}
```

该方法用于处理 /admin/carousels 请求，首先在返回视图前设置 path 字段为 newbee_mall_carousel，然后跳转到 admin 目录下的 newbee_mall_carousel.html 模板页面中。

17.2.3 轮播图管理页面基础样式的实现

接下来就是轮播图管理页面的模板文件制作，在 resources/templates/admin 目录下新建 newbee_mall_carousel.html 模板文件，并引入对应的 JS 文件和 CSS 样式文件，代码如下所示：

```html
<!DOCTYPE html>
<html xmlns:th="http://www.thymeleaf.org">
<header th:replace="admin/header::header-fragment">
</header>
<style>
    .ui-jqgrid tr.jqgrow td {
        white-space: normal !important;
        height: auto;
        vertical-align: text-top;
        padding-top: 2px;
    }
</style>
<body class="hold-transition sidebar-mini">
<div class="wrapper">
    <!-- 引入页面头 header-fragment -->
    <div th:replace="admin/header::header-nav"></div>
    <!-- 引入工具栏 sidebar-fragment -->
```

```html
    <div th:replace="admin/sidebar::sidebar-fragment(${path})"></div>
    <!-- Content Wrapper. Contains（图标content）-->
    <div class="content-wrapper">
        <!-- Content Header（图标header）-->
        <div class="content-header">
            <div class="container-fluid">
            </div><!-- /.container-fluid -->
        </div>
        <!-- Main content -->
        <div class="content">
            <div class="container-fluid">
                <div class="card card-primary card-outline">
                    <div class="card-header">
                        <h3 class="card-title">轮播图管理</h3>
                    </div> <!-- /.card-body -->
                    <div class="card-body">
                        <div class="grid-btn">
                            <button class="btn btn-info" onclick="carouselAdd()"><i
                                    class="fa fa-plus"></i> 新增
                            </button>
                            <button class="btn btn-info" onclick="carouselEdit()"><i
                                    class="fa fa-pencil-square-o"></i> 修改
                            </button>
                            <button class="btn btn-danger" onclick="deleteCarousel()"><i
                                    class="fa fa-trash-o"></i> 删除
                            </button>
                        </div>
                        轮播图管理页面
                    </div><!-- /.card-body -->
                </div>
            </div><!-- /.container-fluid -->
        </div>
    </div>
    <!-- 引入页脚 footer-fragment -->
    <div th:replace="admin/footer::footer-fragment"></div>
</div>
<!-- jQuery -->
<script th:src="@{/admin/plugins/jquery/jquery.min.js}"></script>
<!-- jQuery UI 1.11.4 -->
<script th:src="@{/admin/plugins/jQueryUI/jquery-ui.min.js}"></script>
```

```html
<!-- Bootstrap 4 -->
<script th:src="@{/admin/plugins/bootstrap/js/bootstrap.bundle.min.js}">
</script>
<!-- AdminLTE App -->
<script th:src="@{/admin/dist/js/adminlte.min.js}"></script>
</body>
</html>
```

至此，轮播图管理页面的基础布局和页面的跳转逻辑处理完毕，演示效果如图 17-4 所示。

图 17-4　轮播图配置基础页面

以上只完成了页面的基础内容，在页面上还有三个功能按钮都是静态内容，并没有数据交互，之后会逐步实现这些功能。

17.3　轮播图管理模块后端功能的实现

17.3.1　轮播图表结构设计

在进行接口设计和具体的功能实现前，先要把表结构确定下来。通过对轮播图的介绍，读者应该也可以大致看出两个主要的字段。第一是轮播图片的图片地址，第二是点击轮播图后的跳转链接。其他字段是一些基础的功能字段。新蜂商城轮播图的表结构如下所示：

```sql
USE 'newbee_mall_db ';

DROP TABLE IF EXISTS 'tb_newbee_mall_carousel';
CREATE TABLE 'tb_newbee_mall_carousel'  (
  'carousel_id' int(11) NOT NULL AUTO_INCREMENT COMMENT '首页轮播图主键id',
  'carousel_url' varchar(100) CHARACTER SET utf8 COLLATE utf8_general_ci NOT NULL DEFAULT '' COMMENT '轮播图片url',
  'redirect_url' varchar(100) CHARACTER SET utf8 COLLATE utf8_general_ci NOT NULL DEFAULT '\'##\'' COMMENT '点击后的跳转地址(默认不跳转)',
  'carousel_rank' int(11) NOT NULL DEFAULT 0 COMMENT '排序值(字段越大越靠前)',
  'is_deleted' tinyint(4) NOT NULL DEFAULT 0 COMMENT '删除标识字段(0-未删除 1-已删除)',
  'create_time' datetime(0) NOT NULL DEFAULT CURRENT_TIMESTAMP COMMENT '创建时间',
  'create_user' int(11) NOT NULL DEFAULT 0 COMMENT '创建者id',
  'update_time' datetime(0) NOT NULL DEFAULT CURRENT_TIMESTAMP COMMENT '修改时间',
  'update_user' int(11) NOT NULL DEFAULT 0 COMMENT '修改者id',
  PRIMARY KEY ('carousel_id') USING BTREE
) ENGINE = InnoDB AUTO_INCREMENT = 8 CHARACTER SET = utf8 COLLATE = utf8_general_ci ROW_FORMAT = Dynamic;
```

轮播图表的字段和每个字段对应的含义都在上面的 SQL 中有介绍，读者可以对照 SQL 进行理解，并正确地把该 SQL 导入到数据库中即可。接下来进行编码工作。

17.3.2 轮播图管理模块接口介绍

为了让页面体验更加友好，前后端交互的方式没有采用传统的 MVC 跳转模式，因为一个基础功能对应一个页面这种交互有些浪费资源，翻一页跳转一次也比较烦琐。本书这些功能主要通过 Ajax 异步与后端交互数据实现。管理员用户点击页面上的元素会触发相应的 JS 事件，进而通过 Ajax 的方式向后端请求数据，前端再根据后端返回的数据内容进行响应逻辑的展示。前文个人信息修改功能就是这种实现方式。

轮播图模块在后台管理系统中有 5 个接口，分别如下所示。

①轮播图分页列表接口

②添加轮播图接口

③根据 id 获取单条轮播图记录接口

④修改轮播图接口

⑤批量删除轮播图接口

接下来介绍这些功能具体的实现代码。

17.3.3 新建轮播图实体类和 Mapper 接口

首先在 ltd.newbee.mall.entity 包中创建轮播图实体类，选中 entity 包并右键点击，在弹出的菜单中选择"New→Java Class"，然后在弹出的窗口中输入"Carousel"，最后在 Carousel 类中新增如下代码：

```java
package ltd.newbee.mall.entity;

import com.fasterxml.jackson.annotation.JsonFormat;

import java.util.Date;

public class Carousel {
    private Integer carouselId;

    private String carouselUrl;

    private String redirectUrl;

    private Integer carouselRank;

    private Byte isDeleted;

    @JsonFormat(pattern = "yyyy-MM-dd HH:mm:ss", timezone = "GMT+8")
    private Date createTime;

    private Integer createUser;

    @JsonFormat(pattern = "yyyy-MM-dd HH:mm:ss", timezone = "GMT+8")
    private Date updateTime;

    private Integer updateUser;

    public Integer getCarouselId() {
```

```java
    return carouselId;
}

public void setCarouselId(Integer carouselId) {
    this.carouselId = carouselId;
}

public String getCarouselUrl() {
    return carouselUrl;
}

public void setCarouselUrl(String carouselUrl) {
    this.carouselUrl = carouselUrl == null ? null : carouselUrl.trim();
}

public String getRedirectUrl() {
    return redirectUrl;
}

public void setRedirectUrl(String redirectUrl) {
    this.redirectUrl = redirectUrl == null ? null : redirectUrl.trim();
}

public Integer getCarouselRank() {
    return carouselRank;
}

public void setCarouselRank(Integer carouselRank) {
    this.carouselRank = carouselRank;
}

public Byte getIsDeleted() {
    return isDeleted;
}

public void setIsDeleted(Byte isDeleted) {
    this.isDeleted = isDeleted;
}

public Date getCreateTime() {
    return createTime;
}
```

```java
    public void setCreateTime(Date createTime) {
        this.createTime = createTime;
    }

    public Integer getCreateUser() {
        return createUser;
    }

    public void setCreateUser(Integer createUser) {
        this.createUser = createUser;
    }

    public Date getUpdateTime() {
        return updateTime;
    }

    public void setUpdateTime(Date updateTime) {
        this.updateTime = updateTime;
    }

    public Integer getUpdateUser() {
        return updateUser;
    }

    public void setUpdateUser(Integer updateUser) {
        this.updateUser = updateUser;
    }
}
```

首先在 ltd.newbee.mall.dao 包中新建轮播图实体的 Mapper 接口，选中 dao 包并右键点击，在弹出的菜单中选择"New→Java Class"，在弹出的窗口中输入"CarouselMapper"，然后选中"Interface"选项，最后在 CarouselMapper.java 文件中新增如下代码：

```java
package ltd.newbee.mall.dao;

import ltd.newbee.mall.entity.Carousel;
import ltd.newbee.mall.util.PageQueryUtil;
import org.apache.ibatis.annotations.Param;

import java.util.List;

public interface CarouselMapper {
```

```java
/**
 * 删除一条记录
 * @param carouselId
 * @return
 */
int deleteByPrimaryKey(Integer carouselId);

/**
 * 保存一条新记录
 * @param record
 * @return
 */
int insert(Carousel record);

/**
 * 保存一条新记录
 * @param record
 * @return
 */
int insertSelective(Carousel record);

/**
 * 根据主键查询记录
 * @param carouselId
 * @return
 */
Carousel selectByPrimaryKey(Integer carouselId);

/**
 * 修改记录
 * @param record
 * @return
 */
int updateByPrimaryKeySelective(Carousel record);

/**
 * 修改记录
 * @param record
 * @return
 */
int updateByPrimaryKey(Carousel record);

/**
```

```
 * 查询分页数据
 * @param pageUtil
 * @return
 */
List<Carousel> findCarouselList(PageQueryUtil pageUtil);

/**
 * 查询总数
 * @param pageUtil
 * @return
 */
int getTotalCarousels(PageQueryUtil pageUtil);

/**
 * 批量删除
 * @param ids
 * @return
 */
int deleteBatch(Integer[] ids);

/**
 * 查询固定数量的记录
 * @param number
 * @return
 */
List<Carousel> findCarouselsByNum(@Param("number") int number);
}
```

这里定义了对轮播图操作的数据层方法，包括查询、新增、修改和删除等操作。

17.3.4 创建 CarouselMapper 接口的映射文件

在 resources/mapper 目录下新建 CarouselMapper 接口的映射文件 CarouselMapper.xml，并进行映射文件的编写。

首先，定义映射文件与 Mapper 接口的对应关系。比如在该示例中，需要将 CarouselMapper.xml 文件与对应的 CarouselMapper 接口之间的关系定义出来：

```xml
<mapper namespace="ltd.newbee.mall.dao.CarouselMapper">
```

然后，配置表结构和实体类的对应关系：

```xml
<resultMap id="BaseResultMap" type="ltd.newbee.mall.entity.Carousel">
    <id column="carousel_id" jdbcType="INTEGER" property="carouselId"/>
    <result column="carousel_url" jdbcType="VARCHAR" property="carouselUrl"/>
    <result column="redirect_url" jdbcType="VARCHAR" property="redirectUrl"/>
    <result column="carousel_rank" jdbcType="INTEGER" property="carouselRank"/>
    <result column="is_deleted" jdbcType="TINYINT" property="isDeleted"/>
    <result column="create_time" jdbcType="TIMESTAMP" property="createTime"/>
    <result column="create_user" jdbcType="INTEGER" property="createUser"/>
    <result column="update_time" jdbcType="TIMESTAMP" property="updateTime"/>
    <result column="update_user" jdbcType="INTEGER" property="updateUser"/>
</resultMap>
```

最后，按照对应的接口方法，编写具体的 SQL 语句，最终的 CarouselMapper.xml 文件如下所示：

```xml
<?xml version="1.0" encoding="UTF-8"?>
<!DOCTYPE mapper PUBLIC "-//mybatis.org//DTD Mapper 3.0//EN" "http://mybatis.org/dtd/mybatis-3-mapper.dtd">
<mapper namespace="ltd.newbee.mall.dao.CarouselMapper">
    <resultMap id="BaseResultMap" type="ltd.newbee.mall.entity.Carousel">
        <id column="carousel_id" jdbcType="INTEGER" property="carouselId"/>
        <result column="carousel_url" jdbcType="VARCHAR" property="carouselUrl"/>
        <result column="redirect_url" jdbcType="VARCHAR" property="redirectUrl"/>
        <result column="carousel_rank" jdbcType="INTEGER" property="carouselRank"/>
        <result column="is_deleted" jdbcType="TINYINT" property="isDeleted"/>
        <result column="create_time" jdbcType="TIMESTAMP" property="createTime"/>
        <result column="create_user" jdbcType="INTEGER" property="createUser"/>
        <result column="update_time" jdbcType="TIMESTAMP" property="updateTime"/>
        <result column="update_user" jdbcType="INTEGER" property="updateUser"/>
```

```xml
    </resultMap>
    <sql id="Base_Column_List">
    carousel_id, carousel_url, redirect_url, carousel_rank, is_deleted, create_time,
    create_user, update_time, update_user
    </sql>
    <select id="findCarouselList" parameterType="Map" resultMap="BaseResultMap">
        select
        <include refid="Base_Column_List"/>
        from tb_newbee_mall_carousel
        where is_deleted = 0
        order by carousel_rank desc
        <if test="start!=null and limit!=null">
            limit #{start},#{limit}
        </if>
    </select>
    <select id="getTotalCarousels" parameterType="Map" resultType="int">
        select count(*) from tb_newbee_mall_carousel
        where is_deleted = 0
    </select>
    <select id="selectByPrimaryKey" parameterType="java.lang.Integer" resultMap="BaseResultMap">
        select
        <include refid="Base_Column_List"/>
        from tb_newbee_mall_carousel
        where carousel_id = #{carouselId,jdbcType=INTEGER}
    </select>
    <select id="findCarouselsByNum" parameterType="int" resultMap="BaseResultMap">
        select
        <include refid="Base_Column_List"/>
        from tb_newbee_mall_carousel
        where is_deleted = 0
        order by carousel_rank desc
        limit #{number}
    </select>
    <delete id="deleteByPrimaryKey" parameterType="java.lang.Integer">
        delete from tb_newbee_mall_carousel
        where carousel_id = #{carouselId,jdbcType=INTEGER}
    </delete>
    <insert id="insert" parameterType="ltd.newbee.mall.entity.Carousel">
        insert into tb_newbee_mall_carousel (carousel_id, carousel_url,
```

第 17 章 轮播图管理模块的开发

```xml
            redirect_url,
            carousel_rank, is_deleted, create_time,
            create_user, update_time, update_user
        )
        values (#{carouselId,jdbcType=INTEGER}, #{carouselUrl,jdbcType=VARCHAR}, #{redirectUrl,jdbcType=VARCHAR},
            #{carouselRank,jdbcType=INTEGER}, #{isDeleted,jdbcType=TINYINT}, #{createTime,jdbcType=TIMESTAMP},
            #{createUser,jdbcType=INTEGER}, #{updateTime,jdbcType=TIMESTAMP}, #{updateUser,jdbcType=INTEGER}
        )
    </insert>
    <insert id="insertSelective" parameterType="ltd.newbee.mall.entity.Carousel">
        insert into tb_newbee_mall_carousel
        <trim prefix="(" suffix=")" suffixOverrides=",">
            <if test="carouselId != null">
                carousel_id,
            </if>
            <if test="carouselUrl != null">
                carousel_url,
            </if>
            <if test="redirectUrl != null">
                redirect_url,
            </if>
            <if test="carouselRank != null">
                carousel_rank,
            </if>
            <if test="isDeleted != null">
                is_deleted,
            </if>
            <if test="createTime != null">
                create_time,
            </if>
            <if test="createUser != null">
                create_user,
            </if>
            <if test="updateTime != null">
                update_time,
            </if>
            <if test="updateUser != null">
                update_user,
            </if>
```

```xml
        </trim>
        <trim prefix="values (" suffix=")" suffixOverrides=",">
            <if test="carouselId != null">
                #{carouselId,jdbcType=INTEGER},
            </if>
            <if test="carouselUrl != null">
                #{carouselUrl,jdbcType=VARCHAR},
            </if>
            <if test="redirectUrl != null">
                #{redirectUrl,jdbcType=VARCHAR},
            </if>
            <if test="carouselRank != null">
                #{carouselRank,jdbcType=INTEGER},
            </if>
            <if test="isDeleted != null">
                #{isDeleted,jdbcType=TINYINT},
            </if>
            <if test="createTime != null">
                #{createTime,jdbcType=TIMESTAMP},
            </if>
            <if test="createUser != null">
                #{createUser,jdbcType=INTEGER},
            </if>
            <if test="updateTime != null">
                #{updateTime,jdbcType=TIMESTAMP},
            </if>
            <if test="updateUser != null">
                #{updateUser,jdbcType=INTEGER},
            </if>
        </trim>
    </insert>
    <update id="updateByPrimaryKeySelective" parameterType="ltd.newbee.mall.entity.Carousel">
        update tb_newbee_mall_carousel
        <set>
            <if test="carouselUrl != null">
                carousel_url = #{carouselUrl,jdbcType=VARCHAR},
            </if>
            <if test="redirectUrl != null">
                redirect_url = #{redirectUrl,jdbcType=VARCHAR},
            </if>
            <if test="carouselRank != null">
                carousel_rank = #{carouselRank,jdbcType=INTEGER},
```

```xml
                </if>
                <if test="isDeleted != null">
                    is_deleted = #{isDeleted,jdbcType=TINYINT},
                </if>
                <if test="createTime != null">
                    create_time = #{createTime,jdbcType=TIMESTAMP},
                </if>
                <if test="createUser != null">
                    create_user = #{createUser,jdbcType=INTEGER},
                </if>
                <if test="updateTime != null">
                    update_time = #{updateTime,jdbcType=TIMESTAMP},
                </if>
                <if test="updateUser != null">
                    update_user = #{updateUser,jdbcType=INTEGER},
                </if>
            </set>
            where carousel_id = #{carouselId,jdbcType=INTEGER}
    </update>
    <update id="updateByPrimaryKey" parameterType="ltd.newbee.mall.entity.Carousel">
        update tb_newbee_mall_carousel
        set carousel_url = #{carouselUrl,jdbcType=VARCHAR},
            redirect_url = #{redirectUrl,jdbcType=VARCHAR},
            carousel_rank = #{carouselRank,jdbcType=INTEGER},
            is_deleted = #{isDeleted,jdbcType=TINYINT},
            create_time = #{createTime,jdbcType=TIMESTAMP},
            create_user = #{createUser,jdbcType=INTEGER},
            update_time = #{updateTime,jdbcType=TIMESTAMP},
            update_user = #{updateUser,jdbcType=INTEGER}
        where carousel_id = #{carouselId,jdbcType=INTEGER}
    </update>
    <update id="deleteBatch">
        update tb_newbee_mall_carousel
        set is_deleted=1,update_time=now() where carousel_id in
        <foreach item="id" collection="array" open="(" separator="," close=")">
            #{id}
        </foreach>
    </update>
</mapper>
```

17.3.5 业务层的代码实现

首先在 ltd.newbee.mall.service 包中新建业务处理类，选中 service 包并右键点击，在弹出的菜单中选择 "New→Java Class"，然后在弹出的窗口中输入 "NewBeeMallCarouselService"，并选中 "Interface" 选项，最后在 NewBeeMallCarouselService.java 文件中新增如下代码：

```java
package ltd.newbee.mall.service;

import ltd.newbee.mall.entity.Carousel;
import ltd.newbee.mall.util.PageQueryUtil;
import ltd.newbee.mall.util.PageResult;

public interface NewBeeMallCarouselService {
    /**
     * 查询后台管理系统轮播图分页数据
     *
     * @param pageUtil
     * @return
     */
    PageResult getCarouselPage(PageQueryUtil pageUtil);

    /**
     * 新增一条轮播图记录
     *
     * @param carousel
     * @return
     */
    String saveCarousel(Carousel carousel);

    /**
     * 修改一条轮播图记录
     *
     * @param carousel
     * @return
     */
    String updateCarousel(Carousel carousel);

    /**
     * 根据主键查询轮播图记录
```

```
 *
 * @param id
 * @return
 */
Carousel getCarouselById(Integer id);

/**
 * 批量删除轮播图记录
 *
 * @param ids
 * @return
 */
Boolean deleteBatch(Integer[] ids);
}
```

至此，轮播图模块业务层的方法定义和每个方法的作用都已经编写完成。

首先在 ltd.newbee.mall.service.impl 包中新建 NewBeeMallCarouselService 的实现类，首先选中 impl 包并右键点击，在弹出的菜单中选择"New→Java Class"，然后在弹出的窗口中输入"NewBeeMallCarouselServiceImpl"，最后在 NewBeeMallCarouselServiceImpl 类中新增如下代码：

```
package ltd.newbee.mall.service.impl;

import ltd.newbee.mall.common.ServiceResultEnum;
import ltd.newbee.mall.dao.CarouselMapper;
import ltd.newbee.mall.entity.Carousel;
import ltd.newbee.mall.service.NewBeeMallCarouselService;
import ltd.newbee.mall.util.PageQueryUtil;
import ltd.newbee.mall.util.PageResult;
import org.springframework.beans.factory.annotation.Autowired;
import org.springframework.stereotype.Service;

import java.util.Date;
import java.util.List;

@Service
public class NewBeeMallCarouselServiceImpl implements NewBeeMallCarouselService {

    @Autowired
    private CarouselMapper carouselMapper;

    @Override
```

```java
    public PageResult getCarouselPage(PageQueryUtil pageUtil) {
        List<Carousel> carousels = carouselMapper.findCarouselList(pageUtil);
        int total = carouselMapper.getTotalCarousels(pageUtil);
        PageResult pageResult = new PageResult(carousels, total, pageUtil.getLimit(), pageUtil.getPage());
        return pageResult;
    }

    @Override
    public String saveCarousel(Carousel carousel) {
        if (carouselMapper.insertSelective(carousel) > 0) {
            return ServiceResultEnum.SUCCESS.getResult();
        }
        return ServiceResultEnum.DB_ERROR.getResult();
    }

    @Override
    public String updateCarousel(Carousel carousel) {
        Carousel temp = carouselMapper.selectByPrimaryKey(carousel.getCarouselId());
        if (temp == null) {
            return ServiceResultEnum.DATA_NOT_EXIST.getResult();
        }
        temp.setCarouselRank(carousel.getCarouselRank());
        temp.setRedirectUrl(carousel.getRedirectUrl());
        temp.setCarouselUrl(carousel.getCarouselUrl());
        temp.setUpdateTime(new Date());
        if (carouselMapper.updateByPrimaryKeySelective(temp) > 0) {
            return ServiceResultEnum.SUCCESS.getResult();
        }
        return ServiceResultEnum.DB_ERROR.getResult();
    }

    @Override
    public Carousel getCarouselById(Integer id) {
        return carouselMapper.selectByPrimaryKey(id);
    }

    @Override
    public Boolean deleteBatch(Integer[] ids) {
        if (ids.length < 1) {
            return false;
        }
```

```
        //删除数据
        return carouselMapper.deleteBatch(ids) > 0;
    }
}
```

17.3.6 轮播图管理模块控制层的代码实现

在 NewBeeMallCarouselController 控制器中新增上述接口的实现代码，最终 NewBeeMallCarouselController 类的代码如下所示：

```
package ltd.newbee.mall.controller.admin;

import ltd.newbee.mall.common.ServiceResultEnum;
import ltd.newbee.mall.entity.Carousel;
import ltd.newbee.mall.service.NewBeeMallCarouselService;
import ltd.newbee.mall.util.PageQueryUtil;
import ltd.newbee.mall.util.Result;
import ltd.newbee.mall.util.ResultGenerator;
import org.springframework.stereotype.Controller;
import org.springframework.util.StringUtils;
import org.springframework.web.bind.annotation.*;

import javax.annotation.Resource;
import javax.servlet.http.HttpServletRequest;
import java.util.Map;
import java.util.Objects;

@Controller
@RequestMapping("/admin")
public class NewBeeMallCarouselController {

    @Resource
    NewBeeMallCarouselService newBeeMallCarouselService;

    @GetMapping("/carousels")
    public String carouselPage(HttpServletRequest request) {
        request.setAttribute("path", "newbee_mall_carousel");
        return "admin/newbee_mall_carousel";
    }

    /**
```

```java
 * 列表
 */
@RequestMapping(value = "/carousels/list", method = RequestMethod.GET)
@ResponseBody
public Result list(@RequestParam Map<String, Object> params) {
    if (StringUtils.isEmpty(params.get("page")) || StringUtils.isEmpty(params.get("limit"))) {
        return ResultGenerator.genFailResult("参数异常!");
    }
    PageQueryUtil pageUtil = new PageQueryUtil(params);
    return ResultGenerator.genSuccessResult(newBeeMallCarouselService.getCarouselPage(pageUtil));
}

/**
 * 添加
 */
@RequestMapping(value = "/carousels/save", method = RequestMethod.POST)
@ResponseBody
public Result save(@RequestBody Carousel carousel) {
    if (StringUtils.isEmpty(carousel.getCarouselUrl())
            || Objects.isNull(carousel.getCarouselRank())) {
        return ResultGenerator.genFailResult("参数异常!");
    }
    String result = newBeeMallCarouselService.saveCarousel(carousel);
    if (ServiceResultEnum.SUCCESS.getResult().equals(result)) {
        return ResultGenerator.genSuccessResult();
    } else {
        return ResultGenerator.genFailResult(result);
    }
}

/**
 * 修改
 */
@RequestMapping(value = "/carousels/update", method = RequestMethod.POST)
@ResponseBody
public Result update(@RequestBody Carousel carousel) {
    if (Objects.isNull(carousel.getCarouselId())
            || StringUtils.isEmpty(carousel.getCarouselUrl())
            || Objects.isNull(carousel.getCarouselRank())) {
        return ResultGenerator.genFailResult("参数异常!");
```

```java
        }
        String result = newBeeMallCarouselService.updateCarousel(carousel);
        if (ServiceResultEnum.SUCCESS.getResult().equals(result)) {
            return ResultGenerator.genSuccessResult();
        } else {
            return ResultGenerator.genFailResult(result);
        }
    }

    /**
     * 详情
     */
    @GetMapping("/carousels/info/{id}")
    @ResponseBody
    public Result info(@PathVariable("id") Integer id) {
        Carousel carousel = newBeeMallCarouselService.getCarouselById(id);
        if (carousel == null) {
            return ResultGenerator.genFailResult(ServiceResultEnum.DATA_NOT_EXIST.getResult());
        }
        return ResultGenerator.genSuccessResult(carousel);
    }

    /**
     * 删除
     */
    @RequestMapping(value = "/carousels/delete", method = RequestMethod.POST)
    @ResponseBody
    public Result delete(@RequestBody Integer[] ids) {
        if (ids.length < 1) {
            return ResultGenerator.genFailResult("参数异常!");
        }
        if (newBeeMallCarouselService.deleteBatch(ids)) {
            return ResultGenerator.genSuccessResult();
        } else {
            return ResultGenerator.genFailResult("删除失败");
        }
    }
}
```

首先列表接口负责接收前端传来的分页参数，比如 page、limit 等参数。然后系统将数据总数和对应页面的数据列表查询出来并封装为分页数据返回给前端。接口的映射地址为/admin/carousels/list，请求方法为 GET。业务层的 getCarouselPage()方法在获取响应条数的记录和总数后再进行数据封装。接口根据前端传递的分页参数进行查询并返回分页数据以供前端页面进行数据渲染。

添加接口负责接收前端的 POST 请求并处理其中的参数。接收的参数为 carouselUrl 字段、redirectUrl 字段和 carouselRank 字段。在这个方法里笔者使用了@RequestBody 注解将其转换为 Carousel 对象参数。

删除接口负责接收前端的轮播图删除请求。在处理前端传输过来的数据后，将这些记录从数据库中删除，这里的删除功能并不是真正意义上的删除，而是逻辑删除。将接受的参数设置为一个数组，可以同时删除多条记录，只需要在前端将用户选择的记录 id 封装好再传参到后端即可。接口的请求路径为 /admin/carousels/delete，并使用@RequestBody 注解将前端传过来的参数封装为数组对象。如果数组为空则直接返回异常提醒，如果参数验证通过则调用 deleteBatch()批量删除方法进行数据库操作。

逻辑删除是目前企业开发中比较常用的一种实现方法。比如本章轮播图模块的删除功能，就是将轮播图表中的 is_deleted 字段设置为 1，表示该记录已被删除，而不是执行 DELETE 语句将数据删除。

17.4 轮播图管理模块前端功能的实现

17.4.1 功能按钮和分页信息展示区域

轮播图管理模块有如下功能：轮播图列表、轮播图信息增加、轮播图信息编辑、轮播图信息删除。

其中，列表功能可以使用 jqGrid 分页插件在页面加载时触发实现。另外三个功能则需要在页面中设置功能按钮并设置触发事件。在 17.2 节的页面制作时已经设置了三个功能按钮，分别是添加按钮，对应的触发事件是 carouselAdd()方法；修改按钮，对应的触发事件是 carouselEdit()方法；删除按钮，对应的触发事件是 deleteCarousel()方法。这些方法将在后续前端代码的实现中给出。

关于分页功能的实现，前文已经做了详细的介绍，这里可以直接在页面中引入 jqGrid 的相关静态资源文件，并在页面中展示分页数据的区域增加如下代码：

```
<table id="jqGrid" class="table table-bordered"></table>
```

```
<div id="jqGridPager"></div>
```

此时还没有与后端进行数据交互,接下来将结合 Ajax 和后端接口实现具体的功能。

17.4.2 轮播图管理页面分页功能的实现

在 resources/static/admin/dist/js 目录下新增 newbee_mall_carousel.js 文件,并增加如下代码:

```
$(function () {
    $("#jqGrid").jqGrid({
        url: '/admin/carousels/list',
        datatype: "json",
        colModel: [
            {label: 'id', name: 'carouselId', index: 'carouselId', width: 50, key: true, hidden: true},
            {label: '轮播图', name: 'carouselUrl', index: 'carouselUrl', width: 180, formatter: coverImageFormatter},
            {label: '跳转链接', name: 'redirectUrl', index: 'redirectUrl', width: 120},
            {label: '排序值', name: 'carouselRank', index: 'carouselRank', width: 120},
            {label: '添加时间', name: 'createTime', index: 'createTime', width: 120}
        ],
        height: 560,
        rowNum: 10,
        rowList: [10, 20, 50],
        styleUI: 'Bootstrap',
        loadtext: '信息读取中...',
        rownumbers: false,
        rownumWidth: 20,
        autowidth: true,
        multiselect: true,
        pager: "#jqGridPager",
        jsonReader: {
            root: "data.list",
            page: "data.currPage",
            total: "data.totalPage",
            records: "data.totalCount"
        },
```

```
        prmNames: {
            page: "page",
            rows: "limit",
            order: "order",
        },
        gridComplete: function () {
            //隐藏 grid 底部滚动条
            $("#jqGrid").closest(".ui-jqgrid-bdiv").css({"overflow-x":
"hidden"});
        }
    });

    function coverImageFormatter(cellvalue) {
        return "<img src='" + cellvalue + "' height=\"120\" width=\"160\"
alt='coverImage'/>";
    }

    $(window).resize(function () {
        $("#jqGrid").setGridWidth($(".card-body").width());
    });
});
```

以上代码的主要功能为分页数据展示、字段格式化、jqGrid DOM 宽度的自适应。在页面加载时，首先调用 jqGrid 的初始化方法，将页面中 id 为 jqGrid 的 DOM 渲染为分页表格，并向后端发送请求，请求地址为/admin/carousels/list，然后按照后端返回的 JSON 数据填充表格和表格下方的分页按钮。可以参考前文 jqGrid 分页功能整合进行理解。最后，重启项目验证一下轮播图数据分页功能是否正常。

17.4.3　添加和修改按钮触发事件及 Modal 框的实现

添加和修改两个按钮分别绑定了触发事件。在 newbee_mall_carousel.js 文件中新增 carouselAdd()方法和 carouselEdit()方法，两个方法中的实现为打开信息编辑框。

接下来会介绍 Modal 框的代码和两个按钮的触发事件，代码如下所示：

```
<div class="content">
    <!-- Modal 框 -->
    <div class="modal fade" id="carouselModal" tabindex="-1" role="dialog"
        aria-labelledby="carouselModalLabel">
        <div class="modal-dialog" role="document">
            <div class="modal-content">
```

```html
                <div class="modal-header">
                    <button type="button" class="close" data-dismiss="modal" aria-label="Close"><span aria-hidden="true">&times;</span></button>
                    <h6 class="modal-title" id="carouselModalLabel">Modal</h6>
                </div>
                <div class="modal-body">
                    <form id="indexConfigForm">
                        <div class="form-group">
                            <div class="alert alert-danger" id="edit-error-msg" style="display: none;">
                                错误信息展示栏。
                            </div>
                        </div>
                        <div class="form-group">
                            <div class="col-sm-4">
                                <img id="carouselImg" src="/admin/dist/img/img-upload.png"
                                     style="height: 64px;width: 64px;">
                            </div>
                        </div>
                        <br>
                        <div class="form-group">
                            <div class="col-sm-4">
                                <button class="btn btn-info" style="margin-bottom: 5px;"
                                        id="uploadCarouselImage">
                                    <i class="fa fa-picture-o"></i> 上传轮播图
                                </button>
                            </div>
                        </div>
                        <div class="form-group">
                            <label for="redirectUrl" class="control-label">跳转链接:</label>
                            <input type="text" class="form-control" id="redirectUrl" name="redirectUrl"
                                   placeholder="请输入跳转链接" value="##">
                        </div>
                        <div class="form-group">
                            <label for="carouselRank" class="control-label">排序值:</label>
```

```html
                    <input type="number" class="form-control" id=
"carouselRank" name="carouselRank"
                           placeholder="请输入排序值">
                </div>
            </form>
        </div>
        <div class="modal-footer">
            <button type="button" class="btn btn-default" data-dismiss=
"modal">取消</button>
            <button type="button" class="btn btn-primary" id="save
Button"> 确认</button>
        </div>
    </div>
  </div>
  <!-- /.modal -->
</div>
```

在轮播图模块中需要进行图片上传操作，并获取上传图片的 URL，并整合图片上传相关功能。关于图片上传读者可以结合前文相关的内容进行理解。引入文件上传的插件，并在 Modal 框中实例化文件上传按钮，这部分代码省略，读者可以自行下载本章代码查看。

carouselAdd()方法和 carouselEdit()方法实现代码如下所示：

```javascript
function carouselAdd() {
    reset();
    $('.modal-title').html('轮播图添加');
    $('#carouselModal').modal('show');
}

function carouselEdit() {
    reset();
    var id = getSelectedRow();
    if (id == null) {
        return;
    }
    //请求数据
    $.get("/admin/carousels/info/" + id, function (r) {
        if (r.resultCode == 200 && r.data != null) {
            //填充数据至modal
            $("#carouselImg").attr("src", r.data.carouselUrl);
            $("#carouselImg").attr("style", "height: 64px;width: 64px;
```

```
display:block;");
        $("#redirectUrl").val(r.data.redirectUrl);
        $("#carouselRank").val(r.data.carouselRank);
    }
});
$('.modal-title').html('轮播图编辑');
$('#carouselModal').modal('show');
}
```

添加按钮的作用仅仅是将 Modal 框显示。在修改功能时还有一个步骤，即将选择的记录回显到 Modal 框中以供修改，因此需要请求/admin/carousels/info/{id}轮播图详情接口获取被修改的轮播图数据信息并赋值到对应的 HTML 标签中。

Modal 框显示效果如图 17-5 所示。

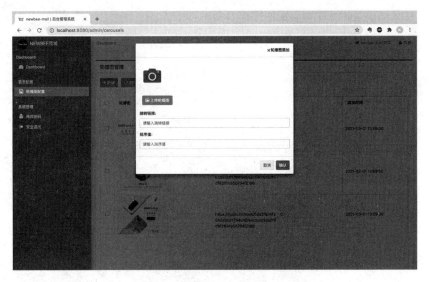

图 17-5　轮播图 Modal 框显示效果

17.4.4　轮播图管理页面添加和编辑功能的实现

在信息录入完成后可以点击信息编辑框下方的"确认"按钮进行数据的交互。JS 实现的代码如下所示：

```
//绑定 Modal 框上的保存按钮
$('#saveButton').click(function () {
```

```javascript
var redirectUrl = $("#redirectUrl").val();
var carouselRank = $("#carouselRank").val();
var carouselUrl = $('#carouselImg')[0].src;
var data = {
    "carouselUrl": carouselUrl,
    "carouselRank": carouselRank,
    "redirectUrl": redirectUrl
};
var url = '/admin/carousels/save';
var id = getSelectedRowWithoutAlert();
if (id != null) {
    url = '/admin/carousels/update';
    data = {
        "carouselId": id,
        "carouselUrl": carouselUrl,
        "carouselRank": carouselRank,
        "redirectUrl": redirectUrl
    };
}
$.ajax({
    type: 'POST',//方法类型
    url: url,
    contentType: 'application/json',
    data: JSON.stringify(data),
    success: function (result) {
        if (result.resultCode == 200) {
            $('#carouselModal').modal('hide');
            swal("保存成功", {
                icon: "success",
            });
            reload();
        } else {
            $('#carouselModal').modal('hide');
            swal(result.message, {
                icon: "error",
            });
        }
        ;
    },
    error: function () {
        swal("操作失败", {
            icon: "error",
        });
```

```
        }
    });
});
```

由于传参和后续处理逻辑类似,为了避免太多重复代码笔者将添加和修改两个触发方法写在了一起,并通过 id 是否大于 0 来确定是修改操作还是添加操作,步骤如下所示。

①前端对用户上传的图片和输入的数据进行简单的正则验证

②封装轮播图实体的参数

③向对应的后端轮播图添加或者修改接口发送 Ajax 请求

④在请求成功后提醒用户请求成功并隐藏当前的轮播图信息 Modal 框,同时刷新轮播图列表数据

⑤请求失败则提醒对应的错误信息

17.4.5　轮播图管理页面删除功能的实现

删除按钮的点击触发事件为 deleteCarousel(),在 newbee_mall_carousel.js 文件中新增如下代码:

```
function deleteCarousel() {
    // 通过 jqGrid 提供的方法获取当前选中列的主键 id
    var ids = getSelectedRows();
    if (ids == null) {
        return;
    }
    swal({
        title: "确认弹框",
        text: "确认要删除数据吗?",
        icon: "warning",
        buttons: true,
        dangerMode: true,
    }).then((flag) => {
        if (flag) {
            $.ajax({
                type: "POST",
                url: "/admin/carousels/delete",
                contentType: "application/json",
                data: JSON.stringify(ids),
                success: function (r) {
```

```
                    if (r.resultCode == 200) {
                        swal("删除成功", {
                            icon: "success",
                        });
                        $("#jqGrid").trigger("reloadGrid");
                    } else {
                        swal(r.message, {
                            icon: "error",
                        });
                    }
                }
            });
        }
    }
)
;
}
```

先获取用户在 jqGrid 表格中选择的需要删除的所有记录的 id。getSelectedRows()方法的实现代码如下所示：

```
/**
 * 获取jqGrid选中的多条记录
 * @returns {*}
 */
function getSelectedRows() {
    var grid = $("#jqGrid");
    var rowKey = grid.getGridParam("selrow");
    if (!rowKey) {
        swal("请选择一条记录", {
            icon: "warning",
        });
        return;
    }
    return grid.getGridParam("selarrrow");
}
```

首先这里调用 jqGrid 内部方法 getGridParam("selrow")获取当前选中列的主键，然后将参数封装并向后端发送请求，请求地址为/admin/carousels/delete。在后端接收到请求后会将对应的记录删除。

17.4.6 功能测试

在编码完成后，启动 Spring Boot 项目。在启动成功后打开浏览器并输入后台管理系统登录页面地址：

http://localhost:8080/admin/login

输入登录信息并登录后台管理系统，点击左侧导航栏中的"轮播图管理"就可以对相关功能进行测试，演示过程如下所示。

（1）列表功能。进入轮播图管理页面就可以看到数据列表，当前列表中有 3 条记录，如图 17-6 所示。

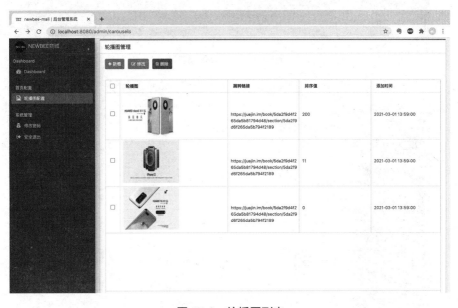

图 17-6　轮播图列表

（2）添加功能。首先点击"新增"按钮会弹出信息编辑框，新增轮播图如图 17-7 所示。

然后上传轮播图片、在 input 框中分别输入设置跳转链接和排序值，最后点击确认按钮。在添加成功后可以看到列表中多了一条数据，如图 17-8 所示。

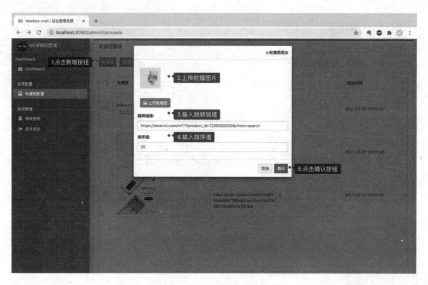

图 17-7　新增轮播图

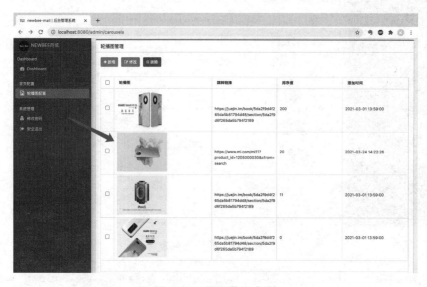

图 17-8　数据添加成功

（3）修改功能。首先选中一条需要修改的记录，并点击"修改"按钮，弹出信息编辑框。此时编辑框中会显示当前记录的所有信息，管理员用户可以修改这些内容，如图 17-9 所示。

第 17 章 轮播图管理模块的开发

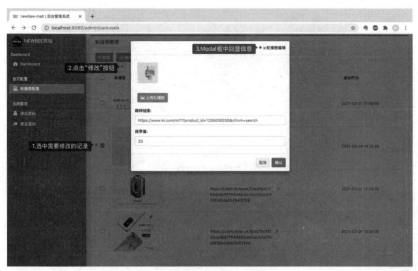

图 17-9 修改轮播图

然后对当前需要选中的轮播图信息进行修改，最后点击"确认"按钮。在修改成功后可以看到之前选中的那一条数据已经修改完成，如图 17-10 所示。

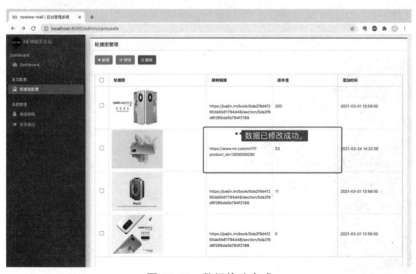

图 17-10 数据修改完成

（4）删除功能。选中需要删除的记录，之后点击"删除"按钮，此时会出现一个确认弹框，如图 17-11 所示。

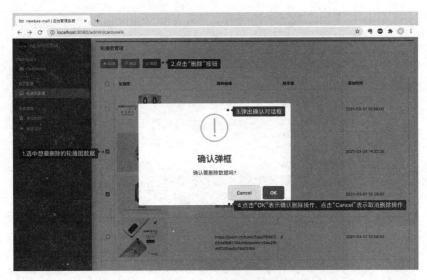

图 17-11　删除轮播图

点击"OK"按钮则表示确认该删除操作，点击"Cancel"按钮则表示取消操作。点击"OK"按钮选中的记录会被删除，如图 17-12 所示。原本列表中有 4 条数据，删除后只剩 2 条数据。

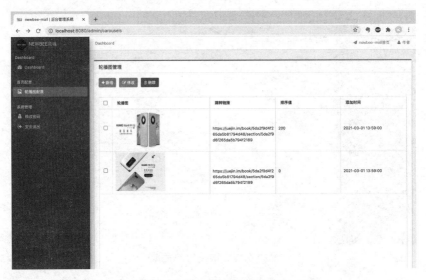

图 17-12　删除数据成功

读者可以按照文中的思路和过程自行测试。至此，轮播图管理模块开发完成！

第 18 章

分类管理模块的开发

除了轮播图数据之外，商品分类数据也是非常重要的。在商城端页面中会展示商品分类信息供用户快速搜索需要的商品。同样地，商城端只需要查询分类数据并展示给用户即可。分类数据的添加和配置则需要在后台管理系统中实现。

18.1 分类管理模块介绍

18.1.1 商品分类

分类是通过比较事物之间的相似性，把具有某些共同点或相似特征的事物归属于一个不确定集合的逻辑方法。而将事物进行分门别类的作用是使一个大集合中的内容条理清楚、层次分明。分类在电商中也叫类目。如果要设计一个商品系统，首先需要把分类系统做好，因为它是商品管理系统非常基础和重要的一个环节。

商品分类就是将商品进行分门别类。一部分商品是衣服，一部分商品是数码产品，另外一部分商品是美妆/护理产品，等等。这样处理的好处就是方便用户筛选和辨别。以天猫商城和京东商城为例，在商城首屏中很大一部分版面都可以进行分类的选择。用户在这里可以通过分类的设置快速进入对应的商品列表页面并进行商品选择。

天猫商城分类显示效果如图 18-1 所示。

图 18-1　天猫商城分类显示效果

京东商城分类显示效果如图 18-2 所示。

图 18-2　京东商城分类显示效果

18.1.2　分类层级

通过观察天猫和京东商城的分类模块，能够看出二者分类层级的设计。在不同的层级下，商城系统需要对商品做进一步的归类操作。因为商品规模和业务的不同，不同层级的展现效果也不同。天猫和京东商城在分类层级的设计思路上是相同的，即三级分层。

如果不设置一定的分类层级，过多的商品类目会造成用户筛选的困难。在设置分类层级后，用户在查找商品时可以遵循"先大类后小类"的原则。比如用户想买一部手机，可以首先在一级分类中筛选并定位到"手机/数码"中，然后在该类目的子分类下再进行筛选。

当然，也有人会提出设计更多层级的分类，比如四级分类、五级分类等。层级太多的话，一是对用户不太友好，层级太多反而不利于搜索；二是对后台管理人员不友好，层级太多不方便管理。目前大部分商城应用选择的分类层级是三级，所以新蜂商城就直接将其设置成三级类目。

18.1.3 分类模块的主要功能

在新蜂商城后台管理系统中，分类模块的主要功能可以整理为以下两点。

①分类数据的设置

②商品数据与分类数据的挂靠和关联

分类数据的存在可以让用户在商城端正常筛选商品，其操作设置包括分类信息的添加、修改等。商品与分类的挂靠和关联是指将商品数据与分类数据建立联系。比如在商品表中设置一个分类 id 的关联字段，使得商品与分类之间产生关联关系。这样就能够通过对应的分类搜索到对应的商品列表。搜索功能的实现会在商城端的开发中讲解，本章主要讲解商品分类后台管理模块的接口实现和页面制作。

18.2 商品类目管理模块前端页面的制作

18.2.1 在导航栏中增加"分类管理"栏目

先在左侧导航栏中新增分类管理页的导航按钮。在 sidebar.html 文件中新增如下代码：

```
<li class="nav-header">管理模块</li>
<li class="nav-item">
    <a th:href="@{/admin/categories?parentId=0&categoryLevel=1&backParentId=0}"
```

```
        th:class="${path}=='newbee_mall_category'?'nav-link active':'nav-
link'">
        <i class="fa fa-list-alt nav-icon" aria-hidden="true"></i>
        <p>
            分类管理
        </p>
    </a>
</li>
```

这里的跳转路径为/admin/categories。在 URL 路径后带有三个参数，分别如下所示。

①parentId：父级分类 id

②categoryLevel：分类级别

③backParentId：返回上一级分类所需的父级分类 id

由于当前设计的商品类目为三级，所以在页面跳转时需要带上这三个参数，并以此确定页面需要展示的分类级数和对应类目的分类数据。如果不设计这些过滤参数的话，每次进入分类管理页面都会显示所有的分类，不利于管理，也不利于查看分类与分类之间的关系。

比如在当前 sidebar.html 中设置的路径：

```
/admin/categories?parentId=0&categoryLevel=1&backParentId=0
```

该路径就表示当前请求的是所有一级分类的列表。在当前系统中默认先显示一级分类的列表，然后根据一级分类的 id 再跳转到二级分类数据的列表页（三级分类同理）。此时，在跳转链接中的三个参数值会发生相应的改变。二级分类和三级分类的数据请求会在后文进行介绍。

最后，新建一个控制器方法来处理该路径并跳转到对应的页面。

18.2.2 控制类处理跳转逻辑

首先在 ltd.newbee.mall.controller.admin 包中新建 NewBeeMallGoodsCategoryController 控制类，选中 admin 包并右键点击，在弹出的菜单中选择"New→Java Class"，然后在弹出的窗口中输入"NewBeeMallGoodsCategoryController"，最后在 NewBeeMallGoodsCategoryController 类中新增如下代码：

```
@GetMapping("/categories")
public String categoriesPage(HttpServletRequest request, @RequestParam
("categoryLevel") Byte categoryLevel, @RequestParam("parentId") Long parentId,
```

```
@RequestParam("backParentId") Long backParentId) {
    if (categoryLevel == null || categoryLevel < 1 || categoryLevel > 3) {
        return "error/error_5xx";
    }
    request.setAttribute("path", "newbee_mall_category");
    request.setAttribute("parentId", parentId);
    request.setAttribute("backParentId", backParentId);
    request.setAttribute("categoryLevel", categoryLevel);
    return "admin/newbee_mall_category";
}
```

该方法用于处理/admin/categories 请求，并接收前端传过来的 URL 参数，并进行简单的参数验证。在返回视图前，程序会设置 path 字段以及其他的参数值到 request 对象中。path 字段主要用于实现导航栏的选中效果，其他参数主要用于分类接口请求的调用。该方法的后面是一个 return 视图语句，即跳转到 admin 目录下的 newbee_mall_category.html 模板页面中。

18.2.3 分类管理页面基础样式的实现

基础样式的实现过程如下。在 resources/templates/admin 目录下新建 newbee_mall_category.html，并引入对应的 JS 文件和 CSS 样式文件，代码如下所示：

```html
<!DOCTYPE html>
<html xmlns:th="http://www.thymeleaf.org">
<header th:replace="admin/header::header-fragment">
</header>
<style>
    .ui-jqgrid tr.jqgrow td {
        white-space: normal !important;
        height: auto;
        vertical-align: text-top;
        padding-top: 2px;
    }
</style>
<body class="hold-transition sidebar-mini">
<div class="wrapper">
    <div th:replace="admin/header::header-nav"></div>
    <div th:replace="admin/sidebar::sidebar-fragment(${path})"></div>
    <div class="content-wrapper">
        <div class="content-header">
```

```html
            <div class="container-fluid">
            </div><!-- /.container-fluid -->
        </div>
        <!-- Main content -->
        <div class="content">
            <div class="container-fluid">
                <div class="card card-primary card-outline">
                    <div class="card-header">
                        <h3 class="card-title">分类管理</h3>
                    </div> <!-- /.card-body -->
                    <div class="card-body">
                        <div class="grid-btn">
                            <button class="btn btn-info" onclick="categoryAdd()"><i
                                    class="fa fa-plus"></i> 新增
                            </button>
                            <button class="btn btn-info" onclick="categoryEdit()"><i
                                    class="fa fa-pencil-square-o"></i> 修改
                            </button>
                            <button class="btn btn-danger" onclick="deleteCagegory()"><i
                                    class="fa fa-trash-o"></i> 删除
                            </button>
                            <button class="btn btn-info" onclick="categoryManage()"><i
                                    class="fa fa-list"></i> 下级分类管理
                            </button>
                            <button class="btn btn-info" onclick="categoryBack()"><i
                                    class="fa fa-backward"></i> 返回
                            </button>
                        </div>
                        <br>
                        <table id="jqGrid" class="table table-bordered">
                        </table>
                        <div id="jqGridPager"></div>
                    </div>
                </div>
            </div>
        </div>
        <div class="content">
            <div class="modal fade" id="categoryModal" tabindex="-1" role=
```

```html
"dialog" aria-labelledby="categoryModalLabel">
            <div class="modal-dialog" role="document">
                <div class="modal-content">
                    <div class="modal-header">
                        <button type="button" class="close" data-dismiss="modal" aria-label="Close"><span
                                aria-hidden="true">&times;</span></button>
                        <h6 class="modal-title" id="categoryModalLabel">Modal</h6>
                    </div>
                    <div class="modal-body">
                        <form id="categoryForm">
                            <div class="form-group">
                                <div class="alert alert-danger" id="edit-error-msg" style="display: none;">
                                    错误信息展示栏。
                                </div>
                            </div>
                            <input type="hidden" class="form-control" id="categoryId" name="categoryId">
                            <input type="hidden" id="categoryLevel" th:value="${categoryLevel}">
                            <input type="hidden" id="parentId" th:value="${parentId}">
                            <input type="hidden" id="backParentId" th:value="${backParentId}">
                            <div class="form-group">
                                <label for="categoryName" class="control-label">分类名称:</label>
                                <input type="text" class="form-control" id="categoryName" name="categoryName"
                                    placeholder="请输入分类名称" required="true">
                            </div>
                            <div class="form-group">
                                <label for="categoryName" class="control-label">排序值:</label>
                                <input type="number" class="form-control" id="categoryRank" name="categoryRank"
                                    placeholder="请输入分类排序值" required="true">
                            </div>
                        </form>
```

```html
                </div>
                <div class="modal-footer">
                    <button type="button" class="btn btn-default" data-dismiss="modal">取消</button>
                    <button type="button" class="btn btn-primary" id="saveButton">确认</button>
                </div>
            </div>
        </div>
    </div>
</div>
<div th:replace="admin/footer::footer-fragment"></div>
</div>
<!-- jQuery -->
<script th:src="@{/admin/plugins/jquery/jquery.min.js}"></script>
<!-- jQuery UI 1.11.4 -->
<script th:src="@{/admin/plugins/jQueryUI/jquery-ui.min.js}"></script>
<!-- Bootstrap 4 -->
<script th:src="@{/admin/plugins/bootstrap/js/bootstrap.bundle.min.js}"></script>
<!-- AdminLTE App -->
<script th:src="@{/admin/dist/js/adminlte.min.js}"></script>
<!-- jqgrid -->
<script th:src="@{/admin/plugins/jqgrid-5.3.0/jquery.jqGrid.min.js}"></script>
<script th:src="@{/admin/plugins/jqgrid-5.3.0/grid.locale-cn.js}"></script>
<!-- sweetalert -->
<script th:src="@{/admin/plugins/sweetalert/sweetalert.min.js}"></script>
<script th:src="@{/admin/dist/js/public.js}"></script>
<script th:src="@{/admin/dist/js/newbee_mall_category.js}"></script>
</body>
</html>
```

18.2.4 功能按钮和分页信息展示区域

分类管理模块设计了常用的几个功能。首先是分类信息增加、分类信息编辑、分类信息删除三个基础的功能。由于商品类目层级的关系，还增加了两个用于不同层级分类间跳转的页面功能。比如由一级分类页面跳转到某一级分类下的所有二级分类数据列表页面；由二级分类页面跳转到某二级分类下的所有三级分类数据列表页面。只有向下跳转也不行，所以这里又设计了返回上一级分类的按钮。

在页面中添加对应的功能按钮和触发事件，代码如下所示：

```html
<div class="grid-btn">
  <button class="btn btn-info" onclick="categoryAdd()"><i class="fa fa-plus"></i> 新增</button>
  <button class="btn btn-info" onclick="categoryEdit()"><i class="fa fa-pencil-square-o"></i> 修改</button>
  <button class="btn btn-danger" onclick="deleteCagegory()"><i class="fa fa-trash-o"></i> 删除</button>
  <button class="btn btn-info" onclick="categoryManage()"><i class="fa fa-list"></i> 下级分类管理</button>
  <button class="btn btn-info" onclick="categoryBack()"><i class="fa fa-backward"></i> 返回</button>
</div>
```

这里共有 5 个功能按钮，分别是新增按钮，对应的触发事件是 categoryAdd()方法；修改按钮，对应的触发事件是 categoryEdit()方法；删除按钮，对应的触发事件是 deleteCagegory()方法；下级分类管理按钮，对应的触发事件是 categoryManage()方法；返回按钮，对应的触发事件是 categoryBack()方法。这些方法将在后续的代码实现中给出。

由于页面已经引入了 jqGrid 的相关静态资源文件，所以需要在页面展示分页数据的区域增加如下代码：

```html
<table id="jqGrid" class="table table-bordered"></table>
<div id="jqGridPager"></div>
```

18.2.5　URL 参数处理

由于分类层级的存在，所以需要在请求 URL 中添加几个参数。在控制器方法中这些参数又放到了 Modal 框中，并传递给了视图对象。这里在前端源码中对这几个参数进行了读取并放在 hidden 模式的 input 框中，代码如下所示：

```html
<input type="hidden" id="categoryLevel" th:value="${categoryLevel}">
<input type="hidden" id="parentId" th:value="${parentId}">
<input type="hidden" id="backParentId" th:value="${backParentId}">
```

这几个参数会在列表接口请求、添加/编辑接口请求的发送中用到，在上下级页面的跳转中也会用到，所以暂时将它们的值放在 input 中以便于读取和使用。

最终的分类管理页面基础效果如图 18-3 所。

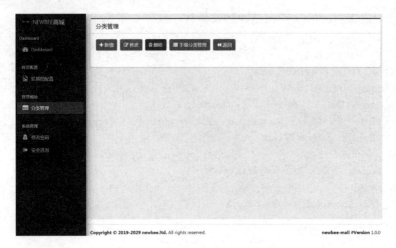

图 18-3　分类管理页面基础效果

这里包括功能按钮、数据列表展示区域以及翻页功能区域,此时只是静态效果展示,并没有与后端进行数据交互,接下来将结合 Ajax 和后端接口实现具体的功能。

18.3　商品分类表的结构设计

在进行分类模块的接口设计和具体的功能实现前,首先要将分类表的结构确定下来。虽然新蜂商城的分类设置成了三个层级,但是在表设计时并没有做成三张表。这是因为表中大部分字段都是一样的,所以就选择增加一个 parentId 字段来区分不同分类层级。同时使用 parentId 字段进行上下级类目之间的关联。分类表的字段设计如下所示:

```sql
USE `newbee_mall_db`;

DROP TABLE IF EXISTS `tb_newbee_mall_goods_category`;
CREATE TABLE `tb_newbee_mall_goods_category` (
  `category_id` bigint(20) NOT NULL AUTO_INCREMENT COMMENT '分类id',
  `category_level` tinyint(4) NOT NULL DEFAULT 0 COMMENT '分类级别(1-一级分类 2-二级分类 3-三级分类)',
  `parent_id` bigint(20) NOT NULL DEFAULT 0 COMMENT '父分类id',
  `category_name` varchar(50) CHARACTER SET utf8 COLLATE utf8_general_ci NOT NULL DEFAULT '' COMMENT '分类名称',
  `category_rank` int(11) NOT NULL DEFAULT 0 COMMENT '排序值(字段越大越靠前)',
  `is_deleted` tinyint(4) NOT NULL DEFAULT 0 COMMENT '删除标识字段(0-未删除 1-
```

```
已删除)',
  'create_time' datetime(0) NOT NULL DEFAULT CURRENT_TIMESTAMP COMMENT '创
建时间',
  'create_user' int(11) NOT NULL DEFAULT 0 COMMENT '创建者id',
  'update_time' datetime(0) NOT NULL DEFAULT CURRENT_TIMESTAMP COMMENT '修
改时间',
  'update_user' int(11) NULL DEFAULT 0 COMMENT '修改者id',
  PRIMARY KEY ('category_id') USING BTREE
) ENGINE = InnoDB CHARACTER SET = utf8 COLLATE = utf8_general_ci ROW_FORMAT
= Dynamic;
```

商品类目表的字段和每个字段对应的含义在上面的 SQL 中都有介绍,读者可以对照理解,并正确地把建表 SQL 导入到数据库中。为了读者在开发时更方便,笔者也准备了一些初始化分类数据以供测试使用,初始化分类数据如下所示:

```
INSERT INTO 'tb_newbee_mall_goods_category' VALUES (15, 1, 0, '家电 数码 手
机', 100, 0, '2021-02-11 18:45:40', 0, '2021-02-11 18:45:40', 0);
INSERT INTO 'tb_newbee_mall_goods_category' VALUES (16, 1, 0, '女装 男装 穿
搭', 99, 0, '2021-02-11 18:46:07', 0, '2021-02-11 18:46:07', 0);
INSERT INTO 'tb_newbee_mall_goods_category' VALUES (17, 2, 15, '家电', 10,
0, '2021-02-11 18:46:32', 0, '2021-02-11 18:46:32', 0);
…省略部分数据
INSERT INTO 'tb_newbee_mall_goods_category' VALUES (103, 3, 83, '腮红', 0,
0, '2021-02-17 18:27:24', 0, '2021-02-17 18:27:24', 0);
INSERT INTO 'tb_newbee_mall_goods_category' VALUES (104, 3, 83, '睫毛膏', 0,
0, '2021-02-17 18:27:47', 0, '2021-02-17 18:27:47', 0);
INSERT INTO 'tb_newbee_mall_goods_category' VALUES (105, 3, 83, '香水', 0,
0, '2021-02-17 18:28:16', 0, '2021-02-17 18:28:16', 0);
INSERT INTO 'tb_newbee_mall_goods_category' VALUES (106, 3, 83, '面膜', 0,
0, '2021-02-17 18:28:21', 0, '2021-02-17 18:28:21', 0);
```

将这部分数据导入数据库中即可。接下来进行分类功能实际的编码工作。

18.4 分类模块后端功能的实现

分类模块在后台管理系统中有 4 个接口,分别是商品类目分页列表接口、添加商品类目接口、修改商品类目接口和批量删除商品类目接口。

接下来讲解每个接口具体的实现代码。

18.4.1 新建分类实体类和 Mapper 接口

首先在 ltd.newbee.mall.entity 包中创建分类实体类，选中 entity 包并右键点击，在弹出的菜单中选择"New→Java Class"，然后在弹出的窗口中输入"GoodsCategory"，最后在 GoodsCategory 类中新增如下代码：

```java
package ltd.newbee.mall.entity;

import com.fasterxml.jackson.annotation.JsonFormat;

import java.util.Date;

public class GoodsCategory {
    private Long categoryId;

    private Byte categoryLevel;

    private Long parentId;

    private String categoryName;

    private Integer categoryRank;

    private Byte isDeleted;

    @JsonFormat(pattern = "yyyy-MM-dd HH:mm:ss", timezone = "GMT+8")
    private Date createTime;

    private Integer createUser;

    @JsonFormat(pattern = "yyyy-MM-dd HH:mm:ss", timezone = "GMT+8")
    private Date updateTime;

    private Integer updateUser;

    public Long getCategoryId() {
        return categoryId;
    }

    public void setCategoryId(Long categoryId) {
        this.categoryId = categoryId;
```

```java
    }

    public Byte getCategoryLevel() {
        return categoryLevel;
    }

    public void setCategoryLevel(Byte categoryLevel) {
        this.categoryLevel = categoryLevel;
    }

    public Long getParentId() {
        return parentId;
    }

    public void setParentId(Long parentId) {
        this.parentId = parentId;
    }

    public String getCategoryName() {
        return categoryName;
    }

    public void setCategoryName(String categoryName) {
        this.categoryName = categoryName == null ? null : categoryName.trim();
    }

    public Integer getCategoryRank() {
        return categoryRank;
    }

    public void setCategoryRank(Integer categoryRank) {
        this.categoryRank = categoryRank;
    }

    public Byte getIsDeleted() {
        return isDeleted;
    }

    public void setIsDeleted(Byte isDeleted) {
        this.isDeleted = isDeleted;
    }

    public Date getCreateTime() {
```

```java
        return createTime;
    }

    public void setCreateTime(Date createTime) {
        this.createTime = createTime;
    }

    public Integer getCreateUser() {
        return createUser;
    }

    public void setCreateUser(Integer createUser) {
        this.createUser = createUser;
    }

    public Date getUpdateTime() {
        return updateTime;
    }

    public void setUpdateTime(Date updateTime) {
        this.updateTime = updateTime;
    }

    public Integer getUpdateUser() {
        return updateUser;
    }

    public void setUpdateUser(Integer updateUser) {
        this.updateUser = updateUser;
    }
}
```

首先在 ltd.newbee.mall.dao 包中新建分类实体的 Mapper 接口，选中 dao 包并右键点击，在弹出的菜单中选择 "New→Java Class"，在弹出的窗口中输入 "GoodsCategoryMapper"，然后选中 "Interface" 选项，最后在 GoodsCategoryMapper.java 文件中新增如下代码：

```java
package ltd.newbee.mall.dao;

import ltd.newbee.mall.entity.GoodsCategory;
import ltd.newbee.mall.util.PageQueryUtil;
import org.apache.ibatis.annotations.Param;
```

```java
import java.util.List;

public interface GoodsCategoryMapper {

    /**
     * 删除一条记录
     *
     * @param categoryId
     * @return
     */
    int deleteByPrimaryKey(Long categoryId);

    /**
     * 保存一条新记录
     *
     * @param record
     * @return
     */
    int insert(GoodsCategory record);

    /**
     * 保存一条新记录
     *
     * @param record
     * @return
     */
    int insertSelective(GoodsCategory record);

    /**
     * 根据主键查询记录
     *
     * @param categoryId
     * @return
     */
    GoodsCategory selectByPrimaryKey(Long categoryId);

    /**
     * 根据分类等级和分类名称查询一条分类记录
     *
     * @param categoryLevel
     * @param categoryName
     * @return
     */
```

```java
    GoodsCategory selectByLevelAndName(@Param("categoryLevel") Byte categoryLevel, @Param("categoryName") String categoryName);

    /**
     * 修改记录
     *
     * @param record
     * @return
     */
    int updateByPrimaryKeySelective(GoodsCategory record);

    /**
     * 修改记录
     *
     * @param record
     * @return
     */
    int updateByPrimaryKey(GoodsCategory record);

    /**
     * 查询分页数据
     *
     * @param pageUtil
     * @return
     */
    List<GoodsCategory> findGoodsCategoryList(PageQueryUtil pageUtil);

    /**
     * 查询总数
     *
     * @param pageUtil
     * @return
     */
    int getTotalGoodsCategories(PageQueryUtil pageUtil);

    /**
     * 批量删除
     *
     * @param ids
     * @return
     */
    int deleteBatch(Integer[] ids);
```

```java
/**
 * 根据父类的分类 id、分类等级和数量查询分类列表
 *
 * @param parentIds
 * @param categoryLevel
 * @param number
 * @return
 */
List<GoodsCategory> selectByLevelAndParentIdsAndNumber(@Param("parentIds") List<Long> parentIds, @Param("categoryLevel") int categoryLevel, @Param("number") int number);
}
```

这里定义了分类数据操作的数据层方法，包括查询、新增、修改和删除等操作。

18.4.2 创建 GoodsCategoryMapper 接口的映射文件

在 resources/mapper 目录下新建 GoodsCategoryMapper 接口的映射文件 GoodsCategoryMapper.xml，并进行映射文件的编写。

首先，定义映射文件与 Mapper 接口的对应关系。比如在该示例中，需要将 GoodsCategoryMapper.xml 文件与对应的 GoodsCategoryMapper 接口之间的关系定义出来：

```xml
<mapper namespace="ltd.newbee.mall.dao.GoodsCategoryMapper">
```

然后，配置表结构和实体类的对应关系：

```xml
<resultMap id="BaseResultMap" type="ltd.newbee.mall.entity.GoodsCategory">
    <id column="category_id" jdbcType="BIGINT" property="categoryId"/>
    <result column="category_level" jdbcType="TINYINT" property="categoryLevel"/>
    <result column="parent_id" jdbcType="BIGINT" property="parentId"/>
    <result column="category_name" jdbcType="VARCHAR" property="categoryName"/>
    <result column="category_rank" jdbcType="INTEGER" property="categoryRank"/>
    <result column="is_deleted" jdbcType="TINYINT" property="isDeleted"/>
    <result column="create_time" jdbcType="TIMESTAMP" property="createTime"/>
    <result column="create_user" jdbcType="INTEGER" property="createUser"/>
    <result column="update_time" jdbcType="TIMESTAMP" property="updateTime"/>
```

```xml
        <result column="update_user" jdbcType="INTEGER" property="updateUser"/>
</resultMap>
```

最后,按照对应的接口方法,编写具体的 SQL 语句,最终的 GoodsCategoryMapper.xml 文件如下所示:

```xml
<?xml version="1.0" encoding="UTF-8"?>
<!DOCTYPE mapper PUBLIC "-//mybatis.org//DTD Mapper 3.0//EN" "http://mybatis.org/dtd/mybatis-3-mapper.dtd">
<mapper namespace="ltd.newbee.mall.dao.GoodsCategoryMapper">
    <resultMap id="BaseResultMap" type="ltd.newbee.mall.entity.GoodsCategory">
        <id column="category_id" jdbcType="BIGINT" property="categoryId"/>
        <result column="category_level" jdbcType="TINYINT" property="categoryLevel"/>
        <result column="parent_id" jdbcType="BIGINT" property="parentId"/>
        <result column="category_name" jdbcType="VARCHAR" property="categoryName"/>
        <result column="category_rank" jdbcType="INTEGER" property="categoryRank"/>
        <result column="is_deleted" jdbcType="TINYINT" property="isDeleted"/>
        <result column="create_time" jdbcType="TIMESTAMP" property="createTime"/>
        <result column="create_user" jdbcType="INTEGER" property="createUser"/>
        <result column="update_time" jdbcType="TIMESTAMP" property="updateTime"/>
        <result column="update_user" jdbcType="INTEGER" property="updateUser"/>
    </resultMap>
    <sql id="Base_Column_List">
        category_id, category_level, parent_id, category_name, category_rank, is_deleted,
        create_time, create_user, update_time, update_user
    </sql>
    <select id="findGoodsCategoryList" parameterType="Map" resultMap="BaseResultMap">
        select
        <include refid="Base_Column_List"/>
        from tb_newbee_mall_goods_category
        <where>
            <if test="categoryLevel!=null and categoryLevel!=''">
                and category_level = #{categoryLevel}
```

```xml
        </if>
        <if test="parentId!=null and parentId!=''">
            and parent_id = #{parentId}
        </if>
        and is_deleted = 0
    </where>
    order by category_rank desc
    <if test="start!=null and limit!=null">
        limit #{start},#{limit}
    </if>
</select>
<select id="getTotalGoodsCategories" parameterType="Map" resultType="int">
    select count(*) from tb_newbee_mall_goods_category
    <where>
        <if test="categoryLevel!=null and categoryLevel!=''">
            and category_level = #{categoryLevel}
        </if>
        <if test="parentId!=null and parentId!=''">
            and parent_id = #{parentId}
        </if>
        and is_deleted = 0
    </where>
</select>
<select id="selectByPrimaryKey" parameterType="java.lang.Long" resultMap="BaseResultMap">
    select
    <include refid="Base_Column_List"/>
    from tb_newbee_mall_goods_category
    where category_id = #{categoryId,jdbcType=BIGINT} and is_deleted=0
</select>
<select id="selectByLevelAndName" resultMap="BaseResultMap">
    select
    <include refid="Base_Column_List"/>
    from tb_newbee_mall_goods_category
    where category_name = #{categoryName,jdbcType=VARCHAR} and category_level = #{categoryLevel,jdbcType=TINYINT}
    and is_deleted = 0 limit 1
</select>
<select id="selectByLevelAndParentIdsAndNumber" resultMap="BaseResultMap">
    select
    <include refid="Base_Column_List"/>
```

```xml
        from tb_newbee_mall_goods_category
        where parent_id in
        <foreach item="parentId" collection="parentIds" open="(" separator="," close=")">
            #{parentId,jdbcType=BIGINT}
        </foreach>
        and category_level = #{categoryLevel,jdbcType=TINYINT}
        and is_deleted = 0
        order by category_rank desc
        <if test="number>0">
            limit #{number}
        </if>
    </select>
    <update id="deleteByPrimaryKey" parameterType="java.lang.Long">
        update tb_newbee_mall_goods_category set is_deleted=1
        where category_id = #{categoryId,jdbcType=BIGINT} and is_deleted=0
    </update>
    <update id="deleteBatch">
        update tb_newbee_mall_goods_category
        set is_deleted=1 where category_id in
        <foreach item="id" collection="array" open="(" separator="," close=")">
            #{id}
        </foreach>
    </update>
    <insert id="insert" parameterType="ltd.newbee.mall.entity.GoodsCategory">
        insert into tb_newbee_mall_goods_category (category_id, category_level, parent_id,
        category_name, category_rank, is_deleted,
        create_time, create_user, update_time,
        update_user)
        values (#{categoryId,jdbcType=BIGINT}, #{categoryLevel,jdbcType=TINYINT}, #{parentId,jdbcType=BIGINT},
        #{categoryName,jdbcType=VARCHAR}, #{categoryRank,jdbcType=INTEGER}, #{isDeleted,jdbcType=TINYINT},
        #{createTime,jdbcType=TIMESTAMP}, #{createUser,jdbcType=INTEGER}, #{updateTime,jdbcType=TIMESTAMP},
        #{updateUser,jdbcType=INTEGER})
    </insert>
    <insert id="insertSelective" parameterType="ltd.newbee.mall.entity.GoodsCategory">
        insert into tb_newbee_mall_goods_category
```

```xml
<trim prefix="(" suffix=")" suffixOverrides=",">
    <if test="categoryId != null">
        category_id,
    </if>
    <if test="categoryLevel != null">
        category_level,
    </if>
    <if test="parentId != null">
        parent_id,
    </if>
    <if test="categoryName != null">
        category_name,
    </if>
    <if test="categoryRank != null">
        category_rank,
    </if>
    <if test="isDeleted != null">
        is_deleted,
    </if>
    <if test="createTime != null">
        create_time,
    </if>
    <if test="createUser != null">
        create_user,
    </if>
    <if test="updateTime != null">
        update_time,
    </if>
    <if test="updateUser != null">
        update_user,
    </if>
</trim>
<trim prefix="values (" suffix=")" suffixOverrides=",">
    <if test="categoryId != null">
        #{categoryId,jdbcType=BIGINT},
    </if>
    <if test="categoryLevel != null">
        #{categoryLevel,jdbcType=TINYINT},
    </if>
    <if test="parentId != null">
        #{parentId,jdbcType=BIGINT},
    </if>
    <if test="categoryName != null">
```

```xml
                #{categoryName,jdbcType=VARCHAR},
            </if>
            <if test="categoryRank != null">
                #{categoryRank,jdbcType=INTEGER},
            </if>
            <if test="isDeleted != null">
                #{isDeleted,jdbcType=TINYINT},
            </if>
            <if test="createTime != null">
                #{createTime,jdbcType=TIMESTAMP},
            </if>
            <if test="createUser != null">
                #{createUser,jdbcType=INTEGER},
            </if>
            <if test="updateTime != null">
                #{updateTime,jdbcType=TIMESTAMP},
            </if>
            <if test="updateUser != null">
                #{updateUser,jdbcType=INTEGER},
            </if>
        </trim>
    </insert>
    <update id="updateByPrimaryKeySelective" parameterType="ltd.newbee.mall.entity.GoodsCategory">
        update tb_newbee_mall_goods_category
        <set>
            <if test="categoryLevel != null">
                category_level = #{categoryLevel,jdbcType=TINYINT},
            </if>
            <if test="parentId != null">
                parent_id = #{parentId,jdbcType=BIGINT},
            </if>
            <if test="categoryName != null">
                category_name = #{categoryName,jdbcType=VARCHAR},
            </if>
            <if test="categoryRank != null">
                category_rank = #{categoryRank,jdbcType=INTEGER},
            </if>
            <if test="isDeleted != null">
                is_deleted = #{isDeleted,jdbcType=TINYINT},
            </if>
            <if test="createTime != null">
                create_time = #{createTime,jdbcType=TIMESTAMP},
```

```xml
        </if>
        <if test="createUser != null">
            create_user = #{createUser,jdbcType=INTEGER},
        </if>
        <if test="updateTime != null">
            update_time = #{updateTime,jdbcType=TIMESTAMP},
        </if>
        <if test="updateUser != null">
            update_user = #{updateUser,jdbcType=INTEGER},
        </if>
    </set>
    where category_id = #{categoryId,jdbcType=BIGINT}
</update>
<update id="updateByPrimaryKey" parameterType="ltd.newbee.mall.entity.GoodsCategory">
    update tb_newbee_mall_goods_category
    set category_level = #{categoryLevel,jdbcType=TINYINT},
      parent_id = #{parentId,jdbcType=BIGINT},
      category_name = #{categoryName,jdbcType=VARCHAR},
      category_rank = #{categoryRank,jdbcType=INTEGER},
      is_deleted = #{isDeleted,jdbcType=TINYINT},
      create_time = #{createTime,jdbcType=TIMESTAMP},
      create_user = #{createUser,jdbcType=INTEGER},
      update_time = #{updateTime,jdbcType=TIMESTAMP},
      update_user = #{updateUser,jdbcType=INTEGER}
    where category_id = #{categoryId,jdbcType=BIGINT}
</update>
</mapper>
```

18.4.3 业务层代码的实现

首先在 ltd.newbee.mall.service 包中新建业务处理类，选中 service 包并右键点击，在弹出的菜单中选择"New→Java Class"，然后在弹出的窗口中输入"NewbeeMallCategoryService"，并选中"Interface"选项，最后在 NewbeeMallCategoryService.java 文件中新增如下代码：

```java
package ltd.newbee.mall.service;

import ltd.newbee.mall.entity.GoodsCategory;
import ltd.newbee.mall.util.PageQueryUtil;
import ltd.newbee.mall.util.PageResult;
```

```java
public interface NewbeeMallCategoryService{

    /**
     * 查询后台管理系统分类分页数据
     *
     * @param pageUtil
     * @return
     */
    PageResult getCategorisPage(PageQueryUtil pageUtil);

    /**
     * 新增一条分类记录
     *
     * @param goodsCategory
     * @return
     */
    String saveCategory(GoodsCategory goodsCategory);

    /**
     * 修改一条分类记录
     *
     * @param goodsCategory
     * @return
     */
    String updateGoodsCategory(GoodsCategory goodsCategory);

    /**
     * 根据主键查询分类记录
     *
     * @param id
     * @return
     */
    GoodsCategory getGoodsCategoryById(Long id);

    /**
     * 批量删除分类数据
     *
     * @param ids
     * @return
     */
    Boolean deleteBatch(Integer[] ids);
```

```
/**
 * 根据parentId和level获取分类列表
 *
 * @param parentIds
 * @param categoryLevel
 * @return
 */
List<GoodsCategory> selectByLevelAndParentIdsAndNumber(List<Long> parentIds, int categoryLevel);
}
```

至此,分类模块的业务层方法定义和每个方法的作用都已经编写完成。

接下来,在 ltd.newbee.mall.service.impl 包中新建 NewbeeMallCategoryService 的实现类,选中 impl 包并右键点击,然后在弹出的菜单中选择"New→Java Class",并在弹出的窗口中输入"NewbeeMallCategoryServiceImpl",最后在 NewbeeMallCategoryServiceImpl 类中新增如下代码:

```java
package ltd.newbee.mall.service.impl;

import ltd.newbee.mall.common.ServiceResultEnum;
import ltd.newbee.mall.dao.GoodsCategoryMapper;
import ltd.newbee.mall.entity.GoodsCategory;
import ltd.newbee.mall.service.NewbeeMallCategoryService;
import ltd.newbee.mall.util.PageQueryUtil;
import ltd.newbee.mall.util.PageResult;
import org.springframework.beans.factory.annotation.Autowired;
import org.springframework.stereotype.Service;

import java.util.Date;
import java.util.List;

@Service
public class NewbeeMallCategoryServiceImpl implements NewbeeMallCategoryService {

    @Autowired
    private GoodsCategoryMapper goodsCategoryMapper;

    @Override
    public PageResult getCategorisPage(PageQueryUtil pageUtil) {
        List<GoodsCategory> goodsCategories = goodsCategoryMapper.findGoodsCategoryList(pageUtil);
        int total = goodsCategoryMapper.getTotalGoodsCategories(pageUtil);
```

```java
        PageResult pageResult = new PageResult(goodsCategories, total,
pageUtil.getLimit(), pageUtil.getPage());
        return pageResult;
    }

    @Override
    public String saveCategory(GoodsCategory goodsCategory) {
        GoodsCategory temp = goodsCategoryMapper.selectByLevelAndName
(goodsCategory.getCategoryLevel(), goodsCategory.getCategoryName());
        if (temp != null) {
            return ServiceResultEnum.SAME_CATEGORY_EXIST.getResult();
        }
        if (goodsCategoryMapper.insertSelective(goodsCategory) > 0) {
            return ServiceResultEnum.SUCCESS.getResult();
        }
        return ServiceResultEnum.DB_ERROR.getResult();
    }

    @Override
    public String updateGoodsCategory(GoodsCategory goodsCategory) {
        GoodsCategory temp = goodsCategoryMapper.selectByPrimaryKey(goods
Category.getCategoryId());
        if (temp == null) {
            return ServiceResultEnum.DATA_NOT_EXIST.getResult();
        }
        GoodsCategory temp2 = goodsCategoryMapper.selectByLevelAndName(goods
Category.getCategoryLevel(), goodsCategory.getCategoryName());
        if (temp2 != null && !temp2.getCategoryId().equals(goodsCategory.
getCategoryId())) {
            //同名且不同id不能继续修改
            return ServiceResultEnum.SAME_CATEGORY_EXIST.getResult();
        }
        goodsCategory.setUpdateTime(new Date());
        if (goodsCategoryMapper.updateByPrimaryKeySelective(goodsCategory)>0) {
            return ServiceResultEnum.SUCCESS.getResult();
        }
        return ServiceResultEnum.DB_ERROR.getResult();
    }

    @Override
    public GoodsCategory getGoodsCategoryById(Long id) {
        return goodsCategoryMapper.selectByPrimaryKey(id);
    }
```

```java
@Override
public Boolean deleteBatch(Integer[] ids) {
    if (ids.length < 1) {
        return false;
    }
    //删除分类数据
    return goodsCategoryMapper.deleteBatch(ids) > 0;
}

@Override
public List<GoodsCategory> selectByLevelAndParentIdsAndNumber(List<Long> parentIds, int categoryLevel) {
    return goodsCategoryMapper.selectByLevelAndParentIdsAndNumber(parentIds, categoryLevel, 0);//0代表查询所有
}
}
```

18.4.4 分类管理模块控制层的代码实现

在 NewBeeMallGoodsCategoryController.java 控制器中新增上述接口的实现代码。最终 NewBeeMallGoodsCategoryController 类的代码如下所示：

```java
package ltd.newbee.mall.controller.admin;

import ltd.newbee.mall.common.newbee_mall_categoryLevelEnum;
import ltd.newbee.mall.common.ServiceResultEnum;
import ltd.newbee.mall.entity.GoodsCategory;
import ltd.newbee.mall.service.NewbeeMallCategoryService;
import ltd.newbee.mall.util.PageQueryUtil;
import ltd.newbee.mall.util.Result;
import ltd.newbee.mall.util.ResultGenerator;
import org.springframework.stereotype.Controller;
import org.springframework.util.CollectionUtils;
import org.springframework.util.StringUtils;
import org.springframework.web.bind.annotation.*;

import javax.annotation.Resource;
import javax.servlet.http.HttpServletRequest;
import java.util.*;
```

```java
@Controller
@RequestMapping("/admin")
public class NewBeeMallGoodsCategoryController {

    @Resource
    private NewbeeMallCategoryService NewbeeMallCategoryService;

    @GetMapping("/categories")
    public String categoriesPage(HttpServletRequest request, @RequestParam("categoryLevel") Byte categoryLevel, @RequestParam("parentId") Long parentId, @RequestParam("backParentId") Long backParentId) {
        if (categoryLevel == null || categoryLevel < 1 || categoryLevel > 3) {
            return "error/error_5xx";
        }
        request.setAttribute("path", "newbee_mall_category");
        request.setAttribute("parentId", parentId);
        request.setAttribute("backParentId", backParentId);
        request.setAttribute("categoryLevel", categoryLevel);
        return "admin/newbee_mall_category";
    }

    /**
     * 列表
     */
    @RequestMapping(value = "/categories/list", method = RequestMethod.GET)
    @ResponseBody
    public Result list(@RequestParam Map<String, Object> params) {
        if (StringUtils.isEmpty(params.get("page")) || StringUtils.isEmpty(params.get("limit")) || StringUtils.isEmpty(params.get("categoryLevel")) || StringUtils.isEmpty(params.get("parentId"))) {
            return ResultGenerator.genFailResult("参数异常！");
        }
        PageQueryUtil pageUtil = new PageQueryUtil(params);
        return ResultGenerator.genSuccessResult(NewbeeMallCategoryService.getCategorisPage(pageUtil));
    }

    /**
     * 添加
     */
    @RequestMapping(value = "/categories/save", method = RequestMethod.POST)
    @ResponseBody
    public Result save(@RequestBody GoodsCategory goodsCategory) {
```

```java
        if (Objects.isNull(goodsCategory.getCategoryLevel())
                || StringUtils.isEmpty(goodsCategory.getCategoryName())
                || Objects.isNull(goodsCategory.getParentId())
                || Objects.isNull(goodsCategory.getCategoryRank())) {
            return ResultGenerator.genFailResult("参数异常！");
        }
        String result = NewbeeMallCategoryService.saveCategory(goodsCategory);
        if (ServiceResultEnum.SUCCESS.getResult().equals(result)) {
            return ResultGenerator.genSuccessResult();
        } else {
            return ResultGenerator.genFailResult(result);
        }
    }


    /**
     * 修改
     */
    @RequestMapping(value = "/categories/update", method = RequestMethod.POST)
    @ResponseBody
    public Result update(@RequestBody GoodsCategory goodsCategory) {
        if (Objects.isNull(goodsCategory.getCategoryId())
                || Objects.isNull(goodsCategory.getCategoryLevel())
                || StringUtils.isEmpty(goodsCategory.getCategoryName())
                || Objects.isNull(goodsCategory.getParentId())
                || Objects.isNull(goodsCategory.getCategoryRank())) {
            return ResultGenerator.genFailResult("参数异常！");
        }
        String result = NewbeeMallCategoryService.updateGoodsCategory(goodsCategory);
        if (ServiceResultEnum.SUCCESS.getResult().equals(result)) {
            return ResultGenerator.genSuccessResult();
        } else {
            return ResultGenerator.genFailResult(result);
        }
    }

    /**
     * 详情
     */
    @GetMapping("/categories/info/{id}")
    @ResponseBody
    public Result info(@PathVariable("id") Long id) {
```

```java
        GoodsCategory goodsCategory = NewbeeMallCategoryService.getGoods
CategoryById(id);
        if (goodsCategory == null) {
            return ResultGenerator.genFailResult("未查询到数据");
        }
        return ResultGenerator.genSuccessResult(goodsCategory);
    }

    /**
     * 分类删除
     */
    @RequestMapping(value = "/categories/delete", method = RequestMethod.POST)
    @ResponseBody
    public Result delete(@RequestBody Integer[] ids) {
        if (ids.length < 1) {
            return ResultGenerator.genFailResult("参数异常！");
        }
        if (NewbeeMallCategoryService.deleteBatch(ids)) {
            return ResultGenerator.genSuccessResult();
        } else {
            return ResultGenerator.genFailResult("删除失败");
        }
    }
}
```

（1）列表接口负责接收前端传来的分页参数，比如 page、limit 等参数。由于分类层级的设计，系统在获取分类列表时会分别获取对应层级的商品类目数据。在传参时需要将类目的层级 categoryLevel 和父级类目的 id 也带上，最后将数据总数和对应页面的数据列表查询出来并封装为分页数据返回给前端。

（2）添加接口负责接收前端的 POST 请求并处理其中的参数。接收的参数分别如下所示。

①类目级别：categoryLevel 字段

②类目的父级 id：parentId 字段

③类目名称：categoryName 字段

④排序值：categoryRank 字段

在这个方法里笔者使用@RequestBody 注解将参数转换为 GoodsCategory 对象参数。在代码实现中，首先会对参数进行校验，然后交给业务层代码进行数据封装并进行数据库 insert 操作。在业务层代码中也有一个逻辑验证，即如果已经存在同级且同名的分类数据就不再继续添加。

（3）修改接口与添加接口类似，只是接受的参数多了一个，接收的参数分别如下所示

①类目 id：categoryId 字段

②类目级别：categoryLevel 字段

③类目的父级 id：parentId 字段

④类目名称：categoryName 字段

⑤排序值：categoryRank 字段

在这个方法里依然使用@RequestBody 注解将参数转换为 GoodsCategory 对象参数。在商品类目修改接口中，依然先会对传递过来的 5 个参数进行基本的校验，再交给业务层代码处理。这里需要进行两个判断。

①根据 id 查询在数据库中的记录判断，如果没有则表示前端传递过来的数据有问题，返回错误信息

②查询是否有同名的类目，如果有则返回错误信息

在校验后封装数据并进行数据库 update 操作。修改商品类目的 SQL 语句也在 GoodsCategoryMapper.xml 文件中，它是标准的 update 修改语句。在执行成功后可以查看数据库中当前的记录是否如预期一样被成功修改。

（4）删除接口负责接收前端商品分类的删除请求。系统在处理前端传输过来的数据后，会将管理员选择需要删除的商品类目记录从数据库中删除。这里的删除功能是逻辑删除。

该接口的参数是一个数组，可以同时删除多条商品类目记录。这里只需要在前端将用户选择的商品类目的记录 id 封装好，再传参到后端即可。接口的请求路径为 /admin/categories/delete，并使用@RequestBody 注解将前端传过来的参数封装为数组对象。如果数组为空则直接返回异常提醒。在参数验证通过后则调用 deleteBatch()批量删除方法进行数据库操作，否则向前端返回错误信息。这里执行的 SQL 语句为 update 语句，将对应记录的 is_deleted 字段值设置为 1 即表示已经删除。

18.5 商品类目管理模块前端功能的实现

18.5.1 分类管理页面分页功能的实现

在 resources/static/admin/dist/js 目录下新增 newbee_mall_category.js 文件，并添加如下代码：

```
$(function () {
    var categoryLevel = $("#categoryLevel").val();
    var parentId = $("#parentId").val();

    $("#jqGrid").jqGrid({
        url: '/admin/categories/list?categoryLevel=' + categoryLevel + '&parentId=' + parentId,
        datatype: "json",
        colModel: [
            {label: 'id', name: 'categoryId', index: 'categoryId', width: 50, key: true, hidden: true},
            {label: '分类名称', name: 'categoryName', index: 'categoryName', width: 240},
            {label: '排序值', name: 'categoryRank', index: 'categoryRank', width: 120},
            {label: '添加时间', name: 'createTime', index: 'createTime', width: 120}
        ],
        height: 400,
        rowNum: 10,
        rowList: [10, 20, 50],
        styleUI: 'Bootstrap',
        loadtext: '信息读取中...',
        rownumbers: false,
        rownumWidth: 20,
        autowidth: true,
        multiselect: true,
        pager: "#jqGridPager",
        jsonReader: {
            root: "data.list",
```

```
            page: "data.currPage",
            total: "data.totalPage",
            records: "data.totalCount"
        },
        prmNames: {
            page: "page",
            rows: "limit",
            order: "order",
        },
        gridComplete: function () {
            //隐藏grid底部滚动条
            $("#jqGrid").closest(".ui-jqgrid-bdiv").css({"overflow-x": "hidden"});
        }
    });

    $(window).resize(function () {
        $("#jqGrid").setGridWidth($(".card-body").width());
    });
});
```

categoryLevel 和 parentId 是分类列表的请求接口必不可少的两个参数。这里读者应该知道左侧菜单栏分类管理的跳转链接设计成 /admin/categories?parentId=0&categoryLevel=1&backParentId=0 和后续处理这几个参数的作用了。因为这些参数会在接口请求中被用到，所以在请求 URL 和页面中都做了处理。在请求列表接口时系统能够通过 jQuery 语法读取 categoryLevel 和 parentId 这两个值，并在列表请求中将它们放入 URL 中。

以上代码的主要功能为分页数据展示、字段格式化、jqGrid 分页 DOM 宽度的自适应。在页面加载时，首先调用 jqGrid 分页插件的初始化方法：将页面中 id 为 jqGrid 的 DOM 渲染为分页表格，在获取 categoryLevel 和 parentId 两个变量值后拼接分类列表的 API 请求地址，并向后端发送请求，然后按照后端返回的 JSON 数据填充商品类目表格和表格下方的分页按钮。这里可以参考前文 jqGrid 分页功能整合进行理解。最后，重启项目验证商品类目数据分页功能是否正常。

18.5.2 上下级分类页面的跳转逻辑处理

从导航页进入分类管理页面，页面上默认展示的是一级分类的数据。由于商品类目层级的关系，管理员用户还需要处理二级分类和三级分类的数据。页面展示的是一级分类列表，想要看到其中一条一级类目下的二级分类数据，比如"家电\数码\手机"这条商品类目下的二级分类，管理员用户首先可以选择这条分类，并点击"下级分类管理"按钮，然后就能够看到二级分类的列表数据了。"下级分类管理"按钮对应的触发事件是 categoryManage()方法，其具体的实现代码如下所示：

```javascript
/**
 * 管理下级分类
 */
function categoryManage() {
    var categoryLevel = parseInt($("#categoryLevel").val());
    var parentId = $("#parentId").val();
    var id = getSelectedRow();
    if (id == undefined || id < 0) {
        return false;
    }
    if (categoryLevel == 1 || categoryLevel == 2) {
        categoryLevel = categoryLevel + 1;
        window.location.href = '/admin/categories?categoryLevel=' + categoryLevel + '&parentId=' + id + '&backParentId=' + parentId;
    } else {
        swal("无下级分类", {
            icon: "warning",
        });
    }
}
```

通过 JS 源码可以看出，在点击了"下级分类管理"按钮后系统还是会跳转到/admin/categories，只是此时的请求参数有所不同。比如当前在一级分类列表中，categoryLevel 变量的值其实是 1，但是在跳转到下级分类时，categoryLevel 变量的值就会变成 2。parentId 变量则是当前选择一级分类的 id，其跳转逻辑与一级分类相似，只是请求参数不同。所以在 jqGrid 请求列表数据时系统也会获取二级分类的数据，此时页面上展示的数据就有了变化。三级分类的展示过程同理。

以上是分类层级的向下跳转，与之对应的是返回上一级分类列表。系统页面上有"返

回"按钮,在点击后执行的方法是 categoryBack(),其代码如下所示:

```
/**
 * 返回上一层级
 */
function categoryBack() {
    var categoryLevel = parseInt($("#categoryLevel").val());
    var backParentId = $("#backParentId").val();
    if (categoryLevel == 2 || categoryLevel == 3) {
        categoryLevel = categoryLevel - 1;
        window.location.href = '/admin/categories?categoryLevel=' +
categoryLevel + '&parentId=' + backParentId + '&backParentId=0';
    } else {
        swal("无上级分类", {
            icon: "warning",
        });
    }
}
```

通过 JS 源码可以看出,在点击"返回"按钮后系统依然会跳转到/admin/categories,只是此时与之前跳转到下级分类时的逻辑相反。categoryLevel 变量的值由"加 1"的逻辑变成"减 1"的逻辑,parentId 变量则是 backParentId 的值。

18.5.3 分类管理页面添加和修改按钮的触发事件

添加和修改两个按钮分别绑定了对应的触发事件。在 newbee_mall_category.js 文件中新增 categoryAdd()方法和 categoryEdit()方法,两个方法中的实现为打开分类数据编辑框。categoryAdd()方法和 categoryEdit()方法实现的代码如下所示:

```
function categoryAdd() {
    reset();
    $('.modal-title').html('分类添加');
    $('#categoryModal').modal('show');
}

function categoryEdit() {
    reset();
    var id = getSelectedRow();
    if (id == null) {
        return;
    }
```

```
    var rowData = $("#jqGrid").jqGrid("getRowData", id);
    $('.modal-title').html('分类编辑');
    $('#categoryModal').modal('show');
    $("#categoryId").val(id);
    $("#categoryName").val(rowData.categoryName);
    $("#categoryRank").val(rowData.categoryRank);
}
```

在点击"添加"按钮后,仅仅是将 Modal 框显示出来,在修改时还有一个步骤,即将选择的记录回显到 Modal 框中以供修改。这里使用了 jqGrid 插件的 getRowData() 方法获取当前选择的所有字段的数据并赋值到对应的 HTML 标签中。

分类管理页面 Modal 框实际效果如图 18-4 所示。

图 18-4 分类管理页面 Modal 框实际效果

18.5.4 分类管理页面添加和编辑功能的实现

在信息录入完成后可以点击信息编辑框下方的"确认"按钮,此时系统会进行数据的交互,其 JS 实现代码如下所示:

```
$('#saveButton').click(function () {
    var categoryName = $("#categoryName").val();
    var categoryLevel = $("#categoryLevel").val();
    var parentId = $("#parentId").val();
    var categoryRank = $("#categoryRank").val();
    if (!validCN_ENString2_18(categoryName)) {
        $('#edit-error-msg').css("display", "block");
        $('#edit-error-msg').html("请输入符合规范的分类名称!");
```

```javascript
} else {
    var data = {
        "categoryName": categoryName,
        "categoryLevel": categoryLevel,
        "parentId": parentId,
        "categoryRank": categoryRank
    };
    var url = '/admin/categories/save';
    var id = getSelectedRowWithoutAlert();
    if (id != null) {
        url = '/admin/categories/update';
        data = {
            "categoryId": id,
            "categoryName": categoryName,
            "categoryLevel": categoryLevel,
            "parentId": parentId,
            "categoryRank": categoryRank
        };
    }
    $.ajax({
        type: 'POST',//方法类型
        url: url,
        contentType: 'application/json',
        data: JSON.stringify(data),
        success: function (result) {
            if (result.resultCode == 200) {
                $('#categoryModal').modal('hide');
                swal("保存成功", {
                    icon: "success",
                });
                reload();
            } else {
                $('#categoryModal').modal('hide');
                swal(result.message, {
                    icon: "error",
                });
            }
            ;
        },
        error: function () {
```

```
            swal("操作失败", {
                icon: "error",
            });
        }
    });
});
```

由于两个方法的传参和后续处理逻辑类似，为了避免太多重复代码这里就将商品分类数据的添加和修改两个触发方法写在一起了，步骤如下所示。

①在 JS 方法中对输入的数据进行简单的正则验证

②封装商品类目参数

③向对应的后端商品类目的添加或者修改接口发送 Ajax 请求

④在请求成功后提醒用户请求成功并隐藏当前的商品类目信息 Modal 框，同时刷新商品类目列表数据

⑤请求失败则提醒对应的错误信息

18.5.5 分类管理页面删除功能的实现

点击"删除"按钮的触发事件为 deleteCagegory()。在 newbee_mall_category.js 文件中新增如下代码：

```
function deleteCagegory() {
    var ids = getSelectedRows();
    if (ids == null) {
        return;
    }
    swal({
        title: "确认弹框",
        text: "确认要删除数据吗?",
        icon: "warning",
        buttons: true,
        dangerMode: true,
    }).then((flag) => {
        if (flag) {
            $.ajax({
                type: "POST",
```

```
            url: "/admin/categories/delete",
            contentType: "application/json",
            data: JSON.stringify(ids),
            success: function (r) {
                if (r.resultCode == 200) {
                    swal("删除成功", {
                        icon: "success",
                    });
                    $("#jqGrid").trigger("reloadGrid");
                } else {
                    swal(r.message, {
                        icon: "error",
                    });
                }
            }
        });
    }
})
;
}
```

这里，获取用户在 jqGrid 表格中选择需要删除的所有记录的 id，并将参数封装并向后端发送 Ajax 请求，请求地址为/admin/categories/delete，后端接收到请求后会将对应的记录删除。

18.5.6 功能测试

在编码完成后，启动 Spring Boot 项目。在启动成功后打开浏览器并输入后台管理系统登录页面地址：

http://localhost:8080/admin/login

在输入登录信息后登录后台管理系统，点击左侧导航栏中的"分类管理"就可以对相关功能进行测试，演示过程如下所示。

（1）列表功能。进入分类管理页面就可以看到数据列表，当前列表中有 9 条记录，分类列表如图 18-5 所示。

图 18-5　分类列表

（2）添加功能。首先，在点击"新增"按钮后会弹出信息编辑框，新增分类如图 18-6 所示。

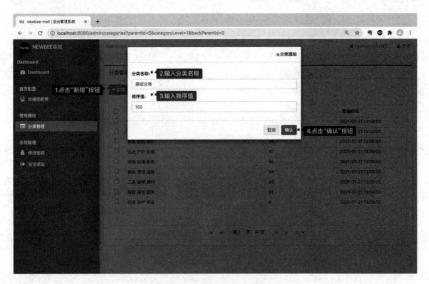

图 18-6　新增分类

然后，在 input 框中分别输入分类名称和排序值。最后，点击"确认"按钮。在添加成功后可以看到列表中多了一条数据，添加成功如图 18-7 所示。

第 18 章　分类管理模块的开发

图 18-7　添加成功

（3）修改功能。首先，选中一条需要修改的记录并点击"修改"按钮，弹出 Modal 框。此时 Modal 框中会显示当前记录的所有信息，管理员用户可以修改这些内容，修改分类如图 18-8 所示。

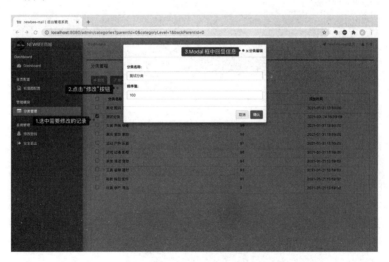

图 18-8　修改分类

然后，对当前需要选中的商品分类信息进行修改，最后点击确认按钮。在修改成功后可以看到列表之前选中的那一条数据已经修改完成，修改完成如图 18-9 所示。

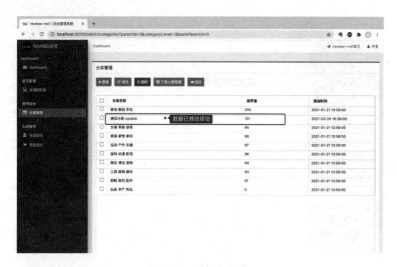

图 18-9　修改完成

（4）删除功能。首先，选中需要删除的记录并点击"删除"按钮，此时会出现一个确认弹框，删除分类如图 18-10 所示。

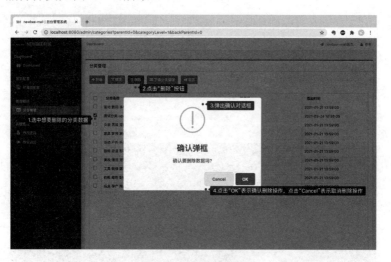

图 18-10　删除分类

点击"OK"表示确认该删除操作，点击"Cancel"表示取消操作。在点击"OK"后，选中的记录会被删除。删除成功如图 18-11 所示，列表中已经不存在之前选中的商品分类记录了。

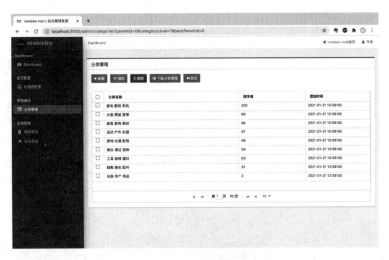

图 18-11　删除成功

读者可以按照文中的思路和过程自行测试。至此，商品类目管理模块开发完成！

18.6　分类数据的三级联动功能开发

18.6.1　多层级数据联动效果的常见场景

二级联动、三级联动或者更多层级的数据联动是互联网产品比较常见的交互方式。这种方式可以提升用户的使用体验，使得本来需要用户逐个输入的文字内容，可以直接借助联动的选择框进行选择。另外，这种基于联动的选择框，可以限制用户随意输入内容，规范用户提交的数据。

这里以电商网站的收货地址编辑场景为例，如图 18-12～图 18-16 所示。

这是一个省市区县的四级联动效果。在新增收货地址时，依次选择浙江、杭州市、西湖区、三墩镇，最终确定了收货地址的所在地区。用户可以直接通过网站提供的四级联动进行区域的选择，避免了手动输入的麻烦，同时，也防止用户随意输入不规范的地址数据。

新蜂商城的商品类目也设计成了多层级的方式，在选择分类信息时也用了这种联动的方式。这一节就来讲一讲这种交互方式，包括它的原理、实现方式。最终会借助新蜂商城三级分类联动的功能实现来完成实践。

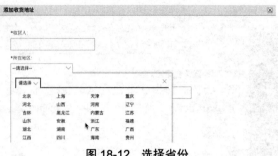

图 18-12 选择省份

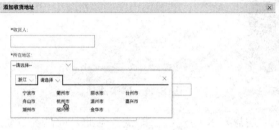

图 18-13 选择城市

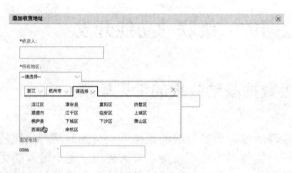

图 18-14 选择区县

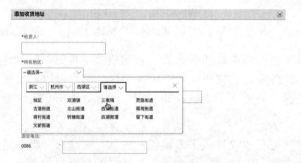

图 18-15 选择街道

图 18-16 所在地区选择完毕

18.6.2 多层级的数据联动实现原理和方式

二级联动就是当下拉框中选择列表 N 的值发生改变时,下一级的下拉框中选择列表 M 也发生改动。下拉列表 M 中的所有数据与下拉列表 N 中的值相对应。这种在交互中的数据变动并不是无关联的变动。比如,在选择浙江省后,下级城市列表中的数据就全是浙江省的城市;选择广东省后,下级城市列表中的数据就全是广东省的城市。城市区域的联动效果也一样。

三级联动或多级联动,就是利用二级联动的原理,在第一级的内容发生改变时,第二级跟着变,第三级跟着第二级变。

数据种类的不同,代码实现的方式也会不同。比如上述所举例的省市区县数据联动,由于这些行政区数据都是固定的且改动较小的数据,就可以直接在前端把所有数据"写死",并通过下拉选择框的 change 事件来实现联动功能。当然,也可以不把所有数据"写死",而是将省市区县的数据存在数据库中,并通过接口的形式返回数据列表。

由于新蜂商城的商品类目数据发生更改的频率较高,增加、编辑、删除等操作都会造成列表数据的变化,所以新蜂商城的三级联动功能在实现时采用动态的三级联动。也就是实时地通过数据库读取数据,并通过下拉选择框的 change 事件来实现联动功能。

接下来笔者会以新蜂商城商品编辑页面的分类选择交互模块为例,来实现一个基础的三级联动功能。

18.6.3 分类三级联动页面基础样式的实现

在 resources/templates/admin 目录下新建 coupling-test.html,并引入基础的 JS 文件和 CSS 样式文件。该测试页面比较简单并没有特别的插件,代码如下所示:

```html
<!DOCTYPE html>
<html xmlns:th="http://www.thymeleaf.org">
<header th:replace="admin/header::header-fragment"></header>
<body class="hold-transition sidebar-mini">
<div class="wrapper">
    <!-- 引入页面头 header-fragment -->
    <div th:replace="admin/header::header-nav"></div>
    <!-- 引入工具栏 sidebar-fragment -->
    <div th:replace="admin/sidebar::sidebar-fragment(${path})"></div>
    <!-- Content Wrapper. Contains page content -->
    <div class="content-wrapper">
        <!-- Content Header (Page header) -->
        <div class="content-header">
            <div class="container-fluid">
            </div><!-- /.container-fluid -->
        </div>
        <!-- Main content -->
        <div class="content">
            <div class="container-fluid">
                <div class="card card-primary card-outline">
                    <div class="card-header">
                        <h3 class="card-title">分类三级联动</h3>
                    </div>
                    <div class="card-body">
                        <form id="coupling-test" onsubmit="return false;">
                            <div class="form-group" style="display:flex;">
                                <label class="control-label">请选择分类:    </label>
                                <select class="form-control col-sm-3" id="levelOne"
                                        data-placeholder="请选择分类...">
                                </select> 
                                <select class="form-control col-sm-3"
                                        id="levelTwo"
                                        data-placeholder="请选择分类...">
                                </select> 
                                <select class="form-control col-sm-3"
                                        id="levelThree"
                                        data-placeholder="请选择分类...">
                                </select>
```

```html
                    </div>
                </form>
            </div>

        </div>
    </div><!-- /.container-fluid -->
  </div>
 </div>
 <!-- /.content-wrapper -->
 <!-- 引入页脚 footer-fragment -->
 <div th:replace="admin/footer::footer-fragment"></div>
</div>
<!-- jQuery -->
<script th:src="@{/admin/plugins/jquery/jquery.min.js}"></script>
<!-- jQuery UI 1.11.4 -->
<script th:src="@{/admin/plugins/jQueryUI/jquery-ui.min.js}"></script>
<!-- Bootstrap 4 -->
<script th:src="@{/admin/plugins/bootstrap/js/bootstrap.bundle.min.js}">
</script>
<!-- AdminLTE App -->
<script th:src="@{/admin/dist/js/adminlte.min.js}"></script>
</body>
</html>
```

这里主要的代码就是三个 select 下拉选择框。它们分别定义了不同的 id 属性，分别是 levelOne、levelTwo 和 levelThree。这里只是页面的实现，并没有涉及任何功能。该页面主要是先把三级分类联动的三个下拉选择框的基础展示效果实现出来。

接下来，在 NewBeeMallGoodsCategoryController 类中新增一个方法，即在访问 admin/coupling-test 时返回该视图，并让页面能够显示出来，代码如下所示：

```java
@GetMapping("/coupling-test")
public String couplingTest(HttpServletRequest request) {
    return "admin/coupling-test";
}
```

在编码完成后，启动 Spring Boot 项目。在启动成功后打开浏览器并输入如下请求地址：

```
http://localhost:8080/admin/coupling-test
```

此时，分类三级联动测试页面效果如图 18-17 所示。

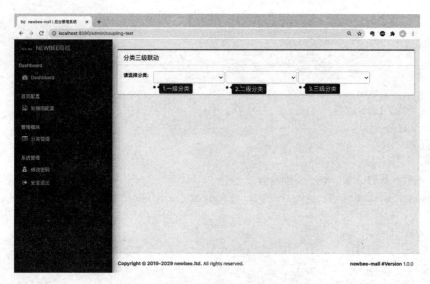

图 18-17 分类三级联动测试页面效果

三个数字框是笔者添加的标注,并不是页面上的内容,页面只有标题和三个下拉选择框。

18.6.4 数据初始化

前文三级联动的基础选择框已经实现。接下来进行第二步,给选择框配置初始数据,让用户可以进行选择。注意,这一步还是对选择框中的数据进行初始化的步骤,并没有进行联动功能的实现。

既然要初始化数据,就先分析页面对应的三个选择框需要显示什么数据。

一级分类的下拉选择框的数据就是显示所有的一级分类。因为一级分类并没有父级分类,不会随着上层分类的变化而发生联动变化,但是二级分类选择框和三级分类选择框中的内容则需要根据上级数据的变化而发生联动变化。

在一级分类发生更改时(注意,这里指一级分类选择框的更改),二级分类下拉选择框的数据发生了更改。此时三级分类下拉选择框中显示的数据是什么呢?应该显示当前二级分类下拉选择框中的第一条数据下的所有三级分类。一级分类下拉选择框中的每一次更改都会使得二级分类下拉选择框的数据发生改变。而三级分类下拉选择框在此时都应该显示二级分类下拉选择框中的第一个选项下的所有三级分类数据。

在初始化时,一级分类下拉选择框的数据是显示所有的一级分类,对应的二级分类下拉选择框中的数据就应该显示当前一级分类下拉选择框中第一条一级分类数据的所有二级分类。至此,一级分类和二级分类两个下拉选择框中的初始化数据就确定了。同理,三级分类下拉选择框中的数据也可以确定下来,就是当前二级分类下拉选择框中第一条二级分类下的所有三级分类数据。

因此,这里需要先修改控制器下的 couplingTest()方法。在转发视图之前,把上面确定下来的初始化数据带上,再转发视图到 coupling-test.html 页面,编码如下所示:

```java
@GetMapping("/coupling-test")
public String couplingTest(HttpServletRequest request) {
    request.setAttribute("path", "coupling-test");
    //查询所有的一级分类
    List<GoodsCategory> firstLevelCategories = NewbeeMallCategoryService.selectByLevelAndParentIdsAndNumber(Collections.singletonList(0L), newbee_mall_categoryLevelEnum.LEVEL_ONE.getLevel());
    if (!CollectionUtils.isEmpty(firstLevelCategories)) {
        //查询一级分类列表中第一个实体的所有二级分类
        List<GoodsCategory> secondLevelCategories = NewbeeMallCategoryService.selectByLevelAndParentIdsAndNumber(Collections.singletonList(firstLevelCategories.get(0).getCategoryId()), newbee_mall_categoryLevelEnum.LEVEL_TWO.getLevel());
        if (!CollectionUtils.isEmpty(secondLevelCategories)) {
            //查询二级分类列表中第一个实体的所有三级分类
            List<GoodsCategory> thirdLevelCategories = NewbeeMallCategoryService.selectByLevelAndParentIdsAndNumber(Collections.singletonList(secondLevelCategories.get(0).getCategoryId()), newbee_mall_categoryLevelEnum.LEVEL_THREE.getLevel());
            request.setAttribute("firstLevelCategories", firstLevelCategories);
            request.setAttribute("secondLevelCategories", secondLevelCategories);
            request.setAttribute("thirdLevelCategories", thirdLevelCategories);
            return "admin/coupling-test";
        }
    }
    return "error/error_5xx";
}
```

该段代码会查询所有的一级分类列表数据 firstLevelCategories。firstLevelCategories 列表对象不为空时,再查询该一级分类列表中第一个实体下的所有二级分类 secondLevelCategories。secondLevelCategories 列表对象不为空时,再查询该二级分类列表中第一个实体下的所有三级分类 thirdLevelCategories。最后,把这些数据都放到 request 对象域中并返回最终视图。

在后端代码实现后，紧接着就是更改前端模板代码。先把后端传过来的数据读取到页面中。这里就通过 th:each 标签读取所有的初始化数据。coupling-test.html 页面代码更改如下所示：

```html
<select class="form-control col-sm-3" id="levelOne" data-placeholder="请选择分类...">
    <th:block th:unless="${null == firstLevelCategories}"><!-- 非空判断 -->
    <th:block th:each="c : ${firstLevelCategories}"><!-- 循环列表 -->
        <option th:value="${c.categoryId}" th:text="${c.categoryName}">
</option>
    </th:block>
    </th:block>
</select> 
<select class="form-control col-sm-3" id="levelTwo" data-placeholder="请选择分类...">
    <th:block th:unless="${null == secondLevelCategories}">
      <th:block th:each="c : ${secondLevelCategories}">
        <option th:value="${c.categoryId}" th:text="${c.categoryName}">
</option>
    </th:block>
    </th:block>
</select> 
<select class="form-control col-sm-3" id="levelThree" data-placeholder="请选择分类...">
    <th:block th:unless="${null == thirdLevelCategories}">
      <th:block th:each="c : ${thirdLevelCategories}">
        <option th:value="${c.categoryId}" th:text="${c.categoryName}">
</option>
    </th:block>
    </th:block>
</select>
```

本来是三个内容为空的 select 下拉选择框，就使用循环语句将返回的列表数据分别读取到对应的下拉选择框中，并给 option 标签分别赋值。value 属性值为每个分类的 id，text 属性值为每个分类的名称。

在编码完成后重启项目，此时的页面效果如图 18-18 所示。

第 18 章 分类管理模块的开发

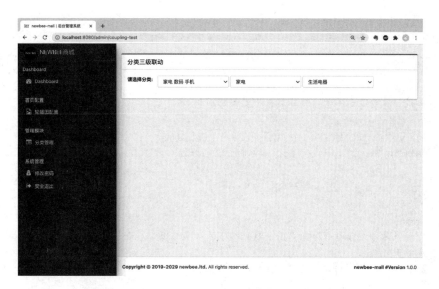

图 18-18 分类三级联动初始化页面

目前只是完成了初始数据的显示，在下拉选择框中选择不同的数据时，下级分类的联动功能并没有实现，还需要继续编码。

18.6.5 数据联动后端接口的实现

这里把注意力放在三个下拉选择框中。在二级分类下拉框中的选中条目发生更改时，三级分类下拉框中的数据就会重新加载；当一级分类下拉框中的选中条目发生更改时，二级分类和三级分类下拉框中的数据都会重新加载。翻译成伪代码语言可以这么理解，当触发上级下拉选择框的 change 事件时，需要重新发起请求获取当前分类的下级分类数据，并将这些内容重新赋值到 select 下拉选择框中。

首先后端需要新增一个接口。该接口的功能就是根据前端传过来的分类 id 来获取该分类下的下级分类数据。如果该分类是一级分类，则返回该一级分类下的所有二级分类列表数据和第一个二级分类下的所有三级分类数据。如果该分类是二级分类，则返回该二级分类下的所有三级分类列表数据。新增接口代码和注释如下所示：

```
@RequestMapping(value = "/categories/listForSelect", method = RequestMethod.GET)
@ResponseBody
public Result listForSelect(@RequestParam("categoryId") Long categoryId) {
```

```java
    if (categoryId == null || categoryId < 1) {
        return ResultGenerator.genFailResult("缺少参数!");
    }
    GoodsCategory category = NewbeeMallCategoryService.getGoodsCategoryById(categoryId);
    //既不是一级分类也不是二级分类则不返回数据
    if (category == null || category.getCategoryLevel() == newbee_mall_categoryLevelEnum.LEVEL_THREE.getLevel()) {
        return ResultGenerator.genFailResult("参数异常!");
    }
    Map categoryResult = new HashMap(2);
    if (category.getCategoryLevel() == newbee_mall_categoryLevelEnum.LEVEL_ONE.getLevel()) {
        //如果是一级分类则返回当前一级分类下的所有二级分类和二级分类列表中第一条数据下的所有三级分类列表
        //查询一级分类列表中第一个实体的所有二级分类
        List<GoodsCategory> secondLevelCategories = NewbeeMallCategoryService.selectByLevelAndParentIdsAndNumber(Collections.singletonList(categoryId), newbee_mall_categoryLevelEnum.LEVEL_TWO.getLevel());
        if (!CollectionUtils.isEmpty(secondLevelCategories)) {
            //查询二级分类列表中第一个实体的所有三级分类
            List<GoodsCategory> thirdLevelCategories = NewbeeMallCategoryService.selectByLevelAndParentIdsAndNumber(Collections.singletonList(secondLevelCategories.get(0).getCategoryId()), newbee_mall_categoryLevelEnum.LEVEL_THREE.getLevel());
            categoryResult.put("secondLevelCategories", secondLevelCategories);
            categoryResult.put("thirdLevelCategories", thirdLevelCategories);
        }
    }
    if (category.getCategoryLevel() == newbee_mall_categoryLevelEnum.LEVEL_TWO.getLevel()) {
        //如果是二级分类则返回当前分类下的所有三级分类列表
        List<GoodsCategory> thirdLevelCategories = NewbeeMallCategoryService.selectByLevelAndParentIdsAndNumber(Collections.singletonList(categoryId), newbee_mall_categoryLevelEnum.LEVEL_THREE.getLevel());
        categoryResult.put("thirdLevelCategories", thirdLevelCategories);
    }
    return ResultGenerator.genSuccessResult(categoryResult);
}
```

18.6.6 监听选择框的 change 事件并实现联动功能

在接口实现后，需要处理前端交互。当用户在更改选择框中的内容时，会触发选择框的 change 事件，并在该事件中进行业务处理和数据的渲染。

当选择框的 change 事件触发时，首先能获取当前选择的分类 id，并且根据该分类 id 请求后端接口获取该分类下的下级分类数据，然后根据后端返回的分类列表数据重新拼装选择框中的 option 列表内容，最终完成联动效果。整个过程如下所示。

①触发下拉选择框的 change 事件

②获取当前选中的分类 id

③向 listForSelect 接口请求下级分类数据

④动态拼装 option 列表

⑤使用新的 option 列表内容并赋值给下拉选择框

虽然过程如此，不过一级分类和二级分类的实现略有不同。一级分类的更改会改变二级分类和三级分类两个下拉选择框中的值，而二级分类的更改只会改变三级分类下拉选择框中的值，其实现代码如下所示：

```javascript
$('#levelOne').on('change', function () {
    $.ajax({
        url: '/admin/categories/listForSelect?categoryId=' + $(this).val(),
        type: 'GET',
        success: function (result) {
            if (result.resultCode == 200) {
                var levelTwoSelect = '';
                var secondLevelCategories = result.data.secondLevelCategories;
                var length2 = secondLevelCategories.length;
                for (var i = 0; i < length2; i++) {
                    levelTwoSelect += '<option value=\"' + secondLevelCategories[i].categoryId + '\">' + secondLevelCategories[i].categoryName + '</option>';
                }
                $('#levelTwo').html(levelTwoSelect);
                var levelThreeSelect = '';
                var thirdLevelCategories = result.data.thirdLevelCategories;
                var length3 = thirdLevelCategories.length;
                for (var i = 0; i < length3; i++) {
                    levelThreeSelect += '<option value=\"' + thirdLevelCategories[i].categoryId + '\">' + thirdLevelCategories[i].categoryName +
```

```javascript
'</option>';
                    }
                    $('#levelThree').html(levelThreeSelect);
                } else {
                    swal(result.message, {
                        icon: "error",
                    });
                }
                ;
            },
            error: function () {
                swal("操作失败", {
                    icon: "error",
                });
            }
        });
    });

    $('#levelTwo').on('change', function () {
        $.ajax({
            url: '/admin/categories/listForSelect?categoryId=' + $(this).val(),
            type: 'GET',
            success: function (result) {
                if (result.resultCode == 200) {
                    var levelThreeSelect = '';
                    var thirdLevelCategories = result.data.thirdLevelCategories;
                    var length = thirdLevelCategories.length;
                    for (var i = 0; i < length; i++) {
                        levelThreeSelect += '<option value=\"' + thirdLevelCategories[i].categoryId + '\">' + thirdLevelCategories[i].categoryName + '</option>';
                    }
                    $('#levelThree').html(levelThreeSelect);
                } else {
                    swal(result.message, {
                        icon: "error",
                    });
                }
                ;
            },
            error: function () {
                swal("操作失败", {
```

```
            icon: "error",
        });
    }
  });
});
```

在编码完成后,启动 Spring Boot 项目。在启动成功后打开浏览器并输入如下请求地址:

```
http://localhost:8080/admin/coupling-test
```

分类三级联动测试如图 18-19 所示。笔者主要是测试了在一级分类更改时的联动效果,每次一级分类更改,二级分类选择框和三级分类选择框都会联动修改。

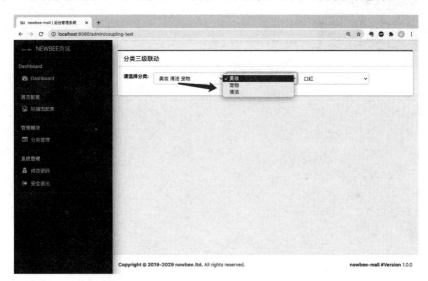

图 18-19　分类三级联动测试

最后测试在二级分类更改时的联动效果。如图 18-20 所示,在选择"家电 数码 手机"中的"手机"二级分类时,三级分类下拉选择框中的数据会对应修改为新的数据。

这里也提醒读者,在测试的时候可以打开浏览器控制台的 Network 面板,观察在每一次切换分类时发送的 Ajax 请求和返回的数据,这样可以更好地理解所开发的各个功能。

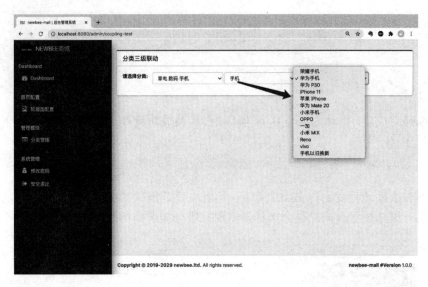

图 18-20 三级联动测试

第 19 章

富文本编辑器介绍及整合

在后台管理系统的商品管理模块中,在进行商品信息编辑时需要使用富文本编辑器来处理商品详情的内容,所以本章会讲解有关富文本编辑器的相关知识。

19.1 富文本编辑器详解

19.1.1 如何处理复杂的文本内容

接下来将从两张图片展开讲述,分别是某品牌的智能电视产品详情页面(如图 19-1 所示)和某新闻平台的新闻详情页面(如图 19-2 所示)。

现在的大部分网站在展示信息时都会采用这种图文并茂的方式。

为什么先介绍这两张图片呢?

这就要从笔者第一次接触富文本编辑器的一段经历讲起。笔者刚入行的时候,在一次需求更改中就负责开发这种类似的页面,功能比较简单,显示一张页面的详情文案即可。页面的展现与图 19-1 和图 19-2 的排版布局类似。而刚入行的笔者就闹了笑话。因为需要实现后端的功能,笔者就开始思考表结构设计的合理性,对照其中一张设计稿开始在表里加字段:主键+介绍信息字段+图片字段+介绍信息字段+图片字段……笔者把页面展示的内容全部拆开成单独的字段。

图 19-1　某品牌的智能电视产品详情页面

图 19-2　某新闻平台的新闻详情页面

但是再看另一份设计稿的时候，笔者瞬间慌了神。因为这张设计稿与第一张设计稿上的内容排版不同，不是文字+图片+文字+图，而是图片+图片+文字。内容布局有了很大的变化就完全不知该如何做，建表的过程也停滞不前。当时还在心里感慨，怎么实际开发项目这么难？

浏览至此的读者可能知道笔者当时为什么会被难住了。因为当时并不知道富文本编辑器的概念，也没有用过类似的富文本编辑器，因此这个简单的需求被想象得过于复杂了。其实需求很简单，建表也不复杂，只需要在对应的表中设置一个内容字段就可以。

而富文本编辑器就可以让开发人员很顺利地实现这种复杂的图文混排需求。

接下来将会讲解商品管理模块的开发，由于在商品编辑时就用到了富文本编辑器，所以本章会对富文本编辑器进行简单的介绍，并整合到实际的项目中，最终实现这种复杂排版内容的录入和读取功能，为之后商品管理模块的开发做铺垫。

19.1.2 富文本编辑器介绍及其优势

在介绍富文本编辑器的之前，先思考一下如何实现 HTML 页面的数据获取。在 form 表单中通常会设置 input 标签或者 textarea 标签，对于一些简单的需求，比如如登录信息的获取，使用 input 标签即可，而文字多的内容使用 textarea 标签获取。但是，当碰到需要排版复杂的图文混合的内容或者更多内容录入的时候，这两个标签就无法满足需求了。

某些产品经理可能会问开发人员：为什么用 Word 或者 WPS 能做的事情，在网站上就不能做呢？

这时候开发人员往往无言以对。这本身就是产品和技术实现的区别。

图文结合且排版良好的内容形式是 input 标签和 textarea 标签都无法完美支持的。因此，出现了富文本编辑器的概念。通过富文本编辑器可以制作出类似文章详情页和商品详情页的良好交互效果。

富文本编辑器是一种可内嵌于浏览器，所见即所得的文本编辑器。富文本编辑器不同于文本编辑器（比如 textarea 标签、input 标签），也叫作图文编辑器。在富文本编辑器里可以编辑非常丰富的内容，比如文字、图片、表情、代码等。

目前的富文本编辑器主要有 Markdown 版本和非 Markdown 版本的编辑器。因为大多数运营人员不懂 Markdown 语法，一般企业开发会采用非 Markdown 版本的富文本编辑器，常见的有 UEditor、wangEditor、KindEditor 等。而很多技术性平台的文章编辑通常使用 Markdown 编辑器，因为这部分人员可以很快掌握 Markdown 语法。掘金、CSDN 等网站都会默认使用 Markdown 编辑器作为用户的文章编辑器。

为什么要使用富文本编辑器，它又有什么优势呢？

①业务方提出的编辑需求越来越复杂

②文本框中需要编辑的内容变得越来越复杂、越来越丰富

③使用富文本编辑器比开发和编写 HTML 文件更灵活

④富文本编辑器功能丰富，可满足大部分需求

19.2 富文本编辑器wangEditor的介绍

本书实战项目所选用的富文本编辑器为开源产品 wangEditor。wangEditor 是一款轻量级 Web 富文本编辑器，配置方便，使用简单。读者可在其官网和 GitHub 仓库进行更多了解。

wangEditor 的功能很多。那么，在网站开发中常用的功能有哪些呢？这里笔者通过实际效果向各位读者进行展示。

（1）图文混排

wangEditor 可以编辑的内容比较丰富，以图 19-2 展示的新闻内容为例，使用 wangEditor 进行图文混排的效果如图 19-3 所示。

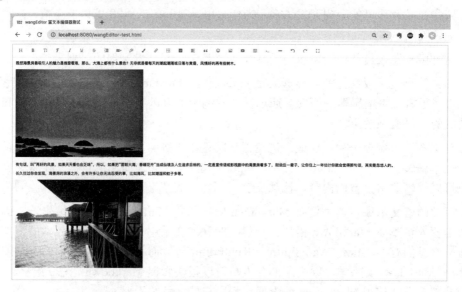

图 19-3　图文混排

（2）字体调整

wangEditor 可以很方便的进行文本内容格式的调整，包括字体、字号大小、字体颜色、背景颜色等内容都可以快速编辑，效果如图 19-4 所示。

图 19-4　字体调整

（3）全屏编辑

wangEditor 在初始化时有默认宽度，通常在页面中只占有部分版面。读者如果觉得不方便编辑，也可以点击全屏编辑的按钮让编辑器最大化，如图 19-5 所示。

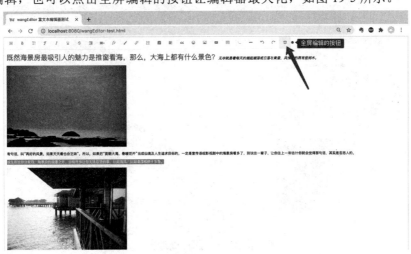

图 19-5　全屏编辑

（4）多图上传

wangEditor 支持图片上传功能。它不仅支持单图上传，也支持多图上传，非常人性化，如图 19-6 所示。

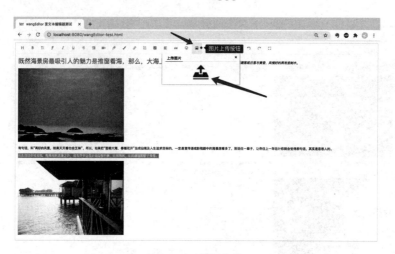

图 19-6　多图上传

（5）图片处理

wangEditor 支持调整图片位置、图片大小等内容。点击需要调整的图片，图片调整的工具栏就会出现，如图 19-7 所示。

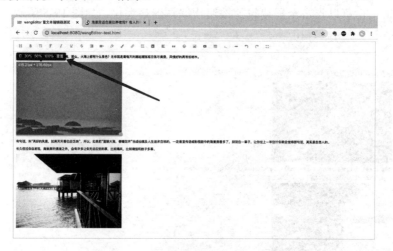

图 19-7　图片处理

以上只演示了与后续商品编辑功能有关联的部分。wangEditor 还有许多其他非常有意思的功能，比如插入表格、插入表情、插入代码等。这里就不再逐一演示了，读者可以点击工具栏中的各个按钮自行体验。

19.3 wangEditor整合编码案例

接下来将通过一个实际的整合案例,介绍如何在项目中使用 wangEditor。

(1) 新建 HTML 测试页面

在 resources/static 目录下新建 wangEditor-test.html 文件,并引入 wangEditor 的 JS 文件,代码如下所示:

```
<script type="text/javascript" src="//unpkg.com/wangeditor/dist/wangEditor.min.js"></script>
```

(2) 创建富文本编辑框 DOM

在 wangEditor-test.html 页面中创建 id 为 wangEditor 的 div 标签。对它进行定义主要是为了后续编辑器的初始化。

这里笔者将它的 id 命名为 wangEditor,该值可以自行修改,代码如下所示:

```
<div id="wangEditor"></div>
```

编辑区域高度默认为 300px。这里也可以在 wangEditor 对象初始化时通过 height 属性进行设置。

(3) 初始化 wangEditor 对象

初始化 wangEditor 对象并对一些配置项进行设置。在 wangEditor-test.html 文件中新增代码如下所示:

```
<script type="text/javascript">
// 初始化富文本编辑器 start

const E = window.wangEditor;
const editorD = new E("#wangEditor");
// 设置编辑区域高度为 640px
editorD.config.height = 640;
// 配置服务端图片上传地址
editorD.config.uploadImgServer = "/uploadFiles";
editorD.config.uploadFileName = "files";
// 限制图片大小 2MB
editorD.config.uploadImgMaxSize = 2 * 1024 * 1024;
// 限制一次最多能传 5 张图片
editorD.config.uploadImgMaxLength = 5;
```

```
// 隐藏插入网络图片的功能
editorD.config.showLinkImg = false;
editorD.create();
</script>
```

这里，相关的配置项和注释都在以上代码中了。

在配置项设置完成后，下一步就可以调用 create()创建 wangEditor 对象了。

（4）获取文档内容

在整理好富文本内容并写入编辑器后，还需要获取在 wangEdito 中输入的内容，并通过请求传给后端进行逻辑处理。wangEditor 提供了对应的方法来获取其中的内容，比如获取输入的商品详情内容，就可以用如下代码实现：

```
var content = editorD.txt.html();
```

在获取成功后，将商品详情内容字段进行封装，并与后端接口进行交互。这部分内容会在后文讲解。

wangEditor-test.html 文件的完整代码如下所示：

```
<!DOCTYPE html>
<html lang="en">
  <head>
    <meta charset="UTF-8" />
    <title>wangEditor 富文本编辑器测试</title>
  </head>
  <body>
    <br>
    wangEditor 富文本编辑器：
        <br>
    <div id="wangEditor"></div>
        <br>
    <input type="button" onclick="getContent()" value="获取文档内容" />
  </body>
  <script
    type="text/javascript"
    src="https://unpkg.com/wangeditor/dist/wangEditor.min.js"
  ></script>
  <script type="text/javascript">
    // 初始化富文本编辑器 start

    const E = window.wangEditor;
    const editorD = new E("#wangEditor");
```

```
// 设置编辑区域高度为 640px
editorD.config.height = 640;
// 配置服务端图片上传地址
editorD.config.uploadImgServer = "/uploadFiles";
editorD.config.uploadFileName = "files";
// 限制图片大小 2MB
editorD.config.uploadImgMaxSize = 2 * 1024 * 1024;
// 限制一次最多能传 5 张图片
editorD.config.uploadImgMaxLength = 5;
// 隐藏插入网络图片的功能
editorD.config.showLinkImg = false;
editorD.create();

function getContent() {
  var content = editorD.txt.html();
  alert(content);
}
</script>
</html>
```

wangEditor-test.html 页面效果如图 19-7 所示。

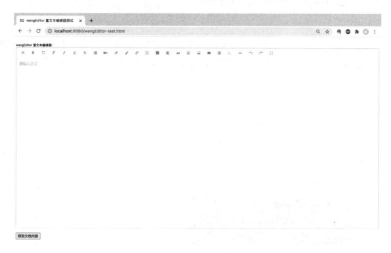

图 19-7　wangEditor-test.html 页面效果

在编辑器中输入内容，点击"获取文档内容"按钮，也可以获取数据，效果如图 19-8 所示。

图 19-8　获取文档内容

19.4　新蜂商城项目wangEditor的应用情况

19.4.1　为什么选择 wangEditor

新蜂商城的商品详情编辑框一开始选择的富文本编辑器并不是 wangEditor，而是 KindEditor。当时商品详情编辑页面的效果如图 19-9 所示。

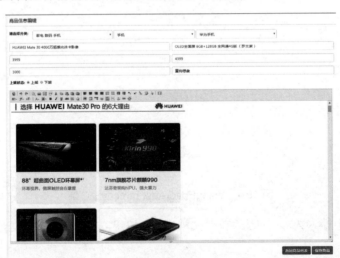

图 19-9　使用 KindEditor 的商品编辑页面

更改为 wangEditor 主要有以下三个原因。

（1）KindEditor 长时间没有维护更新

KindEditor 同样是一款开源产品。仓库在 GitHub 上，但是该仓库已经长时间没有维护更新了。提交记录比较频繁的时期为 2009 年至 2013 年，此后只有零星几条提交记录。

（2）收到了很多关于 KindEditor 的问题反馈

KindEditor 长时间未更新，各种浏览器却一直在更新版本，再加上现在有些浏览器默认不再开启 Flash 插件，导致现在使用 KindEditor 会出现一些兼容、图片上传的问题。

（3）wangEditor 重启更新，V4.0 版本发布

（1）（2）两个问题让笔者动了使用其他富文本编辑器的想法，不过一开始并没有找到特别合适的替代产品。直到 2020 年 wangEditor 重启版本更新，笔者才决定更换为 wangEditor。后来笔者就把 wangEditor 整合到了新蜂商城项目中，在功能测试完成后，于 2020 年 10 月 14 号将所有修改后的代码上传至 GitHub 开源仓库。

19.4.2　wangEditor 整合过程中的问题

因为 wangEditor 使用文档比较清晰，所以整合过程也比较顺利。但有两个问题还是花了一些时间处理。

（1）文件上传处理。wangEditor 在进行多图上传时，input 框中的 name 属性各不同。比如上传 3 张图片，它们对应的 input 框中的 name 属性分别为 files1、files2、files3。请求文件的内容如图 19-1 所示。

在 Spring Boot 项目中，此时的文件上传就需要进行额外的代码处理，具体可以参考 11.4.2 节的内容。

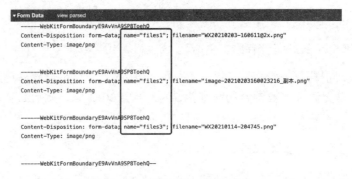

图 19-10　请求文件的内容

（2）wangEditor 在初始化时只会处理带有 html 标签的字符串，无 html 标签的字符串不会被初始化到编辑器中。

测试代码如下所示：

```
<!DOCTYPE html>
<html lang="en">
<head>
    <meta charset="UTF-8" />
    <title>wangEditor 富文本编辑器测试 2</title>
</head>
<body>
<div id="div1">
初始化字符串 1
<p>初始化字符串 2</p>
</div>
</body>
<script type="text/javascript" src="https://unpkg.com/wangeditor/dist/wangEditor.min.js"></script>
<script type="text/javascript">
    // 初始化富文本编辑器 start
    const E = window.wangEditor;
    const editorD = new E("#div1");
    // 设置编辑区域高度为 640px
    editorD.config.height = 640;
    // 配置服务端图片上传地址
    editorD.config.uploadImgServer = "/uploadFiles";
    editorD.config.uploadFileName = "files";
    // 限制图片大小 2MB
    editorD.config.uploadImgMaxSize = 2 * 1024 * 1024;
    // 限制一次最多能传 5 张图片
    editorD.config.uploadImgMaxLength = 5;
```

```
    // 隐藏插入网络图片的功能
    editorD.config.showLinkImg = false;
    editorD.create();
  </script>
</html>
```

该段代码的作用是将 div1 初始化为编辑器模块，div1 有两个字符串，其中一个字符串中带有 p 标签，另外一个字符串不带 p 标签。在初始化后，只有带着 p 标签的字符串才会出现在编辑器中，而不带有 p 标签的字符串则出现在编辑器外，如图 19-11 所示。

图 19-11　wangEditor-test2.html 页面效果

与此对应，新蜂商城的商品详情编辑页面就会出现图 19-12 的情况，即想要编辑商品详情内容，而有些内容会出现在编辑器的外面。

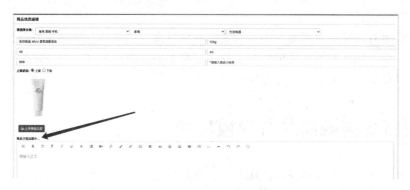

图 19-12　商品信息显示在编辑器外面

出现这种问题可能是因为 wangEditor 不支持纯字符串内容。于是笔者把有问题的字符串包上 p 标签，一切就正常了。

后来笔者咨询过 wangEditor 的作者，他的答复是为了避免出现其他问题，wangEditor 要遵循一些规则，想要将内容渲染进编辑器中，必须要使用标签包裹，否则可能会出现额外的问题。因此在 wangEditor 支持之前，开发人员尽量把需要初始化到编辑器中的内容字段使用标签包裹起来。

以上就是 wangEditor 整合过程中两个需要注意的问题。当然，在新蜂商城项目中这两个问题已经处理完毕。

第 20 章

商品编辑页面及商品管理模块的开发

本章将进行商品编辑页面及商品管理模块的开发,主要涉及新蜂商城商品模块表结构的设计、商品信息编辑页面的制作、商品信息添加功能的实现、商品信息管理接口的实现和商品信息管理页面的设计及编码。

20.1 新蜂商城商品管理模块简介

新蜂商城虽然是购物网站,也属于电子商务产品,但它主要是面向 Java 开发人员的学习资源和练习项目。如果读者想要实际上线供企业商业的产品,还需要进行一定的二次开发工作。

商品是交易的基础,那么商品管理就是电商系统最重要的部分。商品管理模块用于记录与商品有关的数据。虽然商品管理的系统逻辑不复杂,但是操作的数据比较多。订单、营销、支付、物流等环节都需要从商品模块中获取和操作数据。

商品编辑页面肯定不只有编辑器的整合和分类信息的选择,商品信息表也不只是商品类目 id 和商品内容两个字段。接下来笔者会先对商品模块进行分析并进行表结构的设计,然后进行商品编辑页面的制作和功能开发,最后进行商品管理模块功能的页面和功能开发。

20.2 新蜂商城商品信息表结构的设计

为了设计商品信息表的字段，笔者参考了知名线上商城的设计，比如天猫商城和京东商城的商品详情页。

天猫商城商品详情页如图 20-1 所示。

图 20-1 天猫商城商品详情页

京东商城商品详情页如图 20-2 所示。

图 20-2 京东商城商品详情页

由于用户在各大线上商城的长期引导下形成了固有的使用习惯，所以商品信息的设计和模块划分一般差不多。这已经沉淀为一种无形的规则，主要分为如下 4 方面。

①商品基础信息

②商品图片

③商品详情内容

④商品价格信息

商品详情的必要字段包括 6 各方面。

①商品名称

②商品简介

③商品原价

④商品实际售价

⑤商品详情内容

⑥商品主图/商品封面图

以上字段是商品信息实体应该具有的基础字段，而新蜂商城系统在此基础上增加了 3 个字段。

①商品库存，订单出库及仓库数据统计

②分类 id，与商品类目的关联关系

③销售状态，可以控制商品是否能够正常销售

最终，商品信息表的 SQL 设计就好了。在数据库中直接执行以下 SQL 语句即可：

```sql
USE 'newbee_mall_db';

DROP TABLE IF EXISTS 'tb_newbee_mall_goods_info';

CREATE TABLE 'tb_newbee_mall_goods_info' (
  'goods_id' bigint(20) UNSIGNED NOT NULL AUTO_INCREMENT COMMENT '商品表主键id',
  'goods_name' varchar(200) CHARACTER SET utf8 COLLATE utf8_general_ci NOT NULL DEFAULT '' COMMENT '商品名',
  'goods_intro' varchar(200) CHARACTER SET utf8 COLLATE utf8_general_ci NOT NULL DEFAULT '' COMMENT '商品简介',
  'goods_category_id' bigint(20) NOT NULL DEFAULT 0 COMMENT '关联分类id',
  'goods_cover_img' varchar(200) CHARACTER SET utf8 COLLATE utf8_general_ci NOT NULL DEFAULT '/admin/dist/img/no-img.png' COMMENT '商品主图',
  'goods_carousel' varchar(500) CHARACTER SET utf8 COLLATE utf8_general_ci
```

```sql
NOT NULL DEFAULT '/admin/dist/img/no-img.png' COMMENT '商品轮播图',
  'goods_detail_content' text CHARACTER SET utf8 COLLATE utf8_general_ci NOT
NULL COMMENT '商品详情',
  'original_price' int(11) NOT NULL DEFAULT 1 COMMENT '商品价格',
  'selling_price' int(11) NOT NULL DEFAULT 1 COMMENT '商品实际售价',
  'stock_num' int(11) NOT NULL DEFAULT 0 COMMENT '商品库存数量',
  'tag' varchar(20) CHARACTER SET utf8 COLLATE utf8_general_ci NOT NULL
DEFAULT '' COMMENT '商品标签',
  'goods_sell_status' tinyint(4) NOT NULL DEFAULT 0 COMMENT '商品上架状态 0-
下架 1-上架',
  'create_user' int(11) NOT NULL DEFAULT 0 COMMENT '添加者主键id',
  'create_time' datetime(0) NOT NULL DEFAULT CURRENT_TIMESTAMP COMMENT '商
品添加时间',
  'update_user' int(11) NOT NULL DEFAULT 0 COMMENT '修改者主键id',
  'update_time' datetime(0) NOT NULL DEFAULT CURRENT_TIMESTAMP COMMENT '商
品修改时间',
  PRIMARY KEY ('goods_id') USING BTREE
) ENGINE = InnoDB CHARACTER SET = utf8 COLLATE = utf8_general_ci ROW_FORMAT
= Dynamic;
```

商品信息表的字段和每个字段对应的含义在上面的 SQL 中都有介绍，读者可以对照理解，并正确地把建表 SQL 导入数据库中即可。如果有需要，读者也可以自行根据该 SQL 设计进行扩展。

20.3 商品编辑页面的制作

20.3.1 导航栏中增加"商品信息"栏目

在左侧导航栏中新增"商品信息"的导航按钮，在 sidebar.html 文件的"Dashboard"目录下新增如下代码：

```html
<li class="nav-item">
    <a th:href="@{/admin/goods/edit}" th:class="${path}=='goods-edit'?'nav-link active':'nav-link'">
        <i class="nav-icon fa fa-plus"></i>
        <p>
            商品信息
        </p>
```

```
    </a>
</li>
```

在点击"商品信息"后的跳转路径为/admin/goods/edit,新建 Controller 控制类来处理该路径并跳转到对应的页面上。

20.3.2 控制类处理跳转逻辑

在 controller/admin 包下新建 NewBeeMallGoodsController 类,并新增如下代码:

```
package ltd.newbee.mall.controller.admin;

import ltd.newbee.mall.common.NewBeeMallCategoryLevelEnum;
import ltd.newbee.mall.entity.GoodsCategory;
import ltd.newbee.mall.service.NewBeeMallCategoryService;
import org.springframework.stereotype.Controller;
import org.springframework.util.CollectionUtils;
import org.springframework.web.bind.annotation.*;

import javax.annotation.Resource;
import javax.servlet.http.HttpServletRequest;
import java.util.Collections;
import java.util.List;

@Controller
@RequestMapping("/admin")
public class NewBeeMallGoodsController {

    @Resource
    private NewBeeMallCategoryService newBeeMallCategoryService;

    @GetMapping("/goods/edit")
    public String edit(HttpServletRequest request) {
        request.setAttribute("path", "edit");
        //查询所有的一级分类
        List<GoodsCategory> firstLevelCategories = newBeeMallCategoryService.selectByLevelAndParentIdsAndNumber(Collections.singletonList(0L), NewBeeMallCategoryLevelEnum.LEVEL_ONE.getLevel());
        if (!CollectionUtils.isEmpty(firstLevelCategories)) {
            //查询一级分类列表中第一个实体的所有二级分类
            List<GoodsCategory> secondLevelCategories = newBeeMallCategory
```

```
Service.selectByLevelAndParentIdsAndNumber(Collections.singletonList
(firstLevelCategories.get(0).getCategoryId()), NewBeeMallCategoryLevel
Enum.LEVEL_TWO.getLevel());
            if (!CollectionUtils.isEmpty(secondLevelCategories)) {
                //查询二级分类列表中第一个实体的所有三级分类
                List<GoodsCategory> thirdLevelCategories = newBeeMallCategory
Service.selectByLevelAndParentIdsAndNumber(Collections.singletonList
(secondLevelCategories.get(0).getCategoryId()), NewBeeMallCategoryLevel
Enum.LEVEL_THREE.getLevel());
                request.setAttribute("firstLevelCategories", firstLevel
Categories);
                request.setAttribute("secondLevelCategories", secondLevel
Categories);
                request.setAttribute("thirdLevelCategories", thirdLevel
Categories);
                request.setAttribute("path", "goods-edit");
                return "admin/newbee_mall_goods_edit";
            }
        }
        return "error/error_5xx";
    }
}
```

该方法用于处理/admin/goods/edit 请求，并先在返回视图前设置 path 字段和分类数据。因为商品信息添加页面需要进行分类选择，所以需要把分类的三级联动初始化数据也查询并设置到 request 域中。18.6 节已经讲解了三级分类联动的知识点，这里可以直接使用。最后跳转到 admin 目录下的 newbee_mall_goods_edit.html 模板页面中。

20.3.3 商品信息编辑页面的制作

首先在 resources/templates/admin 目录下新建 newbee_mall_goods_edit.html 模板文件，并引入对应的 JS 文件和 CSS 样式文件，然后把商品信息编辑页面按照商品表需要处理的字段进行完善，最后将商品信息表中需要输入内容的字段填充到页面 DOM 中。某些字段只需要一个 input 框即可，比如商品名称字段、商品简介字段。而其他一些字段的输入则需要一些前端插件来完成，比如商品的关联分类、商品主图、商品详情内容。仅仅 input 框肯定是无法满足这些字段的输入，必须借助一些前端插件来完善页面。比如分类选择需要选择框 select，文章详情内容需要富文本编辑器 wangEditor。

根据字段内容,这里新增对应的输入框和 DOM 组件,代码如下所示:

```html
<!-- Copyright (c) 2019-2020 十三 all rights reserved. -->
<!DOCTYPE html>
<html xmlns:th="http://www.thymeleaf.org">
<header th:replace="admin/header::header-fragment"></header>
<body class="hold-transition sidebar-mini">
<div class="wrapper">
    <!-- 引入页面头 header-fragment -->
    <div th:replace="admin/header::header-nav"></div>
    <!-- 引入工具栏 sidebar-fragment -->
    <div th:replace="admin/sidebar::sidebar-fragment(${path})"></div>
    <!-- Content Wrapper. Contains page content -->
    <div class="content-wrapper">
        <!-- Content Header (Page header) -->
        <div class="content-header">
            <div class="container-fluid">
            </div><!-- /.container-fluid -->
        </div>
        <!-- Main content -->
        <div class="content">
            <div class="container-fluid">
                <div class="card card-primary card-outline">
                    <div class="card-header">
                        <h3 class="card-title">商品信息编辑</h3>
                    </div>
                    <div class="card-body">
                        <form id="goodsForm" onsubmit="return false;">
                            <div class="form-group" style="display:flex;">
                                <label class="control-label">请选择分类:    </label>
                                <select class="form-control col-sm-3" id="levelOne"
                                        data-placeholder="请选择分类...">
                                    <th:block th:unless="${null == firstLevelCategories}">
                                        <th:block th:each="c : ${firstLevelCategories}">
                                            <option th:value="${c.categoryId}" th:text="${c.categoryName}">
                                                >
                                            </option>
                                        </th:block>
                                    </th:block>
```

```html
                </select> 
                <select class="form-control col-sm-3"
                        id="levelTwo"
                        data-placeholder="请选择分类...">
                    <th:block th:unless="${null == secondLevelCategories}">
                        <th:block th:each="c : ${secondLevelCategories}">
                            <option th:value="${c.categoryId}" th:text="${c.categoryName}">
                                >
                            </option>
                        </th:block>
                    </th:block>
                </select> 
                <select class="form-control col-sm-3"
                        id="levelThree"
                        data-placeholder="请选择分类...">
                    <th:block th:unless="${null == thirdLevelCategories}">
                        <th:block th:each="c : ${thirdLevelCategories}">
                            <option th:value="${c.categoryId}" th:text="${c.categoryName}">
                                >
                            </option>
                        </th:block>
                    </th:block>
                </select>
            </div>
            <div class="form-group" style="display:flex;">
                <input type="hidden" id="goodsId" name="goodsId">
                <input type="text" class="form-control col-sm-6" id="goodsName" name="goodsName" placeholder="*请输入商品名称(必填)" required="true">
                  
                <input type="text" class="form-control col-sm-6" id="goodsIntro" name="goodsIntro" placeholder="*请输入商品简介(100字以内)" required="true">
            </div>
            <div class="form-group" style="display:flex;">
                <input type="number" class="form-control col-sm-
```

```html
6" id="originalPrice"
                                    name="originalPrice"
                                    placeholder="*请输入商品价格"
                                    required="true">
                           
                         <input type="number" class="form-control col-sm-6" id="sellingPrice"
                                    name="sellingPrice"
                                    placeholder="*请输入商品售卖价"
                                    required="true">
                    </div>
                    <div class="form-group" style="display:flex;">
                         <input type="number" class="form-control col-sm-6" id="stockNum" name="stockNum"
                                    placeholder="*请输入商品库存数"
                                    required="true">
                           
                         <input type="text" class="form-control col-sm-6" id="tag" name="tag"
                                    placeholder="*请输入商品小标签"
                                    required="true">
                    </div>
                    <div class="form-group">
                         <label class="control-label">上架状态: </label>
                         <input name="goodsSellStatus" type="radio" id="goodsSellStatusTrue" checked=true value="0"/> 上架 
                         <input name="goodsSellStatus" type="radio" id="goodsSellStatusFalse" value="1"/> 下架 
                    </div>
                    <div class="form-group">
                         <div class="form-group">
                              <div class="col-sm-4">
                                    <th:block th:if="${null == goods}">
                                         <img id="goodsCoverImg" src="/admin/dist/img/img-upload.png"
                                                style="height: 64px;width: 64px;">
                                    </th:block>
                                    <th:block th:unless="${null == goods}">
                                         <img id="goodsCoverImg"
                                                style="width:160px ;height: 160px;display:block;">
                                    </th:block>
```

```html
                    </div>
                </div>
                <br>
                <div class="form-group">
                    <div class="col-sm-4">
                        <button class="btn btn-info" style="margin-bottom: 5px;"
                                id="uploadGoodsCoverImg">
                            <i class="fa fa-picture-o"></i> 上传商品主图
                        </button>
                    </div>
                </div>
                <div class="form-group" id="wangEditor">
                </div>
                <div class="form-group">
                    <!-- 按钮 -->
                     <button class="btn btn-info float-right" style="margin-left: 5px;"
                                  id="saveButton">保存商品
                    </button> 
                     <button class="btn btn-secondary float-right" style="margin-left: 5px;"
                                  id="cancelButton">返回商品列表
                    </button> 
                </div>
            </form>
        </div>

    </div><!-- /.container-fluid -->
  </div>
</div>
<!-- /.content-wrapper -->
<!-- 引入页脚 footer-fragment -->
<div th:replace="admin/footer::footer-fragment"></div>
</div>
<!-- jQuery -->
<script th:src="@{/admin/plugins/jquery/jquery.min.js}"></script>
<!-- Bootstrap 4 -->
<script th:src="@{/admin/plugins/bootstrap/js/bootstrap.bundle.min.js}">
</script>
```

```html
<!-- AdminLTE App -->
<script th:src="@{/admin/dist/js/adminlte.min.js}"></script>
<!-- sweetalert -->
<script th:src="@{/admin/plugins/sweetalert/sweetalert.min.js}"></script>
<!-- ajaxupload -->
<script th:src="@{/admin/plugins/ajaxupload/ajaxupload.js}"></script>
<!-- wangEditor -->
<script type="text/javascript"
src="//unpkg.com/wangeditor/dist/wangEditor.min.js"></script>
<script th:src="@{/admin/dist/js/public.js}"></script>
<script th:src="@{/admin/dist/js/newbee_mall_goods_edit.js}"></script>
</body>
</html>
```

其中商品标题、商品介绍、商品售价等字段直接使用 input 框。部分字段比较特殊，分类字段使用下拉选择框实现三级联动；商品封面图则使用图片上传插件；商品详情内容的输入使用 wangEditor；销售状态字段字段使用 radio 选择框。代码最下面是两个功能按钮，用于保存商品信息和返回。

20.3.4 初始化插件

在 resources/static/admin/dist/js 目录下新增 newbee_mall_goods_edit.js 文件，并把富文本编辑器和图片上传插件的初始化代码加进来，新增代码如下所示：

```javascript
var editorD;
$(function () {
    //富文本编辑器，用于商品详情编辑
    const E = window.wangEditor;
    editorD = new E('#wangEditor')
    // 设置编辑区域高度为 750px
    editorD.config.height = 750
    //配置服务端图片上传地址
    editorD.config.uploadImgServer = '/admin/upload/files'
    editorD.config.uploadFileName = 'files'
    //限制图片大小 2MB
    editorD.config.uploadImgMaxSize = 2 * 1024 * 1024
    //限制一次最多能传 5 张图片
    editorD.config.uploadImgMaxLength = 5
    //隐藏插入网络图片的功能
    editorD.config.showLinkImg = false
```

```javascript
editorD.config.uploadImgHooks = {
    // 图片上传并返回结果：图片插入已成功
    success: function (xhr) {
        console.log('success', xhr)
    },
    // 图片上传并返回结果：图片插入时出错了
    fail: function (xhr, editor, resData) {
        console.log('fail', resData)
    },
    // 上传图片出错，一般为HTTP请求的错误
    error: function (xhr, editor, resData) {
        console.log('error', xhr, resData)
    },
    // 上传图片超时
    timeout: function (xhr) {
        console.log('timeout')
    },
    customInsert: function (insertImgFn, result) {
        if (result != null && result.resultCode == 200) {
            // insertImgFn 可把图片插入编辑器，传入图片 src 属性，执行函数即可
            result.data.forEach(img => {
                insertImgFn(img)
            });
        } else {
            alert("error");
        }
    }
}
editorD.create();

//图片上传插件的初始化 用于商品预览图上传
new AjaxUpload('#uploadGoodsCoverImg', {
    action: '/admin/upload/file',
    name: 'file',
    autoSubmit: true,
    responseType: "json",
    onSubmit: function (file, extension) {
        if (!(extension && /^(jpg|jpeg|png|gif)$/.test(extension.toLowerCase()))) {
            alert('只支持jpg、png、gif格式的文件！');
            return false;
        }
    },
```

```javascript
            onComplete: function (file, r) {
                if (r != null && r.resultCode == 200) {
                    $("#goodsCoverImg").attr("src", r.data);
                    $("#goodsCoverImg").attr("style", "width: 128px;height: 128px;display:block;");
                    return false;
                } else {
                    alert("error");
                }
            }
        });
    });

    $('#levelOne').on('change', function () {
        $.ajax({
            url: '/admin/categories/listForSelect?categoryId=' + $(this).val(),
            type: 'GET',
            success: function (result) {
                if (result.resultCode == 200) {
                    var levelTwoSelect = '';
                    var secondLevelCategories = result.data.secondLevelCategories;
                    var length2 = secondLevelCategories.length;
                    for (var i = 0; i < length2; i++) {
                        levelTwoSelect += '<option value=\"' + secondLevelCategories[i].categoryId + '\">' + secondLevelCategories[i].categoryName + '</option>';
                    }
                    $('#levelTwo').html(levelTwoSelect);
                    var levelThreeSelect = '';
                    var thirdLevelCategories = result.data.thirdLevelCategories;
                    var length3 = thirdLevelCategories.length;
                    for (var i = 0; i < length3; i++) {
                        levelThreeSelect += '<option value=\"' + thirdLevelCategories[i].categoryId + '\">' + thirdLevelCategories[i].categoryName + '</option>';
                    }
                    $('#levelThree').html(levelThreeSelect);
                } else {
                    swal(result.message, {
                        icon: "error",
                    });
                }
                ;
```

```javascript
        },
        error: function () {
            swal("操作失败", {
                icon: "error",
            });
        }
    });
});

$('#levelTwo').on('change', function () {
    $.ajax({
        url: '/admin/categories/listForSelect?categoryId=' + $(this).val(),
        type: 'GET',
        success: function (result) {
            if (result.resultCode == 200) {
                var levelThreeSelect = '';
                var thirdLevelCategories = result.data.thirdLevelCategories;
                var length = thirdLevelCategories.length;
                for (var i = 0; i < length; i++) {
                    levelThreeSelect += '<option value=\"' + thirdLevelCategories[i].categoryId + '\">' + thirdLevelCategories[i].categoryName + '</option>';
                }
                $('#levelThree').html(levelThreeSelect);
            } else {
                swal(result.message, {
                    icon: "error",
                });
            }
            ;
        },
        error: function () {
            swal("操作失败", {
                icon: "error",
            });
        }
    });
});
```

以上代码中初始化了两个字段的页面 DOM 属性，分别是商品信息编辑框、图片上传框。同时，为了实现分类信息的三级联动，也绑定了分类下拉选择框的 change 事件。

20.3.5 新增控制类处理图片上传

由于商品编辑页面需要上传商品主图，wangEditor 的多图上传模块也需要后端处理，因此需要新增一个控制类来分别处理单图上传和多图上传。新增 UploadController.java 控制类，代码如下所示：

```java
package ltd.newbee.mall.controller.common;

import ltd.newbee.mall.common.Constants;
import ltd.newbee.mall.util.NewBeeMallUtils;
import ltd.newbee.mall.util.Result;
import ltd.newbee.mall.util.ResultGenerator;
import org.springframework.beans.factory.annotation.Autowired;
import org.springframework.stereotype.Controller;
import org.springframework.util.CollectionUtils;
import org.springframework.web.bind.annotation.PostMapping;
import org.springframework.web.bind.annotation.RequestMapping;
import org.springframework.web.bind.annotation.RequestParam;
import org.springframework.web.bind.annotation.ResponseBody;
import org.springframework.web.multipart.MultipartFile;
import org.springframework.web.multipart.MultipartHttpServletRequest;
import org.springframework.web.multipart.support.StandardServletMultipartResolver;

import javax.servlet.http.HttpServletRequest;
import java.io.File;
import java.io.IOException;
import java.net.URI;
import java.net.URISyntaxException;
import java.text.SimpleDateFormat;
import java.util.*;

/**
 * @author 13
 * @qq 交流群 796794009
 * @email 2449207463@qq.com
 * @link https://github.com/newbee-ltd
 */
```

```java
@Controller
@RequestMapping("/admin")
public class UploadController {

    @Autowired
    private StandardServletMultipartResolver standardServletMultipartResolver;

    @PostMapping({"/upload/file"})
    @ResponseBody
    public Result upload(HttpServletRequest httpServletRequest, @RequestParam("file") MultipartFile file) throws URISyntaxException {
        String fileName = file.getOriginalFilename();
        String suffixName = fileName.substring(fileName.lastIndexOf("."));
        //生成文件名称通用方法
        SimpleDateFormat sdf = new SimpleDateFormat("yyyyMMdd_HHmmss");
        Random r = new Random();
        StringBuilder tempName = new StringBuilder();
        tempName.append(sdf.format(new Date())).append(r.nextInt(100)).append(suffixName);
        String newFileName = tempName.toString();
        File fileDirectory = new File(Constants.FILE_UPLOAD_DIC);
        //创建文件
        File destFile = new File(Constants.FILE_UPLOAD_DIC + newFileName);
        try {
            if (!fileDirectory.exists()) {
                if (!fileDirectory.mkdir()) {
                    throw new IOException("文件夹创建失败,路径为: " + fileDirectory);
                }
            }
            file.transferTo(destFile);
            Result resultSuccess = ResultGenerator.genSuccessResult();
            resultSuccess.setData(NewBeeMallUtils.getHost(new URI(httpServletRequest.getRequestURL() + "")) + "/upload/" + newFileName);
            return resultSuccess;
        } catch (IOException e) {
            e.printStackTrace();
            return ResultGenerator.genFailResult("文件上传失败");
        }
    }
}
```

```java
@PostMapping({"/upload/files"})
@ResponseBody
public Result uploadV2(HttpServletRequest httpServletRequest) throws URISyntaxException {
    List<MultipartFile> multipartFiles = new ArrayList<>(8);
    if (standardServletMultipartResolver.isMultipart(httpServletRequest)) {
        MultipartHttpServletRequest multiRequest = (MultipartHttpServletRequest) httpServletRequest;
        Iterator<String> iter = multiRequest.getFileNames();
        int total = 0;
        while (iter.hasNext()) {
            if (total > 5) {
                return ResultGenerator.genFailResult("最多上传5张图片");
            }
            total += 1;
            MultipartFile file = multiRequest.getFile(iter.next());
            multipartFiles.add(file);
        }
    }
    if (CollectionUtils.isEmpty(multipartFiles)) {
        return ResultGenerator.genFailResult("参数异常");
    }
    if (multipartFiles != null && multipartFiles.size() > 5) {
        return ResultGenerator.genFailResult("最多上传5张图片");
    }
    List<String> fileNames = new ArrayList(multipartFiles.size());
    for (int i = 0; i < multipartFiles.size(); i++) {
        String fileName = multipartFiles.get(i).getOriginalFilename();
        String suffixName = fileName.substring(fileName.lastIndexOf("."));
        //生成文件名称通用方法
        SimpleDateFormat sdf = new SimpleDateFormat("yyyyMMdd_HHmmss");
        Random r = new Random();
        StringBuilder tempName = new StringBuilder();
        tempName.append(sdf.format(new Date())).append(r.nextInt(100)).append(suffixName);
        String newFileName = tempName.toString();
        File fileDirectory = new File(Constants.FILE_UPLOAD_DIC);
        //创建文件
        File destFile = new File(Constants.FILE_UPLOAD_DIC + newFileName);
```

```
        try {
            if (!fileDirectory.exists()) {
                if (!fileDirectory.mkdir()) {
                    throw new IOException("文件夹创建失败,路径为: " + fileDirectory);
                }
            }
            multipartFiles.get(i).transferTo(destFile);
            fileNames.add(NewBeeMallUtils.getHost(new URI(httpServlet
Request.getRequestURL() + "")) + "/upload/" + newFileName);
        } catch (IOException e) {
            e.printStackTrace();
            return ResultGenerator.genFailResult("文件上传失败");
        }
    }
    Result resultSuccess = ResultGenerator.genSuccessResult();
    resultSuccess.setData(fileNames);
    return resultSuccess;
    }
}
```

这里映射的路径分别为/admin/upload/file 和/admin/upload/files，在图片上传插件和 wangEditor 初始化时将后端处理路径分别配置即可。

至此，商品信息编辑页面的前端样式和插件初始化就完成了。在编码完成后，启动 Spring Boot 项目。在启动成功后打开浏览器并输入后台管理系统登录页面地址：

```
http://localhost:8080/admin/login
```

在输入登录信息后登录后台管理系统，点击左侧导航栏中的"商品信息"就能够看到商品信息编辑页面，页面效果如图 20-3 所示。

依次选择商品分类、填写商品信息并上传图片后，商品编辑页面的显示效果如图 20-4 所示。

该页面中的内容较多，交互也相对复杂一些，甚至可以说该页面整合了信息编辑页面该有的全部元素：简单输入框、下拉选择框、radio 选择框、图片上传、富文本编辑器。读者可以将本页面与接触过的其他项目的编辑页面进行比较。

不过，此时只完成了前端显示步骤，还没有与后端进行交互。接下来开发后端接口，并把商品编辑页面上的数据保存到数据库中。

图 20-3 商品编辑页面初始效果

图 20-4 商品编辑页面效果

20.4 商品信息添加接口的开发与联调

页面已经制作完成,接下来实现商品信息的添加接口。

商品添加接口负责接收前端的 POST 请求并处理其中的参数。接收的参数为用户在商品信息编辑页面输入的所有字段内容。字段名称与对应的含义如下所示。

①goodsName:商品名称

②goodsIntro:商品简介

③goodsCategoryId:分类 id(在下拉框中选择)

④tag:商品小标签字段

⑤originalPrice:商品原价

⑥sellingPrice:商品实际售价(后续订单流程的计算会使用)

⑦stockNum:商品库存

⑧goodsDetailContent:商品详情内容

⑨goodsCoverImg:商品主图/商品封面图

⑩goodsSellStatus:商品销售状态(上架或下架)

20.4.1 新建商品实体类和 Mapper 接口

首先在 ltd.newbee.mall.entity 包中创建商品实体类,选中 entity 包并右键点击,然后在弹出的菜单中选择"New→Java Class",在弹出的窗口中输入"NewBeeMallGoods",最后在 NewBeeMallGoods 类中新增如下代码:

```
package ltd.newbee.mall.entity;

import com.fasterxml.jackson.annotation.JsonFormat;
import java.util.Date;

public class NewBeeMallGoods {
    private Long goodsId;

    private String goodsName;

    private String goodsIntro;
```

```java
    private Long goodsCategoryId;

    private String goodsCoverImg;

    private String goodsCarousel;

    private Integer originalPrice;

    private Integer sellingPrice;

    private Integer stockNum;

    private String tag;

    private Byte goodsSellStatus;

    private Integer createUser;

    @JsonFormat(pattern = "yyyy-MM-dd HH:mm:ss", timezone = "GMT+8")
    private Date createTime;

    private Integer updateUser;

    @JsonFormat(pattern = "yyyy-MM-dd HH:mm:ss", timezone = "GMT+8")
    private Date updateTime;

    private String goodsDetailContent;

    public Long getGoodsId() {
        return goodsId;
    }

    public void setGoodsId(Long goodsId) {
        this.goodsId = goodsId;
    }

    public String getGoodsName() {
        return goodsName;
    }

    public void setGoodsName(String goodsName) {
        this.goodsName = goodsName == null ? null : goodsName.trim();
```

```java
    }

    public String getGoodsIntro() {
        return goodsIntro;
    }

    public void setGoodsIntro(String goodsIntro) {
        this.goodsIntro = goodsIntro == null ? null : goodsIntro.trim();
    }

    public Long getGoodsCategoryId() {
        return goodsCategoryId;
    }

    public void setGoodsCategoryId(Long goodsCategoryId) {
        this.goodsCategoryId = goodsCategoryId;
    }

    public String getGoodsCoverImg() {
        return goodsCoverImg;
    }

    public void setGoodsCoverImg(String goodsCoverImg) {
        this.goodsCoverImg = goodsCoverImg == null ? null : goodsCoverImg.trim();
    }

    public String getGoodsCarousel() {
        return goodsCarousel;
    }

    public void setGoodsCarousel(String goodsCarousel) {
        this.goodsCarousel = goodsCarousel == null ? null : goodsCarousel.trim();
    }

    public Integer getOriginalPrice() {
        return originalPrice;
    }

    public void setOriginalPrice(Integer originalPrice) {
        this.originalPrice = originalPrice;
    }
```

```java
public Integer getSellingPrice() {
    return sellingPrice;
}

public void setSellingPrice(Integer sellingPrice) {
    this.sellingPrice = sellingPrice;
}

public Integer getStockNum() {
    return stockNum;
}

public void setStockNum(Integer stockNum) {
    this.stockNum = stockNum;
}

public String getTag() {
    return tag;
}

public void setTag(String tag) {
    this.tag = tag == null ? null : tag.trim();
}

public Byte getGoodsSellStatus() {
    return goodsSellStatus;
}

public void setGoodsSellStatus(Byte goodsSellStatus) {
    this.goodsSellStatus = goodsSellStatus;
}

public Integer getCreateUser() {
    return createUser;
}

public void setCreateUser(Integer createUser) {
    this.createUser = createUser;
}

public Date getCreateTime() {
    return createTime;
```

```java
    }

    public void setCreateTime(Date createTime) {
        this.createTime = createTime;
    }

    public Integer getUpdateUser() {
        return updateUser;
    }

    public void setUpdateUser(Integer updateUser) {
        this.updateUser = updateUser;
    }

    public Date getUpdateTime() {
        return updateTime;
    }

    public void setUpdateTime(Date updateTime) {
        this.updateTime = updateTime;
    }

    public String getGoodsDetailContent() {
        return goodsDetailContent;
    }

    public void setGoodsDetailContent(String goodsDetailContent) {
        this.goodsDetailContent = goodsDetailContent == null ? null : goodsDetailContent.trim();
    }
}
```

首先在 ltd.newbee.mall.dao 包中新建商品实体的 Mapper 接口，选中 dao 包并右键点击，然后在弹出的菜单中选择 "New→Java Class"，在弹出的窗口中输入 "NewBeeMallGoodsMapper"，并选中 "Interface" 选项，最后在 NewBeeMallGoodsMapper.java 文件中新增如下代码：

```java
package ltd.newbee.mall.dao;

import ltd.newbee.mall.entity.NewBeeMallGoods;

public interface NewBeeMallGoodsMapper {
```

```
    /**
     * 保存一条新记录
     * @param record
     * @return
     */
    int insertSelective(NewBeeMallGoods record);
}
```

NewBeeMallGoodsMapper 接口用于定义商品操作的数据层方法。此时只添加了 insert 方法，其他方法会在后续功能中介绍。

20.4.2 创建 NewBeeMallGoodsMapper 接口的映射文件

在 resources/mapper 目录下新建 NewBeeMallGoodsMapper 接口的映射文件 NewBeeMallGoodsMapper.xml，并进行映射文件的编写。

首先，定义映射文件与 Mapper 接口的对应关系。比如在该示例中，需要将 NewBeeMallGoodsMapper.xml 文件与对应的 NewBeeMallGoodsMapper 接口之间的关系定义出来：

```
<mapper namespace="ltd.newbee.mall.dao.NewBeeMallGoodsMapper">
```

然后，配置表结构和实体类的对应关系：

```
<resultMap id="BaseResultMap" type="ltd.newbee.mall.entity.NewBeeMallGoods">
  <id column="goods_id" jdbcType="BIGINT" property="goodsId"/>
  <result column="goods_name" jdbcType="VARCHAR" property="goodsName"/>
  <result column="goods_intro" jdbcType="VARCHAR" property="goodsIntro"/>
  <result column="goods_category_id" jdbcType="BIGINT" property="goodsCategoryId"/>
  <result column="goods_cover_img" jdbcType="VARCHAR" property="goodsCoverImg"/>
  <result column="goods_carousel" jdbcType="VARCHAR" property="goodsCarousel"/>
  <result column="original_price" jdbcType="INTEGER" property="originalPrice"/>
  <result column="selling_price" jdbcType="INTEGER" property="sellingPrice"/>
  <result column="stock_num" jdbcType="INTEGER" property="stockNum"/>
  <result column="tag" jdbcType="VARCHAR" property="tag"/>
```

```xml
  <result column="goods_sell_status" jdbcType="TINYINT" property="goodsSellStatus"/>
  <result column="create_user" jdbcType="INTEGER" property="createUser"/>
  <result column="create_time" jdbcType="TIMESTAMP" property="createTime"/>
  <result column="update_user" jdbcType="INTEGER" property="updateUser"/>
  <result column="update_time" jdbcType="TIMESTAMP" property="updateTime"/>
</resultMap>
```

最后,按照对应的接口方法,编写具体的 SQL 语句,最终的 NewBeeMallGoodsMapper.xml 文件如下所示:

```xml
<?xml version="1.0" encoding="UTF-8"?>
<!DOCTYPE mapper PUBLIC "-//mybatis.org//DTD Mapper 3.0//EN" "http://mybatis.org/dtd/mybatis-3-mapper.dtd">
<mapper namespace="ltd.newbee.mall.dao.NewBeeMallGoodsMapper">
    <resultMap id="BaseResultMap" type="ltd.newbee.mall.entity.NewBeeMallGoods">
        <id column="goods_id" jdbcType="BIGINT" property="goodsId"/>
        <result column="goods_name" jdbcType="VARCHAR" property="goodsName"/>
        <result column="goods_intro" jdbcType="VARCHAR" property="goodsIntro"/>
        <result column="goods_category_id" jdbcType="BIGINT" property="goodsCategoryId"/>
        <result column="goods_cover_img" jdbcType="VARCHAR" property="goodsCoverImg"/>
        <result column="goods_carousel" jdbcType="VARCHAR" property="goodsCarousel"/>
        <result column="original_price" jdbcType="INTEGER" property="originalPrice"/>
        <result column="selling_price" jdbcType="INTEGER" property="sellingPrice"/>
        <result column="stock_num" jdbcType="INTEGER" property="stockNum"/>
        <result column="tag" jdbcType="VARCHAR" property="tag"/>
        <result column="goods_sell_status" jdbcType="TINYINT" property="goodsSellStatus"/>
      <result column="create_user" jdbcType="INTEGER" property="createUser"/>
      <result column="create_time" jdbcType="TIMESTAMP" property="createTime"/>
      <result column="update_user" jdbcType="INTEGER" property="updateUser"/>
        <result column="update_time" jdbcType="TIMESTAMP" property="updateTime"/>
    </resultMap>
    <resultMap extends="BaseResultMap" id="ResultMapWithBLOBs" type="ltd.newbee.mall.entity.NewBeeMallGoods">
        <result column="goods_detail_content" jdbcType="LONGVARCHAR" property=
```

```xml
"goodsDetailContent"/>
    </resultMap>
    <insert id="insertSelective" parameterType="ltd.newbee.mall.entity.NewBeeMallGoods">
        insert into tb_newbee_mall_goods_info
        <trim prefix="(" suffix=")" suffixOverrides=",">
            <if test="goodsId != null">
                goods_id,
            </if>
            <if test="goodsName != null">
                goods_name,
            </if>
            <if test="goodsIntro != null">
                goods_intro,
            </if>
            <if test="goodsCategoryId != null">
                goods_category_id,
            </if>
            <if test="goodsCoverImg != null">
                goods_cover_img,
            </if>
            <if test="goodsCarousel != null">
                goods_carousel,
            </if>
            <if test="originalPrice != null">
                original_price,
            </if>
            <if test="sellingPrice != null">
                selling_price,
            </if>
            <if test="stockNum != null">
                stock_num,
            </if>
            <if test="tag != null">
                tag,
            </if>
            <if test="goodsSellStatus != null">
                goods_sell_status,
            </if>
            <if test="createUser != null">
                create_user,
            </if>
            <if test="createTime != null">
```

```xml
            create_time,
        </if>
        <if test="updateUser != null">
            update_user,
        </if>
        <if test="updateTime != null">
            update_time,
        </if>
        <if test="goodsDetailContent != null">
            goods_detail_content,
        </if>
    </trim>
    <trim prefix="values (" suffix=")" suffixOverrides=",">
        <if test="goodsId != null">
            #{goodsId,jdbcType=BIGINT},
        </if>
        <if test="goodsName != null">
            #{goodsName,jdbcType=VARCHAR},
        </if>
        <if test="goodsIntro != null">
            #{goodsIntro,jdbcType=VARCHAR},
        </if>
        <if test="goodsIntro != null">
            #{goodsCategoryId,jdbcType=BIGINT},
        </if>
        <if test="goodsCoverImg != null">
            #{goodsCoverImg,jdbcType=VARCHAR},
        </if>
        <if test="goodsCarousel != null">
            #{goodsCarousel,jdbcType=VARCHAR},
        </if>
        <if test="originalPrice != null">
            #{originalPrice,jdbcType=INTEGER},
        </if>
        <if test="sellingPrice != null">
            #{sellingPrice,jdbcType=INTEGER},
        </if>
        <if test="stockNum != null">
            #{stockNum,jdbcType=INTEGER},
        </if>
        <if test="tag != null">
            #{tag,jdbcType=VARCHAR},
        </if>
```

```xml
            <if test="goodsSellStatus != null">
                #{goodsSellStatus,jdbcType=TINYINT},
            </if>
            <if test="createUser != null">
                #{createUser,jdbcType=INTEGER},
            </if>
            <if test="createTime != null">
                #{createTime,jdbcType=TIMESTAMP},
            </if>
            <if test="updateUser != null">
                #{updateUser,jdbcType=INTEGER},
            </if>
            <if test="updateTime != null">
                #{updateTime,jdbcType=TIMESTAMP},
            </if>
            <if test="goodsDetailContent != null">
                #{goodsDetailContent,jdbcType=LONGVARCHAR},
            </if>
        </trim>
    </insert>
</mapper>
```

20.4.3 业务层的代码实现

首先在 ltd.newbee.mall.service 包中新建业务处理类，选中 service 包并右键点击，然后在弹出的菜单中选择"New→Java Class"，在弹出的窗口中输入"NewBeeMallGoodsService"，并选中"Interface"选项，最后在 NewBeeMallGoodsService.java 文件中新增如下代码：

```java
package ltd.newbee.mall.service;

import ltd.newbee.mall.entity.NewBeeMallGoods;

public interface NewBeeMallGoodsService {

    /**
     * 添加商品
     *
     * @param goods
     * @return
```

```
    */
    String saveNewBeeMallGoods(NewBeeMallGoods goods);
}
```

首先在 ltd.newbee.mall.service.impl 包中新建 NewBeeMallGoodsService 的实现类，选中 impl 包并右键点击击，然后在弹出的菜单中选择"New→Java Class"，在弹出的窗口中输入"NewBeeMallGoodsServiceImpl"，最后在 NewBeeMallGoodsServiceImpl 类中新增如下代码：

```java
package ltd.newbee.mall.service.impl;

import ltd.newbee.mall.common.ServiceResultEnum;
import ltd.newbee.mall.dao.NewBeeMallGoodsMapper;
import ltd.newbee.mall.entity.NewBeeMallGoods;
import ltd.newbee.mall.service.NewBeeMallGoodsService;
import org.springframework.beans.factory.annotation.Autowired;
import org.springframework.stereotype.Service;

@Service
public class NewBeeMallGoodsServiceImpl implements NewBeeMallGoodsService {

    @Autowired
    private NewBeeMallGoodsMapper goodsMapper;

    @Override
    public String saveNewBeeMallGoods(NewBeeMallGoods goods) {
        if (goodsMapper.insertSelective(goods) > 0) {
            return ServiceResultEnum.SUCCESS.getResult();
        }
        return ServiceResultEnum.DB_ERROR.getResult();
    }
}
```

20.4.4　商品添加接口控制层的代码实现

在 NewBeeMallGoodsController 控制器中新增商品添加功能的实现代码，代码如下所示：

```
/**
 * 添加
 */
```

```java
@RequestMapping(value = "/goods/save", method = RequestMethod.POST)
@ResponseBody
public Result save(@RequestBody NewBeeMallGoods newBeeMallGoods) {
    if (StringUtils.isEmpty(newBeeMallGoods.getGoodsName())
            || StringUtils.isEmpty(newBeeMallGoods.getGoodsIntro())
            || StringUtils.isEmpty(newBeeMallGoods.getTag())
            || Objects.isNull(newBeeMallGoods.getOriginalPrice())
            || Objects.isNull(newBeeMallGoods.getGoodsCategoryId())
            || Objects.isNull(newBeeMallGoods.getSellingPrice())
            || Objects.isNull(newBeeMallGoods.getStockNum())
            || Objects.isNull(newBeeMallGoods.getGoodsSellStatus())
            || StringUtils.isEmpty(newBeeMallGoods.getGoodsCoverImg())
            || StringUtils.isEmpty(newBeeMallGoods.getGoodsDetailContent())) {
        return ResultGenerator.genFailResult("参数异常！");
    }
    String result = newBeeMallGoodsService.saveNewBeeMallGoods(newBeeMallGoods);
    if (ServiceResultEnum.SUCCESS.getResult().equals(result)) {
        return ResultGenerator.genSuccessResult();
    } else {
        return ResultGenerator.genFailResult(result);
    }
}
```

在商品信息添加接口中，首先会对参数进行校验，然后对业务层代码进行操作，最后向数据库中插入一条商品数据。

20.4.5 前端调用商品添加接口

在信息录入完成后可以点击商品信息编辑页面中的"保存商品"按钮，此时系统会调用后端接口并进行数据的交互。因此这里需要给这个按钮增加一个绑定事件，在点击按钮后获取管理员用户在页面中输入的数据，并请求后端的商品添加接口完成商品信息的保存功能。

在 newbee_mall_goods_edit.js 文件中新增代码如下：

```javascript
$('#saveButton').click(function () {
    var goodsId = $('#goodsId').val();
    var goodsCategoryId = $('#levelThree option:selected').val();
    var goodsName = $('#goodsName').val();
    var tag = $('#tag').val();
```

```js
var originalPrice = $('#originalPrice').val();
var sellingPrice = $('#sellingPrice').val();
var goodsIntro = $('#goodsIntro').val();
var stockNum = $('#stockNum').val();
var goodsSellStatus = $("input[name='goodsSellStatus']:checked").val();
var goodsDetailContent = editorD.txt.html();
var goodsCoverImg = $('#goodsCoverImg')[0].src;
if (isNull(goodsCategoryId)) {
    swal("请选择分类", {
        icon: "error",
    });
    return;
}
if (isNull(goodsName)) {
    swal("请输入商品名称", {
        icon: "error",
    });
    return;
}
if (!validLength(goodsName, 100)) {
    swal("请输入商品名称", {
        icon: "error",
    });
    return;
}
if (isNull(tag)) {
    swal("请输入商品小标签", {
        icon: "error",
    });
    return;
}
if (!validLength(tag, 100)) {
    swal("标签过长", {
        icon: "error",
    });
    return;
}
if (isNull(goodsIntro)) {
    swal("请输入商品简介", {
        icon: "error",
    });
    return;
}
```

```javascript
if (!validLength(goodsIntro, 100)) {
    swal("简介过长", {
        icon: "error",
    });
    return;
}
if (isNull(originalPrice) || originalPrice < 1) {
    swal("请输入商品价格", {
        icon: "error",
    });
    return;
}
if (isNull(sellingPrice) || sellingPrice < 1) {
    swal("请输入商品售卖价", {
        icon: "error",
    });
    return;
}
if (isNull(stockNum) || sellingPrice < 0) {
    swal("请输入商品库存数", {
        icon: "error",
    });
    return;
}
if (isNull(goodsSellStatus)) {
    swal("请选择上架状态", {
        icon: "error",
    });
    return;
}
if (isNull(goodsDetailContent)) {
    swal("请输入商品介绍", {
        icon: "error",
    });
    return;
}
if (!validLength(goodsDetailContent, 50000)) {
    swal("商品介绍内容过长", {
        icon: "error",
    });
    return;
}
if (isNull(goodsCoverImg) || goodsCoverImg.indexOf('img-upload') != -1)
```

```
{
        swal("封面图片不能为空", {
            icon: "error",
        });
        return;
    }
    var url = '/admin/goods/save';
    var swlMessage = '保存成功';
    var data = {
        "goodsName": goodsName,
        "goodsIntro": goodsIntro,
        "goodsCategoryId": goodsCategoryId,
        "tag": tag,
        "originalPrice": originalPrice,
        "sellingPrice": sellingPrice,
        "stockNum": stockNum,
        "goodsDetailContent": goodsDetailContent,
        "goodsCoverImg": goodsCoverImg,
        "goodsCarousel": goodsCoverImg,
        "goodsSellStatus": goodsSellStatus
    };
    console.log(data);
    $.ajax({
        type: 'POST',//方法类型
        url: url,
        contentType: 'application/json',
        data: JSON.stringify(data),
        success: function (result) {
            if (result.resultCode == 200) {
                swal(swlMessage, {
                    icon: "success",
                });
            } else {
                swal(result.message, {
                    icon: "error",
                });
            }
            ;
        },
        error: function () {
            swal("操作失败", {
                icon: "error",
            });
```

```
        }
    });
});
```

在"保存商品"按钮的点击事件触发后,系统首先会获取所有的输入内容并进行验证,然后封装数据向后端发送请求添加商品信息,最后根据后端返回的数据来提醒管理员用户是否添加成功。

20.4.6 功能测试

在编码完成后,启动 Spring Boot 项目。在启动成功后打开浏览器并输入后台管理系统登录页面地址:

```
http://localhost:8080/admin/login
```

在输入登录信息后登录后台管理系统,点击左侧导航栏中的"商品信息"就可以进行商品信息添加功能的测试了。

在商品编辑页面依次选择商品分类、填写商品信息、上传商品封面图片并在富文本编辑器中输入商品详情信息,最后点击"保存商品"按钮。一切正常的话,页面上会提示"保存成功",如图 20-5 所示。

图 20-5　保存商品成功

这里主要是将用户在商品信息编辑页面输入的所有数据保存到数据库中。在测试过程中也可以对编辑器中的图片上传和封面图上传进行测试，在点击"保存商品"按钮后提示"保存成功"，则一切功能正常。

最后做一个简单的功能总结，读者需要留意。

由于内容比较复杂、知识点的整合较多，建议读者多联系前文的知识点进行理解和学习。如果有不理解的地方，就先拆解知识点，掌握单个知识点，之后再进行整合。这样会事半功倍。

在编码完成进行测试时需要注意两点，一是商品信息数据是否被插入数据库表中；二是图片有没有正确上传。因为在封面图上传时使用了图片上传功能，而默认上传路径为 D:\upload\，因此在功能测试前先检查是否创建了该目录（此目录可根据部署情况自行修改），最后验证图片上传是否正常。

虽然页面中提示保存成功了，但还需要再次验证一下数据库和图片文件。

首先验证在 MySQL 数据库中商品信息表是否有对应的新增记录，直接查询 tb_newbee_mall_goods_info 表中的记录即可，结果如图 20-6 所示。

图 20-6 查看商品表中的记录

可以看到在测试时输入的数据已经正确保存在数据库中。最后，再去图片保存目录查看图片是否已经正确保存在目录中。至此，新增商品信息功能开发完成。

20.5 商品信息编辑页面的完善

接下来继续讲解商品信息修改功能的整个流程和逻辑实现。

想要修改一件商品的信息数据，需要先获取商品信息的所有字段，再回显到编辑页面中。用户根据页面上需要修改的内容，在保存后会向后端发送商品信息修改请求，后端接口在接收请求后会进行参数验证和相应的逻辑操作，最后进行数据的入库操作。这样整个商品信息修改流程完成。

20.5.1 控制类处理跳转的逻辑

根据流程，首先需要获取商品信息详情。由于商品信息编辑页面已经开发完成，所以可以选择与添加商品信息相同的前后端交互方式，将请求转发到编辑页面即可。要获取商品信息详情就需要根据字段来查询，这里选择 id 字段作为传参。在 NewBeeMallGoodsController 中新增如下代码：

```
@GetMapping("/goods/edit/{goodsId}")
public String edit(HttpServletRequest request, @PathVariable("goodsId") Long goodsId) {
  request.setAttribute("path", "edit");
  NewBeeMallGoods newBeeMallGoods = newBeeMallGoodsService.getNewBeeMallGoodsById(goodsId);
  if (newBeeMallGoods == null) {
    return "error/error_400";
  }
  if (newBeeMallGoods.getGoodsCategoryId() > 0) {
    if (newBeeMallGoods.getGoodsCategoryId() != null || newBeeMallGoods.getGoodsCategoryId() > 0) {
      //有分类字段则查询相关分类数据并返回给前端以供分类的三级联动显示
      GoodsCategory currentGoodsCategory = newBeeMallCategoryService.getGoodsCategoryById(newBeeMallGoods.getGoodsCategoryId());
      //商品表中存储的分类id字段为三级分类id,不为三级分类则是错误数据
      if (currentGoodsCategory != null && currentGoodsCategory.getCategoryLevel() == NewBeeMallCategoryLevelEnum.LEVEL_THREE.getLevel()) {
        //查询所有的一级分类
        List<GoodsCategory> firstLevelCategories = newBeeMallCategoryService.selectByLevelAndParentIdsAndNumber(Collections.singletonList(0L), NewBeeMallCategoryLevelEnum.LEVEL_ONE.getLevel());
        //根据parentId查询在当前parentId下所有的三级分类
        List<GoodsCategory> thirdLevelCategories = newBeeMallCategoryService.selectByLevelAndParentIdsAndNumber(Collections.singletonList(currentGoodsCategory.getParentId()), NewBeeMallCategoryLevelEnum.LEVEL_THREE.getLevel());
        //查询当前三级分类的父级二级分类
        GoodsCategory secondCategory = newBeeMallCategoryService.getGoodsCategoryById(currentGoodsCategory.getParentId());
        if (secondCategory != null) {
          //根据parentId查询在当前parentId下所有的二级分类
          List<GoodsCategory> secondLevelCategories = newBeeMallCategory
```

```java
Service.selectByLevelAndParentIdsAndNumber(Collections.singletonList
(secondCategory.getParentId()),
NewBeeMallCategoryLevelEnum.LEVEL_TWO.getLevel());
        //查询当前二级分类的父级一级分类
        GoodsCategory firestCategory = newBeeMallCategoryService.getGoods
CategoryById(secondCategory.getParentId());
        if (firestCategory != null) {
            //在所有分类数据得到后放入 request 对象中供前端读取
            request.setAttribute("firstLevelCategories", firstLevelCategories);
            request.setAttribute("secondLevelCategories", secondLevel
Categories);
            request.setAttribute("thirdLevelCategories", thirdLevel
Categories);
            request.setAttribute("firstLevelCategoryId", firestCategory.get
CategoryId());
            request.setAttribute("secondLevelCategoryId", secondCategory.
getCategoryId());
            request.setAttribute("thirdLevelCategoryId", currentGoods
Category.getCategoryId());
        }
      }
    }
  }
}
if (newBeeMallGoods.getGoodsCategoryId() == 0) {
  //查询所有的一级分类
  List<GoodsCategory> firstLevelCategories = newBeeMallCategoryService.
selectByLevelAndParentIdsAndNumber(Collections.singletonList(0L), NewBee
MallCategoryLevelEnum.LEVEL_ONE.getLevel());
  if (!CollectionUtils.isEmpty(firstLevelCategories)) {
    //查询在一级分类列表中第一个实体的所有二级分类
    List<GoodsCategory> secondLevelCategories = newBeeMallCategoryService.
selectByLevelAndParentIdsAndNumber(Collections.singletonList(firstLevel
Categories.get(0).getCategoryId()), NewBeeMallCategoryLevelEnum.LEVEL_
TWO.getLevel());
    if (!CollectionUtils.isEmpty(secondLevelCategories)) {
      //查询在二级分类列表中第一个实体的所有三级分类
      List<GoodsCategory> thirdLevelCategories = newBeeMallCategoryService.
selectByLevelAndParentIdsAndNumber(Collections.singletonList(secondLevel
Categories.get(0).getCategoryId()), NewBeeMallCategoryLevelEnum.LEVEL_
THREE.getLevel());
      request.setAttribute("firstLevelCategories", firstLevelCategories);
      request.setAttribute("secondLevelCategories", secondLevelCategories);
```

```
            request.setAttribute("thirdLevelCategories", thirdLevelCategories);
        }
    }
}
request.setAttribute("goods", newBeeMallGoods);
request.setAttribute("path", "goods-edit");
return "admin/newbee_mall_goods_edit";
}
```

在访问/admin/goods/edit/{goodsId}路径时，这里会查询商品信息编辑页所需的商品信息详情内容。该方法中有一段代码是关于分类数据列表的查询。除商品信息之外，这里也会查询分类数据并交给前端的三级联动模块进行数据的初始化。

查询分类数据分为两种情况。第一种情况是商品表中关联的分类 id 为 0，表示分类数据并未关联，就需要把分类数据按照在添加时的情况处理。另一种情况是商品表中关联的 id 为正常三级分类的 id。在这种情况下，需要在进入编辑页时将该三级分类和该三级分类关联的二级分类和一级分类设置为选中状态。

因此，这里做了如上代码的处理。首先查询当前三级分类 C，紧接着查询该三级分类的上级分类 B，然后根据 B 的主键查询出与 C 分类同级的三级分类列表。同理，再查出与 B 分类同级的二级分类列表和一级分类列表。当然，这里还要查出 B 分类的父级分类 A。这样在初始化时的分类列表数据就都得到了，需要选中的一级分类 A、二级分类 B、三级分类 C 三个分类数据也都查询到。最后，将对应的数据放入 request 域中并转发到 newbee_mall_goods_edit.html 模板页面中。

20.5.2 商品信息编辑页面数据的回显

首先修改商品信息编辑页面 newbee_mall_goods_edit.html 的代码，并通过 Thymeleaf 语法将请求中携带的数据进行读取，然后显示在编辑页面对应的 DOM 中，修改代码如下所示：

```
<div class="form-group" style="display:flex;">
    <label class="control-label">请选择分类:    </label>
    <select class="form-control col-sm-3" id="levelOne"
            data-placeholder="请选择分类...">
        <th:block th:unless="${null == firstLevelCategories}">
            <th:block th:each="c : ${firstLevelCategories}">
                <option th:value="${c.categoryId}" th:text="${c.categoryName}"
                        th:selected="${null !=firstLevelCategoryId and firstLevel
```

```html
CategoryId==c.categoryId} ?true:false">
            >
          </option>
        </th:block>
      </th:block>
    </select> 
    <select class="form-control col-sm-3"
            id="levelTwo"
            data-placeholder="请选择分类...">
      <th:block th:unless="${null == secondLevelCategories}">
        <th:block th:each="c : ${secondLevelCategories}">
          <option th:value="${c.categoryId}" th:text="${c.categoryName}"
                  th:selected="${null !=secondLevelCategoryId and secondLevelCategoryId==c.categoryId} ?true:false"
            >
          </option>
        </th:block>
      </th:block>
    </select> 
    <select class="form-control col-sm-3"
            id="levelThree"
            data-placeholder="请选择分类...">
      <th:block th:unless="${null == thirdLevelCategories}">
        <th:block th:each="c : ${thirdLevelCategories}">
          <option th:value="${c.categoryId}" th:text="${c.categoryName}"
                  th:selected="${null !=thirdLevelCategoryId and thirdLevelCategoryId==c.categoryId} ?true:false"
            >
          </option>
        </th:block>
      </th:block>
    </select>
</div>
<div class="form-group" style="display:flex;">
    <input type="hidden" id="goodsId" name="goodsId"
           th:value="${goods!=null and goods.goodsId!=null }?${goods.goodsId}: 0">
    <input type="text" class="form-control col-sm-6" id="goodsName" name="goodsName"
           placeholder="*请输入商品名称(必填)"
           th:value="${goods!=null and goods.goodsName!=null }?${goods.goodsName}: ''"
           required="true">
      
```

```html
<input type="text" class="form-control col-sm-6" id="goodsIntro" name="goodsIntro"
       placeholder="*请输入商品简介(100字以内)"
       th:value="${goods!=null and goods.goodsIntro!=null }?${goods.goodsIntro}: ''"
       required="true">
</div>
<div class="form-group" style="display:flex;">
  <input type="number" class="form-control col-sm-6" id="originalPrice"
         name="originalPrice"
         placeholder="*请输入商品价格"
         th:value="${goods!=null and goods.originalPrice!=null }?${goods.originalPrice}: 1"
         required="true">
    
  <input type="number" class="form-control col-sm-6" id="sellingPrice"
         name="sellingPrice"
         placeholder="*请输入商品售卖价"
         th:value="${goods!=null and goods.sellingPrice!=null }?${goods.sellingPrice}: 1"
         required="true">
</div>
<div class="form-group" style="display:flex;">
  <input type="number" class="form-control col-sm-6" id="stockNum" name="stockNum"
         placeholder="*请输入商品库存数"
         th:value="${goods!=null and goods.stockNum!=null }?${goods.stockNum}: 0"
         required="true">
    
  <input type="text" class="form-control col-sm-6" id="tag" name="tag"
         placeholder="*请输入商品小标签"
         th:value="${goods!=null and goods.tag!=null }?${goods.tag}: ''"
         required="true">
</div>
<div class="form-group">
  <label class="control-label">上架状态: </label>
  <input name="goodsSellStatus" type="radio" id="goodsSellStatusTrue" checked=true
         th:checked="${null==goods||(null !=goods and null !=goods.goodsSellStatus and goods.goodsSellStatus==0)} ?true:false"
         value="0"/> 上架 
  <input name="goodsSellStatus" type="radio" id="goodsSellStatusFalse"
```

```
value="1"
        th:checked="${null !=goods and null !=goods.goodsSellStatus and
goods.goodsSellStatus==1} ?true:false"/> 下架 
    </div>
    <div class="form-group">
      <div class="form-group">
        <div class="col-sm-4">
          <th:block th:if="${null == goods}">
            <img id="goodsCoverImg" src="/admin/dist/img/img-upload.png"
                style="height: 64px;width: 64px;">
          </th:block>
          <th:block th:unless="${null == goods}">
            <img id="goodsCoverImg" th:src="${goods.goodsCoverImg}"
                style="width:160px ;height: 160px;display:block;">
          </th:block>
        </div>
      </div>
      <br>
      <div class="form-group">
        <div class="col-sm-4">
          <button class="btn btn-info" style="margin-bottom: 5px;"
              id="uploadGoodsCoverImg">
            <i class="fa fa-picture-o"></i> 上传商品主图
          </button>
        </div>
      </div>
    </div>
    <div class="form-group" id="wangEditor" th:utext="${goods!=null and goods.
goodsDetailContent !=null}?${goods.goodsDetailContent}: ''">
    </div>
```

这里，在原来编辑的 DOM 基础上增加了 Thymeleaf 语法作为读取方法。除了基础的字段回显之外，这里还包括三级联动的回显和富文本编辑器内容的回显。这样，在改造后就完成了编辑功能的前两步，即获取商品信息详情并回显到编辑页面中。

20.6 商品信息修改的开发与联调

修改接口的实现与添加接口类似，唯一的不同点就是修改接口需要知道修改的是哪一条。因此，可以模仿添加接口来实现商品信息修改的接口。商品信息修改接口负责接收前端的 POST 请求并处理其中的参数，接收的参数为管理员用户在商品信息编辑页面输入的所有字段内容和待修改商品信息的主键 id。其中，字段名称与对应的含义如下所示。

①goodsId：商品信息主键

②goodsName：商品名称

③goodsIntro：商品简介

④goodsCategoryId：分类 id（在下拉框中选择）

⑤tag：商品小标签字段

⑥originalPrice：商品原价

⑦sellingPrice：商品实际售价

⑧stockNum：商品库存

⑨goodsDetailContent：商品详情内容

⑩goodsCoverImg：商品主图/商品封面图

⑪odsSellStatus：商品销售状态（上架和下架）

20.6.1 数据层代码的实现

首先，在商品实体 Mapper 接口的 NewBeeMallGoodsMapper.java 文件中新增如下方法：

```
/**
 * 根据主键 id 获取记录
 * @param goodsId
 * @return
 */
NewBeeMallGoods selectByPrimaryKey(Long goodsId);

/**
 * 修改一条记录
 * @param record
 * @return
 */
int updateByPrimaryKeySelective(NewBeeMallGoods record);
```

然后，在映射文件 NewBeeMallGoodsMapper.xml 中按照对应的接口方法，编写具体的 SQL 语句，新增代码如下所示：

```xml
<sql id="Base_Column_List">
goods_id, goods_name, goods_intro,goods_category_id, goods_cover_img,
goods_carousel, original_price,selling_price, stock_num, tag, goods_sell_
status, create_user, create_time, update_user, update_time
</sql>

<sql id="Blob_Column_List">
  goods_detail_content
</sql>

<select id="selectByPrimaryKey" parameterType="java.lang.Long" resultMap=
"ResultMapWithBLOBs">
    select
    <include refid="Base_Column_List"/>
    ,
    <include refid="Blob_Column_List"/>
    from tb_newbee_mall_goods_info
    where goods_id = #{goodsId,jdbcType=BIGINT}
</select>

<update id="updateByPrimaryKeySelective" parameterType="ltd.newbee.mall.
entity.NewBeeMallGoods">
    update tb_newbee_mall_goods_info
    <set>
        <if test="goodsName != null">
            goods_name = #{goodsName,jdbcType=VARCHAR},
        </if>
        <if test="goodsIntro != null">
            goods_intro = #{goodsIntro,jdbcType=VARCHAR},
        </if>
        <if test="goodsCategoryId != null">
            goods_category_id = #{goodsCategoryId,jdbcType=BIGINT},
        </if>
        <if test="goodsCoverImg != null">
            goods_cover_img = #{goodsCoverImg,jdbcType=VARCHAR},
        </if>
        <if test="goodsCarousel != null">
            goods_carousel = #{goodsCarousel,jdbcType=VARCHAR},
        </if>
        <if test="originalPrice != null">
            original_price = #{originalPrice,jdbcType=INTEGER},
```

```xml
        </if>
        <if test="sellingPrice != null">
            selling_price = #{sellingPrice,jdbcType=INTEGER},
        </if>
        <if test="stockNum != null">
            stock_num = #{stockNum,jdbcType=INTEGER},
        </if>
        <if test="tag != null">
            tag = #{tag,jdbcType=VARCHAR},
        </if>
        <if test="goodsSellStatus != null">
            goods_sell_status = #{goodsSellStatus,jdbcType=TINYINT},
        </if>
        <if test="createUser != null">
            create_user = #{createUser,jdbcType=INTEGER},
        </if>
        <if test="createTime != null">
            create_time = #{createTime,jdbcType=TIMESTAMP},
        </if>
        <if test="updateUser != null">
            update_user = #{updateUser,jdbcType=INTEGER},
        </if>
        <if test="updateTime != null">
            update_time = #{updateTime,jdbcType=TIMESTAMP},
        </if>
        <if test="goodsDetailContent != null">
            goods_detail_content = #{goodsDetailContent,jdbcType=LONGVARCHAR},
        </if>
    </set>
    where goods_id = #{goodsId,jdbcType=BIGINT}
</update>
```

20.6.2 业务层代码的实现

首先，在商品业务层的 NewBeeMallGoodsService.java 文件中新增如下方法：

```
/**
 * 修改商品信息
 *
 * @param goods
```

```
 * @return
 */
String updateNewBeeMallGoods(NewBeeMallGoods goods);

/**
 * 获取商品详情
 *
 * @param id
 * @return
 */
NewBeeMallGoods getNewBeeMallGoodsById(Long id);
```

然后，在 NewBeeMallGoodsServiceImpl 类中将上述方法实现，新增如下代码：

```
@Override
public String updateNewBeeMallGoods(NewBeeMallGoods goods) {
    NewBeeMallGoods temp = goodsMapper.selectByPrimaryKey(goods.getGoodsId());
    if (temp == null) {
        return ServiceResultEnum.DATA_NOT_EXIST.getResult();
    }
    goods.setUpdateTime(new Date());
    if (goodsMapper.updateByPrimaryKeySelective(goods) > 0) {
        return ServiceResultEnum.SUCCESS.getResult();
    }
    return ServiceResultEnum.DB_ERROR.getResult();
}

@Override
public NewBeeMallGoods getNewBeeMallGoodsById(Long id) {
    return goodsMapper.selectByPrimaryKey(id);
}
```

updateNewBeeMallGoods()这个方法可以结合商品信息添加的业务层方法来理解，该方法会先判断是否存在当前想要修改的记录，再调用执行 update 语句进行数据库修改。

20.6.3 商品添加接口控制层代码的实现

在 NewBeeMallGoodsController 类中新增 update()方法，接口的映射地址为 /admin/goods/update，请求方法为 POST，新增代码如下所示：

```java
/**
 * 修改
 */
@RequestMapping(value = "/goods/update", method = RequestMethod.POST)
@ResponseBody
public Result update(@RequestBody NewBeeMallGoods newBeeMallGoods) {
    if (Objects.isNull(newBeeMallGoods.getGoodsId())
            || StringUtils.isEmpty(newBeeMallGoods.getGoodsName())
            || StringUtils.isEmpty(newBeeMallGoods.getGoodsIntro())
            || StringUtils.isEmpty(newBeeMallGoods.getTag())
            || Objects.isNull(newBeeMallGoods.getOriginalPrice())
            || Objects.isNull(newBeeMallGoods.getSellingPrice())
            || Objects.isNull(newBeeMallGoods.getGoodsCategoryId())
            || Objects.isNull(newBeeMallGoods.getStockNum())
            || Objects.isNull(newBeeMallGoods.getGoodsSellStatus())
            || StringUtils.isEmpty(newBeeMallGoods.getGoodsCoverImg())
            || StringUtils.isEmpty(newBeeMallGoods.getGoodsDetailContent())) {
        return ResultGenerator.genFailResult("参数异常!");
    }
    String result = newBeeMallGoodsService.updateNewBeeMallGoods(newBeeMallGoods);
    if (ServiceResultEnum.SUCCESS.getResult().equals(result)) {
        return ResultGenerator.genSuccessResult();
    } else {
        return ResultGenerator.genFailResult(result);
    }
}
```

这里先对参数进行校验，业务层代码再操作。这与添加接口不同的是传参，其多了一个主键 id 的字段。系统通过这个字段确定将要修改的商品数据。

20.6.4　前端调用商品修改接口

在对商品信息数据进行修改后，点击信息编辑框下方的"保存商品"按钮，此时会调用后端接口并进行数据的交互，JS 的实现代码如下所示：

```javascript
$('#saveButton').click(function () {
    ... 省略部分代码
    var url = '/admin/goods/save';
    var swlMessage = '保存成功';
```

```js
var data = {
    "goodsName": goodsName,
    "goodsIntro": goodsIntro,
    "goodsCategoryId": goodsCategoryId,
    "tag": tag,
    "originalPrice": originalPrice,
    "sellingPrice": sellingPrice,
    "stockNum": stockNum,
    "goodsDetailContent": goodsDetailContent,
    "goodsCoverImg": goodsCoverImg,
    "goodsCarousel": goodsCoverImg,
    "goodsSellStatus": goodsSellStatus
};
if (goodsId > 0) {
    url = '/admin/goods/update';
    swlMessage = '修改成功';
    data = {
        "goodsId": goodsId,
        "goodsName": goodsName,
        "goodsIntro": goodsIntro,
        "goodsCategoryId": goodsCategoryId,
        "tag": tag,
        "originalPrice": originalPrice,
        "sellingPrice": sellingPrice,
        "stockNum": stockNum,
        "goodsDetailContent": goodsDetailContent,
        "goodsCoverImg": goodsCoverImg,
        "goodsCarousel": goodsCoverImg,
        "goodsSellStatus": goodsSellStatus
    };
}
console.log(data);
$.ajax({
    type: 'POST',//方法类型
    url: url,
    contentType: 'application/json',
    data: JSON.stringify(data),
    success: function (result) {
        if (result.resultCode == 200) {
            swal(swlMessage, {
                icon: "success",
            });
        } else {
```

```
            swal(result.message, {
                icon: "error",
            });
        }
        ;
    },
    error: function () {
        swal("操作失败", {
            icon: "error",
        });
    }
});
});
```

这个方法直接从商品信息添加的 JS 方法中改造而来。在"保存商品"按钮的点击事件处理中,首先要判断 goodsId 是否大于 0,如果大于 0 则证明这是一个修改请求。然后封装数据并向后端发送请求修改商品信息。

20.7 商品信息管理页面的制作

20.7.1 导航栏中增加"商品管理"按钮

在左侧导航栏中新增"商品管理"的导航按钮,就需要在 sidebar.html 文件中的"管理模块"目录下新增如下代码:

```html
<li class="nav-item">
    <a th:href="@{/admin/goods}" th:class="${path}=='newbee_mall_goods'?'nav-link active':'nav-link'">
        <i class="fa fa-archive nav-icon" aria-hidden="true"></i>
        <p>
            商品管理
        </p>
    </a>
</li>
```

这里在点击"商品管理"后的跳转路径为/admin/goods。新建 Controller 控制类来处理该路径并跳转到对应的页面上。

20.7.2　控制类处理跳转逻辑

在 NewBeeMallGoodsController 控制器中新增如下代码：

```
@GetMapping("/goods")
public String goodsPage(HttpServletRequest request) {
    request.setAttribute("path", "newbee_mall_goods");
    return "admin/newbee_mall_goods";
}
```

该方法用于处理/admin/goods 请求，即先在返回视图前设置 path 字段 newbeemallgoods，再跳转到 admin 目录下的 newbee_mall_goods.html 模板页面中。

20.7.3　商品管理页面基础样式的实现

接下来就是商品管理页面的模板文件制作。在 resources/templates/admin 目录下新建 newbee_mall_goods.html 模板文件，并引入对应的 JS 文件和 CSS 样式文件，代码如下所示：

```html
<!-- Copyright (c) 2019-2020 十三 all rights reserved. -->
<!DOCTYPE html>
<html xmlns:th="http://www.thymeleaf.org">
<header th:replace="admin/header::header-fragment">
</header>
<style>
    .ui-jqgrid tr.jqgrow td {
        white-space: normal !important;
        height: auto;
        vertical-align: text-top;
        padding-top: 2px;
    }

    a {
        color: #1baeae;
        text-decoration: none;
        background-color: transparent;
        -webkit-text-decoration-skip: objects;
    }
```

```html
        a:hover {
            color: white;
            background-color: #1baeae;
        }
</style>
<body class="hold-transition sidebar-mini">
<div class="wrapper">
    <!-- 引入页面头 header-fragment -->
    <div th:replace="admin/header::header-nav"></div>
    <!-- 引入工具栏 sidebar-fragment -->
    <div th:replace="admin/sidebar::sidebar-fragment(${path})"></div>
    <!-- Content Wrapper. Contains 图标 content -->
    <div class="content-wrapper">
        <!-- Content Header (图标 header) -->
        <div class="content-header">
            <div class="container-fluid">
            </div><!-- /.container-fluid -->
        </div>
        <!-- Main content -->
        <div class="content">
            <div class="container-fluid">
                <div class="card card-primary card-outline">
                    <div class="card-header">
                        <h3 class="card-title">商品管理</h3>
                    </div> <!-- /.card-body -->
                    <div class="card-body">
                        <div class="grid-btn">
                            <button class="btn btn-info" onclick="addGoods()"><i
                                    class="fa fa-plus"></i> 添加商品
                            </button>
                            <button class="btn btn-info" onclick="editGoods()"><i
                                    class="fa fa-pencil-square-o"></i> 修改商品
                            </button>
                            <button class="btn btn-success" onclick="putUpGoods()"><i
                                    class="fa fa-arrow-up"></i> 上架商品
                            </button>
                            <button class="btn btn-danger" onclick="putDownGoods()"><i
                                    class="fa fa-arrow-down"></i> 下架商品
                            </button>
```

```html
                </div>
                <table id="jqGrid" class="table table-bordered">
                </table>
                <div id="jqGridPager"></div>
            </div><!-- /.card-body -->
        </div>
    </div><!-- /.container-fluid -->
  </div>
</div>
<!-- /.content-wrapper -->
<!-- 引入页脚 footer-fragment -->
<div th:replace="admin/footer::footer-fragment"></div>
</div>
<!-- jQuery -->
<script th:src="@{/admin/plugins/jquery/jquery.min.js}"></script>
<!-- Bootstrap 4 -->
<script th:src="@{/admin/plugins/bootstrap/js/bootstrap.bundle.min.js}">
</script>
<!-- AdminLTE App -->
<script th:src="@{/admin/dist/js/adminlte.min.js}"></script>
<!-- jqgrid -->
<script th:src="@{/admin/plugins/jqgrid-5.5.2/jquery.jqGrid.min.js}">
</script>
<script th:src="@{/admin/plugins/jqgrid-5.5.2/grid.locale-cn.js}">
</script>
<!-- sweetalert -->
<script th:src="@{/admin/plugins/sweetalert/sweetalert.min.js}"></script>
<script th:src="@{/admin/dist/js/public.js}"></script>
<script th:src="@{/admin/dist/js/newbee_mall_goods.js}"></script>
</body>
</html>
```

商品管理页面的基础布局和页面的跳转逻辑处理完毕。不过目前只是实现了页面跳转和商品管理页面的基础内容，功能交互和数据显示接下来实现。

20.8 商品信息管理模块接口的实现

商品信息添加和商品信息修改功能都已经实现，在管理模块中只需要增加列表功能和商品上下架功能了。这里只介绍分页接口和商品上下架接口，其他接口在前文已经开发和联调完成。

20.8.1 数据层代码的实现

首先，在商品实体 Mapper 接口的 NewBeeMallGoodsMapper.java 文件中新增如下方法：

```java
/**
 * 查询分页数据
 * @param pageUtil
 * @return
 */
List<NewBeeMallGoods> findNewBeeMallGoodsList(PageQueryUtil pageUtil);

/**
 * 查询总数
 * @param pageUtil
 * @return
 */
int getTotalNewBeeMallGoods(PageQueryUtil pageUtil);

/**
 * 批量修改记录
 * @param orderIds
 * @param sellStatus
 * @return
 */
int batchUpdateSellStatus(@Param("goodsIds")Long[] goodsIds,@Param("sellStatus") int sellStatus);
```

然后，在映射文件 NewBeeMallGoodsMapper.xml 中按照对应的接口方法，编写具体的 SQL 语句，新增代码如下所示：

```xml
<select id="findNewBeeMallGoodsList" parameterType="Map" resultMap="BaseResultMap">
    select
    <include refid="Base_Column_List"/>
    from tb_newbee_mall_goods_info
    order by goods_id desc
    <if test="start!=null and limit!=null">
        limit #{start},#{limit}
    </if>
```

```xml
</select>

<select id="getTotalNewBeeMallGoods" parameterType="Map" resultType="int">
   select count(*) from tb_newbee_mall_goods_info
</select>

<update id="batchUpdateSellStatus">
    update tb_newbee_mall_goods_info
    set goods_sell_status=#{sellStatus},update_time=now() where goods_id in
    <foreach item="id" collection="goodsIds" open="(" separator="," close=")">
       #{id}
    </foreach>
</update>
```

20.8.2 业务层代码的实现

首先，在商品业务层的 NewBeeMallGoodsService.java 文件中新增如下方法：

```java
/**
 * 后台分页
 *
 * @param pageUtil
 * @return
 */
PageResult getNewBeeMallGoodsPage(PageQueryUtil pageUtil);

/**
 * 批量修改销售状态（上下架）
 *
 * @param ids
 * @return
 */
Boolean batchUpdateSellStatus(Long[] ids,int sellStatus);
```

然后，在 NewBeeMallGoodsServiceImpl 类中将上述方法实现，新增代码如下所示：

```java
@Override
public PageResult getNewBeeMallGoodsPage(PageQueryUtil pageUtil) {
    List<NewBeeMallGoods> goodsList = goodsMapper.findNewBeeMallGoodsList(pageUtil);
    int total = goodsMapper.getTotalNewBeeMallGoods(pageUtil);
    PageResult pageResult = new PageResult(goodsList, total,
```

```
        pageUtil.getLimit(), pageUtil.getPage());
    return pageResult;
}

@Override
public Boolean batchUpdateSellStatus(Long[] ids, int sellStatus) {
    return goodsMapper.batchUpdateSellStatus(ids, sellStatus) > 0;
}
```

20.8.3 控制层代码的实现

在 NewBeeMallGoodsController 类中新增列表接口和商品销售状态批量修改接口的实现方法，代码如下所示：

```
/**
 * 列表
 */
@RequestMapping(value = "/goods/list", method = RequestMethod.GET)
@ResponseBody
public Result list(@RequestParam Map<String, Object> params) {
    if (StringUtils.isEmpty(params.get("page")) || StringUtils.isEmpty
(params.get("limit"))) {
        return ResultGenerator.genFailResult("参数异常！");
    }
    PageQueryUtil pageUtil = new PageQueryUtil(params);
    return ResultGenerator.genSuccessResult(newBeeMallGoodsService.
getNewBeeMallGoodsPage(pageUtil));
}

/**
 * 批量修改销售状态
 */
@RequestMapping(value = "/goods/status/{sellStatus}", method = Request
Method.PUT)
@ResponseBody
public Result delete(@RequestBody Long[] ids, @PathVariable("sellStatus")
int sellStatus) {
    if (ids.length < 1) {
        return ResultGenerator.genFailResult("参数异常！");
    }
    if (sellStatus != Constants.SELL_STATUS_UP && sellStatus != Constants.
```

```
SELL_STATUS_DOWN) {
    return ResultGenerator.genFailResult("状态异常！");
}
if (newBeeMallGoodsService.batchUpdateSellStatus(ids, sellStatus)) {
    return ResultGenerator.genSuccessResult();
} else {
    return ResultGenerator.genFailResult("修改失败");
}
}
```

首先列表接口负责接收前端传来的分页参数，比如 page 、limit 等，并把数据总数和对应页面的数据列表查询出来，然后封装为分页数据返回给前端进行数据渲染。SQL 语句在 NewBeeMallGoodsMapper.xml 文件中，一般的分页就使用 limit 关键字来实现，而系统在获取相应条数的记录和总数后再进行数据封装。

系统根据商品信息表中的销售状态字段来确定商品是否处于正常销售状态。如果是下架商品则需要限制部分功能。因此，在后台管理系统中设置了商品的批量上下架功能。销售状态批量修改接口负责接收前端商品上架或者商品下架的请求，在处理前端传输过来的数据后，系统修改这些商品信息记录的销售状态。这里，将接收的参数设置为一个数组，就可以同时操作多条记录。最后，只需要在前端将用户选择记录的 id 封装好，传参到后端即可。在这个过程中，使用@RequestBody 注解将前端传过来的参数封装为 id 数组，在参数验证通过后则调用 batchUpdateSellStatus()批量修改方法进行数据库操作。以上 SQL 语句的含义就是将这些商品信息记录的 goods_sell_status 字段进行批量修改。

20.9 商品管理模块前端功能的实现

20.9.1 商品管理页面功能按钮的设置

在商品信息管理页面中设置了常用的几个功能按钮：商品信息添加按钮、商品信息修改按钮、商品上架按钮和商品下架按钮。这里添加对应的功能按钮和触发事件，添加按钮对应的触发事件是 addGoods()方法；修改按钮对应的触发事件是 editGoods()方法；商品上架按钮对应的触发事件是 putUpGoods()方法；商品下架按钮按钮对应的触发事件是 putDownGoods()方法。这些方法会在后续的代码实现。

20.9.2 商品管理页面分页功能的实现

在 resources/static/admin/dist/js 目录下新增 newbee_mall_goods.js 文件，并添加如下代码：

```
$(function () {
    $("#jqGrid").jqGrid({
        url: '/admin/goods/list',
        datatype: "json",
        colModel: [
            {label: '商品编号', name: 'goodsId', index: 'goodsId', width: 60, key: true},
            {label: '商品名', name: 'goodsName', index: 'goodsName', width: 120},
            {label: '商品简介', name: 'goodsIntro', index: 'goodsIntro', width: 120},
            {label: '商品图片', name: 'goodsCoverImg', index: 'goodsCoverImg', width: 120, formatter: coverImageFormatter},
            {label: '商品库存', name: 'stockNum', index: 'stockNum', width: 60},
            {label: '商品售价', name: 'sellingPrice', index: 'sellingPrice', width: 60},
            {
                label: '上架状态',
                name: 'goodsSellStatus',
                index: 'goodsSellStatus',
                width: 80,
                formatter: goodsSellStatusFormatter
            },
            {label: '创建时间', name: 'createTime', index: 'createTime', width: 60}
        ],
        height: 760,
        rowNum: 20,
        rowList: [20, 50, 80],
        styleUI: 'Bootstrap',
        loadtext: '信息读取中...',
        rownumbers: false,
        rownumWidth: 20,
        autowidth: true,
        multiselect: true,
        pager: "#jqGridPager",
        jsonReader: {
```

```
            root: "data.list",
            page: "data.currPage",
            total: "data.totalPage",
            records: "data.totalCount"
        },
        prmNames: {
            page: "page",
            rows: "limit",
            order: "order",
        },
        gridComplete: function () {
            //隐藏grid底部滚动条
            $("#jqGrid").closest(".ui-jqgrid-bdiv").css({"overflow-x":
"hidden"});
        }
    });

    $(window).resize(function () {
        $("#jqGrid").setGridWidth($(".card-body").width());
    });

    function goodsSellStatusFormatter(cellvalue) {
        //商品上架状态：0-上架 1-下架
        if (cellvalue == 0) {
            return "<button type=\"button\" class=\"btn btn-block btn-success
btn-sm\" style=\"width: 80%;\">销售中</button>";
        }
        if (cellvalue == 1) {
            return "<button type=\"button\" class=\"btn btn-block btn-
secondary btn-sm\" style=\"width: 80%;\">已下架</button>";
        }
    }

    function coverImageFormatter(cellvalue) {
        return "<img src='" + cellvalue + "' height=\"80\" width=\"80\" alt='
商品主图'/>";
    }
});
```

以上代码的主要功能为分页数据的展示、字段格式化、jqGrid DOM 宽度的自适应。在页面加载时，调用 jqGrid 的初始化方法，将页面 id 为 jqGrid 的 DOM 渲染为分页表格，并向后端发送请求,请求地址为/admin/carousels/list,最后系统按照后端返回的 JSON 数据填充表格和表格下方的分页按钮。这里可以参考前文 jqGrid 分页功能整合进行理解。重启 Spring Boot 项目可以验证商品数据分页功能是否正常。

20.9.3 商品添加和修改按钮的触发事件

由于添加商品和修改商品两个功能已经开发完成,且编辑页面也制作完毕,因此点击添加和修改两个功能按钮跳转到商品编辑页面即可。此处给两个按钮分别绑定了在点击时的触发事件,即在 newbee_mall_goods.js 文件中新增 addGoods()方法和 editGoods()方法。这两个方法的实现均为跳转至商品信息编辑页面。触发事件实现的代码如下所示:

```
/**
 * 添加商品
 */
function addGoods() {
    window.location.href = "/admin/goods/edit";
}

/**
 * 修改商品
 */
function editGoods() {
    var id = getSelectedRow();
    if (id == null) {
        return;
    }
    window.location.href = "/admin/goods/edit/" + id;
}
```

点击添加按钮是直接跳转到商品信息编辑页的。而在点击修改按钮时首先要获取当前选择需要修改的商品信息 id,然后跳转至商品信息编辑页。添加和修改的相关功能的操作是在编辑页面完成的,商品管理页面只负责一次跳转即可。

20.9.4 商品上架和下架功能的实现

商品上架和下架两个按钮的点击触发事件分别为 putUpGoods()和 putDownGoods()。在 newbee_mall_goods.js 文件中新增如下代码:

```javascript
/**
 * 上架
 */
function putUpGoods() {
    var ids = getSelectedRows();
    if (ids == null) {
        return;
    }
    swal({
        title: "确认弹框",
        text: "确认要执行上架操作吗?",
        icon: "warning",
        buttons: true,
        dangerMode: true,
    }).then((flag) => {
        if (flag) {
            $.ajax({
                type: "PUT",
                url: "/admin/goods/status/0",
                contentType: "application/json",
                data: JSON.stringify(ids),
                success: function (r) {
                    if (r.resultCode == 200) {
                        swal("上架成功", {
                            icon: "success",
                        });
                        $("#jqGrid").trigger("reloadGrid");
                    } else {
                        swal(r.message, {
                            icon: "error",
                        });
                    }
                }
            });
        }
    })
    ;
}
/**
 * 下架
 */
```

```js
function putDownGoods() {
    var ids = getSelectedRows();
    if (ids == null) {
        return;
    }
    swal({
        title: "确认弹框",
        text: "确认要执行下架操作吗?",
        icon: "warning",
        buttons: true,
        dangerMode: true,
    }).then((flag) => {
        if (flag) {
            $.ajax({
                type: "PUT",
                url: "/admin/goods/status/1",
                contentType: "application/json",
                data: JSON.stringify(ids),
                success: function (r) {
                    if (r.resultCode == 200) {
                        swal("下架成功", {
                            icon: "success",
                        });
                        $("#jqGrid").trigger("reloadGrid");
                    } else {
                        swal(r.message, {
                            icon: "error",
                        });
                    }
                }
            });
        }
    }
    )
    ;
}
```

这里首先获取用户在 jqGrid 表格中选择需要操作的所有商品信息的 id 字段，然后将参数封装并向后端发送请求修改商品的上下架状态。请求地址分别为 /admin/goods/status/0 和 /admin/goods/status/1，在请求成功后会刷新页面数据。

读者可以结合源码和实际的功能进行学习和理解，接下来进行商品模块的功能测试。

20.9.5 功能测试

在编码完成后,启动 Spring Boot 项目。在启动成功后打开浏览器并输入后台管理系统登录页面地址:

```
http://localhost:8080/admin/login
```

在输入登录信息后登录后台管理系统,点击左侧导航栏中的"商品管理"按钮就可以对相关功能进行测试。

(1) 列表功能

进入商品管理页面就可以看到数据列表,如图 20-7 所示。

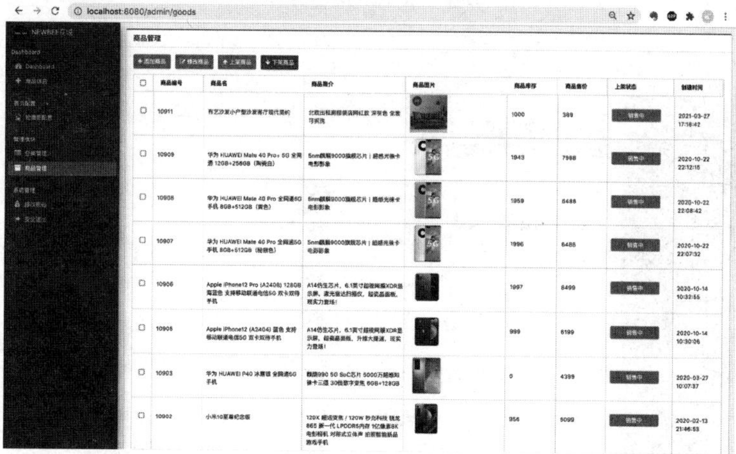

图 20-7 商品列表

笔者已经提供了商品初始化数据 SQL 文件和商品图片的压缩文件,里面有 600 条左右的商品测试数据。将初始化数据和商品图片解压即可使用。

在初始化时,显示的是第一页的数据,列表中是商品的一些基本字段。进行翻页功能测试:翻到第 29 页,每页默认 20 条数据。功能一切正常,商品列表翻页效果如图 20-8 所示。

(2) 商品信息修改

首先选中一条需要修改的记录,并点击"修改商品"按钮,如图 20-9 所示。

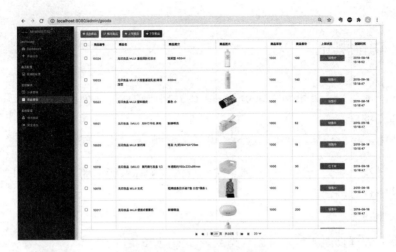

图 20-8　商品列表翻页效果

图 20-9　修改商品信息

然后会跳转到商品编辑页面，管理员用户可以修改商品信息，如图 20-10 所示。

在跳转到商品编辑页面后，首先可以看到商品的所有信息都已经被正确地回显到页面中，商品分类信息也被正确地初始化，且当前商品的分类也在三级分类联动功能中的三个 select 下拉选择框中被正确选中。然后分别更改商品标题、商品简介、库存、商品售价这几个字段，最后点击"保存商品"按钮。

在商品信息更新成功后可以看到列表中的数据已经被修改，如图 20-11 所示。

第 20 章　商品编辑页面及商品管理模块的开发

图 20-10　修改商品信息

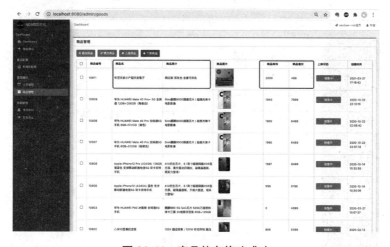

图 20-11　商品信息修改成功

（3）商品下架

首先选中需要下架的记录，然后点击"下架商品"按钮，此时会出现一个确认弹框，如图 20-12 所示。

点击"OK"表示确认操作，点击"Cancel"表示取消操作。点击"OK"后，选中的商品记录会被下架。例如商品编号 10907、10908 和 10909 的三个商品销售状态为"销售中"。在选择这三个商品后点击"下架商品"按钮，并进行确认操作。在请求发送成功后，可以看到这三个商品的销售状态已经改为"已下架"，如图 20-13 所示。

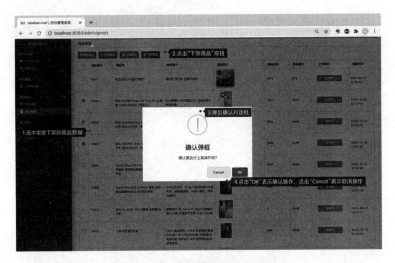

图 20-12　下架商品的确认弹框

图 20-13　商品下架成功

（4）商品上架

与商品下架的操作步骤类似，首先选中需要上架的记录，然后点击"上架商品"按钮即可完成商品的状态修改。如图 20-14 所示，商品编号 10908 和 10909 的两个商品销售状态为"已下架"，选择这两个商品并点击"上架商品"按钮，就会出现商品上架的确认弹框。

最后进行确认操作。在请求发送成功后，可以看到这两条商品的销售状态已经改为"销售中"，如图 20-15 所示。

第 20 章　商品编辑页面及商品管理模块的开发

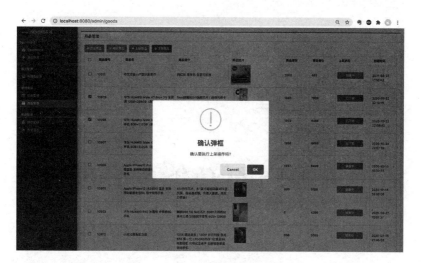

图 20-14　上架商品的确认弹框

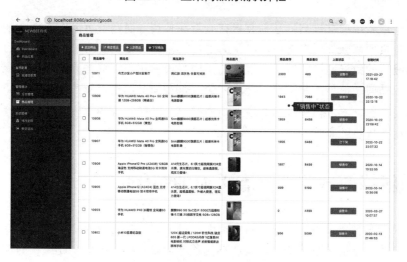

图 20-15　商品上架成功

与 20.4.6 节的功能验证步骤一样，因为都是关于商品信息的功能，所以需要注意两点。第一，在操作完成后，查看在数据库中的记录是否正确。第二，如果有图片上传，查看在 upload 目录下是否有正确的文件。

读者可以按照文中的思路和过程自行测试。至此，商品模块的功能开发完成！

第 21 章

新蜂商城首页功能的开发

后台管理系统通常只是网站拥有者在查看和使用，而商城页面则有较大的不同。商城中的所有页面都会被用户访问，涉及的用户操作行为多为查看。也就是把后台管理系统添加和编辑的数据通过商城各个模块的页面呈现给用户。这些页面偏重于展示，供用户查看，包括商品信息、购物车、商品归类、推荐商品、用户订单等信息的查看。

从本章开始，笔者将会讲解商城端功能模块的开发。本章主要是新蜂商城首页功能的开发和相关模块的完善。

21.1 新蜂商城首页静态页面的制作

相较于后台管理系统，商城端相关页面的设计和制作会更注重用户体验。虽然涉及的操作只是数据查询和数据聚合，但这不代表开发难度就会降低。商城端的页面往往更加注重页面观感和元素设计，做到简洁、美观、实用。

21.1.1 商城首页的设计注意事项

在新蜂商城页面中，首页是最先被用户浏览到的页面，也是非常重要的入口。如果用户在该页面就萌生退意那就证明页面的设计还需要再仔细斟酌。因此，新蜂商城的首页的设计和制作就是重中之重。关于商城首页的设计和制作笔者总结了 4 个设计注意事项。

（1）图文结合

图文相结合最大的优势就是直观、易懂、生动。要让用户在短短几秒钟之内就了解商城的一系列内容，仅凭简短的文字是不够的，还需要使用图片从侧面衬托主题。精准的文案和合适的配图更能吸引用户。尤其是商城系统，想要在首页更完整地将商品内容输出给用户，推荐采取图文结合的方式。

（2）精心设计

在电商网站中，首页是寸土寸金的。因此，首页的文字和图片要精心设计，尽可能用最少的篇幅把信息表达清楚。商品的推荐文案和展示图片都需要再三斟酌。

（3）降低交互难度

视觉体验决定用户的停留时间。用户交互部分的视觉设计必须符合逻辑，比如导航栏、轮播图、商品类目、搜索框等。各个模块一定不要过度设计，可点击部分尽要意图明显，降低用户对首页交互的难度。简单、符合用户习惯即可。

（4）完整的网站导航和清晰的网站 Logo

网站导航一般包括网站首页、项目介绍、公司产品、联系我们、关于我们等分类。比如天猫、京东在页面的顶部和底部都会设计很多导航链接。新蜂商城借鉴了这些设计。导航所占位置有限，所以应该放最重要的内容。

清晰的网站 Logo 能让用户直观地记住商城的名称和形象，也能够给用户展现品牌的特色。一个优秀的 Logo 会给网站带来积极的影响。同时，需要将 Logo 图片设置为带有网站首页的链接，以方便用户点击 Logo 直接返回首页。这也是最常见的做法。

21.1.2 新蜂商城首页的排版设计

后台管理系统的布局和代码实现完全使用 AdminLTE3 模板，而商城端的页面则是自行设计、自行开发的。这是商城端与后台管理系统在开发时最大的不同。

新蜂商城首页布局和排版设计如图 21-1 所示。

由图 21-1 可以看出，新蜂商城首页的整个设计版面被切分成 8 个部分。

①导航及个人信息：该区域主要用于放置与新蜂商城相关的导航链接、登录信息、购物车数据等内容。

②商城 Logo 及搜索框：可以放置 Logo 图片、商品搜索输入框、用户输入需要的商品关键字、跳转到对应的商品搜索列表页。

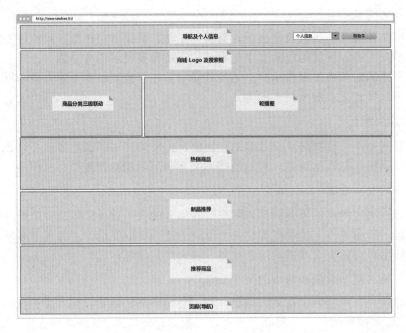

图 21-1 新蜂商城首页布局和排版设计

③商品分类三级联动：展示后台已配置的所有的商品类目数据，包括一级分类、二级分类、三级分类。

④轮播图：以轮播的形式展示后台配置的轮播图。

⑤热销商品：展示后台配置的热销商品数据。

⑥新品推荐：展示后台配置的新品数据。

⑦推荐商品：展示后台配置的推荐商品数据。

⑧页脚（导航）：该区域主要放置备案信息、新蜂商城产品介绍、基础信息、联系方式等。

当然，以上版面设计只是针对新蜂商城这个项目，在开源社区或者企业开发中还有许多其他的商城系统项目。由于前端的设计和实现非常灵活多变，不同的商城系统可能有不同的页面样式和页面布局。

以上是对新蜂商城系统首页布局的大致切分和讲解，接下来是具体的页面和样式编码的实现。

21.1.3 新蜂商城首页基础样式的实现

根据排版设计制作出 HTML 页面，代码如下所示：

```html
<!DOCTYPE html>
<html lang="en">
<head>
    <meta charset="UTF-8">
    <title>新蜂商城 NewBee 商城 newbee-mall</title>
    <link rel="stylesheet" href="css/iconfont.css">
    <link rel="stylesheet" href="css/common.css">
    <link rel="stylesheet" href="styles/header.css">
    <link rel="stylesheet" href="styles/index.css">
</head>
<body>
<header id="header">
    <div class="center">
        <ul class="fl">
            <li><a href="/index">新蜂商城</a></li>
            <li><a href="/admin">后台管理系统</a></li>
            <li><a href="https://edu.csdn.net/course/detail/26258">课程视频</a></li>
            <li><a href="https://juejin.im/book/5da2f9d4f265da5b81794d48/section/5da2f9d6f265da5b794f2189">课程文档</a></li>
            <!-- 省略部分代码 -->
        </ul>
        <div class="fr">
            <ul class="login">
                <li><a href="/login">登录</a></li>
                <li><a href="/register">注册</a></li>
            </ul>
            <div class="shopcart">
                <a href="##" style="color: white;"><i class="iconfont icon-cart"></i>
                    购物车(13)</a>
            </div>
        </div>
    </div>
</header>
<content id="content">
    <nav id="nav">
```

```html
    <div class="banner_x center">
        <a href="/index" class="logo"><h1>新蜂商城</h1></a>
        <div class="fr">
            <div class="search">
                <input class="text" type="text" id="keyword" autocomplete="off">
                <div class="search_hot">
                </div>
            </div>
            <div class="button iconfont icon-search"></div>
        </div>
    </div>
</nav>
    <div id="banner">
        <div class="all-sort-list">
            <div class="item">
                <h3><span>·</span><a href="##">家电 数码 手机</a></h3>
            </div>
            <div class="item">
                <h3><span>·</span><a href="##">女装 男装 穿搭</a></h3>
            </div>
            <div class="item">
                <h3><span>·</span><a href="##">家具 家饰 家纺</a></h3>
            </div>
            <div class="item">
                <h3><span>·</span><a href="##">运动 户外 乐器</a></h3>
            </div>
            <div class="item">
                <h3><span>·</span><a href="##">游戏 动漫 影视</a></h3>
            </div>
            <div class="item">
                <h3><span>·</span><a href="##">美妆 清洁 宠物</a></h3>
            </div>

            <div class="item">
                <h3><span>·</span><a href="##">工具 装修 建材</a></h3>
            </div>

            <div class="item">
                <h3><span>·</span><a href="##">鞋靴 箱包 配件</a></h3>
            </div>

            <div class="item">
                <h3><span>·</span><a href="##">玩具 孕产 用品</a></h3>
            </div>
```

```html
            </div>
            <div class="swiper-container fl">
                <div class="swiper-wrapper">
                        <div class="swiper-slide">
                                <a href="https://juejin.im/book/5da2f9d4f265da5b81794d48/section/5da2f9d6f265da5b794f2189">
                                        <img src="https://newbee-mall.oss-cn-beijing.aliyuncs.com/images/banner1.png" alt="">
                                </a>
                        </div>
                </div>
            </div>
        </div>
        <div id="sub_banner">
                <div class="hot-image">
                    <a href="##">
                        <img src="image/hot3.jpg" alt="热销">
                    </a>
                </div>
                <div class="hot-image">
                    <a href="##">
                        <img src="image/hot3.jpg" alt="热销">
                    </a>
                </div>
                 <div class="hot-image">
                    <a href="##">
                        <img src="image/hot3.jpg" alt="热销">
                    </a>
                </div>
                <div class="hot-image">
                    <a href="##">
                        <img src="image/hot3.jpg" alt="热销">
                    </a>
                </div>
        </div>

        <div id="flash">
            <h2>新品上线</h2>
            <ul>
                    <li>
                        <a href="##">
                            <img src="image/new.png" alt="新品">
                            <p class="name">HUAWEI Mate 30 Pro 双4000万徕卡电影四摄</p>
                            <p class="discount">超曲面OLED环幕屏</p>
                            <p class="item_price">5399</p>
```

```html
                    </a>
                </li>
                <!-- 省略部分代码 -->
        </ul>
    </div>

    <div id="recommend">
        <h2>为你推荐</h2>
        <a href="##" class="more">查看更多>></a>
        <ul>
                <li>
                    <a href="##">
                        <div class="info discount">智能降噪 长久续航</div>
                        <img src="image/recommend.jpg" alt="索尼 WH-1000XM3 头戴式耳机">
                        <p class="name">索尼 WH-1000XM3 头戴式耳机</p>
                        <p class="item_price">2599</p>
                        <p class="counter">猜你喜欢</p>
                        <div class="comment">
                            <p>新蜂精选</p>
                            <p>好物也可以不贵</p>
                        </div>
                    </a>
                </li>
                <!-- 省略部分代码 -->
        </ul>
    </div>
</content>
<div class="site-footer">
    <div class="footer-related">
        <div class="footer-article w1100">
            <dl class="contact clearfix">
                <dt class="fl">
                    <i class="iconfont"></i>
                </dt>
                <dd class="fl">
                    <p class="text">人工客服</p>
                    <p class="tel">400-xxx-xx13</p>
                    <a href="##">在线咨询</a>
                </dd>
            </dl>
            <dl class="col-article">
                <dt>帮助中心</dt>
                <dd><a rel="nofollow" href="https://gitee.com/newbee-ltd/newbee-mall" target="_blank">账户管理</a></dd>
```

```html
            <dd><a rel="nofollow" href="https://gitee.com/newbee-ltd/
newbee-mall" target="_blank">购物指南</a></dd>
            <dd><a rel="nofollow" href="https://gitee.com/newbee-ltd/
newbee-mall" target="_blank">订单操作</a></dd>
          </dl>
      <!-- 省略部分代码 -->
      </div>
      <!-- 省略部分代码 -->
      <div class="footer-info w1100">
         <div class="info-text w1100">
            <p><a href="https://gitee.com/newbee-ltd/newbee-mall"
target="_blank">新蜂商城   |  
               Powered by 十三 
               |  </a>
                QQ 交流群：796794009  
            </p>
            <p><a href="https://gitee.com/newbee-ltd/newbee-mall"
target="_blank">Copyright ©newbee.ltd All
               Rights
               Reserved.  |   浙 ICP 备 17008806 号-5 </a></p>
         </div>
      </div>
   </div>
</div>
</body>
</html>
```

由于篇幅限制，这里直接给出代码文件，分别是样式文件、基础的展示图片及首页的代码，文件名是 index-html.zip，读者可以直接下载并查看源码，文件目录结构如下所示：

```
index-html
    ├── css
    ├── image
    ├── styles
    └── index.html
```

通过首页的静态页面源码，可以看出导航及个人信息、商城 Logo 及搜索框、商品分类三级联动、轮播图、热销商品、新品推荐、推荐商品、页脚（导航）等元素都已经设置在对应的页面区域，只是当前页面中的交互效果还没有实现。此时对应的商品文案、商品图片也都是"假数据"，并非在数据库中读取的数据。后续还要继续对该页面进行完善。

该 HTML 代码是可以直接在本地通过浏览器打开的，页面效果如图 21-2 所示。

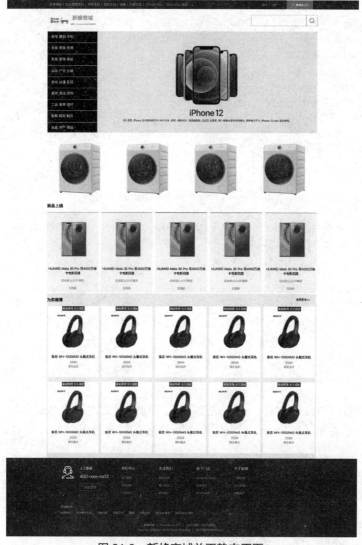

图 21-2　新蜂商城首页静态页面

读者可以对照笔者提供的源码和浏览器呈现的效果，并结合文章分析的页面设计和排版来理解这部分内容。这里只完成了前端页面的样式制作，仅仅实现了静态页面效果。页面上的数据也是固定不变的数据，没有与数据库进行交互并进行数据的读取和渲染。页面的交互和数据的动态获取还没有完成。接下来将会讲解首页展示数据的查询，后台管理系统配置的内容在首页的填充。

21.2 新蜂商城首页功能的实现

为了与后端管理系统区别，商城端的页面实现及相关功能的实现会重新建包和文件目录。

21.2.1 首页跳转逻辑的实现

在 resources/templates 目录下新建 mall 目录，用于存放新蜂商城页面的模板页面，并放入 index.html 页面文件。在 resources/static 目录下新建 mall 目录，用于存放新蜂商城页面相关的静态资源文件，并将前文中涉及的静态资源文件都移到该目录下。

打开 index.html 文件并在该模板文件的标签中导入 Thymeleaf 的名称空间：

```html
<html lang="en" xmlns:th="http://www.thymeleaf.org">
```

导入该名称空间主要是为了 Thymeleaf 的语法提示和 Thymeleaf 标签的使用。接着在模板中使用 th 标签来修改静态资源的引用路径。

21.2.2 Controller 处理跳转

在前端文件制作完成后，接下来新建 Controller 类来处理首页请求路径并跳转到对应的页面中。

首先在 ltd.newbee.mall.controller 包下新建 mall 包，并新建 IndexController.java，然后新增如下代码：

```java
package ltd.newbee.mall.controller.mall;

import org.springframework.stereotype.Controller;
import org.springframework.web.bind.annotation.GetMapping;

import javax.servlet.http.HttpServletRequest;

@Controller
public class IndexController {
```

```
@GetMapping({"/index", "/", "/index.html"})
public String indexPage(HttpServletRequest request) {
    return "mall/index";
}
}
```

该方法用于处理/、/index、index.html 请求路径。这种路径的请求一般为首页请求。如果还需要添加其他路径，也可以在@GetMapping 注解的配置中加上。该方法最后返回的是 mall/index，即访问该方法会跳转到 mall 目录下的 index.html 模板文件中。此时添加首页代码的项目目录结构如图 21-3 所示。

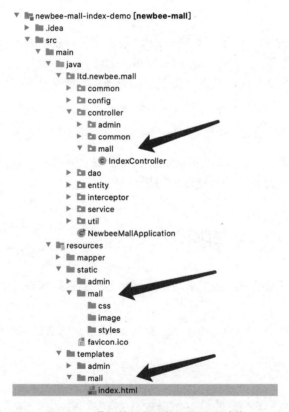

图 21-3　添加首页代码的项目目录结构

至此，商城端首页跳转逻辑处理完毕。

21.2.3 公共页面的抽取

前文讲解了页面设计和基本的页面区域。页面顶部和页面底部两个区域在商城端的页面中是相同的，因此对这两部分进行公共代码的抽取以减少重复编码。在 resources/templates/mall 目录下新建 header.html 和 footer.html 两个模板文件。

header.html 代码如下所示：

```html
<!DOCTYPE html>
<html lang="en" xmlns:th="http://www.thymeleaf.org">
<head th:fragment="head-fragment(title,path)">
    <meta charset="UTF-8">
    <title>新蜂商城 NewBee 商城 newbee-mall</title>
    <link rel="stylesheet" th:href="@{/mall/css/iconfont.css}">
    <link rel="stylesheet" th:href="@{/mall/css/common.css}">
    <link rel="stylesheet" th:href="@{/mall/styles/header.css}">
    <link rel="stylesheet" th:href="@{'/mall/styles/'+${path}+'.css'}">
</head>
<nav id="nav" th:fragment="nav-fragment">
    <div class="banner_x center">
        <a th:href="@{/index}" class="logo"><h1>新蜂商城</h1></a>
        <div class="fr">
            <div class="search">
                <input class="text" type="text" id="keyword" autocomplete="off">
                <div class="search_hot">
                </div>
            </div>
            <div class="button iconfont icon-search" onclick="##"></div>
        </div>
    </div>
</nav>
<header id="header" th:fragment="header-fragment">
    <div class="center">
        <ul class="fl">
            <li><a th:href="@{/index}">新蜂商城</a></li>
            <li><a th:href="@{/admin}">后台管理系统</a></li>
            <li><a href="https://edu.csdn.net/course/detail/26258">课程视频</a></li>
            <li><a href="https://juejin.im/book/5da2f9d4f265da5b81794d48/section/5da2f9d6f2
```

```
65da5b794f2189">课程文档</a></li>
            <li><a href="https://gitee.com/newbee-ltd/newbee-donate">捐赠</a></li>
            <li><a href="https://github.com/newbee-ltd/newbee-mall/issues">问题反馈</a></li>
            <li><a href="https://github.com/newbee-ltd/newbee-mall">GitHub地址</a></li>
            <li><a href="https://gitee.com/newbee-ltd/newbee-mall">码云Gitee地址</a></li>
        </ul>
        <div class="fr">
            <ul class="login">
                <li><a th:href="@{/login}">登录</a></li>
                <li><a th:href="@{/register}">注册</a></li>
            </ul>
            <div class="shopcart">
                <a href="##" style="color: white;"><i class="iconfont icon-cart"></i>
                    购物车(0)</a>
            </div>
        </div>
    </div>
</header>
</html>
```

footer.html 代码如下所示：

```
<!DOCTYPE html>
<html lang="en" xmlns:th="http://www.thymeleaf.org">
<div class="site-footer" th:fragment="footer-fragment">
    <div class="footer-related">
        <div class="footer-article w1100">
            <dl class="contact clearfix">
                <dt class="fl">
                    <i class="iconfont"></i>
                </dt>
                <dd class="fl">
                    <p class="text">人工客服</p>
                    <p class="tel">400-xxx-xx13</p>
                    <a href="//shang.qq.com/wpa/qunwpa?idkey=dc0e028f177932aee2c212a2dd60e0b8342042ac205305803ea801c4eea6727c">在线咨询</a>
                </dd>
```

```html
            </dl>
            ...省略部分代码
        </div>
        <div class="footer-links w1100">
            <p>友情链接: </p>
            <div class="clearfix">
                <a th:href="@{/index}">新蜂商城</a>
                <a th:href="@{/admin}">后台管理系统</a>
                <a href="https://edu.csdn.net/course/detail/26258">课程视频</a>
                <a href="https://juejin.im/book/5da2f9d4f265da5b81794d48/section/5da2f9d6f265da5b794f2189">课程文档</a>
                <a href="https://gitee.com/newbee-ltd/newbee-donate">捐赠</a>
                <a href="https://github.com/newbee-ltd/newbee-mall/issues">问题反馈</a>
                <a href="https://github.com/newbee-ltd/newbee-mall">GitHub 地址</a>
                <a href="https://gitee.com/newbee-ltd/newbee-mall">码云 Gitee 地址</a>
            </div>
        </div>
        <div class="footer-info w1100">
            <div class="info-text w1100">
                <p><a href="https://gitee.com/newbee-ltd/newbee-mall" target="_blank">新蜂商城   |  
                    Powered by 十三 
                    |  </a>
                     QQ 交流群: 796794009   <a
                            href="//shang.qq.com/wpa/qunwpa?idkey=dc0e028f177932aee2c212a2dd60e0b8342042ac205305803ea801c4eea6727c"><img
                            border="0" src="//pub.idqqimg.com/wpa/images/group.png" alt="SpringBoot 技术交流"
                            title="SpringBoot 技术交流"></a>
                </p>
                <p><a href="https://gitee.com/newbee-ltd/newbee-mall" target="_blank">Copyright ©newbee.ltd All
                    Rights
                    Reserved.  |   浙 ICP 备 17008806 号-5 </a></p>
            </div>
        </div>
```

```html
    </div>
  </div>
</html>
```

这里修改 index.html，页面的顶部区域和底部区域通过 th:replace 标签引入，修改的代码如下所示：

```html
<!-- 引入 head 部分 -->
<head th:replace="mall/header::head-fragment('NewBee 商城-首页','index')">

<!-- 引入导航栏和搜索栏 -->
<header th:replace="mall/header::header-fragment"></header>
<nav th:replace="mall/header::nav-fragment"></nav>

<!-- 引入页脚 footer-fragment -->
<div th:replace="mall/footer::footer-fragment"></div>
```

这个知识点在后台管理系统开发时已经介绍过，这里就不再占用太多篇幅。在修改完成后重启项目查看页面效果，一切正常则证明编码成功。

21.3 商城端首页轮播图功能的实现

横跨屏幕的首页轮播图是时下比较流行的网页设计手法，新蜂商城项目也设计了首页轮播图的效果。前文已有相关介绍，这里将介绍轮播图在首页的数据读取和轮播效果的实现。

21.3.1 Swiper 轮播图插件的介绍

新蜂商城首页轮播图的效果通过插件 Swiper 实现。这个插件非常知名，一般的轮播图都会使用到。它的功能也非常强大，只需要一些简单的配置就能实现很多实用的轮播效果。Swiper 也是一款比较轻量级的轮播图插件，即使不引入 JQuery 它也能完成轮播图功能。而且 Swiper 在移动端、PC 端都能够使用，这也是它受欢迎的原因之一。同时 Swiper 完全免费并开源，使用的协议为 MIT 开源协议，无论在个人网站或商业网站均可使用。Swiper 在 GitHub 网站的仓库主页为：
https://github.com/nolimits4web/swiper。

Swiper 源文件的下载地址如下所示：

https://github.com/nolimits4web/swiper/releases

在浏览器中输入该地址就可以看到各个版本 Swipe 代码压缩包。如图 21-4 所示，本书演示代码所选择的 Swiper 版本为 6.4.5，该版本发布于 2020 年 12 月 19 日。

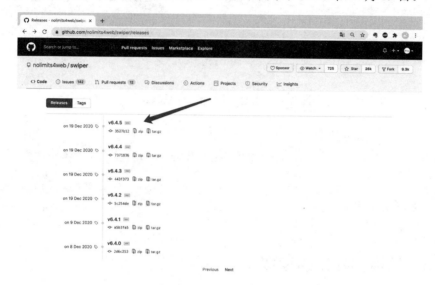

图 21-4　swiper 下载页面

将代码压缩包解压后得到 swiper-master 文件夹。此文件夹包含 Swiper 正式包的目录结构，所有文件都在解压目录中。但是并不需要把这些文件全部放入项目，只需要将 swiper-master 文件夹下的 package 目录中的 swiper-bundle.min.js 和 swiper-bundle.min.css 文件引入即可。

21.3.2　轮播图插件 Swiper 的整合

首先在 static/mall 目录下新建 js 目录，分别将 swiper-bundle.min.js 和 swiper-bundle.min.css 文件放到 static/mall/css 目录和 static/mall/js 目录中。整合后的目录结构如图 21-5 所示。

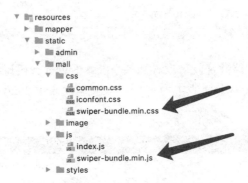

图 21-5　整合后的 Swiper 目录结构

然后将这两个文件引入到页面上，代码如下所示：

```
<link rel="stylesheet" th:href="@{/mall/css/swiper-bundle.min.css}">
<script th:src="@{/mall/js/swiper-bundle.min.js}" type="text/javascript">
</script>
```

21.3.3　轮播图数据的读取

在演示后台管理系统的轮播图管理功能时，此处配置了多条轮播图数据，可以首先在 IndexController 控制类中将这些数据读取出来并放到 request 域中，然后使用 th 语法将这些数据渲染到页面上。

修改 IndexController 控制类中的 indexPage()方法，首先注入 NewBeeMallCarouselService 对象，然后新增代码如下：

```
List<NewBeeMallIndexCarouselVO> carousels = newBeeMallCarouselService.getCarouselsForIndex(Constants.INDEX_CAROUSEL_NUMBER);
request.setAttribute("carousels", carousels);//轮播图
```

读取轮播图列表并将其设置到 request 对象中。NewBeeMallCarouselService 业务类的 getCarouselsForIndex()方法实现代码如下所示：

```
/**
 * 返回固定数量的轮播图对象（首页调用）
 *
 * @param number
```

```
 * @return
 */
public List<NewBeeMallIndexCarouselVO> getCarouselsForIndex(int number) {
    List<NewBeeMallIndexCarouselVO> newBeeMallIndexCarouselVOS = new ArrayList<>
(number);
    List<Carousel> carousels = carouselMapper.findCarouselsByNum(number);
    if (!CollectionUtils.isEmpty(carousels)) {
        newBeeMallIndexCarouselVOS = BeanUtil.copyList(carousels, NewBeeMall
IndexCarouselVO.class);
    }
    return newBeeMallIndexCarouselVOS;
}
```

该方法的作用是返回固定数量的轮播图对象，供首页数据进行渲染。

21.3.4 轮播图数据的渲染

修改前端模板的代码，首先使用 Thymeleaf 语法读取 carousels 轮播图列表对象，然后使用 th:each 循环将轮播图数据渲染到页面上，代码如下所示：

```
<div class="swiper-container fl">
  <div class="swiper-wrapper">
    <th:block th:unless="${#lists.isEmpty(carousels)}">
      <th:block th:each="carousel : ${carousels}">
        <div class="swiper-slide">
          <a th:href="@{${carousel.redirectUrl}}">
            <img th:src="@{${carousel.carouselUrl}}" alt="">
          </a>
        </div>
      </th:block>
    </th:block>
  </div>
  <div class="swiper-pagination"></div>
  <div class="swiper-button-prev"></div>
  <div class="swiper-button-next"></div>
</div>
```

21.3.5 轮播效果的实现

在所有的轮播图片已经渲染到页面中后，在 static/mall/js 目录中新 index.js 并新增 Swiper 插件的初始化方法，代码如下所示：

```
var newbeeSwiper = new Swiper('.swiper-container', {
    //设置自动播放
    autoplay: {
        delay: 2000,
        disableOnInteraction: false
    },
    //设置无限循环播放
    loop: true,
    //设置圆点指示器
    pagination: {
        el: '.swiper-pagination',
    },
    //设置上下页按钮
    navigation: {
        nextEl: '.swiper-button-next',
        prevEl: '.swiper-button-prev',
    }
})
```

通过代码可以看出，该初始化方法首先将对 class 为 swiper-container 的 DOM 对象进行轮播效果的初始化，然后设置自动播放并将设置间隔时间为 2 秒，同时设置手动轮播的按钮。至此，轮播功能的实现完成。

21.4 首页分类效果的制作

商品分类模块的相关内容已经在前文详细讲解过，这里将介绍商品的三级分类在首页的数据读取和联动效果的实现。在后台管理系统中已经配置了很多不同商品的分类数据，也为商品数据与分类数据建立了关联关系，所以这里首先需要读取数据并放入 request 对象中，然后使用 th 语法将这些数据渲染到页面上。

21.4.1 首页商品分类数据的读取

修改 IndexController 中的 indexPage()方法,并注入 NewBeeMallCategoryService 对象,新增代码如下所示:

```
List<NewBeeMallIndexCategoryVO> categories = newBeeMallCategoryService.getCategoriesForIndex();

request.setAttribute("categories", categories);//分类数据
```

分别读取三个级别的商品分类数据,并将这些数据设置到 request 对象中。

NewBeeMallCategoryService 业务类的 getCategoriesForIndex()方法实现代码如下所示:

```
/**
 * 返回分类数据(首页调用)
 *
 * @return
 */
public List<NewBeeMallIndexCategoryVO> getCategoriesForIndex() {
    List<NewBeeMallIndexCategoryVO> newBeeMallIndexCategoryVOS = new ArrayList<>();
    //获取一级分类的固定数量的数据
    List<GoodsCategory> firstLevelCategories = goodsCategoryMapper.selectByLevelAndParentIdsAndNumber(Collections.singletonList(0L), NewBeeMallCategoryLevelEnum.LEVEL_ONE.getLevel(), Constants.INDEX_CATEGORY_NUMBER);
    if (!CollectionUtils.isEmpty(firstLevelCategories)) {
        List<Long> firstLevelCategoryIds = firstLevelCategories.stream().map(GoodsCategory::getCategoryId).collect(Collectors.toList());
        //获取二级分类的数据
        List<GoodsCategory> secondLevelCategories = goodsCategoryMapper.selectByLevelAndParentIdsAndNumber(firstLevelCategoryIds, NewBeeMallCategoryLevelEnum.LEVEL_TWO.getLevel(), 0);
        if (!CollectionUtils.isEmpty(secondLevelCategories)) {
            List<Long> secondLevelCategoryIds = secondLevelCategories.stream().map(GoodsCategory::getCategoryId).collect(Collectors.toList());
            //获取三级分类的数据
            List<GoodsCategory> thirdLevelCategories = goodsCategoryMapper.selectByLevelAndParentIdsAndNumber(secondLevelCategoryIds, NewBeeMallCategoryLevelEnum.LEVEL_THREE.getLevel(), 0);
            if (!CollectionUtils.isEmpty(thirdLevelCategories)) {
                //根据parentId将thirdLevelCategories分组
```

```java
        Map<Long, List<GoodsCategory>> thirdLevelCategoryMap = thirdLevelCategories.stream().collect(groupingBy(GoodsCategory::getParentId));
        List<SecondLevelCategoryVO> secondLevelCategoryVOS = new ArrayList<>();
        //处理二级分类
        for (GoodsCategory secondLevelCategory : secondLevelCategories) {
            SecondLevelCategoryVO secondLevelCategoryVO = new SecondLevelCategoryVO();
            BeanUtil.copyProperties(secondLevelCategory, secondLevelCategoryVO);
            //如果该二级分类下有数据则放入 secondLevelCategoryVOS 对象中
            if (thirdLevelCategoryMap.containsKey(secondLevelCategory.getCategoryId())) {
                //根据二级分类的 id 取出 thirdLevelCategoryMap 分组中的三级分类 list
                List<GoodsCategory> tempGoodsCategories = thirdLevelCategoryMap.get(secondLevelCategory.getCategoryId());
                secondLevelCategoryVO.setThirdLevelCategoryVOS((BeanUtil.copyList(tempGoodsCategories, ThirdLevelCategoryVO.class)));
                secondLevelCategoryVOS.add(secondLevelCategoryVO);
            }
        }
        //处理一级分类
        if (!CollectionUtils.isEmpty(secondLevelCategoryVOS)) {
            //根据 parentId 将 thirdLevelCategories 分组
            Map<Long, List<SecondLevelCategoryVO>> secondLevelCategoryVOMap = secondLevelCategoryVOS.stream().collect(groupingBy(SecondLevelCategoryVO::getParentId));
            for (GoodsCategory firstCategory : firstLevelCategories) {
                NewBeeMallIndexCategoryVO newBeeMallIndexCategoryVO = new NewBeeMallIndexCategoryVO();
                BeanUtil.copyProperties(firstCategory, newBeeMallIndexCategoryVO);
                //如果该一级分类下有数据则放入 newBeeMallIndexCategoryVOS 对象中
                if (secondLevelCategoryVOMap.containsKey(firstCategory.getCategoryId())) {
                    //根据一级分类的 id 取出 secondLevelCategoryVOMap 分组中的二级级分类 list
                    List<SecondLevelCategoryVO> tempGoodsCategories = secondLevelCategoryVOMap.get(firstCategory.getCategoryId());
                    newBeeMallIndexCategoryVO.setSecondLevelCategoryVOS(tempGoodsCategories);
                    newBeeMallIndexCategoryVOS.add(newBeeMallIndexCategoryVO);
                }
            }
        }
    }
    return newBeeMallIndexCategoryVOS;
```

```
    } else {
        return null;
    }
}
```

该方法的作用是返回后台设置的分类数据，供首页数据进行渲染。

首先读取固定数量的一级分类数据，然后获取对应的二级分类并设置到一级分类下，最后获取二级分类下的三级分类数据。新蜂商城首页的分类联动效果实现方式是将所有三级分类数据读取并渲染到页面上的，并通过 JS 代码来实现二级分类和三级分类数据的显示。因此，这里将所有的分类数据都读取出来并封装成一个对象返回给视图层。

21.4.2 首页商品分类数据的渲染

修改前端模板的代码，首先使用 Thymeleaf 语法读取 newBeeMallIndexCategoryVOS 商品分类对象，然后使用三层 th:each 循环将一级分类数据、二级分类数据、三级分类数据分别渲染到页面上，代码如下所示：

```html
<div class="all-sort-list">
  <th:block th:each="category : ${categories}">
    <div class="item">
      <h3><span> </span><a href="##"><th:block th:text="${category.categoryName}"></th:block></a></h3>
      <div class="item-list clearfix">
        <div class="subitem">
          <th:block th:each="secondLevelCategory : ${category.secondLevelCategoryVOS}">
            <dl class="fore1">
              <dt><a href="#"><th:block th:text="${secondLevelCategory.categoryName}"></th:block></a></dt>
              <dd>
                <th:block th:each="thirdLevelCategory : ${secondLevelCategory.thirdLevelCategoryVOS}">
                  <em><a href="#" th:href="@{'/search?goodsCategoryId='+${thirdLevelCategory.categoryId}}">
                    <th:block th:text="${thirdLevelCategory.categoryName}"></th:block>
                  </a></em>
```

```html
                </th:block>
              </dd>
            </dl>
          </th:block>
        </div>
      </div>
    </div>
  </th:block>
</div>
```

21.4.3　首页商品分类联动效果的实现

这里通过 JS 代码实现三级分类的联动效果。监听 hover 事件和窗口的滚动距离对应显示不同的分类数据。在 index.js 添加如下代码：

```javascript
$('.all-sort-list > .item').hover(function () {
    var eq = $('.all-sort-list > .item').index(this),   //获取当前滑过的元素
        h = $('.all-sort-list').offset().top,           //获取当前下拉菜单距离窗口的像素
        s = $(window).scrollTop(),                      //获取游览器滚动的高度
        i = $(this).offset().top,                       //当前滑过元素距离窗口的像素
        item = $(this).children('.item-list').height(), //下拉菜单子类内容容器的高度
        sort = $('.all-sort-list').height();            //父类分类列表容器的高度

    if (item < sort) {                                  //如果子类的高度小于父类的高度
        if (eq == 0) {
            $(this).children('.item-list').css('top', (i - h));
        } else {
            $(this).children('.item-list').css('top', (i - h) + 1);
        }
    } else {
        if (s > h) {   //判断子类的显示位置，如果滚动的高度大于所有分类列表容器的高度
            if (i - s > 0) { //则继续判断当前滑过容器的位置是否有一半超出窗口，一半在窗口内显示的Bug
                $(this).children('.item-list').css('top', (s - h) + 2);
            } else {
                $(this).children('.item-list').css('top', (s - h) - (-(i - s))
```

```
+ 2);
            }
        } else {
            $(this).children('.item-list').css('top', 3);
        }
    }

    $(this).addClass('hover');
    $(this).children('.item-list').css('display', 'block');
}, function () {
    $(this).removeClass('hover');
    $(this).children('.item-list').css('display', 'none');
});
```

通过以上代码就实现了新蜂商城首页三级分类的联动和轮播图功能。

在编码完成后，启动 Spring Boot 项目。在启动成功后打开浏览器并输入商城端首页访问地址：

```
http://localhost:8080
```

实现效果如图 21-6 和图 21-7 所示。此时能在页面上看到所有的一级分类数据。其实对应的二级分类数据和三级分类数据也已经加载了，只是需要手动将鼠标的指针放到一级分类上，它下面的二级分类和三级分类才会显示在浮层中。在三级分类上点击对应分类就可以跳转到该分类的商品列表页面了。该功能会在搜索功能中实现。

图 21-6　首页分类效果和轮播图效果制作 1

图 21-7　首页分类效果和轮播图效果制作 2

21.5　商城首页推荐商品模块的介绍

商城首页中还有三个版块需要进行数据渲染，分别是热销商品、新品上线和推荐商品。它们在首页中的布局如图 21-8 所示。

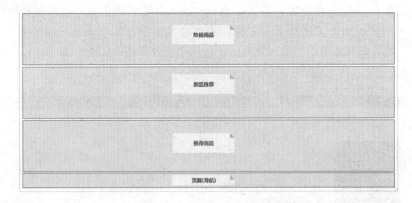

图 21-8　热销商品、新品推荐、推荐商品的布局

在商城首页中设计这三个版块，主要是为了丰富版面布局，使页面不单调。

当然，这部分的设计也参考了当前主流线上商城的商品推荐设计，不过这些线上商城都有大量的正式数据做支撑，做的肯定要比新蜂商城复杂得多，比如热销商品，一定是在大量实际订单的统计下做出来的数据渲染，又比如商品推荐，也一定是在用户的浏览痕迹和下单习惯上计算出来的。目前来说，新蜂商城的开发人员只有笔者一个人，订单也只有模拟数据，如果要做到淘宝、京东那种效果是不现实的，因此新蜂商城中的热销商品、新品上线、推荐商品这三个版块中的数据是在后台中配置的，首页渲染前直接读取数据就可以，这些数据并没有进行实时的数据统计。

由于新蜂商城只是技术实战项目，所以其订单、浏览痕迹和用户习惯等数据都是模拟的，并非实时统计的。这三个版块中的内容是在后台进行配置的，首页配置管理页面如图 21-9 所示。

图 21-9　首页配置管理页面

接下来会介绍具体的实现方式，包括在后台管理系统中首页配置管理模块的开发，以及这些数据在首页的读取和渲染。

21.6　首页配置管理页面的制作

21.6.1　导航栏中增加首页配置相关栏目

首先在左侧导航栏中增加热销商品、新品上线等管理模块的导航按钮。在 sidebar.html 文件中新增如下代码：

```html
<li class="nav-item">
  <a th:href="@{/admin/indexConfigs?configType=3}"
     th:class="${path}=='INDEX_GOODS_HOTS'?'nav-link active':'nav-link'">
    <i class="nav-icon fa fa-hand-o-up"></i>
    <p>
```

```html
        热销商品配置
      </p>
    </a>
</li>
<li class="nav-item">
  <a th:href="@{/admin/indexConfigs?configType=4}"
     th:class="${path}=='INDEX_GOODS_NEW'?'nav-link active':'nav-link'">
    <i class="nav-icon fa fa-hand-o-up"></i>
    <p>
        新品上线配置
      </p>
    </a>
</li>
<li class="nav-item">
  <a th:href="@{/admin/indexConfigs?configType=5}"
     th:class="${path}=='INDEX_GOODS_RECOMMOND'?'nav-link active':'nav-link'">
    <i class="nav-icon fa fa-hand-o-up"></i>
    <p>
        为你推荐配置
      </p>
    </a>
</li>
```

这里的跳转路径为/admin/indexConfigs，链接中包含 configType 参数。该参数主要用于区分首页配置的类型。这种链接设计主要是为了区分不同的内容，比如这里的 configType 参数，不同的值就代表着不同的配置项。最后新建 Controller 控制类来处理该路径并跳转到对应的页面。

21.6.2 控制类处理跳转逻辑

在 controller/admin 包下新建 NewBeeMallGoodsIndexConfigController.java，新增如下代码：

```java
package ltd.newbee.mall.controller.admin;

import org.springframework.stereotype.Controller;
import org.springframework.web.bind.annotation.*;
import javax.servlet.http.HttpServletRequest;
```

```
@Controller
@RequestMapping("/admin")
public class NewBeeMallGoodsIndexConfigController {

    @GetMapping("/indexConfigs")
    public String indexConfigsPage(HttpServletRequest request, @RequestParam("configType") int configType) {
        request.setAttribute("path", indexConfigTypeEnum.getName());
        request.setAttribute("configType", configType);
        return "admin/newbee_mall_index_config";
    }
}
```

该方法用于处理/admin/indexConfigs 请求，在返回视图前首先设置 path 字段和 configType 字段，然后跳转到 admin 目录下的 newbee_mall_index_config.html 模板页面中。

21.6.3 首页配置商品管理页面基础样式的实现

接下来就是首页配置管理页面的模板文件制作。在 resources/templates/admin 目录下新建 newbee_mall_index_config.html 模板文件，并引入对应的 JS 文件和 CSS 样式文件，代码如下所示：

```html
<!DOCTYPE html>
<html xmlns:th="http://www.thymeleaf.org">
<header th:replace="admin/header::header-fragment">
</header>
<style>
    .ui-jqgrid tr.jqgrow td {
        white-space: normal !important;
        height: auto;
        vertical-align: text-top;
        padding-top: 2px;
    }
</style>
<body class="hold-transition sidebar-mini">
<div class="wrapper">
    <!-- 引入页面头 header-fragment -->
    <div th:replace="admin/header::header-nav"></div>
    <!-- 引入工具栏 sidebar-fragment -->
```

```html
    <div th:replace="admin/sidebar::sidebar-fragment(${path})"></div>
    <!-- Content Wrapper. Contains 图标 content -->
    <div class="content-wrapper">
        <!-- Content Header (图标 header) -->
        <div class="content-header">
            <div class="container-fluid">
            </div><!-- /.container-fluid -->
        </div>
        <!-- Main content -->
        <div class="content">
            <div class="container-fluid">
                <div class="card card-primary card-outline">
                    <div class="card-header">
                        <h3 class="card-title">首页配置管理</h3>
                    </div> <!-- /.card-body -->
                    <div class="card-body">
                        <div class="grid-btn">
                            <button class="btn btn-info" onclick="configAdd()"><i
                                    class="fa fa-plus"></i> 新增
                            </button>
                            <button class="btn btn-info" onclick="configEdit()"><i
                                    class="fa fa-pencil-square-o"></i> 修改
                            </button>
                            <button class="btn btn-danger" onclick="deleteConfig()"><i
                                    class="fa fa-trash-o"></i> 删除
                            </button>
                        </div>
                    </div>
                </div>
            </div>
        </div>
    </div>
    <div th:replace="admin/footer::footer-fragment"></div>
</div>
<!-- jQuery -->
<script th:src="@{/admin/plugins/jquery/jquery.min.js}"></script>
<!-- jQuery UI 1.11.4 -->
<script th:src="@{/admin/plugins/jQueryUI/jquery-ui.min.js}"></script>
<!-- Bootstrap 4 -->
<script th:src="@{/admin/plugins/bootstrap/js/bootstrap.bundle.min.js}">
```

```html
</script>
<!-- AdminLTE App -->
<script th:src="@{/admin/dist/js/adminlte.min.js}"></script>
<!-- jqgrid -->
<script
th:src="@{/admin/plugins/jqgrid-5.3.0/jquery.jqGrid.min.js}"></script>
<script
th:src="@{/admin/plugins/jqgrid-5.3.0/grid.locale-cn.js}"></script>

<!-- sweetalert -->
<script th:src="@{/admin/plugins/sweetalert/sweetalert.min.js}"></script>
<script th:src="@{/admin/dist/js/public.js}"></script>
<script th:src="@{/admin/dist/js/newbee_mall_index_config.js}"></script>
</body>
</html>
```

通过以上代码首页配置管理页面的基础内容和页面的跳转逻辑就处理完毕。不同配置项中的 configType 参数值不同，首页配置管理基础页面如图 21-10 所示。

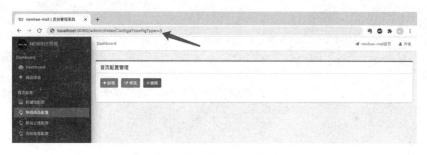

图 21-10　首页配置管理基础页面

此时只完成了管理页面的基础内容。页面上的三个功能按钮还是静态内容，并没有进行数据的交互。

21.7　首页配置管理模块接口的设计及实现

21.7.1　首页配置表结构的设计

商城首页展示内容的主要字段包括配置项的类别和配置项中的商品属性。这里笔者设计了一个商品 id 字段，以此来建立配置项与商品的关系。在实现查询功能时，就可以先查出配置项的数据，然后根据配置项中的商品 id 字段查找对应的商品。其他字段是一些基础的功能字段。首页配置表结构如下所示：

```sql
USE 'newbee_mall_db ';

DROP TABLE IF EXISTS 'tb_newbee_mall_index_config';
CREATE TABLE 'tb_newbee_mall_index_config' (
  'config_id' bigint(20) NOT NULL AUTO_INCREMENT COMMENT '首页配置项主键id',
  'config_name' varchar(50) CHARACTER SET utf8 COLLATE utf8_general_ci NOT NULL DEFAULT '' COMMENT '显示字符(配置搜索时不可为空，其他可为空)',
  'config_type' tinyint(4) NOT NULL DEFAULT 0 COMMENT '1-搜索框热搜 2-搜索下拉框热搜 3-(首页)热销商品 4-(首页)新品上线 5-(首页)为你推荐',
  'goods_id' bigint(20) NOT NULL DEFAULT 0 COMMENT '商品id 默认为0',
  'redirect_url' varchar(100) CHARACTER SET utf8 COLLATE utf8_general_ci NOT NULL DEFAULT '##' COMMENT '点击后的跳转地址(默认不跳转)',
  'config_rank' int(11) NOT NULL DEFAULT 0 COMMENT '排序值(字段越大越靠前)',
  'is_deleted' tinyint(4) NOT NULL DEFAULT 0 COMMENT '删除标识字段(0-未删除 1-已删除)',
  'create_time' datetime(0) NOT NULL DEFAULT CURRENT_TIMESTAMP COMMENT '创建时间',
  'create_user' int(11) NOT NULL DEFAULT 0 COMMENT '创建者id',
  'update_time' datetime(0) NOT NULL DEFAULT CURRENT_TIMESTAMP COMMENT '最新修改时间',
  'update_user' int(11) NULL DEFAULT 0 COMMENT '修改者id',
  PRIMARY KEY ('config_id') USING BTREE
) ENGINE = InnoDB CHARACTER SET = utf8 COLLATE = utf8_general_ci ROW_FORMAT = Dynamic;
```

每个字段对应的含义都在上面的 SQL 中有介绍，读者可以对照理解，正确地把建表 SQL 导入数据库中。接下来进行编码工作。

21.7.2　新建首页配置实体类和 Mapper 接口

首先在 ltd.newbee.mall.entity 包中创建首页配置实体类，选中 entity 包并右键点击，在弹出的菜单中选择"New→Java Class"，然后在弹出的窗口中输入"IndexConfig"，最后在 IndexConfig 类中新增如下代码：

```java
package ltd.newbee.mall.entity;

import com.fasterxml.jackson.annotation.JsonFormat;
```

```java
import java.util.Date;

public class IndexConfig {
    private Long configId;

    private String configName;

    private Byte configType;

    private Long goodsId;

    private String redirectUrl;

    private Integer configRank;

    private Byte isDeleted;

    @JsonFormat(pattern = "yyyy-MM-dd HH:mm:ss", timezone = "GMT+8")
    private Date createTime;

    private Integer createUser;

    @JsonFormat(pattern = "yyyy-MM-dd HH:mm:ss", timezone = "GMT+8")
    private Date updateTime;

    private Integer updateUser;

    public Long getConfigId() {
        return configId;
    }

    public void setConfigId(Long configId) {
        this.configId = configId;
    }

    public String getConfigName() {
        return configName;
    }

    public void setConfigName(String configName) {
        this.configName = configName == null ? null : configName.trim();
    }
```

```java
public Byte getConfigType() {
    return configType;
}

public void setConfigType(Byte configType) {
    this.configType = configType;
}

public Long getGoodsId() {
    return goodsId;
}

public void setGoodsId(Long goodsId) {
    this.goodsId = goodsId;
}

public String getRedirectUrl() {
    return redirectUrl;
}

public void setRedirectUrl(String redirectUrl) {
    this.redirectUrl = redirectUrl == null ? null : redirectUrl.trim();
}

public Integer getConfigRank() {
    return configRank;
}

public void setConfigRank(Integer configRank) {
    this.configRank = configRank;
}

public Byte getIsDeleted() {
    return isDeleted;
}

public void setIsDeleted(Byte isDeleted) {
    this.isDeleted = isDeleted;
}

public Date getCreateTime() {
    return createTime;
```

```java
    public void setCreateTime(Date createTime) {
        this.createTime = createTime;
    }

    public Integer getCreateUser() {
        return createUser;
    }

    public void setCreateUser(Integer createUser) {
        this.createUser = createUser;
    }

    public Date getUpdateTime() {
        return updateTime;
    }

    public void setUpdateTime(Date updateTime) {
        this.updateTime = updateTime;
    }

    public Integer getUpdateUser() {
        return updateUser;
    }

    public void setUpdateUser(Integer updateUser) {
        this.updateUser = updateUser;
    }
}
```

首先在 ltd.newbee.mall.dao 包中新建首页配置实体的 Mapper 接口，选中 dao 包并右键点击，在弹出的菜单中选择"New→Java Class"，然后在弹出的窗口中输入"IndexConfigMapper"，并选中"Interface"选项，最后在 IndexConfigMapper.java 文件中新增如下代码：

```java
package ltd.newbee.mall.dao;

import ltd.newbee.mall.entity.IndexConfig;
import ltd.newbee.mall.util.PageQueryUtil;
import org.apache.ibatis.annotations.Param;

import java.util.List;
```

```java
public interface IndexConfigMapper {

    /**
     * 删除一条记录
     * @param configId
     * @return
     */
    int deleteByPrimaryKey(Long configId);

    /**
     * 保存一条新记录
     * @param record
     * @return
     */
    int insert(IndexConfig record);

    /**
     * 保存一条新记录
     * @param record
     * @return
     */
    int insertSelective(IndexConfig record);

    /**
     * 根据主键查询记录
     * @param configId
     * @return
     */
    IndexConfig selectByPrimaryKey(Long configId);

    /**
     * 修改记录
     * @param record
     * @return
     */
    int updateByPrimaryKeySelective(IndexConfig record);

    /**
     * 修改记录
     * @param record
     * @return
     */
```

```java
    int updateByPrimaryKey(IndexConfig record);
    /**
     * 查询分页数据
     * @param pageUtil
     * @return
     */
    List<IndexConfig> findIndexConfigList(PageQueryUtil pageUtil);

    /**
     * 查询总数
     * @param pageUtil
     * @return
     */
    int getTotalIndexConfigs(PageQueryUtil pageUtil);

    /**
     * 批量删除
     * @param ids
     * @return
     */
    int deleteBatch(Long[] ids);

    /**
     * 根据配置类型查询固定数量的记录
     * @param configType
     * @param number
     * @return
     */
    List<IndexConfig> findIndexConfigsByTypeAndNum(@Param("configType")
int configType, @Param("number") int number);
}
```

这里定义了首页配置实体操作的数据层方法，包括查询、新增、修改和删除等操作。

21.7.3 创建 IndexConfigMapper 接口的映射文件

在 resources/mapper 目录下新建 IndexConfigMapper 接口的映射文件 IndexConfigMapper.xml，并进行映射文件的编写。

首先，定义映射文件与 Mapper 接口的对应关系。比如在该示例中，需要将将 IndexConfigMapper.xml 文件与对应的 IndexConfigMapper 接口之间的关系定义出来：

```xml
<mapper namespace="ltd.newbee.mall.dao.IndexConfigMapper">
```

然后，配置表结构和实体类的对应关系：

```xml
<resultMap id="BaseResultMap" type="ltd.newbee.mall.entity.IndexConfig">
  <id column="config_id" jdbcType="BIGINT" property="configId"/>
  <result column="config_name" jdbcType="VARCHAR" property="configName"/>
  <result column="config_type" jdbcType="TINYINT" property="configType"/>
  <result column="goods_id" jdbcType="BIGINT" property="goodsId"/>
  <result column="redirect_url" jdbcType="VARCHAR" property="redirectUrl"/>
  <result column="config_rank" jdbcType="INTEGER" property="configRank"/>
  <result column="is_deleted" jdbcType="TINYINT" property="isDeleted"/>
  <result column="create_time" jdbcType="TIMESTAMP" property="createTime"/>
  <result column="create_user" jdbcType="INTEGER" property="createUser"/>
  <result column="update_time" jdbcType="TIMESTAMP" property="updateTime"/>
  <result column="update_user" jdbcType="INTEGER" property="updateUser"/>
</resultMap>
```

最后，按照对应的接口方法，编写具体的 SQL 语句，最终的 IndexConfigMapper.xml 文件如下所示：

```xml
<?xml version="1.0" encoding="UTF-8"?>
<!DOCTYPE mapper PUBLIC "-//mybatis.org//DTD Mapper 3.0//EN" "http://mybatis.org/dtd/mybatis-3-mapper.dtd">
<mapper namespace="ltd.newbee.mall.dao.IndexConfigMapper">
    <resultMap id="BaseResultMap" type="ltd.newbee.mall.entity.IndexConfig">
        <id column="config_id" jdbcType="BIGINT" property="configId"/>
        <result column="config_name" jdbcType="VARCHAR" property="configName"/>
        <result column="config_type" jdbcType="TINYINT" property="configType"/>
        <result column="goods_id" jdbcType="BIGINT" property="goodsId"/>
        <result column="redirect_url" jdbcType="VARCHAR" property="redirectUrl"/>
        <result column="config_rank" jdbcType="INTEGER" property="configRank"/>
        <result column="is_deleted" jdbcType="TINYINT" property="isDeleted"/>
        <result column="create_time" jdbcType="TIMESTAMP" property="createTime"/>
        <result column="create_user" jdbcType="INTEGER" property="createUser"/>
        <result column="update_time" jdbcType="TIMESTAMP" property="update
```

```xml
Time"/>
    <result column="update_user" jdbcType="INTEGER" property="updateUser"/>
  </resultMap>
  <sql id="Base_Column_List">
    config_id, config_name, config_type, goods_id, redirect_url, config_rank, is_deleted,
    create_time, create_user, update_time, update_user
  </sql>

  <select id="findIndexConfigList" parameterType="Map" resultMap="BaseResultMap">
    select
    <include refid="Base_Column_List"/>
    from tb_newbee_mall_index_config
    <where>
        <if test="configType!=null and configType!=''">
            and config_type = #{configType}
        </if>
        and is_deleted = 0
    </where>
    order by config_rank desc
    <if test="start!=null and limit!=null">
        limit #{start},#{limit}
    </if>
  </select>
  <select id="getTotalIndexConfigs" parameterType="Map" resultType="int">
    select count(*) from tb_newbee_mall_index_config
    <where>
        <if test="configType!=null and configType!=''">
            and config_type = #{configType}
        </if>
        and is_deleted = 0
    </where>
  </select>
  <select id="findIndexConfigsByTypeAndNum" resultMap="BaseResultMap">
    select
    <include refid="Base_Column_List"/>
    from tb_newbee_mall_index_config
    where config_type = #{configType} and is_deleted = 0
    order by config_rank desc
    limit #{number}
  </select>
```

```xml
    <select id="selectByPrimaryKey" parameterType="java.lang.Long" resultMap="BaseResultMap">
        select
        <include refid="Base_Column_List"/>
        from tb_newbee_mall_index_config
        where config_id = #{configId,jdbcType=BIGINT} and is_deleted=0
    </select>
    <update id="deleteByPrimaryKey" parameterType="java.lang.Long">
        update tb_newbee_mall_index_config set is_deleted=1
        where config_id = #{configId,jdbcType=BIGINT} and is_deleted=0
    </update>
    <update id="deleteBatch">
        update tb_newbee_mall_index_config
        set is_deleted=1,update_time=now() where is_deleted=0 and config_id in
        <foreach item="id" collection="array" open="(" separator="," close=")">
            #{id}
        </foreach>
    </update>
    <insert id="insert" parameterType="ltd.newbee.mall.entity.IndexConfig">
        insert into tb_newbee_mall_index_config (config_id, config_name, config_type,
            goods_id, redirect_url, config_rank,
            is_deleted, create_time, create_user,
            update_time, update_user)
        values (#{configId,jdbcType=BIGINT}, #{configName,jdbcType=VARCHAR}, #{configType,jdbcType=TINYINT},
            #{goodsId,jdbcType=BIGINT}, #{redirectUrl,jdbcType=VARCHAR}, #{configRank,jdbcType=INTEGER},
            #{isDeleted,jdbcType=TINYINT}, #{createTime,jdbcType=TIMESTAMP}, #{createUser,jdbcType=INTEGER},
            #{updateTime,jdbcType=TIMESTAMP}, #{updateUser,jdbcType=INTEGER})
    </insert>
    <insert id="insertSelective" parameterType="ltd.newbee.mall.entity.IndexConfig">
        insert into tb_newbee_mall_index_config
        <trim prefix="(" suffix=")" suffixOverrides=",">
            <if test="configId != null">
                config_id,
            </if>
            <if test="configName != null">
                config_name,
```

```xml
        </if>
        <if test="configType != null">
            config_type,
        </if>
        <if test="goodsId != null">
            goods_id,
        </if>
        <if test="redirectUrl != null">
            redirect_url,
        </if>
        <if test="configRank != null">
            config_rank,
        </if>
        <if test="isDeleted != null">
            is_deleted,
        </if>
        <if test="createTime != null">
            create_time,
        </if>
        <if test="createUser != null">
            create_user,
        </if>
        <if test="updateTime != null">
            update_time,
        </if>
        <if test="updateUser != null">
            update_user,
        </if>
    </trim>
    <trim prefix="values (" suffix=")" suffixOverrides=",">
        <if test="configId != null">
            #{configId,jdbcType=BIGINT},
        </if>
        <if test="configName != null">
            #{configName,jdbcType=VARCHAR},
        </if>
        <if test="configType != null">
            #{configType,jdbcType=TINYINT},
        </if>
        <if test="goodsId != null">
            #{goodsId,jdbcType=BIGINT},
        </if>
        <if test="redirectUrl != null">
```

```xml
                #{redirectUrl,jdbcType=VARCHAR},
            </if>
            <if test="configRank != null">
                #{configRank,jdbcType=INTEGER},
            </if>
            <if test="isDeleted != null">
                #{isDeleted,jdbcType=TINYINT},
            </if>
            <if test="createTime != null">
                #{createTime,jdbcType=TIMESTAMP},
            </if>
            <if test="createUser != null">
                #{createUser,jdbcType=INTEGER},
            </if>
            <if test="updateTime != null">
                #{updateTime,jdbcType=TIMESTAMP},
            </if>
            <if test="updateUser != null">
                #{updateUser,jdbcType=INTEGER},
            </if>
        </trim>
    </insert>
    <update id="updateByPrimaryKeySelective" parameterType="ltd.newbee.mall.entity.IndexConfig">
        update tb_newbee_mall_index_config
        <set>
            <if test="configName != null">
                config_name = #{configName,jdbcType=VARCHAR},
            </if>
            <if test="configType != null">
                config_type = #{configType,jdbcType=TINYINT},
            </if>
            <if test="goodsId != null">
                goods_id = #{goodsId,jdbcType=BIGINT},
            </if>
            <if test="redirectUrl != null">
                redirect_url = #{redirectUrl,jdbcType=VARCHAR},
            </if>
            <if test="configRank != null">
                config_rank = #{configRank,jdbcType=INTEGER},
            </if>
            <if test="isDeleted != null">
                is_deleted = #{isDeleted,jdbcType=TINYINT},
```

```xml
            </if>
            <if test="createTime != null">
                create_time = #{createTime,jdbcType=TIMESTAMP},
            </if>
            <if test="createUser != null">
                create_user = #{createUser,jdbcType=INTEGER},
            </if>
            <if test="updateTime != null">
                update_time = #{updateTime,jdbcType=TIMESTAMP},
            </if>
            <if test="updateUser != null">
                update_user = #{updateUser,jdbcType=INTEGER},
            </if>
        </set>
        where config_id = #{configId,jdbcType=BIGINT}
    </update>
    <update id="updateByPrimaryKey" parameterType="ltd.newbee.mall.entity.IndexConfig">
        update tb_newbee_mall_index_config
        set config_name = #{configName,jdbcType=VARCHAR},
            config_type = #{configType,jdbcType=TINYINT},
            goods_id = #{goodsId,jdbcType=BIGINT},
            redirect_url = #{redirectUrl,jdbcType=VARCHAR},
            config_rank = #{configRank,jdbcType=INTEGER},
            is_deleted = #{isDeleted,jdbcType=TINYINT},
            create_time = #{createTime,jdbcType=TIMESTAMP},
            create_user = #{createUser,jdbcType=INTEGER},
            update_time = #{updateTime,jdbcType=TIMESTAMP},
            update_user = #{updateUser,jdbcType=INTEGER}
        where config_id = #{configId,jdbcType=BIGINT}
    </update>
</mapper>
```

21.7.4　业务层代码的实现

首先在 ltd.newbee.mall.service 包中新建业务处理类，选中 service 包并右键点击，在弹出的菜单中选择 "New→Java Class"，然后在弹出的窗口中输入 "NewBeeMallIndexConfigService"，并选中 "Interface" 选项，最后在 NewBeeMallIndexConfigService.java 文件中新增如下代码：

```
package ltd.newbee.mall.service;
```

```java
import ltd.newbee.mall.controller.vo.NewBeeMallIndexConfigGoodsVO;
import ltd.newbee.mall.entity.IndexConfig;
import ltd.newbee.mall.util.PageQueryUtil;
import ltd.newbee.mall.util.PageResult;

import java.util.List;

public interface NewBeeMallIndexConfigService {
    /**
     * 查询后台管理系统首页配置的分页数据
     *
     * @param pageUtil
     * @return
     */
    PageResult getConfigsPage(PageQueryUtil pageUtil);

    /**
     * 新增一条首页配置记录
     *
     * @param indexConfig
     * @return
     */
    String saveIndexConfig(IndexConfig indexConfig);

    /**
     * 修改一条首页配置记录
     *
     * @param indexConfig
     * @return
     */
    String updateIndexConfig(IndexConfig indexConfig);

    /**
     * 返回固定数量的首页配置商品对象（首页调用）
     *
     * @param number
     * @return
     */
    List<NewBeeMallIndexConfigGoodsVO> getConfigGoodsesForIndex(int configType, int number);

    /**
```

```
     * 批量删除
     *
     * @param ids
     * @return
     */
    Boolean deleteBatch(Long[] ids);
}
```

首页配置模块的业务层方法的定义和每个方法的作用都已经编写完成。

接下来，首先在 ltd.newbee.mall.service.impl 包中新建 NewBeeMallIndexConfig Service 的实现类，选中 impl 包并右键点击，在弹出的菜单中选择"New→Java Class"，然后在弹出的窗口中输入"NewBeeMallIndexConfigServiceImpl"，最后在 NewBeeMallIndex ConfigServiceImpl 类中新增如下代码：

```java
package ltd.newbee.mall.service.impl;

import ltd.newbee.mall.common.ServiceResultEnum;
import ltd.newbee.mall.controller.vo.NewBeeMallIndexConfigGoodsVO;
import ltd.newbee.mall.dao.IndexConfigMapper;
import ltd.newbee.mall.dao.NewBeeMallGoodsMapper;
import ltd.newbee.mall.entity.IndexConfig;
import ltd.newbee.mall.entity.NewBeeMallGoods;
import ltd.newbee.mall.service.NewBeeMallIndexConfigService;
import ltd.newbee.mall.util.BeanUtil;
import ltd.newbee.mall.util.PageQueryUtil;
import ltd.newbee.mall.util.PageResult;
import org.springframework.beans.factory.annotation.Autowired;
import org.springframework.stereotype.Service;
import org.springframework.util.CollectionUtils;

import java.util.ArrayList;
import java.util.List;
import java.util.stream.Collectors;

@Service
public class NewBeeMallIndexConfigServiceImpl implements NewBeeMallIndexConfigService {

    @Autowired
    private IndexConfigMapper indexConfigMapper;

    @Autowired
    private NewBeeMallGoodsMapper goodsMapper;
```

```java
    @Override
    public PageResult getConfigsPage(PageQueryUtil pageUtil) {
        List<IndexConfig> indexConfigs = indexConfigMapper.findIndexConfigList(pageUtil);
        int total = indexConfigMapper.getTotalIndexConfigs(pageUtil);
        PageResult pageResult = new PageResult(indexConfigs, total, pageUtil.getLimit(), pageUtil.getPage());
        return pageResult;
    }

    @Override
    public String saveIndexConfig(IndexConfig indexConfig) {
        if (indexConfigMapper.insertSelective(indexConfig) > 0) {
            return ServiceResultEnum.SUCCESS.getResult();
        }
        return ServiceResultEnum.DB_ERROR.getResult();
    }

    @Override
    public String updateIndexConfig(IndexConfig indexConfig) {
        IndexConfig temp = indexConfigMapper.selectByPrimaryKey(indexConfig.getConfigId());
        if (temp == null) {
            return ServiceResultEnum.DATA_NOT_EXIST.getResult();
        }
        if (indexConfigMapper.updateByPrimaryKeySelective(indexConfig) > 0) {
            return ServiceResultEnum.SUCCESS.getResult();
        }
        return ServiceResultEnum.DB_ERROR.getResult();
    }

    @Override
    public List<NewBeeMallIndexConfigGoodsVO> getConfigGoodsesForIndex(int configType, int number) {
        List<NewBeeMallIndexConfigGoodsVO> newBeeMallIndexConfigGoodsVOS = new ArrayList<>(number);
        List<IndexConfig> indexConfigs = indexConfigMapper.findIndexConfigsByTypeAndNum(configType, number);
        if (!CollectionUtils.isEmpty(indexConfigs)) {
            //取出所有的goodsId
            List<Long> goodsIds = indexConfigs.stream().map(IndexConfig::getGoodsId).collect(Collectors.toList());
```

```
            List<NewBeeMallGoods> newBeeMallGoods = goodsMapper.selectBy
PrimaryKeys(goodsIds);
            newBeeMallIndexConfigGoodsVOS = BeanUtil.copyList(newBeeMall
Goods, NewBeeMallIndexConfigGoodsVO.class);
            for (NewBeeMallIndexConfigGoodsVO newBeeMallIndexConfigGoodsVO :
newBeeMallIndexConfigGoodsVOS) {
                String goodsName = newBeeMallIndexConfigGoodsVO.getGoods
Name();
                String goodsIntro = newBeeMallIndexConfigGoodsVO.getGoods
Intro();
                // 字符串过长导致文字超出的问题
                if (goodsName.length() > 30) {
                    goodsName = goodsName.substring(0, 30) + "...";
                    newBeeMallIndexConfigGoodsVO.setGoodsName(goodsName);
                }
                if (goodsIntro.length() > 22) {
                    goodsIntro = goodsIntro.substring(0, 22) + "...";
                    newBeeMallIndexConfigGoodsVO.setGoodsIntro(goodsIntro);
                }
            }
        }
        return newBeeMallIndexConfigGoodsVOS;
    }

    @Override
    public Boolean deleteBatch(Long[] ids) {
        if (ids.length < 1) {
            return false;
        }
        //删除数据
        return indexConfigMapper.deleteBatch(ids) > 0;
    }
}
```

21.7.5 首页管理模块控制层代码的实现

在 NewBeeMallGoodsIndexConfigController 控制器中新增前文接口的实现代码，最终 NewBeeMallGoodsIndexConfigController 类的代码如下所示：

```
package ltd.newbee.mall.controller.admin;
```

```java
import ltd.newbee.mall.common.IndexConfigTypeEnum;
import ltd.newbee.mall.common.ServiceResultEnum;
import ltd.newbee.mall.entity.GoodsCategory;
import ltd.newbee.mall.entity.IndexConfig;
import ltd.newbee.mall.service.NewBeeMallCategoryService;
import ltd.newbee.mall.service.NewBeeMallIndexConfigService;
import ltd.newbee.mall.util.PageQueryUtil;
import ltd.newbee.mall.util.Result;
import ltd.newbee.mall.util.ResultGenerator;
import org.springframework.stereotype.Controller;
import org.springframework.util.StringUtils;
import org.springframework.web.bind.annotation.*;

import javax.annotation.Resource;
import javax.servlet.http.HttpServletRequest;
import java.util.Map;
import java.util.Objects;

@Controller
@RequestMapping("/admin")
public class NewBeeMallGoodsIndexConfigController {

    @Resource
    private NewBeeMallIndexConfigService newBeeMallIndexConfigService;

    @GetMapping("/indexConfigs")
    public String indexConfigsPage(HttpServletRequest request, @RequestParam("configType") int configType) {
        IndexConfigTypeEnum indexConfigTypeEnum = IndexConfigTypeEnum.getIndexConfigTypeEnumByType(configType);
        if (indexConfigTypeEnum.equals(IndexConfigTypeEnum.DEFAULT)) {
            return "error/error_5xx";
        }
        request.setAttribute("path", indexConfigTypeEnum.getName());
        request.setAttribute("configType", configType);
        return "admin/newbee_mall_index_config";
    }

    /**
     * 列表
     */
    @RequestMapping(value = "/indexConfigs/list", method = RequestMethod.GET)
```

```java
    @ResponseBody
    public Result list(@RequestParam Map<String, Object> params) {
        if (StringUtils.isEmpty(params.get("page")) || StringUtils.isEmpty(params.get("limit"))) {
            return ResultGenerator.genFailResult("参数异常!");
        }
        PageQueryUtil pageUtil = new PageQueryUtil(params);
        return ResultGenerator.genSuccessResult(newBeeMallIndexConfigService.getConfigsPage(pageUtil));
    }

    /**
     * 添加
     */
    @RequestMapping(value = "/indexConfigs/save", method = RequestMethod.POST)
    @ResponseBody
    public Result save(@RequestBody IndexConfig indexConfig) {
        if (Objects.isNull(indexConfig.getConfigType())
                || StringUtils.isEmpty(indexConfig.getConfigName())
                || Objects.isNull(indexConfig.getConfigRank())) {
            return ResultGenerator.genFailResult("参数异常!");
        }
        String result = newBeeMallIndexConfigService.saveIndexConfig(indexConfig);
        if (ServiceResultEnum.SUCCESS.getResult().equals(result)) {
            return ResultGenerator.genSuccessResult();
        } else {
            return ResultGenerator.genFailResult(result);
        }
    }

    /**
     * 修改
     */
    @RequestMapping(value = "/indexConfigs/update", method = RequestMethod.POST)
    @ResponseBody
    public Result update(@RequestBody IndexConfig indexConfig) {
        if (Objects.isNull(indexConfig.getConfigType())
                || Objects.isNull(indexConfig.getConfigId())
                || StringUtils.isEmpty(indexConfig.getConfigName())
```

```java
            || Objects.isNull(indexConfig.getConfigRank())) {
        return ResultGenerator.genFailResult("参数异常！");
    }
    String result = newBeeMallIndexConfigService.updateIndexConfig(indexConfig);
    if (ServiceResultEnum.SUCCESS.getResult().equals(result)) {
        return ResultGenerator.genSuccessResult();
    } else {
        return ResultGenerator.genFailResult(result);
    }
}

/**
 * 删除
 */
@RequestMapping(value = "/indexConfigs/delete", method = RequestMethod.POST)
@ResponseBody
public Result delete(@RequestBody Long[] ids) {
    if (ids.length < 1) {
        return ResultGenerator.genFailResult("参数异常！");
    }
    if (newBeeMallIndexConfigService.deleteBatch(ids)) {
        return ResultGenerator.genSuccessResult();
    } else {
        return ResultGenerator.genFailResult("删除失败");
    }
}
}
```

列表接口负责接收前端传来的分页参数（比如 page、limit 等），并将查询数据总数和对应页面的数据列表并封装为分页数据返回给前端。其中一个重要的参数是 configType，这个参数能够区分不同类型的首页配置项。实际的查询 SQL 语句在 IndexConfigMapper.xml 文件中。除了分页参数的过滤之外，针对 config_type 字段系统也进行了过滤。前端请求的 configType 参数不同，执行的 SQL 也会对应查询不同类型的分页记录，并获取响应条数的记录和总数并进行数据封装。列表接口是根据前端传递的分页参数进行查询并返回分页数据以供前端页面进行数据渲染的。

添加接口负责接收前端的 POST 请求并处理其中的参数。接收的参数为 configName 字段、configType 字段、redirectUrl 字段、goodsId 字段和 configRank 字段。在里使用 @RequestBody 注解将这些字段转换为 IndexConfig 对象。需要注意的一点是，这里不会存储和设置任何关于商品表的其他字段，也不会对商品表进行操作，只是通过设置 goodsId 字段将配置项和对应的商品记录之间的关联关系存储下来。

删除接口负责接收前端的首页删除请求，在处理前端传输过来的数据后，它将这些记录从数据库中删除。这里的删除功能并不是真正意义上的删除，而是逻辑删除。将接受的参数设置为一个数组，可以同时删除多条记录。只需要在前端将用户选择的记录 id 封装好再传参到后端即可。接口的请求路径为 /admin/indexConfigs/delete，并使用 @RequestBody 注解将前端传过来的参数封装为数组对象。如果数组为空则直接返回异常提醒。在参数验证通过后则调用 deleteBatch() 批量删除方法进行数据库操作。

21.8 首页配置管理模块前端功能的实现

21.8.1 功能按钮和分页信息展示区域

首页配置管理模块包括首页配置列表、首页配置信息增加、首页配置信息编辑和首页配置信息删除。

其中，列表功能可以在页面加载时使用 jqGrid 触发实现，另外三个功能则需要在页面中设置功能按钮并设置触发事件。在 21.6 节页面制作时已经设置了这三个功能按钮，分别是添加按钮，对应的触发事件是 configAdd() 方法；修改按钮，对应的触发事件是 configEdit() 方法；删除按钮，对应的触发事件是 deleteConfig() 方法。

关于分页功能的实现，前文已经做了详细的介绍，这里可以直接在页面中引入 jqGrid 的相关静态资源文件，并在页面中展示分页数据的区域增加如下代码：

```
<table id="jqGrid" class="table table-bordered"></table>
<div id="jqGridPager"></div>
```

此时还没有与后端进行数据交互。接下来将结合 Ajax 和后端接口实现具体的功能。

21.8.2 首页配置管理页面分页功能的实现

在 resources/static/admin/dist/js 目录下新增 newbee_mall_index_config.js 文件，并添加如下代码：

```
$(function () {
```

```javascript
$("#jqGrid").jqGrid({
    url: '/admin/carousels/list',
    datatype: "json",
    colModel: [
        {label: 'id', name: 'carouselId', index: 'carouselId', width: 50, key: true, hidden: true},
        {label: '轮播图', name: 'carouselUrl', index: 'carouselUrl', width: 180, formatter: coverImageFormatter},
        {label: '跳转链接', name: 'redirectUrl', index: 'redirectUrl', width: 120},
        {label: '排序值', name: 'carouselRank', index: 'carouselRank', width: 120},
        {label: '添加时间', name: 'createTime', index: 'createTime', width: 120}
    ],
    height: 560,
    rowNum: 10,
    rowList: [10, 20, 50],
    styleUI: 'Bootstrap',
    loadtext: '信息读取中...',
    rownumbers: false,
    rownumWidth: 20,
    autowidth: true,
    multiselect: true,
    pager: "#jqGridPager",
    jsonReader: {
        root: "data.list",
        page: "data.currPage",
        total: "data.totalPage",
        records: "data.totalCount"
    },
    prmNames: {
        page: "page",
        rows: "limit",
        order: "order",
    },
    gridComplete: function () {
        //隐藏grid底部滚动条
        $("#jqGrid").closest(".ui-jqgrid-bdiv").css({"overflow-x": "hidden"});
    }
});
```

```
    function coverImageFormatter(cellvalue) {
        return "<img src='" + cellvalue + "' height=\"120\" width=\"160\" alt='coverImage'/>";
    }

    $(window).resize(function () {
        $("#jqGrid").setGridWidth($(".card-body").width());
    });
});
```

在跳转到对应的首页配置管理页面时,这里都会带上 configType 参数。该参数在接口请求中会用到,所以在请求 URL 和页面中都做了处理。在请求列表接口时,系统就能够通过 jQuery 语法取 configType 参数的值并在列表请求中将它们放在请求的 URL 中。

以上代码的主要功能为分页数据展示、字段格式化、jqGrid 的 DOM 宽度自适应。在页面加载时,系统首先会调用 jqGrid 的初始化方法,将页面中 id 为 jqGrid 的 DOM 渲染为分页表格,在获取 configType 变量值后拼接首页配置列表的请求地址,并向后端发送请求,然后按照后端返回的 JSON 数据填充分页表格和表格下方的分页按钮。这里可以参考前文 jqGrid 分页功能整合进行理解。重启项目可以验证首页配置数据分页功能是否正常。

21.8.3　添加和修改按钮触发事件及 Modal 框实现

添加和修改两个功能按钮分别绑定了触发事件,即在 newbee_mall_index_config.js 文件中新增 configAdd()方法和 configEdit()方法。两个方法的实现为打开首页配置信息编辑框。接下来实现信息编辑框和两个触发事件,代码如下所示:

```
<div class="content">
        <!-- Modal 框 -->
        <div class="modal fade" id="indexConfigModal" tabindex="-1" role="dialog"
            aria-labelledby="indexConfigModalLabel">
            <div class="modal-dialog" role="document">
                <div class="modal-content">
                    <div class="modal-header">
                        <button type="button" class="close" data-dismiss="modal" aria-label="Close"><span
                                aria-hidden="true">&times;</span></button>
                        <h6 class="modal-title" id="categoryModalLabel">Modal</h6>
```

```html
            </div>
            <div class="modal-body">
                <form id="indexConfigForm">
                    <div class="form-group">
                        <div class="alert alert-danger" id="edit-error-msg" style="display: none;">
                            错误信息展示栏
                        </div>
                    </div>
                    <input type="hidden" class="form-control" id="indexConfigModaconfigId" name="indexConfigModaconfigId">
                    <input type="hidden" id="configType" th:value="${configType}">
                    <div class="form-group">
                        <label for="configName" class="control-label">配置项显示字符:</label>
                        <input type="text" class="form-control" id="configName" name="configName"
                               placeholder="请输入配置项显示字符" required="true">
                    </div>
                    <div class="form-group">
                        <label for="redirectUrl" class="control-label">跳转链接:</label>
                        <input type="text" class="form-control" id="redirectUrl" name="redirectUrl"
                               placeholder="请输入跳转链接" value="##">
                    </div>
                    <div class="form-group">
                        <label for="goodsId" class="control-label">关联商品编号:</label>
                        <input type="number" class="form-control" id="goodsId" name="goodsId"
                               placeholder="关联商品编号" value="0">
                    </div>
                    <div class="form-group">
                        <label for="configRank" class="control-label">排序值:</label>
                        <input type="number" class="form-control" id="configRank" name="configRank"
                               placeholder="请输入排序值">
                    </div>
                </form>
```

```html
                    </div>
                    <div class="modal-footer">
                        <button type="button" class="btn btn-default" data-dismiss="modal">取消</button>
                        <button type="button" class="btn btn-primary" id="saveButton">确认</button>
                    </div>
                </div>
            </div>
        </div>
        <!-- /.modal -->
    </div>
```

configAdd()方法和 configEdit()方法实现的代码如下所示：

```javascript
function configAdd() {
    reset();
    $('.modal-title').html('首页配置项添加');
    $('#indexConfigModal').modal('show');
}

function configEdit() {
    reset();
    var id = getSelectedRow();
    if (id == null) {
        return;
    }
    var rowData = $("#jqGrid").jqGrid("getRowData", id);
    $('.modal-title').html('首页配置项编辑');
    $('#indexConfigModal').modal('show');
    $("#configId").val(id);
    $("#configName").val(rowData.configName);
    $("#redirectUrl").val(rowData.redirectUrl);
    $("#goodsId").val(rowData.goodsId);
    $("#configRank").val(rowData.configRank);
}
```

添加按钮的作用仅仅是显示 Modal 框。修改功能则多了一个步骤，需要将选择的记录回显到 Modal 框中以供修改。这里使用了 jqGrid 插件中的 getRowData()方法获取当前选择行上的所有字段的数据并赋值到对应的输入框中。

首页配置管理页面 Modal 框效果如图 21-11 所示。

图 21-11　首页配置管理页面 Modal 框效果

21.8.4　首页配置管理页面添加和编辑功能的实现

在信息录入完成后可以点击信息编辑框下方的"确认"按钮，此时会进行数据的交互，JS 实现代码如下所示：

```
//绑定 Modal 框上的保存按钮
$('#saveButton').click(function () {
    var configName = $("#configName").val();
    var configType = $("#configType").val();
    var redirectUrl = $("#redirectUrl").val();
    var goodsId = $("#goodsId").val();
    var configRank = $("#configRank").val();
    if (!validCN_ENString2_18(configName)) {
        $('#edit-error-msg').css("display", "block");
        $('#edit-error-msg').html("请输入符合规范的配置项名称！");
    } else {
        var data = {
            "configName": configName,
            "configType": configType,
            "redirectUrl": redirectUrl,
            "goodsId": goodsId,
            "configRank": configRank
        };
        var url = '/admin/indexConfigs/save';
        var id = getSelectedRowWithoutAlert();
        if (id != null) {
            url = '/admin/indexConfigs/update';
            data = {
```

```
            "configId": id,
            "configName": configName,
            "configType": configType,
            "redirectUrl": redirectUrl,
            "goodsId": goodsId,
            "configRank": configRank
        };
    }
    $.ajax({
        type: 'POST',//方法类型
        url: url,
        contentType: 'application/json',
        data: JSON.stringify(data),
        success: function (result) {
            if (result.resultCode == 200) {
                $('#indexConfigModal').modal('hide');
                swal("保存成功", {
                    icon: "success",
                });
                reload();
            } else {
                $('#indexConfigModal').modal('hide');
                swal(result.message, {
                    icon: "error",
                });
            }
            ;
        },
        error: function () {
            swal("操作失败", {
                icon: "error",
            });
        }
    });
}
});
```

添加和修改两个方法的传参和后续处理逻辑类似。为了避免太多重复代码的编写，因此将两个方法写在一起了，并通过 id 是否大于 0 来确定是修改操作还是添加操作。步骤如下所示。

①封装首页配置实体的参数

②向对应的后端首页配置的添加或者修改接口发送请求

③在请求成功后提醒用户请求成功并隐藏当前的首页配置信息 Modal 框，同时刷新首页配置列表数据

④请求失败则提醒对应的错误信息

21.8.5　首页配置管理页面删除功能的实现

删除按钮的点击触发事件为 deleteConfig()。在 newbee_mall_index_config.js 文件中新增如下代码：

```javascript
function deleteConfig () {
    var ids = getSelectedRows();
    if (ids == null) {
        return;
    }
    swal({
        title: "确认弹框",
        text: "确认要删除数据吗?",
        icon: "warning",
        buttons: true,
        dangerMode: true,
    }).then((flag) => {
        if (flag) {
            $.ajax({
                type: "POST",
                url: "/admin/indexConfigs/delete",
                contentType: "application/json",
                data: JSON.stringify(ids),
                success: function (r) {
                    if (r.resultCode == 200) {
                        swal("删除成功", {
                            icon: "success",
                        });
                        $("#jqGrid").trigger("reloadGrid");
                    } else {
                        swal(r.message, {
                            icon: "error",
```

```
                });
            }
        }
    });
                }
            }
        )
    ;
}
```

这里首先获取用户在 jqGrid 表格中选择的需要删除的所有记录的 id，然后将参数封装并向后端发送请求，请求地址为/admin/carousels/delete，后端接收到请求后会将对应的记录删除。

21.9　商城首页功能完善

接下来，读取并渲染首页最后三个版块上的数据。

21.9.1　首页推荐商品数据的读取

修改 IndexController 中的 indexPage()方法，并注入 NewBeeMallIndexConfigService 对象中，新增代码如下所示：

```
List<NewBeeMallIndexConfigGoodsVO> hotGoodses = newBeeMallIndexConfig
Service.getConfigGoodsesForIndex(IndexConfigTypeEnum.INDEX_GOODS_HOT.
getType(), Constants.INDEX_GOODS_HOT_NUMBER);
List<NewBeeMallIndexConfigGoodsVO> newGoodses = newBeeMallIndexConfig
Service.getConfigGoodsesForIndex(IndexConfigTypeEnum.INDEX_GOODS_NEW.
getType(), Constants.INDEX_GOODS_NEW_NUMBER);
List<NewBeeMallIndexConfigGoodsVO> recommendGoodses = newBeeMallIndex
ConfigService.getConfigGoodsesForIndex(IndexConfigTypeEnum.INDEX_GOODS_
RECOMMOND.getType(), Constants.INDEX_GOODS_RECOMMOND_NUMBER);

request.setAttribute("hotGoodses", hotGoodses);//热销商品
request.setAttribute("newGoodses", newGoodses);//新品推荐
request.setAttribute("recommendGoodses", recommendGoodses);//推荐商品
```

读取一定数量的首页配置商品数据，并分别将其设置到 request 对象中。

NewBeeMallIndexConfigService 业务类的 getConfigGoodsesForIndex()方法实现代码如下所示：

```java
/**
 * 返回固定数量的首页配置商品对象（首页调用）
 */
public List<NewBeeMallIndexConfigGoodsVO> getConfigGoodsesForIndex(int configType, int number) {
    List<NewBeeMallIndexConfigGoodsVO> newBeeMallIndexConfigGoodsVOS = new ArrayList<>(number);
    List<IndexConfig> indexConfigs = indexConfigMapper.findIndexConfigsByTypeAndNum(configType, number);
    if (!CollectionUtils.isEmpty(indexConfigs)) {
        //读取所有的 goodsId
        List<Long> goodsIds = indexConfigs.stream().map(IndexConfig::getGoodsId).collect(Collectors.toList());
        List<NewBeeMallGoods> newBeeMallGoods = goodsMapper.selectByPrimaryKeys(goodsIds);
        newBeeMallIndexConfigGoodsVOS = BeanUtil.copyList(newBeeMallGoods, NewBeeMallIndexConfigGoodsVO.class);
        for (NewBeeMallIndexConfigGoodsVO newBeeMallIndexConfigGoodsVO : newBeeMallIndexConfigGoodsVOS) {
            String goodsName = newBeeMallIndexConfigGoodsVO.getGoodsName();
            String goodsIntro = newBeeMallIndexConfigGoodsVO.getGoodsIntro();
            // 字符串过长导致文字超出的问题
            if (goodsName.length() > 30) {
                goodsName = goodsName.substring(0, 30) + "...";
                newBeeMallIndexConfigGoodsVO.setGoodsName(goodsName);
            }
            if (goodsIntro.length() > 22) {
                goodsIntro = goodsIntro.substring(0, 22) + "...";
                newBeeMallIndexConfigGoodsVO.setGoodsIntro(goodsIntro);
            }
        }
    }
    return newBeeMallIndexConfigGoodsVOS;
}
```

该方法的作用是返回数量的配置项对象，以供首页进行数据渲染。

首先根据 configType 参数读取固定数量的首页配置数据，然后获取配置项中关联的商品记录属性，再对字符串进行处理并封装到 VO 对象里，最后设置到 request 对象中。

这里主要是将需要的数据读取出来并封装成一个对象返回给视图层。

注意，VO 对象就是视图层使用的对象，与 entity 对象有一点区别。entity 对象中的字段与数据库表字段逐一对应，而 VO 对象中的字段则是根据视图层需要来设置。也可以不新增 VO 对象，直接返回 entity 对象，这取决于各个开发人员的开发习惯。

21.9.2 首页推荐商品数据的渲染

前端模板代码的修改，首先使用 Thymeleaf 语法依次读取热销商品 hotGoodses 对象、新品 newGoodses 对象和推荐商品 recommendGoodses 对象，然后使用 th:each 循环将这些商品数据分别渲染到页面上，代码如下所示：

```html
<div id="sub_banner">
  <th:block th:unless="${#lists.isEmpty(hotGoodses)}">
    <th:block th:each="hotGoodse : ${hotGoodses}">
      <div class="hot-image">
        <a th:href="@{'/goods/detail/'+${hotGoodse.goodsId}}">
          <img th:src="@{${hotGoodse.goodsCoverImg}}" th:alt="${hotGoodse.goodsName}">
        </a>
      </div>
    </th:block>
  </th:block>
</div>

<div id="flash">
  <h2>新品上线</h2>
  <ul>
    <th:block th:unless="${#lists.isEmpty(newGoodses)}">
      <th:block th:each="newGoods : ${newGoodses}">
        <li>
          <a th:href="@{'/goods/detail/'+${newGoods.goodsId}}">
            <img th:src="@{${newGoods.goodsCoverImg}}" th:alt="${newGoods.goodsName}">
            <p class="name" th:text="${newGoods.goodsName}">NewBeeMall</p>
            <p class="discount" th:text="${newGoods.goodsIntro}">NewBeeMall</p>
            <p class="item_price" th:text="${newGoods.sellingPrice}">NewBeeMall</p>
          </a>
        </li>
      </th:block>
```

```html
    </th:block>
  </ul>
</div>

<div id="recommend">
  <h2>为你推荐</h2>
  <a href="##" class="more">查看更多>></a>
  <ul>
    <th:block th:unless="${#lists.isEmpty(recommendGoodses)}">
      <th:block th:each="recommendGoods : ${recommendGoodses}">
        <li>
          <a th:href="@{'/goods/detail/'+${recommendGoods.goodsId}}">
            <div class="info discount" th:text="${recommendGoods.tag}">
              推荐
            </div>
            <img th:src="@{${recommendGoods.goodsCoverImg}}" th:alt="${recommendGoods.goodsName}">
            <p class="name" th:text="${recommendGoods.goodsName}">NewBeeMall</p>
            <p class="item_price" th:text="${recommendGoods.sellingPrice}">NewBeeMall</p>
            <p class="counter">猜你喜欢</p>
            <div class="comment">
              <p>新蜂精选</p>
              <p>好物也可以不贵</p>
            </div>
          </a>
        </li>
      </th:block>
    </th:block>
  </ul>
</div>
```

　　首页配置中的这三个部分并没有进行复杂的 Thymeleaf 语法运算，只是对后端传输过来的数据进行了解析和读取。这样，新蜂商城中的热销商品、新品上线、推荐商品这三个版块中的数据就读取并渲染完成。

　　商城端首页整个页面的数据读取和渲染步骤至此也完成了。打开新蜂商城项目的首页，可以看到新蜂商城首页中所有的数据都已经是后台管理系统中动态配置的数据了。数据读取并渲染成功，轮播图、分类数据、新品上线、推荐商品等内容都将正确地显示在页面上。读者可以自行对这些数据进行新增和修改，并在后台管理系统中进行配置，并通过刷新首页页面来验证数据是否正常。

第 22 章

商城端用户登录和注册功能的开发

如果不是商城的注册用户，会被限制使用商城的很多功能，比如个人信息管理、购物车、下单等。在商城端浏览的用户可以选择在注册页面填写注册信息成为商城用户，就能够登录账号并使用商城端的所有功能了。在介绍和开发商城端后续的功能模块前，需要先把新蜂商城用户模块的登录、注册等功能开发完成。

商城端的登录状态依然是通过 session 对象来保存的。用户在登录成功后，用户信息被放到 session 对象中。在访问项目时，通过拦截器判断在 session 对象中是否有用户信息的数据。有用户信息数据则表示正常登录，需要放行请求；没有用户信息数据则表示未登录，需要跳转到新蜂商城的登录页面。关于登录功能的介绍、用户登录状态的保持和验证、登录流程的设计，前文已有介绍，本章主要介绍商城端登录和注册功能及其实现，包括前端页面、数据库设计及后端业务代码的编写。

22.1 商城端用户表结构的设计

商城端用户表结构设计的主要字段包括用户的基本信息、收货地址、是否注销等。新蜂商城第一版的收货地址的字段设计是单个收货地址，且存放在用户表中，并没有设计成多收货地址。字段代码如下所示：

```
USE `newbee_mall_db`;

DROP TABLE IF EXISTS `tb_newbee_mall_user`;
CREATE TABLE `tb_newbee_mall_user` (
```

```sql
  'user_id' bigint(20) NOT NULL AUTO_INCREMENT COMMENT '用户主键id',
  'nick_name' varchar(50) CHARACTER SET utf8 COLLATE utf8_general_ci NOT NULL DEFAULT '' COMMENT '用户昵称',
  'login_name' varchar(11) CHARACTER SET utf8 COLLATE utf8_general_ci NOT NULL DEFAULT '' COMMENT '登录名称(默认为手机号)',
  'password_md5' varchar(32) CHARACTER SET utf8 COLLATE utf8_general_ci NOT NULL DEFAULT '' COMMENT 'MD5加密后的密码',
  'introduce_sign' varchar(100) CHARACTER SET utf8 COLLATE utf8_general_ci NOT NULL DEFAULT '' COMMENT '个性签名',
  'address' varchar(100) CHARACTER SET utf8 COLLATE utf8_general_ci NOT NULL DEFAULT '' COMMENT '收货地址',
  'is_deleted' tinyint(4) NOT NULL DEFAULT 0 COMMENT '注销标识字段(0-正常 1-已注销)',
  'locked_flag' tinyint(4) NOT NULL DEFAULT 0 COMMENT '锁定标识字段(0-未锁定 1-已锁定)',
  'create_time' datetime(0) NOT NULL DEFAULT CURRENT_TIMESTAMP COMMENT '注册时间',
  PRIMARY KEY ('user_id') USING BTREE
) ENGINE = InnoDB AUTO_INCREMENT = 1 CHARACTER SET = utf8 COLLATE = utf8_general_ci ROW_FORMAT = Dynamic;

INSERT INTO 'tb_newbee_mall_user' VALUES (1, '十三', '13700002703', 'e10adc3949ba59abbe56e057f20f883e', '我不怕千万人阻挡,只怕自己投降', '杭州市西湖区 xx 小区 x 幢 419 十三 137xxxx2703', 0, 0, '2021-03-22 08:44:57');
INSERT INTO 'tb_newbee_mall_user' VALUES (6, '测试用户1', '13711113333', 'dda01dc6d334badcd031102be6bee182', '测试用户1', '上海浦东新区XX路XX号 999 137xxxx7797', 0, 0, '2021-03-29 10:51:39');
INSERT INTO 'tb_newbee_mall_user' VALUES (7, '测试用户2测试用户2测试用户2测试用户2', '13811113333', 'dda01dc6d334badcd031102be6bee182', '测试用户2', '杭州市西湖区 xx 小区 x 幢 419 十三 137xxxx2703', 0, 0, '2021-03-29 10:55:08');
INSERT INTO 'tb_newbee_mall_user' VALUES (8, '测试用户3', '13911113333', 'dda01dc6d334badcd031102be6bee182', '测试用户3', '杭州市西湖区 xx 小区 x 幢 419 十三 137xxxx2703', 0, 0, '2021-03-29 10:55:16');
```

这里,在数据库中新建一张 tb_newbee_mall_user 表,并在表中新增几条用户数据,在之后演示登录功能时会用到。

22.2 商城端用户登录和注册页面的制作

22.2.1 商城端登录页面基础样式的实现

通过数据库的设计和前文开发完成的后台管理系统登录功能可以看出，类似的登录页面的主要字段就是登录信息和验证码字段。商城端的登录功能也是这样设计的。登录功能页面展示效果如图 22-1 所示。

图 22-1　新蜂商城登录页面

新蜂商城的商城端登录页面的设计比较简洁，主要元素就是新蜂商城的 Logo 图片和 slogan。这里主要是取了"新蜂"两个字的谐音：随新所欲，蜂富多彩。登录功能需要用到的三个用户输入字段为手机号、密码和验证码。

在 templates/mall 目录中新建 login.html 模板页面。模板引擎选择的是 Thymeleaf，代码如下所示：

```
<!-- Copyright (c) 2019-2020 十三 all rights reserved. -->
<!DOCTYPE html>
<html lang="en" xmlns:th="http://www.thymeleaf.org">
<head>
    <meta charset="UTF-8">
    <title>NewBee商城-登录</title>
    <link rel="stylesheet" th:href="@{mall/css/common.css}">
```

```html
        <link rel="stylesheet" th:href="@{mall/styles/login.css}">
        <link rel="stylesheet" th:href="@{mall/styles/header.css}">
        <link rel="stylesheet" th:href="@{/admin/plugins/sweetalert/sweetalert.css}"/>
</head>
<body>
<div class="top center">
    <div class="logo center">
        <a href="./index.html" target="_blank"><img src="mall/image/login-logo-2.png" alt=""></a>
    </div>
</div>
<div class="form center">
    <div class="login">
        <div class="login_center">
            <div class="login_top">
                <div class="left fl">会员登录</div>
                <div class="right fr">您还不是我们的会员?<a href="register.html" target="_self">立即注册</a></div>
                <div class="clear"></div>
                <div class="under-line center"></div>
            </div>
            <form id="loginForm" onsubmit="return false;" action="##">
                <div class="login_main center">
                    <div class="login-info">手机号: <input class="login-info-input" type="text" name="loginName"
                                                            id="loginName"
                                                            placeholder="请输入你的手机号"/>
                    </div>
                    <div class="login-info">密    码: <input class="login-info-input"
                                                                                     id="password"
                                                                                     type="password"
                                                                                     name="password"
                                                                                     placeholder="请输入你的密码"/></div>
                    <div class="login-info">
                        验证码: 
                        <input class="login-info-input verify-code" type="text" name="verifyCode"
                               placeholder="请输入验证码" id="verifyCode"/>
                        <img alt="点击图片刷新!" style="top: 16px;position:
```

```
relative;" th:src="@{/common/mall/kaptcha}"
                    onclick="this.src='/common/mall/kaptcha?d='+new
Date()*1">
                </div>
            </div>
            <div class="login_submit">
                <input class="submit" type="submit" onclick="login()"
value="立即登录">
            </div>
        </form>
    </div>
</div>
</body>
<!-- jQuery -->
<script th:src="@{/admin/plugins/jquery/jquery.min.js}"></script>
<script th:src="@{/admin/dist/js/public.js}"></script>
<script th:src="@{/admin/plugins/sweetalert/sweetalert.min.js}"></script>
</html>
```

22.2.2 商城端注册页面基础样式的实现

注册页面与登录页面非常类似，只是二者的功能有些区别，注册页面效果如图 22-2 所示。

图 22-2 新蜂商城注册页面

在 templates/mall 目录中新建 register.html 模板页面。该页面代码与登录页面基本一致，只是部分文案有修改，此处不再赘述，读者可查看源码。

22.2.3 控制类处理跳转逻辑

在 controller/mall 包下新建 PersonalController.java，新增如下代码：

```java
@Controller
public class PersonalController {

    /**
     * 登录页面跳转
     * @return
     */
    @GetMapping({"/login", "login.html"})
    public String loginPage() {
        return "mall/login";
    }

    /**
     * 注册页面跳转
     * @return
     */
    @GetMapping({"/register", "register.html"})
    public String registerPage() {
        return "mall/register";
    }
}
```

该段代码用于处理/login 和/register 两个请求，请求方法为 GET，分别是登录页面和注册页面的跳转处理方法。在发起请求后会分别跳转到 templates/mall 目录下的 login.html 模板页面和 register.html 模板页面中。

22.3 商城端用户登录和注册模块接口的实现

22.3.1 新建商城端用户实体类和 Mapper 接口

首先在 ltd.newbee.mall.entity 包中创建用户实体类，选中 entity 包并右键点击，在弹出的菜单中选择"New→Java Class",然后在弹出的窗口中输入"MallUser",最后在 MallUser 类中新增如下代码:

```java
package ltd.newbee.mall.entity;

import com.fasterxml.jackson.annotation.JsonFormat;

import java.util.Date;

public class MallUser {
    private Long userId;

    private String nickName;

    private String loginName;

    private String passwordMd5;

    private String introduceSign;

    private String address;

    private Byte isDeleted;

    private Byte lockedFlag;

    @JsonFormat(pattern = "yyyy-MM-dd HH:mm:ss", timezone = "GMT+8")
    private Date createTime;

    public Long getUserId() {
        return userId;
    }

    public void setUserId(Long userId) {
        this.userId = userId;
    }

    public String getNickName() {
        return nickName;
    }

    public void setNickName(String nickName) {
        this.nickName = nickName == null ? null : nickName.trim();
    }
```

```java
    public String getLoginName() {
        return loginName;
    }

    public void setLoginName(String loginName) {
        this.loginName = loginName == null ? null : loginName.trim();
    }

    public String getPasswordMd5() {
        return passwordMd5;
    }

    public void setPasswordMd5(String passwordMd5) {
        this.passwordMd5 = passwordMd5 == null ? null : passwordMd5.trim();
    }

    public String getIntroduceSign() {
        return introduceSign;
    }

    public void setIntroduceSign(String introduceSign) {
        this.introduceSign = introduceSign == null ? null : introduceSign.trim();
    }

    public String getAddress() {
        return address;
    }

    public void setAddress(String address) {
        this.address = address;
    }

    public Byte getIsDeleted() {
        return isDeleted;
    }

    public void setIsDeleted(Byte isDeleted) {
        this.isDeleted = isDeleted;
    }

    public Byte getLockedFlag() {
```

```
        return lockedFlag;
    }

    public void setLockedFlag(Byte lockedFlag) {
        this.lockedFlag = lockedFlag;
    }

    public Date getCreateTime() {
        return createTime;
    }

    public void setCreateTime(Date createTime) {
        this.createTime = createTime;
    }
}
```

首先在 ltd.newbee.mall.dao 包中新建用户实体的 Mapper 接口,选中 dao 包并右键点击,在弹出的菜单中选择"New→Java Class",然后在弹出的窗口中输入"MallUserMapper",并选中"Interface"选项,最后在 MallUserMapper.java 文件中新增如下代码:

```
package ltd.newbee.mall.dao;

import ltd.newbee.mall.entity.MallUser;
import org.apache.ibatis.annotations.Param;

public interface MallUserMapper {

    /**
     * 保存一条新记录
     * @param record
     * @return
     */
    int insertSelective(MallUser record);

    /**
     * 根据 loginName 查询记录
     * @param loginName
     * @return
     */
    MallUser selectByLoginName(String loginName);

    /**
```

```
 * 根据loginName和密码字段查询记录
 * @param loginName
 * @return
 */
MallUser selectByLoginNameAndPasswd(@Param("loginName") String loginName,
@Param("password") String password);
}
```

这里定义了商城用户实体店数据层的操作，主要是新增和查询方法。

22.3.2　创建 MallUserMapper 接口的映射文件

在 resources/mapper 目录下新建 MallUserMapper 接口的映射文件 MallUserMapper.xml，并进行映射文件的编写。

首先，定义映射文件与 Mapper 接口的对应关系。比如在该示例中，需要将 MallUserMapper.xml 文件与对应的 MallUserMapper 接口之间的关系定义出来：

```xml
<mapper namespace="ltd.newbee.mall.dao.MallUserMapper">
```

然后，配置表结构和实体类的对应关系：

```xml
<resultMap id="BaseResultMap" type="ltd.newbee.mall.entity.MallUser">
  <id column="user_id" jdbcType="BIGINT" property="userId"/>
  <result column="nick_name" jdbcType="VARCHAR" property="nickName"/>
  <result column="login_name" jdbcType="VARCHAR" property="loginName"/>
  <result column="password_md5" jdbcType="VARCHAR" property="passwordMd5"/>
  <result column="introduce_sign" jdbcType="VARCHAR" property="introduceSign"/>
  <result column="address" jdbcType="VARCHAR" property="address"/>
  <result column="is_deleted" jdbcType="TINYINT" property="isDeleted"/>
  <result column="locked_flag" jdbcType="TINYINT" property="lockedFlag"/>
  <result column="create_time" jdbcType="TIMESTAMP" property="createTime"/>
</resultMap>
```

最后，按照对应的接口方法，编写具体的 SQL 语句，最终的 MallUserMapper.xml 文件如下所示：

```xml
<?xml version="1.0" encoding="UTF-8"?>
<!DOCTYPE mapper PUBLIC "-//mybatis.org//DTD Mapper 3.0//EN" "http://mybatis.org/dtd/mybatis-3-mapper.dtd">
<mapper namespace="ltd.newbee.mall.dao.MallUserMapper">
```

```xml
<resultMap id="BaseResultMap" type="ltd.newbee.mall.entity.MallUser">
    <id column="user_id" jdbcType="BIGINT" property="userId"/>
    <result column="nick_name" jdbcType="VARCHAR" property="nickName"/>
    <result column="login_name" jdbcType="VARCHAR" property="loginName"/>
    <result column="password_md5" jdbcType="VARCHAR" property="passwordMd5"/>
    <result column="introduce_sign" jdbcType="VARCHAR" property="introduceSign"/>
    <result column="address" jdbcType="VARCHAR" property="address"/>
    <result column="is_deleted" jdbcType="TINYINT" property="isDeleted"/>
    <result column="locked_flag" jdbcType="TINYINT" property="lockedFlag"/>
    <result column="create_time" jdbcType="TIMESTAMP" property="createTime"/>
</resultMap>

<sql id="Base_Column_List">
  user_id, nick_name, login_name, password_md5, introduce_sign, address, is_deleted,
  locked_flag, create_time
</sql>

<select id="selectByLoginName" parameterType="java.lang.String" resultMap="BaseResultMap">
    select
    <include refid="Base_Column_List"/>
    from tb_newbee_mall_user
    where login_name = #{loginName} and is_deleted = 0
</select>

<select id="selectByLoginNameAndPasswd" resultMap="BaseResultMap">
    select
    <include refid="Base_Column_List"/>
    from tb_newbee_mall_user
    where login_name = #{loginName} and password_md5 = #{password} and is_deleted = 0
</select>

<insert id="insertSelective" parameterType="ltd.newbee.mall.entity.MallUser">
    insert into tb_newbee_mall_user
```

```xml
<trim prefix="(" suffix=")" suffixOverrides=",">
    <if test="userId != null">
        user_id,
    </if>
    <if test="nickName != null">
        nick_name,
    </if>
    <if test="loginName != null">
        login_name,
    </if>
    <if test="passwordMd5 != null">
        password_md5,
    </if>
    <if test="introduceSign != null">
        introduce_sign,
    </if>
    <if test="address != null">
        address,
    </if>
    <if test="isDeleted != null">
        is_deleted,
    </if>
    <if test="lockedFlag != null">
        locked_flag,
    </if>
    <if test="createTime != null">
        create_time,
    </if>
</trim>
<trim prefix="values (" suffix=")" suffixOverrides=",">
    <if test="userId != null">
        #{userId,jdbcType=BIGINT},
    </if>
    <if test="nickName != null">
        #{nickName,jdbcType=VARCHAR},
    </if>
    <if test="loginName != null">
        #{loginName,jdbcType=VARCHAR},
    </if>
    <if test="passwordMd5 != null">
        #{passwordMd5,jdbcType=VARCHAR},
    </if>
    <if test="introduceSign != null">
```

```xml
                #{introduceSign,jdbcType=VARCHAR},
            </if>
            <if test="address != null">
                #{address,jdbcType=VARCHAR},
            </if>
            <if test="isDeleted != null">
                #{isDeleted,jdbcType=TINYINT},
            </if>
            <if test="lockedFlag != null">
                #{lockedFlag,jdbcType=TINYINT},
            </if>
            <if test="createTime != null">
                #{createTime,jdbcType=TIMESTAMP},
            </if>
        </trim>
    </insert>
</mapper>
```

22.3.3 业务层代码的实现

首先在 ltd.newbee.mall.service 包中新建业务处理类，选中 service 包并右键点击，在弹出的菜单中选择"New→Java Class"，然后在弹出的窗口中输入"NewBeeMallUserService"，并选中"Interface"选项，最后在 NewBeeMallUserService.java 文件中新增如下代码：

```java
package ltd.newbee.mall.service;

import javax.servlet.http.HttpSession;

public interface NewBeeMallUserService {

    /**
     * 用户注册
     *
     * @param loginName
     * @param password
     * @return
     */
    String register(String loginName, String password);

    /**
```

```
 * 登录
 *
 * @param loginName
 * @param passwordMD5
 * @param httpSession
 * @return
 */
String login(String loginName, String passwordMD5, HttpSession httpSession);
}
```

这里分别定义了登录和注册两个业务层方法。

首先在 ltd.newbee.mall.service.impl 包中新建 NewBeeMallUserService 的实现类，选中 impl 包并右键点击，在弹出的菜单中选择"New→Java Class"，然后在弹出的窗口中输入"NewBeeMallUserServiceImpl"，最后在 NewBeeMallUserServiceImpl 类中新增如下代码：

```java
package ltd.newbee.mall.service.impl;

import ltd.newbee.mall.common.Constants;
import ltd.newbee.mall.common.ServiceResultEnum;
import ltd.newbee.mall.controller.vo.NewBeeMallUserVO;
import ltd.newbee.mall.dao.MallUserMapper;
import ltd.newbee.mall.entity.MallUser;
import ltd.newbee.mall.service.NewBeeMallUserService;
import ltd.newbee.mall.util.BeanUtil;
import ltd.newbee.mall.util.MD5Util;
import org.springframework.beans.factory.annotation.Autowired;
import org.springframework.stereotype.Service;

import javax.servlet.http.HttpSession;

@Service
public class NewBeeMallUserServiceImpl implements NewBeeMallUserService {

    @Autowired
    private MallUserMapper mallUserMapper;


    @Override
    public String register(String loginName, String password) {
        if (mallUserMapper.selectByLoginName(loginName) != null) {
            return ServiceResultEnum.SAME_LOGIN_NAME_EXIST.getResult();
        }
```

```java
        MallUser registerUser = new MallUser();
        registerUser.setLoginName(loginName);
        registerUser.setNickName(loginName);
        String passwordMD5 = MD5Util.MD5Encode(password, "UTF-8");
        registerUser.setPasswordMd5(passwordMD5);
        if (mallUserMapper.insertSelective(registerUser) > 0) {
            return ServiceResultEnum.SUCCESS.getResult();
        }
        return ServiceResultEnum.DB_ERROR.getResult();
    }

    @Override
    public String login(String loginName, String passwordMD5, HttpSession httpSession) {
        MallUser user = mallUserMapper.selectByLoginNameAndPasswd(loginName, passwordMD5);
        if (user != null && httpSession != null) {
            if (user.getLockedFlag() == 1) {
                return ServiceResultEnum.LOGIN_USER_LOCKED.getResult();
            }
            //昵称太长，影响页面展示
            if (user.getNickName() != null && user.getNickName().length() > 7) {
                String tempNickName = user.getNickName().substring(0, 7) + "..";
                user.setNickName(tempNickName);
            }
            NewBeeMallUserVO newBeeMallUserVO = new NewBeeMallUserVO();
            BeanUtil.copyProperties(user, newBeeMallUserVO);
            httpSession.setAttribute(Constants.MALL_USER_SESSION_KEY, newBeeMallUserVO);
            return ServiceResultEnum.SUCCESS.getResult();
        }
        return ServiceResultEnum.LOGIN_ERROR.getResult();
    }
}
```

注册方法 register() 的业务层代码逻辑：查询数据库中是否有相同登录名的记录，如果存在则不执行后续逻辑，返回"用户名已存在"的错误信息。如果不存在则证明可以新增这条记录，将密码进行 MD5 转换后执行 insert 语句，再将用户信息插入到数据库中，并返回成功的业务信息。主要 SQL 方法为 selectByLoginName()和 insertSelective()。selectByLoginName()方法根据用户名查询用户记录信息，insertSelective()方法是一条标准的 insert 组装语句，用于将记录保存到数据库中。

登录方法 login() 的业务层代码逻辑：根据用户名和密码查询数据库中是否存在对应的用户记录，若不存在则返回错误信息。如果存在则验证该用户是否已被禁用，若已被禁用则返回登录失败。用户状态正常再对昵称字段进行处理，然后返回登录成功的业务信息。主要的 SQL 方法为 selectByLoginNameAndPasswd()，即根据用户名和密码查询用户记录信息。

22.3.4 商城端用户登录和注册控制层代码的实现

在 PersonalController 控制器中新增登录和注册接口的实现代码，最终 PersonalController 类的代码如下所示：

```java
package ltd.newbee.mall.controller.mall;

import ltd.newbee.mall.common.Constants;
import ltd.newbee.mall.common.ServiceResultEnum;
import ltd.newbee.mall.service.NewBeeMallUserService;
import ltd.newbee.mall.util.MD5Util;
import ltd.newbee.mall.util.Result;
import ltd.newbee.mall.util.ResultGenerator;
import org.springframework.stereotype.Controller;
import org.springframework.util.StringUtils;
import org.springframework.web.bind.annotation.GetMapping;
import org.springframework.web.bind.annotation.PostMapping;
import org.springframework.web.bind.annotation.RequestParam;
import org.springframework.web.bind.annotation.ResponseBody;

import javax.annotation.Resource;
import javax.servlet.http.HttpSession;

@Controller
public class PersonalController {

    @Resource
    private NewBeeMallUserService newBeeMallUserService;

    @GetMapping({"/login", "login.html"})
    public String loginPage() {
        return "mall/login";
    }
```

第22章 商城端用户登录和注册功能的开发

```java
@GetMapping({"/register", "register.html"})
public String registerPage() {
    return "mall/register";
}

@PostMapping("/login")
@ResponseBody
public Result login(@RequestParam("loginName") String loginName,
                    @RequestParam("verifyCode") String verifyCode,
                    @RequestParam("password") String password,
                    HttpSession httpSession) {
    if (StringUtils.isEmpty(loginName)) {
        return ResultGenerator.genFailResult(ServiceResultEnum.LOGIN_NAME_NULL.getResult());
    }
    if (StringUtils.isEmpty(password)) {
        return ResultGenerator.genFailResult(ServiceResultEnum.LOGIN_PASSWORD_NULL.getResult());
    }
    if (StringUtils.isEmpty(verifyCode)) {
        return ResultGenerator.genFailResult(ServiceResultEnum.LOGIN_VERIFY_CODE_NULL.getResult());
    }
    String kaptchaCode = httpSession.getAttribute(Constants.MALL_VERIFY_CODE_KEY) + "";
    if (StringUtils.isEmpty(kaptchaCode) || !verifyCode.toLowerCase().equals(kaptchaCode)) {
        return ResultGenerator.genFailResult(ServiceResultEnum.LOGIN_VERIFY_CODE_ERROR.getResult());
    }
    String loginResult = newBeeMallUserService.login(loginName, MD5Util.MD5Encode(password, "UTF-8"), httpSession);
    //登录成功
    if (ServiceResultEnum.SUCCESS.getResult().equals(loginResult)) {
        return ResultGenerator.genSuccessResult();
    }
    //登录失败
    return ResultGenerator.genFailResult(loginResult);
}

@PostMapping("/register")
@ResponseBody
public Result register(@RequestParam("loginName") String loginName,
```

```java
                    @RequestParam("verifyCode") String verifyCode,
                    @RequestParam("password") String password,
                    HttpSession httpSession) {
        if (StringUtils.isEmpty(loginName)) {
            return ResultGenerator.genFailResult(ServiceResultEnum.LOGIN_NAME_NULL.getResult());
        }
        if (StringUtils.isEmpty(password)) {
            return ResultGenerator.genFailResult(ServiceResultEnum.LOGIN_PASSWORD_NULL.getResult());
        }
        if (StringUtils.isEmpty(verifyCode)) {
            return ResultGenerator.genFailResult(ServiceResultEnum.LOGIN_VERIFY_CODE_NULL.getResult());
        }
        String kaptchaCode = httpSession.getAttribute(Constants.MALL_VERIFY_CODE_KEY) + "";
        if (StringUtils.isEmpty(kaptchaCode) || !verifyCode.toLowerCase().equals(kaptchaCode)) {
            return ResultGenerator.genFailResult(ServiceResultEnum.LOGIN_VERIFY_CODE_ERROR.getResult());
        }
        String registerResult = newBeeMallUserService.register(loginName, password);
        //注册成功
        if (ServiceResultEnum.SUCCESS.getResult().equals(registerResult)) {
            return ResultGenerator.genSuccessResult();
        }
        //注册失败
        return ResultGenerator.genFailResult(registerResult);
    }

    @GetMapping("/logout")
    public String logout(HttpSession httpSession) {
        httpSession.removeAttribute(Constants.MALL_USER_SESSION_KEY);
        return "mall/login";
    }
}
```

注册接口负责接收前端传来的用户注册相关参数，包括手机号、密码、验证码。首先对参数进行校验。验证码的显示和校验的逻辑在前文中已经有介绍，这里就直接整合到了商城端用户的登录和注册功能中。然后在对参数与存储在 session 对象中的验证码值进行比较后，调用 newBeeMallUserService 业务层代码在数据库中新增一条新蜂商城用户记录，最后根据业务层的结果返回对应的 Result 对象。

在注册成功后，数据库中就有了用户记录，接下来就可以进行登录操作了。登录接口负责接收前端传来的用户登录相关参数，包括手机号、密码、验证码。首先对参数进行校验，这里使用 verifyCode 参数与存储在 session 中的验证码值进行比较。然后在校验成功后调用 newBeeMallUserService 业务层代码查询用户输入的登录信息是否正确，最后根据业务层的结果返回对应的 Result 对象。

退出登录的功能比较简单，就是将保存在 session 中的用户信息清除即可。在清除 session 中的用户信息后页面跳转到登录页面。

22.4 商城端用户登录注册模块前端功能的实现

在相关接口实现后，需要在前端进行接口的请求和回调处理。

22.4.1 注册功能的实现

在注册信息输入完成后，可以点击信息编辑框下方的"立即注册"按钮。该按钮已经定义了触发事件，点击后会执行 register()方法，并在该方法中进行数据的交互。在 register.html 文件中新增如下代码：

```javascript
<script type="text/javascript">
    function register() {
        var loginName = $("#loginName").val();
        if (!validPhoneNumber(loginName)) {
            swal('请输入正确的登录名(即手机号)', {
                icon: "error",
            });
            return false;
        }
        var password = $("#password").val();
        if (!validPassword(password)) {
            swal('请输入正确的密码格式(6~20位字符和数字组合)', {
                icon: "error",
            });
            return false;
        }
```

```javascript
        var verifyCode = $("#verifyCode").val();
        if (!validLength(verifyCode, 5)) {
            swal('请输入正确的验证码', {
                icon: "error",
            });
            return false;
        }
        //验证
        var params = $("#registerForm").serialize();
        var url = '/register';
        $.ajax({
            type: 'POST',//方法类型
            url: url,
            data: params,
            success: function (result) {
                if (result.resultCode == 200) {
                    swal({
                        title: "注册成功",
                        text: "是否跳转至登录页?",
                        icon: "success",
                        buttons: true,
                        dangerMode: true,
                    }).then((flag) => {
                        if (flag) {
                            window.location.href = '/login';
                        }
                    }
                    )
                    ;
                } else {
                    swal(result.message, {
                        icon: "error",
                    });
                }
                ;
            },
            error: function () {
                swal("操作失败", {
                    icon: "error",
                });
            }
        });
    }
```

```
</script>
```

register()方法的执行步骤如下所示。

①使用 jQuery 语法获取用户输入的注册信息字段

②使用正则表达式验证用户的输入字段是否符合规范

③封装数据并向注册接口发送请求

④请求成功则提醒用户注册成功

⑤请求失败则提醒对应的错误信息

22.4.2　登录功能的实现

在登录信息输入完成后，可以点击信息编辑框下方的"立即登录"按钮。该按钮已经定义了触发事件，点击后会执行 login()方法，并在该方法中进行数据的交互。在 login.html 文件中新增如下代码：

```
<script type="text/javascript">
    function login() {
        var loginName = $("#loginName").val();
        if (!validPhoneNumber(loginName)) {
            swal('请输入正确的登录名(即手机号)', {
                icon: "error",
            });
            return false;
        }
        var password = $("#password").val();
        if (!validPassword(password)) {
            swal('请输入正确的密码格式(6~20 位字符和数字组合)', {
                icon: "error",
            });
            return false;
        }
        var verifyCode = $("#verifyCode").val();
        if (!validLength(verifyCode, 5)) {
            swal('请输入正确的验证码', {
                icon: "error",
            });
            return false;
        }
```

```javascript
    //验证
    var params = $("#loginForm").serialize();
    var url = '/login';
    $.ajax({
        type: 'POST',//方法类型
        url: url,
        data: params,
        success: function (result) {
            if (result.resultCode == 200) {
                window.location.href = '/index';
            } else {
                swal(result.message, {
                    icon: "error",
                });
            }
            ;
        },
        error: function () {
            swal("操作失败", {
                icon: "error",
            });
        }
    });
}
</script>
```

login()方法的执行步骤如下所示。

①使用 jQuery 语法获取用户输入的登录信息字段

②使用正则表达式验证用户的输入字段是否符合规范

③封装数据并向登录接口发送请求

④在请求成功后会跳转至新蜂商城首页（如果想跳转到其他页面可以更改代码）

⑤请求失败则提醒对应的错误信息

22.5 商城端用户登录拦截器的实现

拦截器的相关知识点已经在前文介绍过，这里就直接进行编码实现了。

22.5.1 定义拦截器

在 interceptor 包中新建 NewBeeMallLoginInterceptor 类，该类需要实现 HandlerInterceptor 接口，代码如下所示：

```java
package ltd.newbee.mall.interceptor;

import ltd.newbee.mall.common.Constants;
import org.springframework.stereotype.Component;
import org.springframework.web.servlet.HandlerInterceptor;
import org.springframework.web.servlet.ModelAndView;

import javax.servlet.http.HttpServletRequest;
import javax.servlet.http.HttpServletResponse;

/**
 * newbee-mall 系统身份验证拦截器
 */
@Component
public class NewBeeMallLoginInterceptor implements HandlerInterceptor {

    @Override
    public boolean preHandle(HttpServletRequest request, HttpServletResponse response, Object o) throws Exception {
        if (null == request.getSession().getAttribute(Constants.MALL_USER_SESSION_KEY)) {
            response.sendRedirect(request.getContextPath() + "/login");
            return false;
        } else {
            return true;
        }
    }

    @Override
    public void postHandle(HttpServletRequest httpServletRequest, HttpServletResponse httpServletResponse, Object o, ModelAndView modelAndView) throws Exception {
    }

    @Override
    public void afterCompletion(HttpServletRequest httpServletRequest,
```

```
HttpServletResponse httpServletResponse, Object o, Exception e) throws
Exception {

    }
}
```

这里需要完善 preHandle 方法，同时在类声明上方添加@Component 注解使其注册到 IOC 容器中。

通过上面代码可以看出，这里在请求的预处理过程中读取并判断了当前 session 对象中是否存在 newBeeMallUser 对象。如果不存在则返回 false 并跳转至商城端登录页面。如果已经存在则返回 true，继续后续处理流程。

22.5.2 配置拦截器

在实现拦截器的相关方法后，需要对该拦截器进行配置以使其生效。由于已经对后台登录拦截器进行了配置，新蜂商城的拦截器配置只需要在 NeeBeeMallWebMvcConfigurer 类中添加关于商城页面的登录拦截即可，最终代码如下所示：

```
package ltd.newbee.mall.config;

import ltd.newbee.mall.common.Constants;
import ltd.newbee.mall.interceptor.AdminLoginInterceptor;
import ltd.newbee.mall.interceptor.NewBeeMallLoginInterceptor;
import org.springframework.beans.factory.annotation.Autowired;
import org.springframework.context.annotation.Configuration;
import org.springframework.web.servlet.config.annotation.InterceptorRegistry;
import org.springframework.web.servlet.config.annotation.ResourceHandlerRegistry;
import org.springframework.web.servlet.config.annotation.WebMvcConfigurer;

@Configuration
public class NeeBeeMallWebMvcConfigurer implements WebMvcConfigurer {

    @Autowired
    private AdminLoginInterceptor adminLoginInterceptor;
    @Autowired
    private NewBeeMallLoginInterceptor newBeeMallLoginInterceptor;
```

```java
public void addInterceptors(InterceptorRegistry registry) {
    // 添加一个拦截器，拦截以/admin为前缀的URL路径（后台登录拦截）
    registry.addInterceptor(adminLoginInterceptor)
            .addPathPatterns("/admin/**")
            .excludePathPatterns("/admin/login")
            .excludePathPatterns("/admin/coupling-test")
            .excludePathPatterns("/admin/categories/listForSelect")
            .excludePathPatterns("/admin/dist/**")
            .excludePathPatterns("/admin/plugins/**");
    // 商城页面登录拦截
    registry.addInterceptor(newBeeMallLoginInterceptor)
            .excludePathPatterns("/admin/**")
            .excludePathPatterns("/register")
            .excludePathPatterns("/login")
            .excludePathPatterns("/logout")
            .addPathPatterns("/goods/detail/**")
            .addPathPatterns("/shop-cart")
            .addPathPatterns("/shop-cart/**")
            .addPathPatterns("/saveOrder")
            .addPathPatterns("/orders")
            .addPathPatterns("/orders/**")
            .addPathPatterns("/personal")
            .addPathPatterns("/personal/updateInfo")
            .addPathPatterns("/selectPayType")
            .addPathPatterns("/payPage");
}

public void addResourceHandlers(ResourceHandlerRegistry registry) {
    registry.addResourceHandler("/upload/**").addResourceLocations("file:" + Constants.FILE_UPLOAD_DIC);
    registry.addResourceHandler("/goods-img/**").addResourceLocations("file:" + Constants.FILE_UPLOAD_DIC);
}
}
```

这里使用@Autowired 注解注入商城端登录拦截器 NewBeeMallLoginInterceptor，并对该拦截器所拦截的路径进行配置。从代码中可以看出该拦截器只对用户操作的功能路径生效，比如个人中心相关功能、购物车模块功能、订单模块功能等。这些功能将会在后文逐一实现。

22.6 功能测试

在编码完成后,启动 Spring Boot 项目。在启动成功后打开浏览器并输入商城端注册页面地址:

`http://localhost:8080/register`

在注册页面填写注册信息,如果有不符合规范的输入则会收到对应的错误提示。然后点击"立即注册"按钮,在注册成功后页面弹框会提醒用户注册成功,效果如图 22-3 所示。

图 22-3 新蜂商城账号注册成功

当然,在功能测试完成后需检查数据库中是否已经存在对应的用户记录,并使用已注册的账号进行登录功能的测试。商城端用户登录页面地址如下所示:

`http://localhost:8080/login`

新蜂商城账号登录效果如图 22-4 所示。

第 22 章　商城端用户登录和注册功能的开发

图 22-4　新蜂商城账号登录

在页面中依次输入登录信息并点击"立即登录"按钮。如果登录信息错误，页面会出现提示信息。如果登录成功，页面会跳转至商城端首页。此时可以看到顶部导航栏已经出现了商城用户的基本身份信息，首页用户信息展示效果如图 22-5 所示。

图 22-5　首页用户信息展示效果

读者可以按照文中的思路和过程自行测试。至此，新蜂商城的登录和注册相关功能就开发完成。

第 23 章

商城端搜索商品功能的开发

至此,商城端首页所有的功能基本已经开发完成,页面上大部分显示区域都实现了数据交互,但是搜索框还没有做任何的数据交互。搜索框是首页还没有实现的一个功能,本章将会讲解商城端的商品搜索功能,包括搜索框的搜索功能和按照分类信息搜索商品列表的功能。

23.1 搜索页面的设计和数据格式的定义

23.1.1 搜索页面的设计

商城端搜索页面最终的展现效果如图 23-1 所示。

该页面的布局与首页非常相似,主要含有顶部导航栏区域、商品列表区域和底部页脚区域。这里重点关注图 23-1 中标注的三个地方。

(1)搜索页副标题区域:这个区域处于整个搜索页面功能区的顶部,用于显示搜索信息。因为分类搜索和关键字搜索都需要使用该页面,所以在显示搜索信息时需要对搜索功能进行区分。如果是根据关键字搜索,则这个区域只会显示用户搜索的关键字。如果是根据分类搜索,那么这里还会显示分类信息的筛选框。

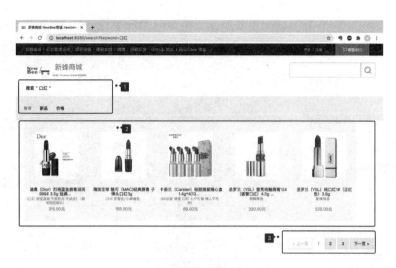

图 23-1　商城端商品搜索页面

（2）商品列表区域：用于展示商品列表，显示商品的概览信息。这是该页面最主要的部分，包括商品图片、商品价格、商品简介等内容。这些信息在页面上主要供用户查看和筛选。在点击单个商品后会跳转到对应的商品详情页面。

（3）分页导航区域：这里放置分页按钮，用于分页跳转。这是分页功能必不可少的一部分。

23.1.2　数据格式的定义

商品分页列表数据肯定是一个 List 对象。同时，因为有分页功能，还需要返回分页字段，所以最终返回的数据格式为 PageResult 对象。而在列表单项对象中的字段则需要通过搜索结果页的内容进行确认。如图 23-2 所示是商品列表展示的字段，即为商品分页列表中需要渲染的内容。

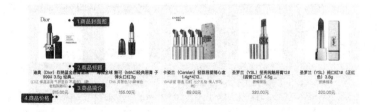

图 23-2　商品列表展示的字段

通过图 23-2 可以看到页面中主要展示了商品封面图字段、商品标题字段、商品简介字段、商品价格字段。这里的字段通常会设计成可跳转的形式，即在点击商品标题或者商品封面图后会跳转到对应的商品详情页面中。因此这里还需要一个商品实体的 id 字段。返回数据的格式编码如下所示：

```java
package ltd.newbee.mall.controller.vo;
import java.io.Serializable;
/**
 * 搜索列表页的商品VO
 */
public class NewBeeMallSearchGoodsVO implements Serializable {

    private Long goodsId;

    private String goodsName;

    private String goodsIntro;

    private String goodsCoverImg;

    private Integer sellingPrice;

    public Long getGoodsId() {
        return goodsId;
    }

    public void setGoodsId(Long goodsId) {
        this.goodsId = goodsId;
    }
    public String getGoodsName() {
        return goodsName;
    }
    public void setGoodsName(String goodsName) {
        this.goodsName = goodsName;
    }
    public String getGoodsIntro() {
        return goodsIntro;
    }
    public void setGoodsIntro(String goodsIntro) {
        this.goodsIntro = goodsIntro;
    }
```

```java
    public String getGoodsCoverImg() {
        return goodsCoverImg;
    }

    public void setGoodsCoverImg(String goodsCoverImg) {
        this.goodsCoverImg = goodsCoverImg;
    }

    public Integer getSellingPrice() {
        return sellingPrice;
    }
    public void setSellingPrice(Integer sellingPrice) {
        this.sellingPrice = sellingPrice;
    }
}
```

23.2 发起搜索请求

接下来讲解的是整个搜索功能的第一个操作：发起搜索请求。该操作有两个途径来实现，分别是通过提交搜索框中的内容来实现商品关键字的搜索，以及点击分类名称来实现按照分类信息搜索商品。

23.2.1 商品的关键字搜索

首先，需要对搜索框所在的 div 进行代码修改，这部分内容在 header.html 中，代码如下所示：

```html
<div class="fr">
  <div class="search">
    <input class="text" type="text" id="keyword" autocomplete="off">
    <div class="search_hot">
    </div>
  </div>
  <div class="button iconfont icon-search" onclick="search()"></div>
</div>
```

然后，给搜索输入框旁边的搜索图片绑定 onclick 事件，在用户点击搜索按钮后执行 search()方法。相应的搜索逻辑会在 JS 方法里进行编码实现。在 resources/static/mall/js

目录下新增 search.js 文件并新增如下代码:

```javascript
$(function () {
    $('#keyword').keypress(function (e) {
        var key = e.which; //e.which是按键的值
        if (key == 13) {
            var q = $(this).val();
            if (q && q != '') {
                window.location.href = '/search?keyword=' + q;
            }
        }
    });
});

function search() {
    var q = $('#keyword').val();
    if (q && q != '') {
        window.location.href = '/search?keyword=' + q;
    }
}
```

该段代码定义了 search()方法的实现,在用户点击搜索按钮后会将用户输入的字段取出并放入搜索请求的路径中进行请求。同时也定义了输入框的 keypress()事件,在用户在输入框中输入需要搜索的字段后按 Enter 键,也可以发起搜索请求。

最后,修改 index.html 模板文件,增加 search.js 的引用代码:

```html
<script th:src="@{/mall/js/search.js}"></script>
```

根据关键字搜索的整个交互过程如图 23-3 所示。先在输入框中输入关键字并点击搜索按钮。

图 23-3 根据关键字搜索

在搜索后,这里会跳转到搜索结果页面。该页面会展示出该关键字所关联的商品内容,如图 23-4 所示。

第 23 章 商城端搜索商品功能的开发

图 23-4 关键字搜索结果页面

23.2.2 商品的分类搜索功能

如果根据分类进行搜索，就需要把三级分类的 id 传给后台进行处理。这个交互在首页分类功能展示时候就已经做了处理，代码如下所示：

```html
<dd><th:block th:each="thirdLevelCategory : ${secondLevelCategory.thirdLevelCategoryVOS}">
<em><a href="#" th:href="@{'/search?goodsCategoryId='+${thirdLevelCategory.categoryId}}">
<th:block th:text="${thirdLevelCategory.categoryName}"></th:block></a>
</em>
</th:block>
</dd>
```

在循环处理三级分类的信息展示时，这些文字信息分别添加了对应的超链接，点击之后会跳转到商品搜索页面，并带上分类搜索所需的参数 goodsCategoryId（即当前三级分类的 id）。

根据商品类目搜索的整个交互过程如图 23-5 所示，选择对应的商品分类信息并点击即可。

图 23-5 根据商品类目搜索

在点击商品分类信息后,跳转到搜索结果页面,该页面会展示出与该分类所关联的商品内容,如图 23-6 所示。

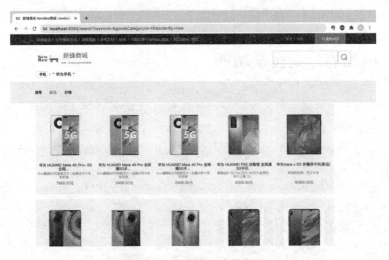

图 23-6 分类搜索结果页面

23.3 商品数据查询的实现代码

接下来是数据查询功能的实现。商品搜索分页列表中的字段可以通过直接查询 tb_newbee_mall_goods_info 商品表来获取。同时需要注意分页功能的实现,在传参时需要传入关键字和页码。

23.3.1 数据层代码的实现

首先，在商品实体 Mapper 接口 NewBeeMallGoodsMapper.java 的文件中新增如下方法：

```
/**
* 根据搜索字段查询分页数据
* @param pageUtil
* @return
*/
List<NewBeeMallGoods> findNewBeeMallGoodsListBySearch(PageQueryUtil pageUtil);

/**
* 根据搜索字段查询总数
* @param pageUtil
* @return
*/
int getTotalNewBeeMallGoodsBySearch(PageQueryUtil pageUtil);
```

然后，在映射文件 NewBeeMallGoodsMapper.xml 中添加具体的 SQL 语句，新增代码如下所示：

```
<select id="findNewBeeMallGoodsListBySearch" parameterType="Map" resultMap="BaseResultMap">
    select
    <include refid="Base_Column_List"/>
    from tb_newbee_mall_goods_info
    <where>
        <if test="keyword!=null and keyword!=''">
            and (goods_name like CONCAT('%',#{keyword},'%') or goods_intro like CONCAT('%',#{keyword},'%'))
        </if>
        <if test="goodsCategoryId!=null and goodsCategoryId!=''">
            and goods_category_id = #{goodsCategoryId}
        </if>
        <if test="goodsSellStatus!=null">
            and goods_sell_status = #{goodsSellStatus}
        </if>
    </where>
```

```xml
    <if test="orderBy!=null and orderBy!=''">
        <choose>
            <when test="orderBy == 'new'">
                <!-- 按照发布时间倒序排列 -->
                order by goods_id desc
            </when>
            <when test="orderBy == 'price'">
                <!-- 按照售价从小到大排列 -->
                order by selling_price asc
            </when>
            <otherwise>
                <!-- 默认按照库存数量从大到小排列 -->
                order by stock_num desc
            </otherwise>
        </choose>
    </if>
    <if test="start!=null and limit!=null">
        limit #{start},#{limit}
    </if>
</select>

<select id="getTotalNewBeeMallGoodsBySearch" parameterType="Map" resultType="int">
    select count(*) from tb_newbee_mall_goods_info
    <where>
        <if test="keyword!=null and keyword!=''">
            and (goods_name like CONCAT('%',#{keyword},'%') or goods_intro like CONCAT('%',#{keyword},'%'))
        </if>
        <if test="goodsCategoryId!=null and goodsCategoryId!=''">
            and goods_category_id = #{goodsCategoryId}
        </if>
        <if test="goodsSellStatus!=null">
            and goods_sell_status = #{goodsSellStatus}
        </if>
    </where>
</select>
```

根据前端传输过来的关键字和商品类目 id，系统会对商品记录进行检索，并使用 MySQL 数据库的 LIKE 语法对关键字进行过滤。首先，根据 goods_category_id 字段对商品类目进行过滤，然后根据 orderBy 字段进行商品搜索分页结果的排序，最后根据 start 和 limit 两个分页所必需的参数过滤出对应页码中的列表数据。

注意，为了避免给读者造成误导，这里解释一下，开发人员可以自行决定字段的命名，只要功能正常实现即可。start 和 limit 两个字段是本项目中定义并命名的，也可以命名为其他名称，比如 startNum 和 size 等。

23.3.2 业务层代码的实现

首先，在商品业务层代码 NewBeeMallGoodsService.java 中新增如下方法：

```
/**
 * 商品搜索
 *
 * @param pageUtil
 * @return
 */
PageResult searchNewBeeMallGoods(PageQueryUtil pageUtil);
```

然后，在 NewBeeMallGoodsServiceImpl 类中将上述方法实现，新增如下代码：

```
@Override
public PageResult searchNewBeeMallGoods(PageQueryUtil pageUtil) {
    List<NewBeeMallGoods> goodsList = goodsMapper.findNewBeeMallGoodsListBySearch(pageUtil);
    int total = goodsMapper.getTotalNewBeeMallGoodsBySearch(pageUtil);
    List<NewBeeMallSearchGoodsVO> newBeeMallSearchGoodsVOS = new ArrayList<>();
    if (!CollectionUtils.isEmpty(goodsList)) {
        newBeeMallSearchGoodsVOS = BeanUtil.copyList(goodsList, NewBeeMallSearchGoodsVO.class);
        for (NewBeeMallSearchGoodsVO newBeeMallSearchGoodsVO : newBeeMallSearchGoodsVOS) {
            String goodsName = newBeeMallSearchGoodsVO.getGoodsName();
            String goodsIntro = newBeeMallSearchGoodsVO.getGoodsIntro();
            // 字符串过长导致文字超出的问题
            if (goodsName.length() > 28) {
                goodsName = goodsName.substring(0, 28) + "...";
                newBeeMallSearchGoodsVO.setGoodsName(goodsName);
            }
            if (goodsIntro.length() > 30) {
                goodsIntro = goodsIntro.substring(0, 30) + "...";
                newBeeMallSearchGoodsVO.setGoodsIntro(goodsIntro);
            }
```

```
    }
  }
  PageResult pageResult = new PageResult(newBeeMallSearchGoodsVOS, total,
pageUtil.getLimit(), pageUtil.getPage());
  return pageResult;
}
```

这里定义了 searchNewBeeMallGoods()方法并传入 PageQueryUtil 对象作为参数。商品类目 id 、关键字 keyword 字段、分页所需的 page 字段、排序字段等都作为属性放在了这个对象中。关键字或者商品类目 id 用于过滤想要的商品列表，page 字段用于确定查询第几页的数据。这里通过 SQL 查询出对应的分页数据，再填充数据。某些字段太长会导致页面上的展示效果不好，所以对这些字段进行了简单的字符串处理并设置到 NewBeeMallSearchGoodsVO 对象中。最终返回的数据类型为 PageResult 对象。

23.4 商品搜索结果页面数据的渲染

23.4.1 参数封装及分页数据的获取

想要将数据通过 Thymeleaf 语法渲染到前端页面上，需要先将数据获取并转发到对应的模板页面中。这就需要在 Controller 方法中将查询到的数据放入 request 对象中，在 ltd.newbee.mall.controller.mall 包中新建 GoodsController 控制类，并新增 searchPage()方法，代码如下所示：

```
package ltd.newbee.mall.controller.mall;

import ltd.newbee.mall.common.Constants;
import ltd.newbee.mall.controller.vo.SearchPageCategoryVO;
import ltd.newbee.mall.service.NewBeeMallCategoryService;
import ltd.newbee.mall.service.NewBeeMallGoodsService;
import ltd.newbee.mall.util.PageQueryUtil;
import org.springframework.stereotype.Controller;
import org.springframework.util.StringUtils;
import org.springframework.web.bind.annotation.GetMapping;
import org.springframework.web.bind.annotation.RequestParam;

import javax.annotation.Resource;
import javax.servlet.http.HttpServletRequest;
import java.util.Map;
```

```java
@Controller
public class GoodsController {

    @Resource
    private NewBeeMallGoodsService newBeeMallGoodsService;
    @Resource
    private NewBeeMallCategoryService newBeeMallCategoryService;

    @GetMapping({"/search", "/search.html"})
    public String searchPage(@RequestParam Map<String, Object> params, HttpServletRequest request) {
        if (StringUtils.isEmpty(params.get("page"))) {
            params.put("page", 1);
        }
        params.put("limit", Constants.GOODS_SEARCH_PAGE_LIMIT);
        //封装分类数据
        if (params.containsKey("goodsCategoryId") && !StringUtils.isEmpty(params.get("goodsCategoryId") + "")) {
            Long categoryId = Long.valueOf(params.get("goodsCategoryId") + "");
            SearchPageCategoryVO searchPageCategoryVO = newBeeMallCategoryService.getCategoriesForSearch(categoryId);
            if (searchPageCategoryVO != null) {
                request.setAttribute("goodsCategoryId", categoryId);
                request.setAttribute("searchPageCategoryVO", searchPageCategoryVO);
            }
        }
        //封装参数供前端回显
        if (params.containsKey("orderBy") && !StringUtils.isEmpty(params.get("orderBy") + "")) {
            request.setAttribute("orderBy", params.get("orderBy") + "");
        }
        String keyword = "";
        //对 keyword 做过滤
        if (params.containsKey("keyword") && !StringUtils.isEmpty((params.get("keyword") + "").trim())) {
            keyword = params.get("keyword") + "";
        }
        request.setAttribute("keyword", keyword);
        params.put("keyword", keyword);
        //封装商品数据
        PageQueryUtil pageUtil = new PageQueryUtil(params);
```

```
            request.setAttribute("pageResult", newBeeMallGoodsService.search
NewBeeMallGoods(pageUtil));
        return "mall/search";
    }
}
```

该方法的映射路径为/search，请求方法为 GET，所有的传参都用 Map 对象来接收。前端传过来的参数主要有 page、keyword、goodsCategoryId 和 orderBy。

page 参数是分页所必需的字段，如果不传参的话默认为第 1 页。keyword 参数是关键字，用来过滤商品名和商品简介。goodsCategoryId 参数是用来过滤商品分类 id 的字段。orderBy 参数则是排序字段，不同的排序方式，返回的数据也会不同。

首先根据以上字段封装查询参数 PageQueryUtil 对象，然后通过 SQL 查询出对应的分页数据 pageResult 并放入 request 对象中，最后跳转到 mall 目录下的 search.html 模板页面进行数据渲染。需要注意的是 keyword、orderBy 等字段也会返回给页面，主要是为了搜索页副标题区域的显示和翻页链接的拼接。

23.4.2 搜索结果页面渲染的逻辑实现

在 resources/templates/mall 目录中新增搜索页 search.html，模板代码如下所示：

```html
<!-- Copyright (c) 2019-2020 十三 all rights reserved. -->
<!DOCTYPE html>
<html lang="en" xmlns:th="http://www.thymeleaf.org">
<head th:replace="mall/header::head-fragment('NewBee 商城-搜索','search')">
</head>
<body>
<header th:replace="mall/header::header-fragment"></header>
<!-- nav -->
<nav th:replace="mall/header::nav-fragment"></nav>

<!--分类筛选-->
<div class="classify">
    <div class="category">
        <div class="category_bar">
            <th:block th:if="${searchPageCategoryVO!=null}">
                <div class="fm c">
                    <a href="##" class="qqq" th:text="${searchPageCategoryVO.secondLevelCategoryName}">newbee-mall</a>
```

```html
                <div>
                    <th:block th:each="thirdLevelCategory : ${searchPageCategoryVO.thirdLevelCategoryList}">
                        <a th:href="@{${'/search?goodsCategoryId='+thirdLevelCategory.categoryId}}"
                           th:text="${thirdLevelCategory.categoryName}">newbee-mall</a>
                    </th:block>
                </div>
            </div>
            <i><img src="mall/image/right-@1x.png" alt=""></i>
            <div class="findword">"
                <th:block th:text="${searchPageCategoryVO.CurrentCategoryName}"></th:block>
                "
            </div>
        </th:block>
        <th:block th:if="${searchPageCategoryVO==null}">
            <th:block th:if="${keyword!=null and keyword !=''}">
                <div class="findword">搜索 "
                    <th:block th:text="${keyword}"></th:block>
                    "
                </div>
            </th:block>
        </th:block>
    </div>
  </div>
</div>

<!--排序-->
<div class="sort">
    <div class="list">
        <a th:href="@{'/search?keyword='+${keyword==null?'':keyword}+'&goodsCategoryId='+${goodsCategoryId==null?'':goodsCategoryId}+'&orderBy=default'}">
            <div th:class="${orderBy==null || orderBy=='default'?'active':''}">推荐</div>
        </a>
        <a th:href="@{'/search?keyword='+${keyword==null?'':keyword}+'&goodsCategoryId='+${goodsCategoryId==null?'':goodsCategoryId}+'&orderBy=new'}">
            <div th:class="${orderBy=='new'?'active':''}">新品</div>
        </a>
```

```html
                <a th:href="@{'/search?keyword='+${keyword==null?'':keyword}+
'&goodsCategoryId='+${goodsCategoryId==null?'':goodsCategoryId}+'&order
By=price'}">
                    <div th:class="${orderBy=='price'?'active':''}">价格</div>
                </a>
            </div>
</div>

<div class="goods_item center">
    <div class="main center">
        <th:block th:if="${#lists.isEmpty(pageResult.list)}">
            <img style="margin-top: 16px;padding: 16px 20px;" th:src="@{/
mall/image/null-content.png}">
        </th:block>
        <th:block th:unless="${#lists.isEmpty(pageResult.list)}">
            <th:block th:each="goods : ${pageResult.list}">
                <div class="item_card_frame">
                    <div class="item_card"><a th:href="@{'/goods/detail/'
+${goods.goodsId}}" target="_blank"><img
                            th:src="@{${goods.goodsCoverImg}}" th:alt="${goods.
goodsName}"></a></div>
                    <div class="item_brand"><a th:href="@{'/goods/detail/
'+${goods.goodsId}}" target="_blank"
                                            th:text="${goods.goodsName}">
newbee.ltd</a></div>
                    <div class="item_sub_intro" th:text="${goods.goods
Intro}">newbee.ltd</div>
                    <div class="item_price" th:text="${goods.sellingPrice+
'.00元'}">1299.00元</div>
                </div>
            </th:block>
        </th:block>
        <div class="clear"></div>
    </div>
    <div class="pages">
        <div class="page_wrap">
            <th:block th:if="${null != pageResult and !#lists.isEmpty(page
Result.list)}">
                <span class="page_span1">
                    <a th:href="@{${pageResult.currPage==1}?'##':'/search?
keyword='+${keyword==null?'':keyword}+'&page=' + ${pageResult.currPage-
1}+'&goodsCategoryId='+${goodsCategoryId==null?'':goodsCategoryId}+'&ord
erBy='+${orderBy==null?'':orderBy}}">
```

```html
                                    < 上一页
                                </a>
                                <th:block th:if="${pageResult.currPage-2 >=1}"><a
                                    th:href="@{'/search?keyword='+${keyword==null?'':
keyword}+'&page=' + ${pageResult.currPage-2}+'&goodsCategoryId='+${goods
CategoryId==null?'':goodsCategoryId}+'&orderBy='+${orderBy==null?'':
orderBy}}"
                                    th:text="${pageResult.currPage -2}">1</a></th:block>
                                <th:block th:if="${pageResult.currPage-1 >=1}"><a
                                    th:href="@{'/search?keyword='+${keyword==null?'':
keyword}+'&page=' + ${pageResult.currPage-1}+'&goodsCategoryId= '+${goods
CategoryId==null?'':goodsCategoryId}+'&orderBy='+${orderBy==null?'':
orderBy}}"
                                    th:text="${pageResult.currPage -1}">1</a></th:block>
                                <a href="##" class="active" th:text="${pageResult.currPage}">1</a>
                                <th:block th:if="${pageResult.currPage+1<=pageResult.total
Page}"><a
                                    th:href="@{'/search?keyword='+${keyword==null?'':
keyword}+'&page=' + ${pageResult.currPage+1}+'&goodsCategoryId=
'+${goodsCategoryId==null?'':goodsCategoryId}+'&orderBy='+${orderBy==
null?'':orderBy}}"
                                    th:text="${pageResult.currPage +1}">1</a></th:block>
                                <th:block th:if="${pageResult.currPage+2<=pageResult.total
Page}"><a
                                    th:href="@{'/search?keyword='+${keyword==null?'':
keyword}+'&page=' + ${pageResult.currPage+2}+'&goodsCategoryId='+${goods
CategoryId==null?'':goodsCategoryId}+'&orderBy='+${orderBy==null?'':
orderBy}}"
                                    th:text="${pageResult.currPage +2}">1</a></th:block>
                                <a th:href="@{${pageResult.currPage>=pageResult.
totalPage}?'##':'/search?keyword='+${keyword==null?'':keyword}+'&page=' +
${pageResult.currPage+1}+'&goodsCategoryId='+${goodsCategoryId==null?'':
goodsCategoryId}+'&orderBy='+${orderBy==null?'':orderBy}}">
                                    下一页 >
                                </a>
                            </span>
                        </th:block>
                    </div>
                </div>
            </div>

            <div th:replace="mall/footer::footer-fragment"></div>
            <!-- jQuery -->
```

```
<script th:src="@{/admin/plugins/jquery/jquery.min.js}"></script>
<script th:src="@{/mall/js/search.js}" type="text/javascript"></script>
</body>
</html>
```

通过上述代码可以看到页面中主要进行了四个逻辑运算,从上至下依次为:

①分类筛选和关键字回显

②排序字段样式渲染和 URL 拼接

③商品列表渲染

④分页区域渲染

关键字回显及排序字段的处理比较简单,通过后台返回的数据直接运用 Thymeleaf 语法即可完成。

这里重点讲一下商品列表渲染和分页区域的生成和显示。在商品分页列表区域和分页功能区域对应的位置读取 pageResult 对象中的 list 数据和分页数据。list 数据为商品分页列表数据,首先使用 th:each 循环语法将商品预览图字段、商品标题字段、商品简介字段、商品价格字段渲染出来。然后根据分页字段 currPage 当前页码、totalPage 总页码以及其他回显字段将下方的分页按钮渲染出来。这里的分页并没有借助插件来实现,而是通过自行编码的方式来生成分页按钮的。

通过页面分页区域的代码可以看出,最多会有 7 个翻页按钮,分别是"上一页""下一页",以及具体的 5 个数字页码翻页按钮。"上一页"和"下一页"两个按钮是固定的。当前页为第 1 页,点击"上一页"不进行跳转,因为此时没有上一页。当前页是最后一页,点击"下一页"不进行跳转,因为此时没有下一页。而中间的具体页码的生成,则根据 currPage 来实现。

这里具体看一下生成逻辑,通过当前页变量和总页码进行判断,如果当前页之前应该有数据,则生成当前页码前面的页码,如果当前页码之后有数据则生成当前页码后面的页码。以当前页码为中点,前后最多分别生成两个分页按钮。这是新蜂商城项目中的默认实现方式。读者也可以进行扩展,增加或者减少生成的翻页按钮数量。

第 24 章

商品详情页及购物车功能的开发

在商品功能模块和订单结算模块之间还有一个功能模块,即购物车模块。它处于整个购物环节由商品到订单转化过程的中间状态,负责打通商品和订单这两个模块。

本章将讲解购物车相关功能模块的开发,同时也会介绍一下商品详情功能的开发。

24.1 商城端商品详情页面的制作

商品详情页面可以让用户看到更多的商品信息,以便他们更好地进行选择和比对。商品详情页的制作并不复杂,实现逻辑就是根据商品 id 查询到商品表中的记录,并使用 Thymeleaf 语法将数据渲染到页面中即可。因为商品详情页与购物车功能模块有一定的关联,所以笔者放在这里讲解。商品详情页中包含"加入购物车"的按钮,如果想在购物车中添加数据需要进入商品详情页面。

24.1.1 商品详情页跳转逻辑的实现

详情页通常通过点击首页或者搜索列表页中的单个商品卡片的链接跳转而来。详情页的路径定义为/goods/detail/{goodsId},在 GoodsController 中添加 detailPage()对这个路径请求进行处理,新增如下代码:

```
@GetMapping("/goods/detail/{goodsId}")
public String detailPage(@PathVariable("goodsId") Long goodsId, HttpServlet
```

```
Request request) {
    if (goodsId < 1) {
        return "error/error_5xx";
    }
    NewBeeMallGoods goods = newBeeMallGoodsService.getNewBeeMallGoodsById
(goodsId);
    if (goods == null) {
        return "error/error_404";
    }
    if (Constants.SELL_STATUS_UP != goods.getGoodsSellStatus()) {
        return "error/error_404";
    }
    NewBeeMallGoodsDetailVO goodsDetailVO = new NewBeeMallGoodsDetailVO();
    BeanUtil.copyProperties(goods, goodsDetailVO);
    goodsDetailVO.setGoodsCarouselList(goods.getGoodsCarousel(). split(","));
    request.setAttribute("goodsDetail", goodsDetailVO);
    return "mall/detail";
}
```

goodsId 这个参数就是商品主键 id。这里通过@PathVariable 注解读取路径中的这个字段值，并根据该值调用 NewBeeMallGoodsService 中的 getNewBeeMallGoodsById()方法获取 NewBeeMallGoods 对象。getNewBccMallGoodsById()方法的实现方式就是根据主键 id 查询数据库中的商品表并返回商品实体数据，最后将查询到的商品详情数据放到 request 请求中并跳转到 detail.html 模板页面中。

24.1.2　商品详情页面数据的渲染

在 templates/mall 目录中新建 detail.html 模板页面，代码如下所示：

```
<!-- Copyright (c) 2019-2020 十三 all rights reserved. -->
<!DOCTYPE html>
<html lang="en" xmlns:th="http://www.thymeleaf.org">
<head th:replace="mall/header::head-fragment('NewBee 商城-商品详情
','detail')">
</head>
<body>
<header th:replace="mall/header::header-fragment"></header>

<div id="detail">
    <!-- nav -->
```

```html
<nav th:replace="mall/header::nav-fragment"></nav>

<div class="dc">
    <div class="content w">
        <div class="title fl">商品详情</div>
        <div class="clear"></div>
    </div>
</div>

<div class="intro mt20 w clearfix">
    <div class="left fl" style="position: relative;">
        <div class="swiper-container fl">
            <img th:src="@{${goodsDetail.goodsCoverImg}}">
        </div>
    </div>
    <div class="right fr">
        <div class="h3 ml20 mt20" th:text="${goodsDetail.goodsName}">NewBeeMall</div>
        <div class="sub_title mr40 ml20 mt10" th:text="${goodsDetail.goodsIntro}">NewBeeMall</div>
        <div class="item_price mr40 ml20 mt10">
            <th:block th:text="${goodsDetail.sellingPrice}+'.00 元'"></th:block>
            <del>
                <th:block th:text="${goodsDetail.originalPrice}+'.00 元'"></th:block>
            </del>
        </div>

        <div class="order">
            <input class="car" type="button" value="立即选购"/>
            <input class="car" type="button" value="加入购物车"/>
        </div>
        <div class="tb-extra ml20" id="J_tbExtra">
            <dl>
                <dt>承诺</dt>
                <dd><a class="J_Cont" title="满足 7 天无理由退换货申请的前提下，包邮商品需要买家承担退货邮费，非包邮商品需要买家承担发货和退货邮费。" href="#" target="_blank"><img th:src="@{/mall/image/7d.jpg}">7 天无理由</a></dd>
            </dl>
            <dl>
                <dt>支付</dt>
```

```html
            <dd><a href="##" target="_blank"><img th:src="@{/mall/image/hua.png}">蚂蚁花呗</a><a href="##"
target="_blank"><img
                    th:src="@{/mall/image/card.png}">信用卡支付</a><a
href="##" target="_blank"><img
                    th:src="@{/mall/image/ji.png}">集分宝</a></dd>
        </dl>
        <dl>
            <dt>支持</dt>
            <dd>折旧变现,买新更省钱。<a style="float:none;text-decoration:underline;" href="##">详情</a></dd>
        </dl>
    </div>
  </div>
  <div class="clear"></div>
</div>
<!-- 这里使用的是th:utext标签,用th:text不会解析html,而用th:utext会解析html -->
<div class="goods mt20 w clearfix" th:utext="${goodsDetail.goodsDetailContent}">
</div>
</div>

<div th:replace="mall/footer::footer-fragment"></div>

<!-- jQuery -->
<script th:src="@{/admin/plugins/jquery/jquery.min.js}"></script>
<script th:src="@{/mall/js/search.js}" type="text/javascript"></script>
<script th:src="@{/admin/plugins/sweetalert/sweetalert.min.js}"></script>
</body>
</html>
```

这里根据返回的 goodsDetail 对象依次将数据渲染到商品标题区域、商品的基础信息区域和商品详情区域。使用到的 Thymeleaf 语法也比较简单,直接读取对应的字段即可。

唯一需要注意的是在详情内容渲染时用到的 th 标签是 th:utext。因为在 MySQL 数据库中存储的商品详情字段包含排版和样式,字段中带有 HTML 标签。如果使用 th:text 标签是无法在页面中显示样式和排版内容的,而是一段源码字符串。因此使用 th:utext 标签来渲染,这样才能够正确地解析和渲染商品详情。

这里笔者做了一次对比,在编码完成后,启动 Spring Boot 项目。在启动成功后打开浏览器并进入商城端首页，点击对应的商品进入该商品的详情页面。th:utext 标签和 th:text 标签的显示效果如图 24-1 和图 24-2 所示。读者可以分别使用这两个标签来渲染，并自行比对效果。

图 24-1　使用 th:utext 标签渲染的详情页效果

图 24-2　使用 th:text 标签渲染的详情页效果

24.2 购物车模块简介及表结构设计

24.2.1 购物车模块简介

为了让读者更好地理解购物车这个功能点，这里先介绍线下超市或者商场的购物流程。

①进入商场

②获取购物车或者购物篮

③在商城中四处逛

④在不同的区域选择不同的商品

⑤经过一番筛选后，将想要购买的商品放入购物车/购物篮

⑥某些商品需要称重或者其他处理

⑦到收银台清点商品并计算价格

⑧结账

⑨离开商场并回家

大部分线上商城中的购物车功能模块，都是将线下购物车进行了抽象而开发的一个功能。新蜂商城也是如此。与线下实体的购物车不同，线上购物车模块的作用是存放商城用户挑选的商品数据。

24.2.2 购物车表结构设计

商城系统的购物车功能模块用到的购物项表结构主要字段如下所示。

①user_id：用户的 id，根据这个字段确定用户购物车中的数据

②goods_id：关联的商品 id，根据这个字段查询对应的商品信息并显示到页面上

③goods_count：购物车中某件商品的数量

④create_time：商品添加到购物车中的时间

因此，购物车表结构设计如下所示：

```
USE 'newbee_mall_db ';
```

```sql
DROP TABLE IF EXISTS 'tb_newbee_mall_shopping_cart_item';

CREATE TABLE 'tb_newbee_mall_shopping_cart_item' (
  'cart_item_id' bigint(20) NOT NULL AUTO_INCREMENT COMMENT '购物项主键id',
  'user_id' bigint(20) NOT NULL COMMENT '用户主键id',
  'goods_id' bigint(20) NOT NULL DEFAULT '0' COMMENT '关联商品id',
  'goods_count' int(11) NOT NULL DEFAULT 1 COMMENT '数量(最大为5)',
  'is_deleted' tinyint(4) NOT NULL DEFAULT '0' COMMENT '删除标识字段(0-未删除 1-已删除)',
  'create_time' datetime NOT NULL DEFAULT CURRENT_TIMESTAMP COMMENT '创建时间',
  'update_time' datetime NOT NULL DEFAULT CURRENT_TIMESTAMP COMMENT '最新修改时间',
  PRIMARY KEY ('cart_item_id')
) ENGINE=InnoDB DEFAULT CHARSET=utf8;
```

每个字段对应的含义在上面的 SQL 中都有介绍，读者可以对照理解，并正确地把建表 SQL 导入到数据库中。上述代码中的购物车模块用来存储用户选择的商品数据，为订单结算做准备。这也是距离结算环节最近的一个步骤和功能点。接下来讲一下它相关功能的实现。

24.3 将商品加入购物车功能的实现

24.3.1 新建购物项实体类和 Mapper 接口

首先在 ltd.newbee.mall.entity 包中创建购物项实体类，选中 entity 包并右键点击，在弹出的菜单中选择 "New→Java Class"，然后在弹出的窗口中输入 "NewBeeMallShoppingCartItem"，最后在 NewBeeMallShoppingCartItem 类中新增如下代码：

```java
package ltd.newbee.mall.entity;

import java.util.Date;

public class NewBeeMallShoppingCartItem {
    private Long cartItemId;

    private Long userId;
```

```java
    private Long goodsId;

    private Integer goodsCount;

    private Byte isDeleted;

    private Date createTime;

    private Date updateTime;

    public Long getCartItemId() {
        return cartItemId;
    }

    public void setCartItemId(Long cartItemId) {
        this.cartItemId = cartItemId;
    }

    public Long getUserId() {
        return userId;
    }

    public void setUserId(Long userId) {
        this.userId = userId;
    }

    public Long getGoodsId() {
        return goodsId;
    }

    public void setGoodsId(Long goodsId) {
        this.goodsId = goodsId;
    }

    public Integer getGoodsCount() {
        return goodsCount;
    }

    public void setGoodsCount(Integer goodsCount) {
        this.goodsCount = goodsCount;
    }
```

```java
    public Byte getIsDeleted() {
        return isDeleted;
    }

    public void setIsDeleted(Byte isDeleted) {
        this.isDeleted = isDeleted;
    }

    public Date getCreateTime() {
        return createTime;
    }

    public void setCreateTime(Date createTime) {
        this.createTime = createTime;
    }

    public Date getUpdateTime() {
        return updateTime;
    }

    public void setUpdateTime(Date updateTime) {
        this.updateTime = updateTime;
    }
}
```

接下来，首先在 ltd.newbee.mall.dao 包中新建购物项实体的 Mapper 接口，选中 dao 包并右键点击，在弹出的菜单中选择"New→Java Class"，然后在弹出的窗口中输入 "NewBeeMallShoppingCartItemMapper"，并选中"Interface"选项，最后在 NewBeeMallShoppingCartItemMapper.java 文件中新增如下代码：

```java
package ltd.newbee.mall.dao;

import ltd.newbee.mall.entity.NewBeeMallShoppingCartItem;
import org.apache.ibatis.annotations.Param;

public interface NewBeeMallShoppingCartItemMapper {

    /**
     * 保存一条新记录
     *
     * @param record
     * @return
     */
```

```
    int insertSelective(NewBeeMallShoppingCartItem record);

    /**
     * 根据userId和goodsId查询记录
     *
     * @param newBeeMallUserId
     * @param goodsId
     * @return
     */
    NewBeeMallShoppingCartItem selectByUserIdAndGoodsId(@Param("newBee
MallUserId") Long newBeeMallUserId, @Param("goodsId") Long goodsId);

    /**
     * 根据userId查询当前用户已添加了多少条记录
     *
     * @param newBeeMallUserId
     * @return
     */
    int selectCountByUserId(Long newBeeMallUserId);
}
```

24.3.2 创建 NewBeeMallShoppingCartItemMapper 接口的映射文件

在 resources/mapper 目录下新建 NewBeeMallShoppingCartItemMapper 接口的映射文件 NewBeeMallShoppingCartItemMapper.xml，并进行映射文件的编写。

首先，定义映射文件与 Mapper 接口的对应关系。比如在该示例中，需要将 NewBeeMallShoppingCartItemMapper.xml 文件与对应的 NewBeeMallShoppingCartItemMapper 接口之间的关系定义出来：

```xml
<mapper namespace="ltd.newbee.mall.dao.NewBeeMallShoppingCartItemMapper">
```

然后，配置表结构和实体类的对应关系：

```xml
<resultMap id="BaseResultMap" type="ltd.newbee.mall.entity.NewBeeMall
ShoppingCartItem">
  <id column="cart_item_id" jdbcType="BIGINT" property="cartItemId"/>
  <result column="user_id" jdbcType="BIGINT" property="userId"/>
  <result column="goods_id" jdbcType="BIGINT" property="goodsId"/>
  <result column="goods_count" jdbcType="INTEGER" property="goodsCount"/>
```

```xml
        <result column="is_deleted" jdbcType="TINYINT" property="isDeleted"/>
        <result column="create_time" jdbcType="TIMESTAMP" property="createTime"/>
        <result column="update_time" jdbcType="TIMESTAMP" property="updateTime"/>
    </resultMap>
```

最后,按照对应的接口方法,编写具体的 SQL 语句,最终的 NewBeeMallShoppingCartItemMapper.xml 文件如下所示:

```xml
<?xml version="1.0" encoding="UTF-8"?>
<!DOCTYPE mapper PUBLIC "-//mybatis.org//DTD Mapper 3.0//EN" "http://mybatis.org/dtd/mybatis-3-mapper.dtd">
<mapper namespace="ltd.newbee.mall.dao.NewBeeMallShoppingCartItemMapper">
    <resultMap id="BaseResultMap" type="ltd.newbee.mall.entity.NewBeeMallShoppingCartItem">
        <id column="cart_item_id" jdbcType="BIGINT" property="cartItemId"/>
        <result column="user_id" jdbcType="BIGINT" property="userId"/>
        <result column="goods_id" jdbcType="BIGINT" property="goodsId"/>
        <result column="goods_count" jdbcType="INTEGER" property="goodsCount"/>
        <result column="is_deleted" jdbcType="TINYINT" property="isDeleted"/>
        <result column="create_time" jdbcType="TIMESTAMP" property="createTime"/>
        <result column="update_time" jdbcType="TIMESTAMP" property="updateTime"/>
    </resultMap>
    <sql id="Base_Column_List">
    cart_item_id, user_id, goods_id, goods_count, is_deleted, create_time, update_time
  </sql>

    <select id="selectByUserIdAndGoodsId" resultMap="BaseResultMap">
        select
        <include refid="Base_Column_List"/>
        from tb_newbee_mall_shopping_cart_item
        where user_id = #{newBeeMallUserId,jdbcType=BIGINT} and goods_id=#{goodsId,jdbcType=BIGINT} and is_deleted = 0
        limit 1
    </select>

    <select id="selectCountByUserId" resultType="int">
        select
        count(*)
        from tb_newbee_mall_shopping_cart_item
```

```xml
        where user_id = #{newBeeMallUserId,jdbcType=BIGINT} and is_deleted = 0
    </select>

    <insert id="insertSelective" parameterType="ltd.newbee.mall.entity.NewBeeMallShoppingCartItem">
        insert into tb_newbee_mall_shopping_cart_item
        <trim prefix="(" suffix=")" suffixOverrides=",">
            <if test="cartItemId != null">
                cart_item_id,
            </if>
            <if test="userId != null">
                user_id,
            </if>
            <if test="goodsId != null">
                goods_id,
            </if>
            <if test="goodsCount != null">
                goods_count,
            </if>
            <if test="isDeleted != null">
                is_deleted,
            </if>
            <if test="createTime != null">
                create_time,
            </if>
            <if test="updateTime != null">
                update_time,
            </if>
        </trim>
        <trim prefix="values (" suffix=")" suffixOverrides=",">
            <if test="cartItemId != null">
                #{cartItemId,jdbcType=BIGINT},
            </if>
            <if test="userId != null">
                #{userId,jdbcType=BIGINT},
            </if>
            <if test="goodsId != null">
                #{goodsId,jdbcType=BIGINT},
            </if>
            <if test="goodsCount != null">
                #{goodsCount,jdbcType=INTEGER},
            </if>
            <if test="isDeleted != null">
```

```xml
                #{isDeleted,jdbcType=TINYINT},
            </if>
            <if test="createTime != null">
                #{createTime,jdbcType=TIMESTAMP},
            </if>
            <if test="updateTime != null">
                #{updateTime,jdbcType=TIMESTAMP},
            </if>
        </trim>
    </insert>
</mapper>
```

24.3.3 业务层代码的实现

首先在 ltd.newbee.mall.service 包中新建业务处理类,选中 service 包并右键点击,在弹出的菜单中选择"New→Java Class",然后在弹出的窗口中输入"NewBeeMallShoppingCartService",并选中"Interface"选项,最后在 NewBeeMallShoppingCartService.java 文件中新增如下代码:

```java
package ltd.newbee.mall.service;

import ltd.newbee.mall.entity.NewBeeMallShoppingCartItem;

public interface NewBeeMallShoppingCartService {

    /**
     * 保存商品至购物车中
     *
     * @param newBeeMallShoppingCartItem
     * @return
     */
    String saveNewBeeMallCartItem(NewBeeMallShoppingCartItem newBeeMallShoppingCartItem);
}
```

目前这里只定义了一个保存商品至购物车中的 saveNewBeeMallCartItem()方法,其他方法后续会逐一讲解。

接下来在 ltd.newbee.mall.service.impl 包中新建 NewBeeMallShoppingCartService 的实现类,选中 impl 包并右键点击,在弹出的菜单中选择"New→Java Class",然后在弹

出的窗口中输入 "NewBeeMallShoppingCartServiceImpl"，最后在 NewBeeMallShoppingCartServiceImpl 类中新增如下代码：

```java
package ltd.newbee.mall.service.impl;

import ltd.newbee.mall.common.Constants;
import ltd.newbee.mall.common.ServiceResultEnum;
import ltd.newbee.mall.dao.NewBeeMallGoodsMapper;
import ltd.newbee.mall.dao.NewBeeMallShoppingCartItemMapper;
import ltd.newbee.mall.entity.NewBeeMallGoods;
import ltd.newbee.mall.entity.NewBeeMallShoppingCartItem;
import ltd.newbee.mall.service.NewBeeMallShoppingCartService;
import org.springframework.beans.factory.annotation.Autowired;
import org.springframework.stereotype.Service;

@Service
public class NewBeeMallShoppingCartServiceImpl implements NewBeeMallShoppingCartService {

    @Autowired
    private NewBeeMallShoppingCartItemMapper newBeeMallShoppingCartItemMapper;

    @Autowired
    private NewBeeMallGoodsMapper newBeeMallGoodsMapper;

    @Override
    public String saveNewBeeMallCartItem(NewBeeMallShoppingCartItem newBeeMallShoppingCartItem) {
        NewBeeMallShoppingCartItem temp = newBeeMallShoppingCartItemMapper.selectByUserIdAndGoodsId(newBeeMallShoppingCartItem.getUserId(), newBeeMallShoppingCartItem.getGoodsId());
        if (temp != null) {
            return "购物车中已存在";
        }
        NewBeeMallGoods newBeeMallGoods = newBeeMallGoodsMapper.selectByPrimaryKey(newBeeMallShoppingCartItem.getGoodsId());
        //商品为空
        if (newBeeMallGoods == null) {
            return ServiceResultEnum.GOODS_NOT_EXIST.getResult();
        }
        int totalItem = newBeeMallShoppingCartItemMapper.selectCountByUserId(newBeeMallShoppingCartItem.getUserId()) + 1;
```

```
        //超出单个商品的最大数量
        if (newBeeMallShoppingCartItem.getGoodsCount() > Constants.SHOPPING_
CART_ITEM_LIMIT_NUMBER) {
            return ServiceResultEnum.SHOPPING_CART_ITEM_LIMIT_NUMBER_ERROR.
getResult();
        }
        //超出最大数量
        if (totalItem > Constants.SHOPPING_CART_ITEM_TOTAL_NUMBER) {
            return ServiceResultEnum.SHOPPING_CART_ITEM_TOTAL_NUMBER_ERROR.
getResult();
        }
        //保存记录
        if (newBeeMallShoppingCartItemMapper.insertSelective(newBeeMall
ShoppingCartItem) > 0) {
            return ServiceResultEnum.SUCCESS.getResult();
        }
        return ServiceResultEnum.DB_ERROR.getResult();
    }
}
```

这里先对参数进行校验,校验步骤如下所示。

①根据用户信息和商品信息查询购物项表中是否已存在相同的记录,如果存在则进行修改操作,不存在则进行后续操作

②判断商品数据是否正确

③判断用户的购物车中的商品数量是否已超出最大限制

在校验通过后再进行新增操作,将该记录保存到数据库中。以上操作都需要调用 SQL 语句来完成。

24.3.4 将商品加入购物车接口的实现

在 controller/mall 包下新建 ShoppingCartController.java 类,并新增如下代码:

```
package ltd.newbee.mall.controller.mall;

import ltd.newbee.mall.common.Constants;
import ltd.newbee.mall.common.ServiceResultEnum;
import ltd.newbee.mall.controller.vo.NewBeeMallUserVO;
import ltd.newbee.mall.entity.NewBeeMallShoppingCartItem;
```

```java
import ltd.newbee.mall.service.NewBeeMallShoppingCartService;
import ltd.newbee.mall.util.Result;
import ltd.newbee.mall.util.ResultGenerator;
import org.springframework.stereotype.Controller;
import org.springframework.web.bind.annotation.PostMapping;
import org.springframework.web.bind.annotation.RequestBody;
import org.springframework.web.bind.annotation.ResponseBody;

import javax.annotation.Resource;
import javax.servlet.http.HttpSession;

@Controller
public class ShoppingCartController {

    @Resource
    private NewBeeMallShoppingCartService newBeeMallShoppingCartService;

    @PostMapping("/shop-cart")
    @ResponseBody
    public Result saveNewBeeMallShoppingCartItem(@RequestBody NewBeeMallShoppingCartItem newBeeMallShoppingCartItem,
                                                 HttpSession httpSession) {
        NewBeeMallUserVO user = (NewBeeMallUserVO) httpSession.getAttribute(Constants.MALL_USER_SESSION_KEY);
        newBeeMallShoppingCartItem.setUserId(user.getUserId());
        String saveResult = newBeeMallShoppingCartService.saveNewBeeMallCartItem(newBeeMallShoppingCartItem);
        //添加成功
        if (ServiceResultEnum.SUCCESS.getResult().equals(saveResult)) {
            return ResultGenerator.genSuccessResult();
        }
        //添加失败
        return ResultGenerator.genFailResult(saveResult);
    }
}
```

该接口首先负责接收前端的 POST 请求并处理其中的参数，接收的参数为 goodsId 字段和 goodsCount 字段，然后调用业务层方法对数据进行验证并保存到数据库中。

24.3.5 前端功能的实现

添加商品到购物车这个功能的交互是在商品详情页实现的。在该页面中有"加入购物车"的按钮。这里给该按钮绑定一个触发事件，在点击该按钮后执行 saveToCart() 方法。修改 detail.html 模板文件中的按钮代码如下所示：

```html
<input class="car" type="button" th:onclick="'saveToCart('+${goodsDetail.goodsId}+')'" value="加入购物车"/>
```

在点击该按钮后就能够使用 JS 代码调用后台的接口了。点击触发事件 saveToCart() 方法的实现代码如下所示：

```html
<script type="text/javascript">
/**
 * 添加到购物车
 */
function saveToCart(id) {
var goodsCount = 1;
var data = {
  "goodsId": id,
  "goodsCount": goodsCount
};
$.ajax({
  type: 'POST',
  url: '/shop-cart',
  contentType: 'application/json',
  data: JSON.stringify(data),
  success: function (result) {
    if (result.resultCode == 200) {
      swal({
        title: "添加成功",
        text: "确认框",
        icon: "success",
        buttons: true,
        dangerMode: true,
      }).then((flag) => {
        window.location.reload();
      }
          );
    } else {
      swal(result.message, {
        icon: "error",
      });
```

```
        }
      },
      error: function () {
        swal("操作失败", {
          icon: "error",
        });
      }
    });
}
</script>
```

用户在商品详情页点击"加入购物车"按钮后，系统就会执行 saveToCart()方法，执行逻辑如下所示。

①封装参数，主要是商品 id 和商品数量，商品数量默认为 1

②向后端添加购物车接口发送请求

③根据返回数据弹出提示信息

24.4 购物车列表功能的实现

24.4.1 数据格式的定义

购物车页面的商品项列表肯定是一个 List 对象，因此后台在返回数据时需要一个购物项列表对象，以及一些总览性的字段。这些总览性字段和列表中单项对象的字段可以通过购物车页面中所展示的内容进行确认。图 24-4 即为新蜂商城购物车页面中需要渲染的字段。

图 24-4 购物车列表页面字段

图 24-4 中包括购物项列表数据和底部总览性数据。购物项列表数据中的商品标题字段、商品预览图字段、商品价格字段可以通过购物项表中的 goods_id 来关联和查询。而其商品数量字段通过购物项表来查询。列表还有一个删除按钮，因此还需要把购物项的 id 字段返回给前端。总览性数据包括加购总量字段和总价字段。这样，返回数据的格式就得出来了：购物项列表数据+加购总量字段+总价字段。小计的两个字段需要单独处理。购物项 VO 对象编码如下所示：

```java
package ltd.newbee.mall.controller.vo;

import java.io.Serializable;

/**
 * 购物车页面购物项VO
 */
public class NewBeeMallShoppingCartItemVO implements Serializable {

    private Long cartItemId;

    private Long goodsId;

    private Integer goodsCount;

    private String goodsName;

    private String goodsCoverImg;

    private Integer sellingPrice;

    public Long getGoodsId() {
        return goodsId;
    }

    public void setGoodsId(Long goodsId) {
        this.goodsId = goodsId;
    }

    public String getGoodsName() {
        return goodsName;
    }

    public void setGoodsName(String goodsName) {
        this.goodsName = goodsName;
    }
```

```java
    public String getGoodsCoverImg() {
        return goodsCoverImg;
    }

    public void setGoodsCoverImg(String goodsCoverImg) {
        this.goodsCoverImg = goodsCoverImg;
    }

    public Integer getSellingPrice() {
        return sellingPrice;
    }

    public void setSellingPrice(Integer sellingPrice) {
        this.sellingPrice = sellingPrice;
    }

    public Long getCartItemId() {
        return cartItemId;
    }

    public void setCartItemId(Long cartItemId) {
        this.cartItemId = cartItemId;
    }

    public Integer getGoodsCount() {
        return goodsCount;
    }

    public void setGoodsCount(Integer goodsCount) {
        this.goodsCount = goodsCount;
    }
}
```

24.4.2 购物车列表数据的获取

接下来是数据查询的功能实现。购物车列表中的字段可以分别通过查询 tb_newbee_mall_shopping_cart_item 购物项表和 tb_newbee_mall_goods_info 商品表来获取。

首先，在购物项实体 Mapper 接口的 NewBeeMallShoppingCartItemMapper.java 文件

第 24 章 商品详情页及购物车功能的开发

中新增如下方法：

```java
/**
 * 根据 userId 和 number 字段获取固定数量的购物项列表数据
 * @param newBeeMallUserId
 * @param number
 * @return
 */
List<NewBeeMallShoppingCartItem> selectByUserId(@Param("newBeeMallUserId") Long newBeeMallUserId, @Param("number") int number);
```

然后，在映射文件 NewBeeMallShoppingCartItemMapper.xml 中添加具体的 SQL 语句，新增代码如下所示：

```xml
<select id="selectByUserId" resultMap="BaseResultMap">
    select
    <include refid="Base_Column_List"/>
    from tb_newbee_mall_shopping_cart_item
    where user_id = #{newBeeMallUserId,jdbcType=BIGINT} and is_deleted = 0
    limit #{number}
</select>
```

再在业务类 NewBeeMallShoppingCartService.java 中新增如下方法：

```java
/**
 * 获取我的购物车中的列表数据
 *
 * @param newBeeMallUserId
 * @return
 */
List<NewBeeMallShoppingCartItemVO> getMyShoppingCartItems(Long newBeeMallUserId);
```

最后根据 userId 获取所有的购物项数据，并在 NewBeeMallShoppingCartServiceImpl 类中将上述方法实现，新增如下代码：

```java
@Override
public List<NewBeeMallShoppingCartItemVO> getMyShoppingCartItems(Long newBeeMallUserId) {
    List<NewBeeMallShoppingCartItemVO> newBeeMallShoppingCartItemVOS = new ArrayList<>();
    List<NewBeeMallShoppingCartItem> newBeeMallShoppingCartItems = newBeeMallShoppingCartItemMapper.selectByUserId(newBeeMallUserId, Constants.SHOPPING_CART_ITEM_TOTAL_NUMBER);
```

```java
        if (!CollectionUtils.isEmpty(newBeeMallShoppingCartItems)) {
            //查询商品信息并做数据转换
            List<Long> newBeeMallGoodsIds = newBeeMallShoppingCartItems.stream().map(NewBeeMallShoppingCartItem::getGoodsId).collect(Collectors.toList());
            List<NewBeeMallGoods> newBeeMallGoods = newBeeMallGoodsMapper.selectByPrimaryKeys(newBeeMallGoodsIds);
            Map<Long, NewBeeMallGoods> newBeeMallGoodsMap = new HashMap<>();
            if (!CollectionUtils.isEmpty(newBeeMallGoods)) {
                newBeeMallGoodsMap = newBeeMallGoods.stream().collect(Collectors.toMap(NewBeeMallGoods::getGoodsId, Function.identity(), (entity1, entity2) -> entity1));
            }
            for (NewBeeMallShoppingCartItem newBeeMallShoppingCartItem : newBeeMallShoppingCartItems) {
                NewBeeMallShoppingCartItemVO newBeeMallShoppingCartItemVO = new NewBeeMallShoppingCartItemVO();
                BeanUtil.copyProperties(newBeeMallShoppingCartItem, newBeeMallShoppingCartItemVO);
                if (newBeeMallGoodsMap.containsKey(newBeeMallShoppingCartItem.getGoodsId())) {
                    NewBeeMallGoods newBeeMallGoodsTemp = newBeeMallGoodsMap.get(newBeeMallShoppingCartItem.getGoodsId());
                    newBeeMallShoppingCartItemVO.setGoodsCoverImg(newBeeMallGoodsTemp.getGoodsCoverImg());
                    String goodsName = newBeeMallGoodsTemp.getGoodsName();
                    // 字符串过长导致文字超出的问题
                    if (goodsName.length() > 28) {
                        goodsName = goodsName.substring(0, 28) + "...";
                    }
                    newBeeMallShoppingCartItemVO.setGoodsName(goodsName);
                    newBeeMallShoppingCartItemVO.setSellingPrice(newBeeMallGoodsTemp.getSellingPrice());
                    newBeeMallShoppingCartItemVOS.add(newBeeMallShoppingCartItemVO);
                }
            }
        }
        return newBeeMallShoppingCartItemVOS;
    }
```

这里首先定义了 getMyShoppingCartItems() 方法并传入 userId 字段作为参数,然后通过 SQL 查询出当前 userId 下的购物项列表数据。因为购物车页面需要展示商品信息,所以通过购物项表中的 goods_id 获取每个购物项对应的商品信息。接着是填充数据,即将相关字段封装到 NewBeeMallShoppingCartItemVO 对象中。某些字段太长会导致页面的展示效果不好，所以对这些字段进行了简单的字符串处理。最后将封装好的 List 对象返回。

24.4.3 购物车列表数据的渲染

想要将数据通过 Thymeleaf 语法渲染到前端页面上，首先需要将数据获取并转发到对应的模板页面中。这就需要在 Controller 方法中将查询到的数据放入 request 请求中。这里在 ShoppingCartController 中新增 cartListPage()方法，代码如下所示：

```
@GetMapping("/shop-cart")
public String cartListPage(HttpServletRequest request,
                           HttpSession httpSession) {
    NewBeeMallUserVO user = (NewBeeMallUserVO) httpSession.getAttribute
(Constants.MALL_USER_SESSION_KEY);
    int itemsTotal = 0;
    int priceTotal = 0;
    List<NewBeeMallShoppingCartItemVO> myShoppingCartItems = newBeeMall
ShoppingCartService.getMyShoppingCartItems(user.getUserId());
    if (!CollectionUtils.isEmpty(myShoppingCartItems)) {
        //购物项总数
        itemsTotal = myShoppingCartItems.stream().mapToInt(NewBeeMall
ShoppingCartItemVO::getGoodsCount).sum();
        if (itemsTotal < 1) {
            return "error/error_5xx";
        }
        //总价
        for (NewBeeMallShoppingCartItemVO newBeeMallShoppingCartItemVO :
myShoppingCartItems) {
            priceTotal += newBeeMallShoppingCartItemVO.getGoodsCount() *
newBeeMallShoppingCartItemVO.getSellingPrice();
        }
        if (priceTotal < 1) {
            return "error/error_5xx";
        }
    }
    request.setAttribute("itemsTotal", itemsTotal);
    request.setAttribute("priceTotal", priceTotal);
    request.setAttribute("myShoppingCartItems", myShoppingCartItems);
    return "mall/cart";
```

}

在该方法中首先会调用业务层的方法，把当前用户添加到购物车中的购物项数据全部读取出来，然后计算商品总数和总价，并将三个对象都放到 request 对象中，最后跳转到 mall 目录下的 cart.html 模板页面进行数据的渲染。

在 resources/templates/mall 目录中新增搜索页 cart.html ，模板代码如下所示：

```html
<!-- Copyright (c) 2019-2020 十三 all rights reserved. -->
<!DOCTYPE html>
<html lang="en" xmlns:th="http://www.thymeleaf.org">
<head>
    <meta charset="UTF-8">
    <title>NewBee 商城-购物车</title>
    <link rel="stylesheet" th:href="@{mall/css/iconfont.css}">
    <link rel="stylesheet" th:href="@{mall/css/common.css}">
    <link rel="stylesheet" th:href="@{mall/styles/header.css}">
    <link rel="stylesheet" th:href="@{mall/styles/cart.css}">
    <link rel="stylesheet" th:href="@{/admin/plugins/sweetalert/sweetalert.css}"/>
</head>
<body>

<div id="cart">
    <div class="banner_x center">
        <a th:href="@{/index}" target="_blank">
            <div class="logo fl">
                <img src="mall/image/new-bee-logo-3.png"/>
            </div>
        </a>

        <div class="wdgwc fl ml20">购物车</div>
        <div class="wxts fl ml20">温馨提示：产品是否购买成功，以最终下单为准哦，请尽快结算</div>
        <div class="clear"></div>
    </div>
    <div class="cart_line"></div>
    <div class="cart_bg">
        <th:block th:if="${#lists.isEmpty(myShoppingCartItems)}">
            <div class="list center">
                <img style="position: absolute;margin-top: 16px;left: 45%;" th:src="@{/mall/image/null-content.png}">
            </div>
        </th:block>
```

```html
<th:block th:unless="${#lists.isEmpty(myShoppingCartItems)}">
    <div class="list center">
        <div class="top2 center">
            <div class="sub_top fl">
            </div>
            <div class="sub_top fl">商品名称</div>
            <div class="sub_top fl">单价</div>
            <div class="sub_top fl">数量</div>
            <div class="sub_top fl">小计</div>
            <div class="sub_top fr">操作</div>
            <div class="clear"></div>
        </div>
        <th:block th:each="item : ${myShoppingCartItems}">
            <div class="content2 center">
                <div class="sub_content fl ">
                </div>
                <div class="sub_content cover fl"><img th:src="@{${item.goodsCoverImg}}"></div>
                <div class="sub_content fl ft20" th:text="${item.goodsName}">商品名称</div>
                <div class="sub_content fl" th:text="${item.sellingPrice+'元'}">1299 元</div>
                <div class="sub_content fl">
                    <input class="goods_count" th:id="${'goodsCount'+item.cartItemId}" type="number"
                           th:onblur="'updateItem('+${item.cartItemId}+')'"
                           th:value="${item.goodsCount}" step="1" min="1"
                           max="5">
                </div>
                <div class="sub_content fl" th:text="${item.goodsCount*item.sellingPrice+'元'}">1299 元</div>
                <div class="sub_content fl"><a href="##" th:onclick="'deleteItem('+${item.cartItemId}+')'">×</a>
                </div>
                <div class="clear"></div>
            </div>
        </th:block>
    </div>
</th:block>
<div class="pre_order mt20 center">
    <div class="tips fl ml20">
        <ul>
```

```html
            <li><a th:href="@{/index}">继续购物</a></li>
            <li>|</li>
            <li>共<span th:text="${itemsTotal}">13</span>件商品</li>
            <div class="clear"></div>
        </ul>
    </div>
    <div class="order_div fr">
        <div class="order_total fl">合计（不含运费）：<span th:text="${priceTotal}+'.00元'">1299.00元</span></div>
        <div class="order_button fr">
            <th:block th:if="${itemsTotal == 0}">
                <input class="order_button_c" type="button" name="tip"
                    onclick="tip()"
                    value="去结算"/>
            </th:block>
            <th:block th:unless="${itemsTotal == 0}">
                <input class="order_button_d" type="button" name="settle"
                    onclick="settle()"
                    value="去结算"/>
            </th:block>
        </div>
        <div class="clear"></div>
    </div>
    <div class="clear"></div>
</div>
</div>
<div th:replace="mall/footer::footer-fragment"></div>

</body>
<!-- jQuery -->
<script th:src="@{/admin/plugins/jquery/jquery.min.js}"></script>
</html>
```

这里，在购物项列表区域对应的位置读取 myShoppingCartItems 数据，首先使用 th:each 循环语法将商品标题字段、商品预览图字段、商品价格字段、商品数量字段、单个商品的总价字段渲染出来，然后读取底部两个小计字段并渲染至页面中。至此，购物车列表页面开发完成。

24.5 编辑购物项功能的实现

在购物车列表中可以编辑需要购买的商品数量，也可以将商品从购物车中删除，接下来就实现这两个功能。

24.5.1 数据层代码的实现

首先，在购物项实体 Mapper 接口的 NewBeeMallShoppingCartItemMapper.java 文件中新增如下方法：

```java
/**
 * 根据主键查询记录
 *
 * @param cartItemId
 * @return
 */
NewBeeMallShoppingCartItem selectByPrimaryKey(Long cartItemId);

/**
 * 修改记录
 *
 * @param record
 * @return
 */
int updateByPrimaryKeySelective(NewBeeMallShoppingCartItem record);

/**
 * 删除一条记录
 *
 * @param cartItemId
 * @return
 */
int deleteByPrimaryKey(Long cartItemId);
```

然后，在映射文件 NewBeeMallShoppingCartItemMapper.xml 中添加具体的 SQL 语句，新增代码如下所示：

```xml
<select id="selectByPrimaryKey" parameterType="java.lang.Long"
```

```xml
    resultMap="BaseResultMap">
    select
    <include refid="Base_Column_List"/>
    from tb_newbee_mall_shopping_cart_item
    where cart_item_id = #{cartItemId,jdbcType=BIGINT} and is_deleted = 0
</select>

<update id="updateByPrimaryKeySelective" parameterType="ltd.newbee.mall.entity.NewBeeMallShoppingCartItem">
    update tb_newbee_mall_shopping_cart_item
    <set>
        <if test="userId != null">
            user_id = #{userId,jdbcType=BIGINT},
        </if>
        <if test="goodsId != null">
            goods_id = #{goodsId,jdbcType=BIGINT},
        </if>
        <if test="goodsCount != null">
            goods_count = #{goodsCount,jdbcType=INTEGER},
        </if>
        <if test="isDeleted != null">
            is_deleted = #{isDeleted,jdbcType=TINYINT},
        </if>
        <if test="createTime != null">
            create_time = #{createTime,jdbcType=TIMESTAMP},
        </if>
        <if test="updateTime != null">
            update_time = #{updateTime,jdbcType=TIMESTAMP},
        </if>
    </set>
    where cart_item_id = #{cartItemId,jdbcType=BIGINT}
</update>

<update id="deleteByPrimaryKey" parameterType="java.lang.Long">
    update tb_newbee_mall_shopping_cart_item set is_deleted = 1
    where cart_item_id = #{cartItemId,jdbcType=BIGINT} and is_deleted = 0
</update>
```

24.5.2 业务层代码的实现

首先，在业务类 NewBeeMallShoppingCartService.java 中新增如下方法：

```
/**
 * 修改购物车中的属性
 *
 * @param newBeeMallShoppingCartItem
 * @return
 */
String updateNewBeeMallCartItem(NewBeeMallShoppingCartItem newBeeMall
ShoppingCartItem);

/**
 * 删除购物车中的商品
 *
 * @param newBeeMallShoppingCartItemId
 * @return
 */
Boolean deleteById(Long newBeeMallShoppingCartItemId);
```

然后，在 NewBeeMallShoppingCartServiceImpl 类中将上述方法实现，新增如下代码：

```
@Override
public String updateNewBeeMallCartItem(NewBeeMallShoppingCartItem newBee
MallShoppingCartItem) {
    NewBeeMallShoppingCartItem newBeeMallShoppingCartItemUpdate = newBee
MallShoppingCartItemMapper.selectByPrimaryKey(newBeeMallShoppingCartItem.
getCartItemId());
    if (newBeeMallShoppingCartItemUpdate == null) {
        return ServiceResultEnum.DATA_NOT_EXIST.getResult();
    }
    //超出单个商品的最大数量
    if (newBeeMallShoppingCartItem.getGoodsCount() > Constants.SHOPPING_
CART_ITEM_LIMIT_NUMBER) {
        return ServiceResultEnum.SHOPPING_CART_ITEM_LIMIT_NUMBER_ERROR.
getResult();
    }
    newBeeMallShoppingCartItemUpdate.setGoodsCount(newBeeMallShoppingCart
Item.getGoodsCount());
    newBeeMallShoppingCartItemUpdate.setUpdateTime(new Date());
    //修改记录
    if (newBeeMallShoppingCartItemMapper.updateByPrimaryKeySelective(new
BeeMallShoppingCartItemUpdate) > 0) {
        return ServiceResultEnum.SUCCESS.getResult();
    }
    return ServiceResultEnum.DB_ERROR.getResult();
}
```

```
@Override
public Boolean deleteById(Long newBeeMallShoppingCartItemId) {
    //userId 不同则不能删除
    return newBeeMallShoppingCartItemMapper.deleteByPrimaryKey(newBeeMallShoppingCartItemId) > 0;
}
```

这里，updateNewBeeMallCartItem()方法首先对参数进行校验，校验步骤如下所示。

① 首先根据前端传参的购物项主键 id 查询购物项表中是否存在该记录，如果不存在则返回错误信息，存在则进行后续操作

② 判断用户购物车中的商品数量是否已超出最大限制

在校验通过后再进行修改操作，将该购物项记录的数量和时间进行修改。以上操作都需要调用 SQL 语句来完成。

deleteById()方法的实现逻辑不复杂，即调用数据层的方法将当前购物项的 is_deleted 字段修改为 1。这里并不是真正的删除，而是逻辑删除。

24.5.3 控制层代码的实现

在 ShoppingCartController 中修改购物项和删除购物项两个方法，代码如下所示：

```
/**
 * 修改购物项
 */
@PutMapping("/shop-cart")
@ResponseBody
public Result updateNewBeeMallShoppingCartItem(@RequestBody NewBeeMallShoppingCartItem newBeeMallShoppingCartItem,
                                               HttpSession httpSession) {
    NewBeeMallUserVO user = (NewBeeMallUserVO) httpSession.getAttribute(Constants.MALL_USER_SESSION_KEY);
    newBeeMallShoppingCartItem.setUserId(user.getUserId());
    String updateResult = newBeeMallShoppingCartService.updateNewBeeMallCartItem(newBeeMallShoppingCartItem);
    //修改成功
    if (ServiceResultEnum.SUCCESS.getResult().equals(updateResult)) {
        return ResultGenerator.genSuccessResult();
```

```java
    }
    //修改失败
    return ResultGenerator.genFailResult(updateResult);
}

/**
 * 删除购物项
 */
@DeleteMapping("/shop-cart/{newBeeMallShoppingCartItemId}")
@ResponseBody
public Result updateNewBeeMallShoppingCartItem(@PathVariable("newBeeMall
ShoppingCartItemId") Long newBeeMallShoppingCartItemId,
                                                HttpSession httpSession) {
    NewBeeMallUserVO user = (NewBeeMallUserVO) httpSession.getAttribute
(Constants.MALL_USER_SESSION_KEY);
    Boolean deleteResult = newBeeMallShoppingCartService.deleteById(new
BeeMallShoppingCartItemId);
    //删除成功
    if (deleteResult) {
        return ResultGenerator.genSuccessResult();
    }
    //删除失败
    return ResultGenerator.genFailResult(ServiceResultEnum.OPERATE_ERROR.
getResult());
}
```

修改购物项接口的映射地址为/shop-cart，请求方法为 PUT。该 Controller 类中的几个方法的路径都是/shop-cart。不同的请求方法就是接口的区分方式，POST 方法是新增接口，GET 方法是购物车列表页面显示，PUT 方法是修改接口。编辑功能主要是修改当前购物项的数量。后端的编辑接口负责接收前端的 PUT 请求并进行处理，接收的参数为 cartItemId 字段和 goodsCount 字段。通过这两个字段就可以确定修改哪一条记录和商品的数量值。在这个方法里使用@RequestBody 注解将其转换为 NewBeeMallShoppingCartItem 对象参数并进行后续的操作。

如果购物车中一些商品不进行后续的结算购买操作，可以选择将其删除。删除购物项接口负责接收前端的 DELETE 请求并进行处理，接收的参数为 cartItemId 字段，最后调用删除方法即可完成删除操作。接口的映射地址为/shop-cart/{newBeeMallShoppingCartItemId}，请求方法为 DELETE。

24.5.4 前端调用修改和删除购物项的接口

修改数量的按钮的实现逻辑通过列表中商品数量 input 框的 onblur()事件类实现。首先给商品数量输入框绑定 onblur()事件，代码如下所示：

```html
<input class="goods_count" th:id="${'goodsCount'+item.cartItemId}" type="number"
                          th:onblur="'updateItem('+${item.cartItemId}+')'"
                          th:value="${item.goodsCount}" step="1" min="1"
                          max="5">
```

当用户的鼠标焦点离开 input 输入框时就会触发执行 updateItem()方法。该方法会向后端发送修改购物项数量的请求。在 cart.html 模板文件中新增代码如下所示：

```js
/**
 * 更新购物项
 *
 */
function updateItem(id) {
    var domId = 'goodsCount' + id;
    var goodsCount = $("#" + domId).val();
    if (goodsCount > 5) {
        swal("单个商品最多可购买5个", {
            icon: "error",
        });
        return;
    }
    if (goodsCount < 1) {
        swal("数量异常", {
            icon: "error",
        });
        return;
    }
    var data = {
        "cartItemId": id,
        "goodsCount": goodsCount
    };
    $.ajax({
        type: 'PUT',
```

```
            url: '/shop-cart',
            contentType: 'application/json',
            data: JSON.stringify(data),
            success: function (result) {
                if (result.resultCode == 200) {
                    window.location.reload();
                } else {
                    swal("操作失败", {
                        icon: "error",
                    });
                }
            },
            error: function () {
                swal("操作失败", {
                    icon: "error",
                });
            }
        });
}
```

该方法具体执行步骤如下所示。

①获取购物项主键 id 和修改后的数量

②进行基本的正则验证，修改数量不能大于 5 也不能小于 1

注意，这个数值的验证以及 5 和 1 的设置都是新蜂商城项目中的逻辑。读者可以自行更改，比如修改为不进行数量限制，或者单个商品的最大数量是 20。

③向后端购物项修改接口发送请求

④根据接口回调进行后续操作

每个购物项的显示区域都有一个删除按钮，删除按钮也绑定了触发事件，绑定的 JS 方法为 deleteItem()。在 cart.html 模板文件中新增代码如下所示：

```
/**
 * * 删除购物项
 * @param id
 */
function deleteItem(id) {
    swal({
        title: "确认弹框",
        text: "确认要删除数据吗?",
        icon: "warning",
        buttons: true,
```

```
            dangerMode: true,
        }).then((flag) => {
            if (flag) {
                $.ajax({
                    type: 'DELETE',
                    url: '/shop-cart/' + id,
                    success: function (result) {
                        if (result.resultCode == 200) {
                            window.location.reload();
                        } else {
                            swal("操作失败", {
                                icon: "error",
                            });
                        }
                    },
                    error: function () {
                        swal("操作失败", {
                            icon: "error",
                        });
                    }
                });
            }
        })
    ;
}
```

这里点击删除按钮后会执行 deleteItem()方法，在用户确认后，获取需要删除的购物项 id 并向后端发送删除请求，最后购物车中就不存在这条记录了。

24.6 功能测试

在编码完成后，启动 Spring Boot 项目。在启动成功后打开浏览器并输入商城端登录页面地址：

```
http://localhost:8080/login
```

在正确登录后就可以对相关功能进行测试，演示过程如下所示。

（1）添加商品到购物车

首先进入商品详情页面，点击"加入购物车"按钮就可以添加商品。比如笔者点进了商品 id 为 10909 的商品详情页面。如图 24-5 所示，在点击"加入购物车"按钮后，系统成功将 id 为 10909 的商品添加到购物车中了。

图 24-5　添加商品到购物车

然后可以进入其他商品页面，并将这些商品添加到购物车中。

（2）列表功能

点击页面顶部右侧的"购物车"按钮，就可以进入购物车页面并查看数据列表。当前购物车中有 3 条记录，且加购数量都为 1，如图 24-6 所示。

图 24-6　购物车列表

（3）修改功能

可以通过数字调整按钮来增加或者减少商品的购买数量，也可以直接在输入框中输入一个数字完成修改。如图 24-7 所示，在修改完成后，对应的小计价格和页面底部的总价也会同步更改。

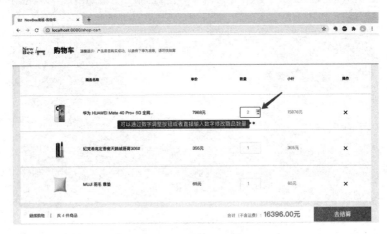

图 24-7　修改商品数量

如果数量超过限制，或者输入了负数或者不规范的数字，都会收到错误提示。比如"单个商品最多可购买 5 个""数量异常"，如图 24-8 所示。

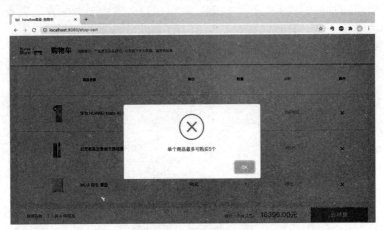

图 24-8　错误提示

（4）删除功能

如果不想结算某件商品，可以直接点击商品信息旁边的删除按钮，此时会出现一个确认弹框，如图 24-9 所示。

图 24-9　删除购物项

点击"OK"则表示确认该删除操作，点击"Cancel"则表示取消操作。在点击"OK"后，选中的记录会被删除。如图 24-10 所示，之前选中删除的商品记录不存在了。原本列表中有 3 条数据，删除后只有 2 条数据。

图 24-10　删除后的列表

第 25 章

订单模块功能开发及讲解

在把心仪的商品加到购物车并确定需要购买的商品和对应的数量后,就可以执行提交订单的操作。此时也就由购物车模块切换到了另外一个电商流程中——订单流程。接下来主要介绍订单模块相关功能的开发。关于订单的生成和后续处理流程,不同公司或者不同商城项目具体的需求与业务场景会有一些差异。但是从订单的生成阶段到订单完成阶段大体的流程还是类似的。

新蜂商城中订单的生成到处理结束,主要有以下几个流程,如图 25-1 所示。

图 25-1 订单流程

① 提交订单(由新蜂商城用户发起)

② 订单入库(后台逻辑,用户无感知)

③ 支付订单(由新蜂商城用户发起)

④ 订单处理(包括确认订单、取消订单、修改订单信息等操作,新蜂商城用户和管理员都可以对订单进行处理)

第 25 章　订单模块功能开发及讲解

25.1　订单确认页面的功能开发

25.1.1　商城中的订单确认步骤

订单确认步骤是订单生成功能中的一个很重要的功能，日常使用到的商城项目基本都有这个步骤。在线上商城购物时也肯定有"订单确认"的环节，以淘宝商城的订单确认页为例，如图 25-2 所示。

图 25-2　淘宝网确认订单页面

这个页面中包含在购物车中选择的商品信息，还有收获地址信息、运费信息、优惠信息等。在购物车中只有商品信息，在确认订单页面则是多种信息的集合。只有全部信息齐全才能够生成订单数据。该页面的设置可以理解为一个信息确认的过程。在该页面将所有信息都确认无误后，就可以进行提交订单的操作，并生成一条订单记录。

25.1.2 订单确认的前置步骤

通过购物车页面的结算按钮进入订单确认页面。订单确认步骤是在购物车页面发起的。点击"去结算"按钮如图 25-3 所示。

图 25-3 "去结算"按钮

根据购物车中的待结算商品数量来判断"去结算"按钮是否正常展示。如果购物车中无数据，点击"去结算"按钮则提示"购物车中无数据"。如果数据正常则执行 settle() 方法并进行跳转，跳转路径为/shop-cart/settle，并在该控制器方法中进行订单确认页面中数据的查询和整合。其代码如下所示：

```html
<div class="order_button fr">
  <th:block th:if="${itemsTotal == 0}">
    <input class="order_button_c" type="button" name="tip"
        onclick="tip()"
        value="去结算"/>
  </th:block>
  <th:block th:unless="${itemsTotal == 0}">
    <input class="order_button_d" type="button" name="settle"
        onclick="settle()"
        value="去结算"/>
  </th:block>
</div>
```

其中 tip()方法和 settle()方法代码如下所示：

```
/**
 * 购物车中数量为 0 时提示
 */
function tip() {
```

```
swal("购物车中无数据，无法结算", {
    icon: "error",
  });
}

/**
 * 跳转至结算页面
 */
function settle() {
  window.location.href = '/shop-cart/settle'
}
```

25.1.3 订单确认页面的数据整合

订单确认页展示的数据与购物车列表中展示的商品数据类似。不过该页面不仅仅只有商品数据，还有用户数据和支付数据。所以该页面主要的数据是"商品数据+用户数据+支付数据"。这里还需要把购买者和购买者的收货信息记录下来，其他的内容还有运费金额、优惠金额、实际支付金额等。新蜂商城的订单确认页信息如图 25-4 所示。

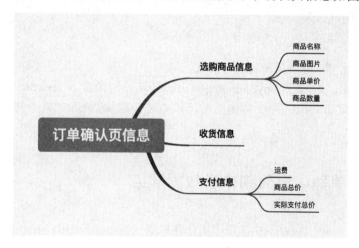

图 25-4 订单确认页信息

在 ShoppingCartController 类中新增 settlePage()方法，该方法用于处理/shop-cart/settle 请求，并将数据带到订单确认页面进行渲染，代码如下所示：

```
@GetMapping("/shop-cart/settle")
public String settlePage(HttpServletRequest request,
```

```java
                            HttpSession httpSession) {
    int priceTotal = 0;
    NewBeeMallUserVO user = (NewBeeMallUserVO) httpSession.getAttribute
(Constants.MALL_USER_SESSION_KEY);
    List<NewBeeMallShoppingCartItemVO> myShoppingCartItems = newBeeMall
ShoppingCartService.getMyShoppingCartItems(user.getUserId());
    if (CollectionUtils.isEmpty(myShoppingCartItems)) {
        //无数据则不跳转至结算页
        return "/shop-cart";
    } else {
        //总价
        for (NewBeeMallShoppingCartItemVO newBeeMallShoppingCartItemVO :
myShoppingCartItems) {
            priceTotal += newBeeMallShoppingCartItemVO.getGoodsCount() *
newBeeMallShoppingCartItemVO.getSellingPrice();
        }
        if (priceTotal < 1) {
            return "error/error_5xx";
        }
    }
    request.setAttribute("priceTotal", priceTotal);
    request.setAttribute("myShoppingCartItems", myShoppingCartItems);
    return "mall/order-settle";
}
```

在新蜂商城第一版中并没有运费和优惠券功能，所以这里只是查询商品信息并计算总价，然后将 priceTotal 和 myShoppingCartItems 两个对象放入 request 请求中，再跳转到 order-settle 页面。用户的收货信息是可以通过读取 session 对象中的 User 对象来获取的。

25.1.4 订单确认页面制作及数据渲染

在 resources/templates/mall 目录中新增订单确认页 order-settle.html，模板代码如下所示：

```html
<!DOCTYPE html>
<html lang="en" xmlns:th="http://www.thymeleaf.org">
<head th:replace="mall/header::head-fragment('NewBee 商城-订单结算','order-detail')">
</head>
```

第25章 订单模块功能开发及讲解

```html
<link th:href="@{/mall/css/bootstrap-modal.css}" rel="stylesheet">
<body>
<header th:replace="mall/header::header-fragment"></header>
<!-- nav -->
<nav th:replace="mall/header::nav-fragment"></nav>

<!-- personal -->
<div id="personal">
    <div class="self-info center">

        <!-- sidebar -->
        <div th:replace="mall/personal-sidebar::sidebar-fragment"></div>

        <div class="intro fr">
            <div class="uc-box uc-main-box">
                <div class="uc-content-box order-view-box">
                    <div class="box-hd">
                        <h1 class="title">填写并核对订单信息</h1>
                        <div class="more clearfix">
                            <div class="actions">
                                <a id="saveOrder" class="btn btn-small btn-primary" title="提交订单">提交订单</a>
                            </div>
                        </div>
                    </div>
                    <div class="box-bd">
                        <div class="uc-order-item uc-order-item-pay">
                            <div class="order-detail">

                                <div class="order-summary">
                                    <div class="order-progress">
                                        <ol class="progress-list clearfix progress-list-5">
                                            <li class="step step-done">
                                                <div class="progress"><span class="text">购物车</span></div>
                                                <div class="info"></div>
                                            </li>
                                            <li class="step step-active">
                                                <div class="progress"><span class="text">下单</span></div>
                                                <div class="info"></div>
                                            </li>
```

```html
                        <li class="step">
                            <div class="progress"><span class="text">付款</span></div>
                            <div class="info"></div>
                        </li>
                        <li class="step">
                            <div class="progress"><span class="text">出库</span></div>
                            <div class="info"></div>
                        </li>
                        <li class="step">
                            <div class="progress"><span class="text">交易成功</span></div>
                            <div class="info"></div>
                        </li>
                    </ol>
                </div>
            </div>
            <table class="order-items-table">
                <tbody>
                <th:block th:each="item : ${myShoppingCartItems}">
                    <tr>
                        <td class="col col-thumb">
                            <div class="figure figure-thumb">
                                <a target="_blank" th:href="@{'/goods/detail/'+${item.goodsId}}">
                                    <img th:src="@{${item.goodsCoverImg}}"
                                         width="80" height="80" alt="">
                                </a>
                            </div>
                        </td>
                        <td class="col col-name">
                            <p class="name">
                                <a target="_blank" th:href="@{'/goods/detail/'+${item.goodsId}}"
                                   th:text="${item.goodsName}">newbee</a>
                            </p>
                        </td>
```

```html
                            <td class="col col-price"><p class="price"                                                                                          th:text="${item.sellingPrice+'元 x '+item.goodsCount}">
                                1299元 × 1</p></td>
                            <td class="col col-actions">
                            </td>
                        </tr>
                    </th:block>
                    </tbody>
                </table>
            </div>
            <div id="editAddr" class="order-detail-info">
                <h3>收货信息</h3>
                <table class="info-table">
                    <tbody>
                    <tr>
                        <th>收货地址：</th>
                        <td class="user_address_label"
                            th:text="${session.newBeeMallUser.address==''?'无':session.newBeeMallUser.address}">
                            newbee
                        </td>
                    </tr>
                    </tbody>
                </table>
                <div class="actions">
                    <a class="btn btn-small btn-line-gray J_editAddr"
                       href="javascript:openUpdateModal();">修改</a>
                </div>
            </div>
            <div id="editTime" class="order-detail-info">
                <h3>支付方式</h3>
                <table class="info-table">
                    <tbody>
                    <tr>
                        <th>支付方式：</th>
                        <td>在线支付</td>
                    </tr>
                    </tbody>
                </table>
```

```html
                    <div class="actions">
                    </div>
                </div>
                <div class="order-detail-total">
                    <table class="total-table">
                        <tbody>
                            <tr>
                                <th>商品总价: </th>
                                <td><span class="num" th:text="${priceTotal+'.00'}">1299.00</span>元</td>
                            </tr>
                            <tr>
                                <th>运费: </th>
                                <td><span class="num">0</span>元</td>
                            </tr>
                            <tr>
                                <th class="total">应付金额: </th>
                                <td class="total"><span class="num" th:text="${priceTotal+'.00'}">1299.00</span>元
                                </td>
                            </tr>
                        </tbody>
                    </table>
                </div>
            </div>
        </div>
    </div>
    <div class="modal fade" id="personalInfoModal" tabindex="-1" role="dialog"
         aria-labelledby="personalInfoModalLabel">
        <div class="modal-dialog" role="document">
            <div class="modal-content">
                <div class="modal-header">
                    <button type="button" class="close" data-dismiss="modal" aria-label="Close"><span
                            aria-hidden="true">&times;</span></button>
                    <h6 class="modal-title" id="personalInfoModalLabel">地址修改</h6>
                </div>
                <div class="modal-body">
                    <form id="personalInfoForm">
```

```html
                    <div class="form-group">
                        <input type="hidden" id="userId" th:value="${session.newBeeMallUser.userId}">
                        <label for="address" class="control-label">收货地址:</label>
                        <input type="text" class="form-control" id="address" name="address"
                               placeholder="请输入收货地址" th:value="${session.newBeeMallUser.address}"
                               required="true">
                    </div>
                </form>
            </div>
            <div class="modal-footer">
                <button type="button" class="btn btn-default" data-dismiss="modal">取消</button>
                <button type="button" class="btn btn-primary" id="saveButton">确认</button>
            </div>
        </div>
    </div>
</div>
    <div class="clear"></div>
  </div>
</div>

<div th:replace="mall/footer::footer-fragment"></div>

<!-- jQuery -->
<script th:src="@{/admin/plugins/jquery/jquery.min.js}"></script>
<script th:src="@{/mall/js/search.js}" type="text/javascript"></script>
<script th:src="@{/admin/plugins/sweetalert/sweetalert.min.js}"></script>
<script th:src="@{/mall/js/bootstrap3.js}"></script>
<script type="text/javascript">
    $('#saveButton').click(function () {
        var address = $("#address").val();
        var userId = $("#userId").val();
        var data = {
            "userId": userId,
            "address": address
        };
        $.ajax({
```

```
            type: 'POST',//方法类型
            url: '/personal/updateInfo',
            contentType: 'application/json',
            data: JSON.stringify(data),
            success: function (result) {
                if (result.resultCode == 200) {
                    $('#personalInfoModal').modal('hide');
                    window.location.reload();
                } else {
                    $('#personalInfoModal').modal('hide');
                    swal(result.message, {
                        icon: "error",
                    });
                }
                ;
            },
            error: function () {
                swal('操作失败', {
                    icon: "error",
                });
            }
        });
    });
</script>
</body>
</html>
```

该页面中需要渲染的数据有收货信息、商品信息、价格信息。

收货信息直接读取 session 对象中的数据，编码实现为 session.newBeeMallUser.address。如果该字段为空或者需要修改可以点击修改按钮进行数据的修改。

商品数据区域的渲染与购物车中的商品数据渲染类似，读取 myShoppingCartItems 数据，使用 th:each 循环语法将商品标题字段、商品预览图字段、商品价格字段、商品数量字段渲染即可。

价格信息中运费为 0，总价字段直接读取 priceTotal 字段。

在修改完成后重启项目查看数据是否正常。如图 25-5 所示即为订单确认页面的显示效果。购物车中已经有 2 条已添加的商品记录，点击结算按钮后，就会跳转到订单确认页面，表示数据一切正常。

第 25 章 订单模块功能开发及讲解

图 25-5 订单确认页面

25.2 订单模块中的表结构设计

25.2.1 订单主表和订单项关联表设计

新蜂商城系统的订单功能模块主要涉及数据库中的两张表。一次下单行为可能购买一件商品也可能购买多件商品,所以除了订单主表 tb_newbee_mall_order 之外,还有一个订单项关联表 tb_newbee_mall_order_item,二者是一对多的关系。订单主表中存储关于订单的相关信息,而订单项表中主要存储关联的商品字段。

订单主表 tb_newbee_mall_order 表结构设计的主要字段如下所示。

①user_id:用户的 id。根据这个字段来确定是哪个用户下的订单。

②order_no:订单号。订单号是唯一的标识订单和是在后续查询订单时用的。这是每个电商系统都会有的设计。

③paystatus、paytype、pay_time:支付信息字段。包括支付状态、支付方式、支付时间。

④order_status:订单状态字段。

⑤username、userphone、user_address:收件信息字段,最好在订单表中设置这几个

字段。有些商城的订单设计,收货地址只会关联一个地址 id 字段,这样做是不合理的。因为该 id 关联的地址表中的记录是可以被修改和删除的。也就是说如果修改或者删除,订单中的收件信息就不是下单时的数据了。因此需要把这些字段放到订单表中,并记录下单当时的收件信息。新蜂商城第一版中只处理了 user_address 字段,后续迭代中会进行优化。

⑥create_time:订单生成时间。

订单主表的主要字段代码如下所示:

```sql
DROP TABLE IF EXISTS 'tb_newbee_mall_order';
CREATE TABLE 'tb_newbee_mall_order'  (
  'order_id' bigint(20) NOT NULL AUTO_INCREMENT COMMENT '订单表主键id',
  'order_no' varchar(20) CHARACTER SET utf8 COLLATE utf8_general_ci NOT NULL DEFAULT '' COMMENT '订单号',
  'user_id' bigint(20) NOT NULL DEFAULT 0 COMMENT '用户主键id',
  'total_price' int(11) NOT NULL DEFAULT 1 COMMENT '订单总价',
  'pay_status' tinyint(4) NOT NULL DEFAULT 0 COMMENT '支付状态:0.未支付,1.支付成功,-1:支付失败',
  'pay_type' tinyint(4) NOT NULL DEFAULT 0 COMMENT '0.无 1.支付宝支付 2.微信支付',
  'pay_time' datetime(0) NULL DEFAULT NULL COMMENT '支付时间',
  'order_status' tinyint(4) NOT NULL DEFAULT 0 COMMENT '订单状态:0.待支付 1.已支付 2.配货完成 3:出库成功 4.交易成功 -1.手动关闭 -2.超时关闭 -3.商家关闭',
  'extra_info' varchar(100) CHARACTER SET utf8 COLLATE utf8_general_ci NOT NULL DEFAULT '' COMMENT '订单body',
  'user_name' varchar(30) CHARACTER SET utf8 COLLATE utf8_general_ci NOT NULL DEFAULT '' COMMENT '收货人姓名',
  'user_phone' varchar(11) CHARACTER SET utf8 COLLATE utf8_general_ci NOT NULL DEFAULT '' COMMENT '收货人手机号',
  'user_address' varchar(100) CHARACTER SET utf8 COLLATE utf8_general_ci NOT NULL DEFAULT '' COMMENT '收货人收货地址',
  'is_deleted' tinyint(4) NOT NULL DEFAULT 0 COMMENT '删除标识字段(0-未删除 1-已删除)',
  'create_time' datetime(0) NOT NULL DEFAULT CURRENT_TIMESTAMP COMMENT '创建时间',
  'update_time' datetime(0) NOT NULL DEFAULT CURRENT_TIMESTAMP COMMENT '最新修改时间',
  PRIMARY KEY ('order_id') USING BTREE
) ENGINE = InnoDB AUTO_INCREMENT = 1 CHARACTER SET = utf8 COLLATE =
```

utf8_general_ci ROW_FORMAT = Dynamic;

订单项表 tb_newbee_mall_order_item 表结构设计的主要字段如下所示。

①order_id：关联的订单主键 id，标识该订单项是哪个订单中的数据。

②goodsid、goodsname、goodscoverimg、sellingprice、goodscount：订单中的商品信息，记录单当时的商品信息。

③create_time：记录生成时间。

订单项表的字段代码如下所示：

```sql
DROP TABLE IF EXISTS 'tb_newbee_mall_order_item';
CREATE TABLE 'tb_newbee_mall_order_item'  (
  'order_item_id' bigint(20) NOT NULL AUTO_INCREMENT COMMENT '订单关联购物项主键id',
  'order_id' bigint(20) NOT NULL DEFAULT 0 COMMENT '订单主键id',
  'goods_id' bigint(20) NOT NULL DEFAULT 0 COMMENT '关联商品id',
  'goods_name' varchar(200) CHARACTER SET utf8 COLLATE utf8_general_ci NOT NULL DEFAULT '' COMMENT '下单时商品的名称(订单快照)',
  'goods_cover_img' varchar(200) CHARACTER SET utf8 COLLATE utf8_general_ci NOT NULL DEFAULT '' COMMENT '下单时商品的主图(订单快照)',
  'selling_price' int(11) NOT NULL DEFAULT 1 COMMENT '下单时商品的价格(订单快照)',
  'goods_count' int(11) NOT NULL DEFAULT 1 COMMENT '数量(订单快照)',
  'create_time' datetime(0) NOT NULL DEFAULT CURRENT_TIMESTAMP COMMENT '创建时间',
  PRIMARY KEY ('order_item_id') USING BTREE
) ENGINE = InnoDB AUTO_INCREMENT = 1 CHARACTER SET = utf8 COLLATE = utf8_general_ci ROW_FORMAT = Dynamic;
```

每个字段对应的含义在上面的 SQL 中都有介绍，读者可以对照理解，并正确地把建表 SQL 导入数据库中即可。关于两张表中的快照字段，包括收件信息字段和商品信息字段，读者可以参考淘宝商城的订单快照来理解。这些信息都是可以更改的，因此不能只关联一个主键 id。比如订单中存储的是在下单时的数据，而商品信息是可以随时更改的。如果没有这几个字段只用商品 id 关联的话，商品信息一旦被更改，订单信息也随之被更改，从而不再是下单时的数据，这就不符合逻辑了。

25.2.2 订单项表的设计思路

接下来主要介绍订单项表 tb_newbee_mall_orderitem 和购物项表 tb_newbee_mall_shopping_cart_ite 的差异，以及单独设计一张订单项表的原因。

购物车模块的购物项和订单模块的订单项是很多商城项目都存在的设计。只是有一些商城项目为了简化开发，在实现的时候选择将两者等同。本来应该设计两张表减少为只设计一张表，用购物项表来替代两张表，那么订单在生成后会在购物项表中增加与订单主键 id 的关联。

其实 OrderItem 和 CartItem 二者的区别是很明显的。它们是相似却完全不同的两个对象，购物项对象是商品与购物车之间抽象出的一个对象，而订单项对象是商品与订单之间抽象出的一个对象。可能他们都与商品相关，而且在页面数据展示时也类似，所以会有简化为一张表的实现方式。但是笔者并不赞同这种实现方式。虽然二者很相似，但是依然有很多不同的地方。购物项的相关操作在购物车中，而在生成订单后该购物项就不再存在了，即该对象已经被删除了。它的生命周期也就到此为止。既然生命周期已经终结再与订单做关联就说不通了。

以订单快照为例，它需要记录下单时的商品内容和订单信息。如果想要查看下单时的商品信息和收货地址，与订单相关的数据都要把下单时的内容保存。比如，淘宝商城的订单设计，可以看到下单时的订单快照数据就是下单时的商品数据，而不是最新的商品数据。而购物车就不需要快照，直接读取最新的商品相关信息即可。

购物车模块的购物项和订单模块的订单项是两个不同的对象，因此笔者选择分别设计两张表。

25.3 订单生成功能的实现

25.3.1 新蜂商城订单生成的流程

在订单确认页面处理完毕后，紧接着就是生成订单的环节。此时用户点击"提交订单"按钮，商城系统就对应生成一笔订单数据并保存在数据库中。

第 25 章 订单模块功能开发及讲解

在点击"提交订单"按钮后,此时后台会进行一系列的操作,包括数据查询、数据判断、数据整合等。订单生成流程如图 25-6 所示。

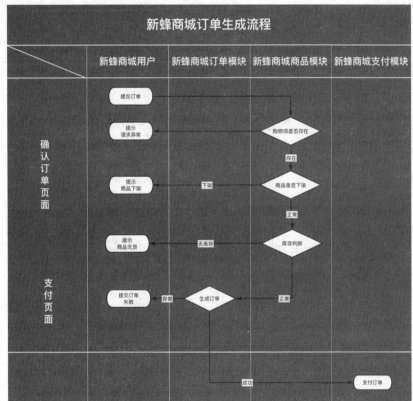

图 25-6 订单生成流程

读者可以仔细看一下这张流程图,并结合流程图更好地理解代码的实现。

25.3.2 发起订单生成请求

修改 order-settle.html 页面代码,给"提交订单"按钮增加点击事件,代码如下所示:

```
$('#saveOrder').click(function () {
  var userAddress = $(".user_address_label").html();
  if (userAddress == '' || userAddress == '无') {
    swal("请填写收货信息", {
```

```
      icon: "error",
    });
    return;
  }
  if (userAddress.trim().length < 10) {
    swal("请输入正确的收货信息", {
      icon: "error",
    });
    return;
  }
  window.location.href = '/saveOrder';
});
```

这里，前端主要验证收获信息是否正确。字段是能够由用户主动输入的，所以需要进行一次判断，如果字段验证通过，则向/saveOrder 请求地址发起订单生成请求。

25.3.3 订单生成请求处理

首先在 controller 包中新建 OrderController 类用于处理订单模块的相关请求，然后新增 saveOrder()方法用于处理订单生成的请求，代码如下所示：

```
package ltd.newbee.mall.controller.mall;

import ltd.newbee.mall.common.Constants;
import ltd.newbee.mall.common.NewBeeMallException;
import ltd.newbee.mall.common.ServiceResultEnum;
import ltd.newbee.mall.controller.vo.NewBeeMallShoppingCartItemVO;
import ltd.newbee.mall.controller.vo.NewBeeMallUserVO;
import ltd.newbee.mall.service.NewBeeMallOrderService;
import ltd.newbee.mall.service.NewBeeMallShoppingCartService;
import org.springframework.stereotype.Controller;
import org.springframework.util.CollectionUtils;
import org.springframework.util.StringUtils;
import org.springframework.web.bind.annotation.GetMapping;

import javax.annotation.Resource;
import javax.servlet.http.HttpSession;
import java.util.List;

@Controller
public class OrderController {
```

```java
    @Resource
    private NewBeeMallShoppingCartService newBeeMallShoppingCartService;
    @Resource
    private NewBeeMallOrderService newBeeMallOrderService;

    @GetMapping("/saveOrder")
    public String saveOrder(HttpSession httpSession) {
        NewBeeMallUserVO user = (NewBeeMallUserVO) httpSession.getAttribute(Constants.MALL_USER_SESSION_KEY);
        List<NewBeeMallShoppingCartItemVO> myShoppingCartItems = newBeeMallShoppingCartService.getMyShoppingCartItems(user.getUserId());
        if (StringUtils.isEmpty(user.getAddress().trim())) {
            //无收货地址
            NewBeeMallException.fail(ServiceResultEnum.NULL_ADDRESS_ERROR.getResult());
        }
        if (CollectionUtils.isEmpty(myShoppingCartItems)) {
            //购物车中无数据则跳转至错误页
            NewBeeMallException.fail(ServiceResultEnum.SHOPPING_ITEM_ERROR.getResult());
        }
        //保存订单并返回订单号
        String saveOrderResult = newBeeMallOrderService.saveOrder(user, myShoppingCartItems);
        //跳转到订单详情页
        return "redirect:/orders/" + saveOrderResult;
    }
}
```

该方法处理的映射地址为/saveOrder，请求方法为 GET，过程如下所示。

①验证收货地址信息，有则继续走后续流程，无则返回异常信息

②验证购物车中是否有数据，有则继续走后续流程，无则返回异常信息

③将购物项数据和用户信息作为参数传给业务层的 saveOrder()方法进行订单生成的业务逻辑操作

④如果订单生成成功，业务层的 saveOrder()方法会返回订单号，直接跳转到订单详情页

25.3.4 订单生成逻辑的实现

接下来在 service 包中新建订单模块的业务实现类，并实现订单生成的业务逻辑，代码如下所示：

```java
@Override
@Transactional
public String saveOrder(NewBeeMallUserVO user, List<NewBeeMallShoppingCartItemVO> myShoppingCartItems) {
    List<Long> itemIdList = myShoppingCartItems.stream().map(NewBeeMallShoppingCartItemVO::getCartItemId).collect(Collectors.toList());
    List<Long> goodsIds = myShoppingCartItems.stream(). map(NewBeeMallShoppingCartItemVO::getGoodsId).collect(Collectors.toList());
    List<NewBeeMallGoods> newBeeMallGoods = newBeeMallGoodsMapper.selectByPrimaryKeys(goodsIds);
    //检查是否包含已下架商品
    List<NewBeeMallGoods> goodsListNotSelling = newBeeMallGoods.stream()
        .filter(newBeeMallGoodsTemp -> newBeeMallGoodsTemp.getGoodsSellStatus() != Constants.SELL_STATUS_UP)
        .collect(Collectors.toList());
    if (!CollectionUtils.isEmpty(goodsListNotSelling)) {
        //goodsListNotSelling 对象非空则表示有下架商品
        NewBeeMallException.fail(goodsListNotSelling.get(0).getGoodsName() + "已下架，无法生成订单");
    }
    Map<Long, NewBeeMallGoods> newBeeMallGoodsMap = newBeeMallGoods.stream().collect(Collectors.toMap(NewBeeMallGoods::getGoodsId, Function.identity(), (entity1, entity2) -> entity1));
    //判断商品库存
    for (NewBeeMallShoppingCartItemVO shoppingCartItemVO : myShoppingCartItems) {
        //查出商品中不存在购物车中的这条关联商品数据，则直接返回错误提醒
        if (!newBeeMallGoodsMap.containsKey(shoppingCartItemVO.getGoodsId())) {
            NewBeeMallException.fail(ServiceResultEnum.SHOPPING_ITEM_ERROR. GetResult());
        }
        //存在数量大于库存的情况，直接返回错误提醒
        if (shoppingCartItemVO.getGoodsCount() > newBeeMallGoodsMap.get(shoppingCartItemVO.getGoodsId()).getStockNum()) {
            NewBeeMallException.fail(ServiceResultEnum.SHOPPING_ITEM_COUNT_ERROR.
```

```java
getResult());
        }
    }
    //删除购物项
    if (!CollectionUtils.isEmpty(itemIdList) && !CollectionUtils.isEmpty(goodsIds) && !CollectionUtils.isEmpty(newBeeMallGoods)) {
        if (newBeeMallShoppingCartItemMapper.deleteBatch(itemIdList) > 0) {
            List<StockNumDTO> stockNumDTOS = BeanUtil.copyList(myShoppingCartItems, StockNumDTO.class);
            int updateStockNumResult = newBeeMallGoodsMapper.updateStockNum(stockNumDTOS);
            if (updateStockNumResult < 1) {
                NewBeeMallException.fail(ServiceResultEnum.SHOPPING_ITEM_COUNT_ERROR.getResult());
            }
            //生成订单号
            String orderNo = NumberUtil.genOrderNo();
            int priceTotal = 0;
            //保存订单
            NewBeeMallOrder newBeeMallOrder = new NewBeeMallOrder();
            newBeeMallOrder.setOrderNo(orderNo);
            newBeeMallOrder.setUserId(user.getUserId());
            newBeeMallOrder.setUserAddress(user.getAddress());
            //总价
            for (NewBeeMallShoppingCartItemVO newBeeMallShoppingCartItemVO : myShoppingCartItems) {
                priceTotal += newBeeMallShoppingCartItemVO.getGoodsCount() * newBeeMallShoppingCartItemVO.getSellingPrice();
            }
            if (priceTotal < 1) {
                NewBeeMallException.fail(ServiceResultEnum.ORDER_PRICE_ERROR.getResult());
            }
            newBeeMallOrder.setTotalPrice(priceTotal);
            String extraInfo = "";
            newBeeMallOrder.setExtraInfo(extraInfo);
            //生成订单项并保存订单项纪录
            if (newBeeMallOrderMapper.insertSelective(newBeeMallOrder) > 0) {
                //生成所有的订单项快照,并保存至数据库
                List<NewBeeMallOrderItem> newBeeMallOrderItems = new ArrayList<>();
                for (NewBeeMallShoppingCartItemVO newBeeMallShoppingCartItemVO : myShoppingCartItems) {
                    NewBeeMallOrderItem newBeeMallOrderItem = new NewBeeMallOrder
```

```
Item();
        //使用 BeanUtil 工具类将 newBeeMallShoppingCartItemVO 中的属性复制到
newBeeMallOrderItem 对象中
        BeanUtil.copyProperties(newBeeMallShoppingCartItemVO, newBeeMall
OrderItem);
        //在 NewBeeMallOrderMappe 文件 insert()方法中使用了 useGeneratedKeys,
因此 orderId 可以获取
        newBeeMallOrderItem.setOrderId(newBeeMallOrder.getOrderId());
        newBeeMallOrderItems.add(newBeeMallOrderItem);
    }
    //保存至数据库
    if (newBeeMallOrderItemMapper.insertBatch(newBeeMallOrderItems) > 0) {
        //在所有操作成功后,将订单号返回,以供 Controller 方法跳转到订单详情
        return orderNo;
    }
    NewBeeMallException.fail(ServiceResultEnum.ORDER_PRICE_ERROR.
getResult());
    }
    NewBeeMallException.fail(ServiceResultEnum.DB_ERROR.getResult());
  }
  NewBeeMallException.fail(ServiceResultEnum.DB_ERROR.getResult());
 }
 NewBeeMallException.fail(ServiceResultEnum.SHOPPING_ITEM_ERROR.
getResult());
 return ServiceResultEnum.SHOPPING_ITEM_ERROR.getResult();
}
```

订单生成的方法中共 80 行代码,就是先验证,再进行订单数据封装,最后将订单数据和订单项数据保存到数据库中。

这里,结合订单生成流程图来理解,订单生成的详细过程如下所示。

①检查在结算商品中是否包含已下架商品,如果有则抛出一个异常,无下架商品则继续后续流程

②判断商品数据和商品库存,如果商品数据有误或者商品库存不足则抛出异常,一切正常则继续走后续流程

③对象的非空判断

④在生成订单后,购物项数据需要删除,这里调用 NewBeeMallShoppingCartItemMapper.deleteBatch()方法将这些数据批量删除

⑤更新商品库存记录

⑥判断订单价格,如果所有购物项加起来的数据为 0 或者小于 0 则不继续生成订单

⑦生成订单号并封装 NewBeeMallOrder 对象,保存订单记录到数据库中

⑧封装订单项数据并保存订单项数据到数据库中

⑨返回新订单的订单号字段值

这里在 saveOrder()方法中同时修改了多张表中的记录,为了保证事务的一致性,在该方法上添加了@Transactional 注解。一旦该方法在执行过程中发生异常,立刻回滚事务,不然可能会出现库存已扣除,但是订单没生成的场景。

至此,订单生成的逻辑就完成了。读者在功能测试时可以关注一下数据库中的相关记录,在功能完成时购物项是否成功删除、商品库存是否成功修改、订单和订单项是否成功生成。订单即商品和用户信息的结合,需要商品信息、需要用户信息。当然,订单生成只是订单模块中的第一步,后续还有一些步骤需要完成。

25.4 订单详情页面功能的实现

在订单成功生成后还有支付、商家确认订单、取消订单等操作。新蜂商城在订单生成后会跳转到详情页。商城端订单的后续步骤也是在详情页操作的,比如支付的发起、取消订单等。当然,这个交互也是可以修改的。比如在订单生成后跳转到支付流程。目前的交互设计主要是为了合理地安排本书的内容。

25.4.1 订单详情跳转处理

详情页面的请求地址定义为/orders/{orderNo},接收的参数为订单号。一般没有商城系统会直接把订单 id 暴露出去,所以在 OrderController 中新增 orderDetailPage()来对这个路径请求进行处理,新增如下代码:

```
@GetMapping("/orders/{orderNo}")
public String orderDetailPage(HttpServletRequest request, @PathVariable
("orderNo") String orderNo, HttpSession httpSession) {
  NewBeeMallUserVO user = (NewBeeMallUserVO) httpSession.getAttribute
(Constants.MALL_USER_SESSION_KEY);
  NewBeeMallOrderDetailVO orderDetailVO = newBeeMallOrderService.getOrder
DetailByOrderNo(orderNo, user.getUserId());
  if (orderDetailVO == null) {
    return "error/error_5xx";
```

```
}
request.setAttribute("orderDetailVO", orderDetailVO);
return "mall/order-detail";
}
```

orderNo 这个参数是订单记录的唯一订单号，通过@PathVariable 注解读取路径中的这个字段值，并根据这个值调用 NewBeeMallOrderService 业务类中的 getOrderDetailByOrderNo()方法并获取 NewBeeMallOrderDetailVO 对象。getOrderDetailByOrderNo()方法的实现方式：根据主键订单号查询数据库中的订单表并返回订单详情所需的数据，最后将查询到的商品详情数据放到 request 请求中并跳转到 order-detail.html 模板页面。

25.4.2 订单详情数据的渲染

在 templates/mall 目录中新建 order-detail.html，代码如下所示：

```
<!-- Copyright (c) 2019-2020 十三 all rights reserved. -->
<!DOCTYPE html>
<html lang="en" xmlns:th="http://www.thymeleaf.org">
<head th:replace="mall/header::head-fragment('NewBee 商城-订单详情','order-detail')">
</head>
<link rel="stylesheet" th:href="@{/admin/plugins/sweetalert/sweetalert.css}"/>
<body>
<header th:replace="mall/header::header-fragment"></header>
<!-- nav -->
<nav th:replace="mall/header::nav-fragment"></nav>

<!-- personal -->
<div id="personal">
    <div class="self-info center">

        <!-- sidebar -->
        <div th:replace="mall/personal-sidebar::sidebar-fragment"></div>

        <div class="intro fr">
            <div class="uc-box uc-main-box">
                <div class="uc-content-box order-view-box">
                    <div class="box-hd">
                        <h1 class="title">订单详情
```

第25章 订单模块功能开发及讲解

```html
                    <small>请谨防钓鱼链接或诈骗电话, <a href="##">了解更多
&gt;</a>
                    </small>
                </h1>
                <div class="more clearfix">
                    <h2 class="subtitle">订单号:
                        <th:block th:text="${orderDetailVO.orderNo}">
</th:block>
                        <span class="tag tag-subsidy"></span>
                    </h2>
                    <div class="actions">
                        <input type="hidden" id="orderNoValue" th:value=
"${orderDetailVO.orderNo}">
                        <th:block th:if="${orderDetailVO.orderStatus>0
and orderDetailVO.orderStatus<3}">
                            <a onclick="cancelOrder()"
                                class="btn btn-small btn-line-gray" title=
"取消订单">取消订单</a>
                        </th:block>
                        <th:block th:if="${orderDetailVO.order
Status==0}">
                            <a onclick="payOrder()"
                                class="btn btn-small btn-primary" title="
去支付">去支付</a>
                        </th:block>
                        <th:block th:if="${orderDetailVO.order
Status==3}">
                            <a onclick="finishOrder()"
                                class="btn btn-small btn-primary" title="
确认收货">确认收货</a>
                        </th:block>
                    </div>
                </div>
            </div>
            <div class="box-bd">
                <div class="uc-order-item uc-order-item-pay">
                    <div class="order-detail">

                        <div class="order-summary">
                            <div class="order-status" th:text="${order
DetailVO.orderStatusString}">
                                newbee
                            </div>
```

· 679 ·

```html
                                    <div class="order-desc">
                                        <th:block th:if="${orderDetailVO.orderStatus==0}">请尽快完成支付哦~</th:block>
                                        <th:block th:if="${orderDetailVO.orderStatus==1}">newbee 商城订单确认中~</th:block>
                                        <th:block th:if="${orderDetailVO.orderStatus==2}">newbee 仓库正在紧急配货中~</th:block>
                                        <th:block th:if="${orderDetailVO.orderStatus==3}">订单已出库正在快马加鞭向您奔来~</th:block>
                                        <th:block th:if="${orderDetailVO.orderStatus==4}">交易成功，感谢您对 newbee 商城的支持~
                                        </th:block>
                                        <th:block th:if="${orderDetailVO.orderStatus<0}">交易已关闭~</th:block>
                                    </div>
                                    <th:block th:if="${orderDetailVO.orderStatus>=0}">
                                        <div class="order-progress">
                                            <ol class="progress-list clearfix progress-list-5">
                                                <th:block th:if="${orderDetailVO.orderStatus==0}">
                                                    <li class="step step-active">
                                                </th:block>
                                                <th:block th:if="${orderDetailVO.orderStatus>0}">
                                                    <li class="step step-first">
                                                </th:block>
                                                <div class="progress"><span class="text">下单</span></div>
                                                <div class="info"
                                                     th:text="${#dates.format(orderDetailVO.createTime, 'yyyy-MM-dd HH:mm:ss')}">
                                                    02 月 07 日
                                                </div>
                                                </li>
                                                <th:block th:if="${orderDetailVO.orderStatus<1}">
                                                    <li class="step">
                                                </th:block>
                                                <th:block th:if="${orderDetailVO.orderStatus==1}">
```

```
                                                                <li class="step step-active">
                                                            </th:block>
                                                            <th:block th:if="${orderDetailVO.
orderStatus>1}">
                                                                <li class="step step-done">
                                                            </th:block>
                                                            <div class="progress"><span class=
"text">付款</span></div>

                                                            <div class="info"></div>
                                                        </li>
                                                        <th:block th:if="${orderDetailVO.
orderStatus<2}">
                                                                <li class="step">
                                                            </th:block>
                                                            <th:block th:if="${orderDetailVO.
orderStatus==2}">
                                                                <li class="step step-active">
                                                            </th:block>
                                                            <th:block th:if="${orderDetailVO.
orderStatus>2}">
                                                                <li class="step step-done">
                                                            </th:block>
                                                            <div class="progress"><span class=
"text">配货</span></div>

                                                            <div class="info"></div>
                                                        </li>
                                                        <th:block th:if="${orderDetailVO.
orderStatus<3}">
                                                                <li class="step">
                                                            </th:block>
                                                            <th:block th:if="${orderDetailVO.
orderStatus==3}">
                                                                <li class="step step-active">
                                                            </th:block>
                                                            <th:block th:if="${orderDetailVO.
orderStatus>3}">
                                                                <li class="step step-done">
                                                            </th:block>
                                                            <div class="progress"><span class=
"text">出库</span></div>

                                                            <div class="info"></div>
                                                        </li>
                                                        <th:block th:if="${orderDetailVO.
```

```html
                                orderStatus<4}">
                                                        <li class="step">
                                                    </th:block>
                                                    <th:block th:if="${orderDetailVO.
orderStatus==4}">
                                                        <li class="step step-active">
                                                    </th:block>
                                                    <th:block th:if="${orderDetailVO.
orderStatus>4}">
                                                        <li class="step step-last">
                                                    </th:block>
                                                    <div class="progress"><span class=
"text">交易成功</span></div>
                                                    <div class="info"></div>
                                                </li>
                                            </ol>
                                        </div>
                                    </th:block>
                                </div>
                                <table class="order-items-table">
                                    <tbody>
                                    <th:block th:each="item : ${orderDetailVO.
newBeeMallOrderItemVOS}">
                                        <tr>
                                            <td class="col col-thumb">
                                                <div class="figure figure-thumb">
                                                    <a target="_blank" th:href=
"@{'/goods/detail/'+${item.goodsId}}">
                                                        <img th:src="@{${item.
goodsCoverImg}}"
                                                             width="80" height="80"
alt="">
                                                    </a>
                                                </div>
                                            </td>
                                            <td class="col col-name">
                                                <p class="name">
                                                    <a target="_blank" th:href=
"@{'/goods/detail/'+${item.goodsId}}"
                                                       th:text="${item.goods
Name}">newbee</a>
                                                </p>
                                            </td>
```

```html
                                        <td class="col col-price"><p class="price"                                                                    th:text="${item.sellingPrice+'元 x '+item.goodsCount}">
                                        1299元 × 1</p></td>
                                        <td class="col col-actions">
                                        </td>
                                    </tr>
                                </th:block>
                            </tbody>
                        </table>
                    </div>
                    <div id="editAddr" class="order-detail-info">
                        <h3>收货信息</h3>
                        <table class="info-table">
                            <tbody>
                                <tr>
                                    <td th:text="${orderDetailVO.userAddress}">newbee</td>
                                </tr>
                            </tbody>
                        </table>
                        <div class="actions">
                        </div>
                    </div>
                    <div id="editTime" class="order-detail-info">
                        <h3>支付方式</h3>
                        <table class="info-table">
                            <tbody>
                                <tr>
                                    <th>支付方式：</th>
                                    <td th:text="${orderDetailVO.payTypeString==null?'在线支付':orderDetailVO.payTypeString}">
                                        在线支付
                                    </td>
                                </tr>
                            </tbody>
                        </table>
                        <div class="actions">
                        </div>
                    </div>
                    <div class="order-detail-total">
                        <table class="total-table">
```

```html
                    <tbody>
                        <tr>
                            <th>运费：</th>
                            <td><span class="num">0</span>元</td>
                        </tr>
                        <tr>
                            <th class="total">商品总价：</th>
                            <td class="total"><span class="num"
                                                    th:text=
"${orderDetailVO.totalPrice+'.00'}">1299.00</span>元
                            </td>
                        </tr>
                    </tbody>
                </table>
            </div>
        </div>
      </div>
     </div>
    </div>
    <div class="clear"></div>
  </div>
</div>
<div th:replace="mall/footer::footer-fragment"></div>
<!-- jQuery -->
<script th:src="@{/admin/plugins/jquery/jquery.min.js}"></script>
<script th:src="@{/admin/plugins/sweetalert/sweetalert.min.js}"></script>
</body>
</html>
```

　　这里根据返回的 orderDetailVO 对象依次将订单数据渲染到订单基本信息区域、订单项展示区域和订单收货信息、价格区域等地方。这里使用的 Thymeleaf 语法也比较简单，直接读取对应的字段即可。在功能实现后，订单详情效果如图 25-7 所示。

　　关于订单状态的显示文案这里也做了逻辑处理。不同的状态下会显示不同的文案。订单进度条也是根据订单状态来显示的，比如当前订单是刚生成的订单，只有"下单"是绿色的。随着后续步骤的完成，订单进度条也会对应更新并一步一步把订单状态的进度条打满。这两个部分的页面渲染用到了一些 Thymeleaf 的逻辑判断语法，读者在阅读代码时可以关注一下。

当然，很多人会认为订单详情页面与订单确认页面非常类似。这里简单地解释一下，虽然两个页面的整体布局类似，也都有商品数据的展示，但是展示的数据并不相同。两个页面展示的字段差别还是非常大的，订单详情页面读取的数据都是订单表和订单项表中的数据，订单确认页面中读取的则是购物项表中的数据。

图 25-7　订单详情页面

25.5　商城端订单列表功能

订单生成后就能够在个人中心的订单列表中看到相关数据了。各种状态的订单都会在这个列表中显示，且商城端的订单列表也支持分页功能。

25.5.1　订单列表数据格式的定义

图 25-8 是新蜂商城订单列表页面中需要渲染的内容。订单列表是一个 List 对象，因此后端返回数据时需要一个订单列表对象。对象中的字段有订单表中的字段，也有订单项表中的字段，这些字段以及列表中单项对象中的字段可以通过图 25-8 中的内容进行确认。

图 25-8　订单列表数据格式

在列表数据中包括订单状态、订单交易时间、订单总价、商品标题字段、商品预览图字段、商品价格字段、商品购买数量字段。一个订单中可能会有多个订单项，所以订单 VO 对象中也有一个订单项 VO 的列表对象，订单列表中返回的 VO 对象的编码如下所示：

```java
package ltd.newbee.mall.controller.vo;

import java.io.Serializable;
import java.util.Date;
import java.util.List;

/**
 * 订单列表页面VO
 */
public class NewBeeMallOrderListVO implements Serializable {

    private Long orderId;

    private String orderNo;

    private Integer totalPrice;

    private Byte payType;

    private Byte orderStatus;
```

```java
    private String orderStatusString;

    private String userAddress;

    private Date createTime;

    private List<NewBeeMallOrderItemVO> newBeeMallOrderItemVOS;

    public Long getOrderId() {
        return orderId;
    }

    public void setOrderId(Long orderId) {
        this.orderId = orderId;
    }

    public String getOrderNo() {
        return orderNo;
    }

    public void setOrderNo(String orderNo) {
        this.orderNo = orderNo;
    }

    public Integer getTotalPrice() {
        return totalPrice;
    }

    public void setTotalPrice(Integer totalPrice) {
        this.totalPrice = totalPrice;
    }

    public Byte getPayType() {
        return payType;
    }

    public void setPayType(Byte payType) {
        this.payType = payType;
    }

    public Byte getOrderStatus() {
        return orderStatus;
    }
```

```java
    public void setOrderStatus(Byte orderStatus) {
        this.orderStatus = orderStatus;
    }

    public String getOrderStatusString() {
        return orderStatusString;
    }

    public void setOrderStatusString(String orderStatusString) {
        this.orderStatusString = orderStatusString;
    }

    public String getUserAddress() {
        return userAddress;
    }

    public void setUserAddress(String userAddress) {
        this.userAddress = userAddress;
    }

    public Date getCreateTime() {
        return createTime;
    }

    public void setCreateTime(Date createTime) {
        this.createTime = createTime;
    }

    public List<NewBeeMallOrderItemVO> getNewBeeMallOrderItemVOS() {
        return newBeeMallOrderItemVOS;
    }

    public void setNewBeeMallOrderItemVOS(List<NewBeeMallOrderItemVO> newBeeMallOrderItemVOS) {
        this.newBeeMallOrderItemVOS = newBeeMallOrderItemVOS;
    }
}
```

订单项 VO 对象的编码如下所示：

```java
package ltd.newbee.mall.controller.vo;

import java.io.Serializable;
```

```java
/**
 * 订单详情页页面订单项 VO
 */
public class NewBeeMallOrderItemVO implements Serializable {

    private Long goodsId;

    private Integer goodsCount;

    private String goodsName;

    private String goodsCoverImg;

    private Integer sellingPrice;

    public Long getGoodsId() {
        return goodsId;
    }

    public void setGoodsId(Long goodsId) {
        this.goodsId = goodsId;
    }

    public String getGoodsName() {
        return goodsName;
    }

    public void setGoodsName(String goodsName) {
        this.goodsName = goodsName;
    }

    public String getGoodsCoverImg() {
        return goodsCoverImg;
    }

    public void setGoodsCoverImg(String goodsCoverImg) {
        this.goodsCoverImg = goodsCoverImg;
    }

    public Integer getSellingPrice() {
        return sellingPrice;
    }
```

```java
    public void setSellingPrice(Integer sellingPrice) {
        this.sellingPrice = sellingPrice;
    }

    public Integer getGoodsCount() {
        return goodsCount;
    }

    public void setGoodsCount(Integer goodsCount) {
        this.goodsCount = goodsCount;
    }
}
```

25.5.2 订单列表页面数据的获取

接下来是数据查询的功能实现。在上述订单列表中的字段，可以通过分别查询 tb_newbee_mall_order 订单表和 tb_newbee_mall_order_item 订单项表来获取。

在订单业务类中新增如下代码：

```java
/**
 * 我的订单列表
 */
public PageResult getMyOrders(PageQueryUtil pageUtil) {
    int total = newBeeMallOrderMapper.getTotalNewBeeMallOrders(pageUtil);
    List<NewBeeMallOrder> newBeeMallOrders = newBeeMallOrderMapper.findNewBeeMallOrderList(pageUtil);
    List<NewBeeMallOrderListVO> orderListVOS = new ArrayList<>();
    if (total > 0) {
        //数据转换，将实体类转成 VO
        orderListVOS = BeanUtil.copyList(newBeeMallOrders, NewBeeMallOrderListVO.class);
        //设置订单状态中文显示值
        for (NewBeeMallOrderListVO newBeeMallOrderListVO : orderListVOS) {
            newBeeMallOrderListVO.setOrderStatusString(NewBeeMallOrderStatusEnum.getNewBeeMallOrderStatusEnumByStatus(newBeeMallOrderListVO.getOrderStatus()).getName());
        }
        List<Long> orderIds = newBeeMallOrders.stream().map(NewBeeMallOrder::getOrderId).collect(Collectors.toList());
```

```java
    if (!CollectionUtils.isEmpty(orderIds)) {
        List<NewBeeMallOrderItem> orderItems = newBeeMallOrderItemMapper.selectByOrderIds(orderIds);
        Map<Long, List<NewBeeMallOrderItem>> itemByOrderIdMap = orderItems.stream().collect(groupingBy(NewBeeMallOrderItem::getOrderId));
        for (NewBeeMallOrderListVO newBeeMallOrderListVO : orderListVOS) {
            //封装每个订单列表对象的订单项数据
            if (itemByOrderIdMap.containsKey(newBeeMallOrderListVO.getOrderId())) {
                List<NewBeeMallOrderItem> orderItemListTemp = itemByOrderIdMap.get(newBeeMallOrderListVO.getOrderId());
                //将NewBeeMallOrderItem对象列表转换成NewBeeMallOrderItemVO对象列表
                List<NewBeeMallOrderItemVO> newBeeMallOrderItemVOS = BeanUtil.copyList(orderItemListTemp, NewBeeMallOrderItemVO.class);
                newBeeMallOrderListVO.setNewBeeMallOrderItemVOS(newBeeMallOrderItemVOS);
            }
        }
    }
    PageResult pageResult = new PageResult(orderListVOS, total, pageUtil.getLimit(), pageUtil.getPage());
    return pageResult;
}
```

这里首先定义了 getMyOrders() 方法并传入 PageUtil 对象作为参数。该对象中会有分页参数和用户的 userId，并且通过 SQL 查询出当前 userId 下的订单列表数据和每个订单所关联的订单项列表数据。然后是填充数据的环节，即将相关字段封装到 NewBeeMallOrderListVO 对象中，并将封装好的 List 对象返回。

想要将数据通过 Thymeleaf 语法渲染到前端页面上，需要将数据获取并转发到对应的模板页面中。在 OrderController 的订单列表方法中将查询到的数据放入 request 请求中，并在 OrderController 中新增 orderListPage() 方法，代码如下所示：

```java
@GetMapping("/orders")
public String orderListPage(@RequestParam Map<String, Object> params, HttpServletRequest request, HttpSession httpSession) {
    NewBeeMallUserVO user = (NewBeeMallUserVO) httpSession.getAttribute(Constants.MALL_USER_SESSION_KEY);
    params.put("userId", user.getUserId());
    if (StringUtils.isEmpty(params.get("page"))) {
        params.put("page", 1);
```

```
    }
    params.put("limit", Constants.ORDER_SEARCH_PAGE_LIMIT);
    //封装我的订单数据
    PageQueryUtil pageUtil = new PageQueryUtil(params);
    request.setAttribute("orderPageResult",
newBeeMallOrderService.getMyOrders(pageUtil));
    request.setAttribute("path", "orders");
    return "mall/my-orders";
}
```

该方法首先将分页参数和当前用户的 userId 封装到 PageQueryUtil 对象中并调用了业务层的方法。然后把当前用户的订单数据按照不同的分页参数查询出来并将返回结果放到 request 对象中。最后跳转到 mall 目录下的 my-orders.html 模板页面进行数据渲染。

25.5.3 订单列表页面渲染的逻辑

在 resources/templates/mall 目录中新增"我的订单"页面模板 my-orders.html，模板代码如下所示：

```html
<!-- Copyright (c) 2019-2020 十三 all rights reserved. -->
<!DOCTYPE html>
<html lang="en" xmlns:th="http://www.thymeleaf.org">
<head th:replace="mall/header::head-fragment('NewBee商城-订单列表','my-orders')">
</head>
<body>
<header th:replace="mall/header::header-fragment"></header>
<!-- nav -->
<nav th:replace="mall/header::nav-fragment"></nav>

<!-- personal -->
<div id="personal">
    <div class="self-info center">

        <!-- sidebar -->
        <div th:replace="mall/personal-sidebar::sidebar-fragment"></div>

        <div class="intro fr">
            <div class="uc-box uc-main-box">
                <div class="uc-content-box order-list-box">
                    <div class="box-hd">
```

```html
                <h1 class="title">我的订单
                    <small>请谨防钓鱼链接或诈骗电话，<a href="##">了解更多&gt;</a>
                    </small>
                </h1>
            </div>
            <div class="box-bd">
                <div id="J_orderList">
                    <ul class="order-list">
                        <th:block th:if="${#lists.isEmpty(orderPageResult.list)}">
                            <img style="margin-top: 16px;"
                                 th:src="@{/mall/image/null-content.png}">
                            <small><a th:href="@{/index}">    去购物&gt;</a>
                            </small>
                        </th:block>
                        <th:block th:unless="${#lists.isEmpty(orderPageResult.list)}">
                            <th:block th:each="order : ${orderPageResult.list}">
                                <li class="uc-order-item uc-order-item-list">
                                    <div class="order-detail">
                                        <div class="order-summary">
                                            <th:block th:if="${order.orderStatus<0}">
                                                <div class="order-status-closed">
                                            </th:block>
                                            <th:block th:if="${order.orderStatus==0}">
                                                <div class="order-status-need-pay">
                                            </th:block>
                                            <th:block th:if="${order.orderStatus>0}">
                                                <div class="order-status-success">
                                            </th:block>
                                            <th:block th:text="${' '+order.orderStatusString}"></th:block>
```

```html
                                                </div>
                                            </div>
                                            <table class="order-detail-table">
                                                <thead>
                                                <tr>
                                                    <th class="col-main"><p class="caption-info"
                                                                            th:text="${#dates.format(order.createTime, 'yyyy-MM-dd HH:mm:ss')}">
                                                        02月07日<span class="sep">|</span>订单号: <a
                                                            href="##"
                                                            th:text="${order.orderNo}">201908121807</a><span
                                                            class="sep">|</span>在线支付</p></th>
                                                    <th class="col-sub"><p class="caption-price">应付金额: <span
                                                            class="num"
                                                            th:text="${order.totalPrice+'.00'}">1299.00</span>元
                                                    </p></th>
                                                </tr>
                                                </thead>
                                                <tbody>
                                                <tr>
                                                    <td class="order-items">
                                                        <ul class="goods-list">
                                                            <th:block
                                                                    th:each="item : ${order.newBeeMallOrderItemVOS}">
                                                                <li>
                                                                    <div class="figure figure-thumb">
                                                                        <a target="_blank"
                                                                           th:href="@{'/goods/detail/'+${item.goodsId}}">
                                                                            <img th:src= "@{${item.goodsCoverImg}}"
                                                                                 width="80" height="80" alt="">
                                                                        </a>
```

```html
                            </div>
                            <p class="name">
<a target="_blank"
th:href="@{'/goods/detail/'+${item.goodsId}}"
th:text="${item.goodsName}">newbee</a>
                            </p>
                            <p class="price"
                                th:text=
"${item.sellingPrice+'元 x '+item.goodsCount}">
                                13元 × 1</p>
                        </th:block>
                    </ul>
                </td>
                <td class="order-actions"><a
class="btn btn-small btn-line-gray"
                                                                th:
href="@{${'/orders/'+order.orderNo}}">订单详情</a>
                </td>
            </tr>
        </tbody>
    </table>
            </div>
        </li>
    </th:block>
    </th:block>
    </ul>
</div>
                <th:block th:unless="${#lists.isEmpty(orderPageResult.list)}">
                    <div id="J_orderListPages">
                        <div class="newbee-pagenavi">
                            <th:block
th:if="${orderPageResult.currPage-2 >=1}"><a
                                        class="numbers"
                                        th:href="@{'/orders?page=' + ${orderPageResult.currPage-2}}"
                                        th:text="${orderPageResult.currPage-2}">1</a>
                            </th:block>
                            <th:block th:if="${orderPageResult.currPage-1 >=1}"><a
```

```html
                                        class="numbers"
                                        th:href="@{'/orders?page=' + ${order
PageResult.currPage-1}}"
                                        th:text="${orderPageResult.currPage
-1}">1</a>
                            </th:block>
                            <a href="##" class="numbers current" th:text=
"${orderPageResult.currPage}">1</a>
                            <th:block th:if="${orderPageResult.currPage+
1<=orderPageResult.totalPage}"><a
                                    class="numbers"
                                    th:href="@{'/orders?page=' +
${orderPageResult.currPage+1}}"
                                    th:text="${orderPageResult.currPage
+1}">1</a></th:block>
                            <th:block th:if="${orderPageResult.currPage+
2<=orderPageResult.totalPage}"><a
                                    class="numbers"
                                    th:href="@{'/orders?page=' + ${orderPage
Result.currPage+2}}"
                                    th:text="${orderPageResult.currPage
+2}">1</a></th:block>
                        </div>
                    </div>
                </th:block>
            </div>
        </div>
      </div>
   </div>
   <div class="clear"></div>
</div>
</div>

<div th:replace="mall/footer::footer-fragment"></div>

<!-- jQuery -->
<script th:src="@{/admin/plugins/jquery/jquery.min.js}"></script>
<script th:src="@{/mall/js/search.js}" type="text/javascript"></script>
</body>
</html>
```

通过代码可以看出，在该页面中主要进行了两部分的数据读取和渲染，分别是订单数据列表渲染和分页区域的渲染。在订单列表区域和分页功能区域对应的位置读取 orderPageResult 对象中的 list 数据和分页数据。list 数据为订单列表数据。使用 th:each 循环语法将订单状态、订单交易时间、订单总价、商品标题字段、商品预览图字段、商品价格字段、商品购买数量字段渲染出来。在渲染商品信息的时候用到了一个 th:each 循环语法和一个 a 标签。在点击对应按钮后会跳转到订单详情页面。

接下来根据分页字段当前页码 currPage、总页码 totalPage 将下方的分页按钮渲染出来。这里的分页并没有借助插件来实现，而是通过自行编码的方式来生成，逻辑与前文的分页按钮渲染一致，只是 CSS 样式有一些区别。

25.6 订单处理流程及订单状态的介绍

25.6.1 订单处理流程

订单模块是整个电商系统的重中之重，甚至可以说它就是电商系统的心脏。因为订单往往决定了一个电商系统的生死，而且订单模块贯穿了整个电商系统的大部分流程。各个环节都与它密不可分，从用户点击提交订单并成功生成订单开始，后续的整个流程都是围绕着订单模块进行的，包括支付成功到确认收货的正常订单流程，也包括订单取消、订单退款等一系列的异常订单流程。

正常订单流程如图 25-9 所示。

图 25-9　正常订单流程

在订单生成后，用户正常进行支付操作，商家正常进行订单确认和订单发货操作，最后由用户进行最后一个步骤：确认收货。这样整个订单流程就正常走完。

异常订单流程如图 25-10 所示。

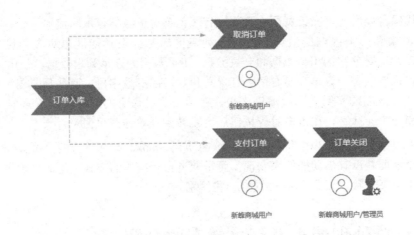

图 25-10 异常订单流程

在订单入库后,用户选择不支付而直接取消订单,或者用户正常支付但是在后续流程中选择取消订单,至此订单就不是正常状态的订单了。因为它的流程并没有如预想的一样。不只是用户可以关闭订单,如果流程中出现了意外事件,商城管理员用户也可以在后台管理系统中关闭订单。

25.6.2 订单状态的介绍

接下来讲解订单流程中各个订单状态和各个状态的转换。订单流程完善的编码实践都是围绕着订单状态的改变来做的功能实现。只要理解了订单状态以及如何发生状态转变的逻辑,对于读者理解代码、理解商城业务是一个很大的帮助。

订单表中的 order_status 字段就是订单状态字段,新蜂商城订单状态的设计如下所示。

① 0:待支付

② 1:已支付

③ 2:配货完成

④ 3:出库成功

⑤ 4:交易成功

⑥ -1:手动关闭

⑦ -2:超时关闭

⑧-3：商家关闭

以上是新蜂商城的订单状态存储的值以及这个值对应的含义。这与主流的商城设计类似，可能文案上有些小差别。比如状态 0，新蜂商城用"待支付"表示，其他商城可能会用"待付款"。数字的使用也可能有差异，比如新蜂商城中将订单的初始状态用数字 0 表示，其他的商城在实现时可能用数字 1 表示。

接下来详细介绍一下这些状态。

（1）待支付/待付款

新蜂商城用数字 0 来表示这个状态。

在用户提交订单后，会进行订单的入库、商品库存修改等操作。此时是订单的初始状态。目前主流的商城或者常用的外卖平台，基本上在订单生成后就会唤起支付操作。所以订单的初始状态就被称为"待支付"或"待付款"。其实它的含义是订单成功入库，也就是初始状态。新蜂商城选用"待支付"表示。

（2）已支付/已付款/待确认

新蜂商城用数字 1 来表示这个状态。

用户完成订单支付，系统需要记录订单支付时间及支付方式等信息。此时成功付款，等待商家进行订单确认以便进行后续操作。这个状态被称为"已支付"或"已付款"，也可以被称为"待商家确认"。这些称谓一般由产品经理或者项目负责人来决定。新蜂商城选用"已支付"表示。

（3）已确认/配货完成/待发货

新蜂商城用数字 2 来表示这个状态。

商家确认订单正常，并且可以进行发货操作，就将订单修改为这个状态。此时商家确认了订单的有效性，接下来就是发货，所以此时的状态可以被称为"已确认"、"待发货"或"配货完成"。新蜂商城选用"配货完成"表示。

（4）出库成功/待收货/已发货

新蜂商城用数字 3 来表示这个状态。

订单中的商品在出库并交给物流系统后就进入了这个状态。对于仓库来说是"出库成功"，对于用户来说是"待收货"，而对于商家来说是"已发货"。新蜂商城选用"出库成功"表示。

（5）交易成功/订单完成

新蜂商城用数字 4 来表示这个状态。

用户收到此次购买的商品，并点击商城订单系统页面中的"确认收货"按钮，这就表示订单已经完成了所有的正向步骤，此次交易就成功了。这个状态被称为"交易成功"或"订单完成"。新蜂商城选用"交易成功"表示。

（5）订单关闭/已取消

新蜂商城用数字-1、-2、-3来表示"手动关闭"、"超时关闭"和"商家关闭"这种订单状态。

这属于订单异常的状态，在付款之前取消订单或者在其他状态下选择主动取消订单都会进入这种状态。也可以统一称为"订单关闭"或"已取消"。

当然，订单流程还会牵涉客服功能、订单售后、订单退款等逻辑，但这些内容并不在本次讲解之内。

25.7 订单状态转换的讲解

25.7.1 订单支付

新蜂商城的支付功能是模拟实现的。在成功生成订单后就可以跳转到支付模块的相关页面。或者通过订单详情页面的"去支付"按钮跳转到支付模块的相关页面。支付步骤最终是为了把订单状态改为支付成功状态，同时记录支付的相关信息。支付信息记录了支付时间和支付方式，还记录了支付的参数和第三方支付公司的回调数据。

新蜂商城中的支付步骤如下所示。

（1）选择支付方式

选择支付方式页面如图 25-11 所示。

图 25-11 选择支付方式页面

（2）跳转至支付页面

模拟的支付页面如图 25-12 所示。

图 25-12　模拟的支付页面

（3）支付成功

点击支付页面的"支付成功"按钮，就会向支付回调地址发送请求。该按钮模拟了正式的第三方支付接口回调，表示已经支付成功，可以修改订单状态了。

这里的处理代码如下所示：

```
@GetMapping("/paySuccess")
@ResponseBody
public Result paySuccess(@RequestParam("orderNo") String orderNo, @RequestParam("payType") int payType) {
    String payResult = newBeeMallOrderService.paySuccess(orderNo, payType);
    if (ServiceResultEnum.SUCCESS.getResult().equals(payResult)) {
        return ResultGenerator.genSuccessResult();
    } else {
        return ResultGenerator.genFailResult(payResult);
    }
}
```

这是 OrderController 类中的代码，负责接收支付回调数据，参数为订单号和支付方式。这里根据这两个参数对订单的状态进行修改，调用的 service 层的方法为 paySuccess()，代码如下所示：

```
public String paySuccess(String orderNo, int payType) {
    NewBeeMallOrder newBeeMallOrder = newBeeMallOrderMapper.selectByOrderNo(orderNo);
    if (newBeeMallOrder != null) {
```

```
        newBeeMallOrder.setOrderStatus((byte) NewBeeMallOrderStatusEnum.
OREDER_PAID.getOrderStatus());
        newBeeMallOrder.setPayType((byte) payType);
        newBeeMallOrder.setPayStatus((byte) PayStatusEnum.PAY_SUCCESS.
getPayStatus());
        newBeeMallOrder.setPayTime(new Date());
        newBeeMallOrder.setUpdateTime(new Date());
        if (newBeeMallOrderMapper.updateByPrimaryKeySelective(newBeeMall
Order) > 0) {
            return ServiceResultEnum.SUCCESS.getResult();
        } else {
            return ServiceResultEnum.DB_ERROR.getResult();
        }
    }
    return ServiceResultEnum.ORDER_NOT_EXIST_ERROR.getResult();
}
```

如此，再根据订单号查询订单，将相关状态进行修改，并调用数据层的方法进行实际的入库操作。其他状态的转换也与此类似。

支付成功后的订单详情页面如图 25-13 所示。

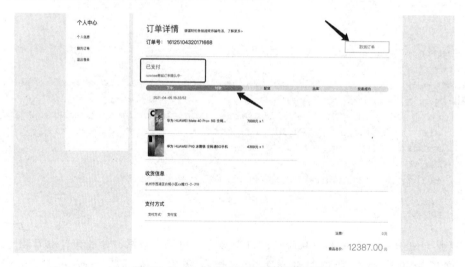

图 25-13　支付成功后的订单详情页面

订单已支付成功，订单状态已更改，下方的订单进度条"付款"环节完成，支付方式也由"无"变成了"支付宝"。

25.7.2 订单确认

订单确认这个步骤是在后台管理系统中操作的，操作按钮为"配货完成"。这里以支付成功的订单（订单号 16125104320171668）的数据为例来操作配货完成的功能。订单管理页面的"配货完成"如图 25-14 所示。

图 25-14 订单管理页面的"配货完成"

如果没有支付成功或者已经过了配货完成的状态，就无法进行"订单确认"的操作。这里只有在支付成功的状态下可以操作，否则会弹出提示框，无法进行后续步骤。在点击"配货完成"按钮后完成订单确认步骤，此时订单配货完成，订单状态已更改，下方的订单进度条"配货"环节完成，订单详情页面如图 25-15 所示。

图 25-15 配货完成后的订单详情页面

25.7.3 订单出库

该步骤与订单确认类似，也是在后台管理系统中操作，订单管理页面的"出库"如图 25-16 所示。

图 25-16 订单管理页面的"出库"

选择对应的订单并点击"出库"按钮，在请求成功后订单的状态也会相应改变，出库后的订单详情页面 15-17 所示。

图 25-17 出库后的订单详情页面

订单出库完成，订单状态已更改，下方的订单进度条"出库"环节完成。

25.7.4 确认收货

确认收货是一个只有用户才能操作的步骤。该操作步骤的触发按钮在订单详情页面上。用户可以点击"确认收货"按钮完成订单的最后一步操作。在请求成功后，订单的状态即转换为"交易成功"。确认收货后的订单详情页面如图 25-18 所示。

图 25-18　确认收货后的订单详情页面

此时，订单详情页面的进度条已经全部完成并"拉满"。

25.7.5 取消订单

前文四个状态是数据订单正向流程的状态。在实际操作中，用户可能会在某个环节选择将订单关闭，因此也需要实现取消订单的操作。该操作可以被用户触发，也可以在后台管理系统中由管理员触发。

用户可以点击订单详情页面中"取消订单"按钮来关闭订单，如图 25-19 所示。

在点击"取消订单"按钮后会出现一个确认弹框，如果点击确认则该订单将会变成关闭状态，订单详情页面中显示的内容也会被修改。取消订单后的订单详情页面如图 25-20 所示。

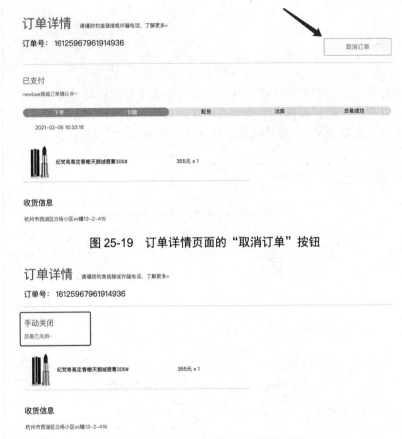

图 25-19 订单详情页面的"取消订单"按钮

图 25-20 取消订单后的订单详情页面

在后台管理系统中也可以进行关闭订单的操作，操作按钮为"关闭订单"。选中需要关闭的订单记录，点击"关闭订单"按钮即可，如图 25-21 所示。

图 25-21 订单管理页面的"关闭订单"

可以看到选中订单的状态已变为"商家关闭",如图 25-22 所示。

图 25-22 选中订单的状态已变为"商家关闭"

在商城端和后台管理系统中进行关闭订单的操作,完成后无法再对订单进行其他的操作。

至此,新蜂商城所有的功能模块及相关知识点都介绍完毕。

读者可以把本文的源码下载到本地并启动项目,结合源码和实际的操作理解。订单生成和各个状态的转换涉及多张表的数据更改,在测试时一定要注意数据库中商品、购物项、订单等数据是否正确被修改。

25.8 商城系统的展望

虽然新蜂商城的功能模块已经全部讲解完成,但是新蜂商城的优化和迭代工作不会停止。在技术栈上也会不断增加新的实用技术。

软件的需求是不断变化的,技术的更新迭代也越来越快。笔者会继续开发商城管理系统 Vue3 版本,使用 Vue3+Element UI 重构新蜂商城后台管理系统。同时,笔者会对新蜂商城 v1 版本进行升级,增加优惠券、接入支付宝、增加秒杀模块,并扩展更多的后端技术栈。新蜂商城版本记录及开发计划如图 25-23 所示。

笔者不仅会继续开发新蜂商城各个版本的功能,也会不断整理相关的开发文档和知识点。

行文至此,笔者万般不舍。不过,对于 Spring Boot 技术栈和大型商城项目开发实战的讲解就要告一段落了。在本书的最后,诚心祝愿各位读者能够在编程道路上寻找到属于自己的精彩!

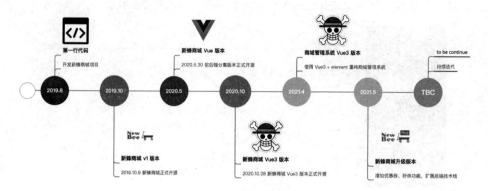

图 25-23 新蜂商城版本记录及开发计划